**Iridium Complexes in
Organic Synthesis**

Edited by
Luis A. Oro and
Carmen Claver

Further Reading

Hashmi, A. S. K., Toste, D. F. (Eds.)

Modern Gold Catalyzed Synthesis

2009
ISBN: 978-3-527-31952-7

Yamamoto, H., Ishihara, K. (Eds.)

Acid Catalysis in Modern Organic Synthesis

2 Volumes
2008
ISBN: 978-3-527-31724-0

Ding, K., Uozumi, Y. (Eds.)

Handbook of Asymmetric Heterogeneous Catalysis

2008
ISBN: 978-3-527-31913-8

Toru, T., Bolm, C. (Eds.)

Organosulfur Chemistry in Asymmetric Synthesis

2008
ISBN: 978-3-527-31854-2

Bolm, C., Hahn, F. E. (Eds.)

Activating Unreactive Substrates
The Role of Secondary Interactions

2009
ISBN: 978-3-527-31823-0

Iridium Complexes in Organic Synthesis

Edited by
Luis A. Oro and Carmen Claver

WILEY-VCH Verlag GmbH & Co. KGaA

The Editors

Prof. Dr. Luis A. Oro
University of Zaragoza-CSIC
Dep. Inorganic Chemistry, Faculty of
Science-ICMA
Instituto Universitario de Catálisis Homogénea
Pedro Cerbuna 12
50009 Zaragoza
Spain

Prof. Dr. Carmen Claver
Universitat Rovira i Virgili
Dept. de Química Física i Inorgànica
Facultat de Química
c/ Marcel.lí Domingo, s/n
43007 Tarragona
Spain

■ All books published by Wiley-VCH are carefully produced. Nevertheless, authors, editors, and publisher do not warrant the information contained in these books, including this book, to be free of errors. Readers are advised to keep in mind that statements, data, illustrations, procedural details or other items may inadvertently be inaccurate.

Library of Congress Card No.: applied for

British Library Cataloguing-in-Publication Data
A catalogue record for this book is available from the British Library.

Bibliographic information published by the Deutsche Nationalbibliothek
The Deutsche Nationalbibliothek lists this publication in the Deutsche Nationalbibliografie; detailed bibliographic data are available on the Internet at <http://dnb.d-nb.de>.

© 2009 WILEY-VCH Verlag GmbH & Co. KGaA, Weinheim

All rights reserved (including those of translation into other languages). No part of this book may be reproduced in any form – by photoprinting, microfilm, or any other means – nor transmitted or translated into a machine language without written permission from the publishers. Registered names, trademarks, etc. used in this book, even when not specifically marked as such, are not to be considered unprotected by law.

Composition   SNP Best-set Typesetter Ltd., Hong Kong
Printing   Strauss GmbH, Mörlenbach
Bookbinding   Litges & Dopf GmbH, Heppenheim

Printed in the Federal Republic of Germany
Printed on acid-free paper

ISBN: 978-3-527-31996-1

Contents

Preface *XIII*
List of Contributors *XV*

1 Application of Iridium Catalysts in the Fine Chemicals Industry 1
Hans-Ulrich Blaser
1.1 Introduction *1*
1.2 Industrial Requirements for Applying Catalysts *1*
1.2.1 Characteristics of the Manufacture of Enantiomerically Pure Products *1*
1.2.2 Process Development: Critical Factors for the Application of Catalysts *2*
1.2.3 Requirements for Practically Useful Catalysts *3*
1.2.3.1 Preparation Methods *3*
1.2.3.2 Catalysts Cost *3*
1.2.3.3 Availability of the Catalysts *3*
1.2.3.4 Catalytic Performance *3*
1.2.3.5 Separation *4*
1.3 Enantioselective Hydrogenation of C=N Bonds *4*
1.3.1 Catalysts and Scope *4*
1.3.2 Industrial Applications *6*
1.4 Enantioselective Hydrogenation of C=C Bonds *8*
1.4.1 Catalysts and Scope *8*
1.4.2 Industrial Applications *9*
1.5 Miscellaneous Catalytic Applications with Industrial Potential *10*
1.6 Conclusions and Outlook *13*
References *13*

2 Dihydrido Iridium Triisopropylphosphine Complexes: From Organometallic Chemistry to Catalysis 15
Luis A. Oro
2.1 Introduction *15*
2.2 [Ir(COD)(NCMe)(PR$_3$)]BF$_4$ (PR$_3$ = P^{i}Pr$_3$, PMe$_3$) and Related Complexes as Catalyst Precursors: Is 1,5-Cyclo-Octadiene an Innocent and Removable Ligand? *16*

Iridium Complexes in Organic Synthesis. Edited by Luis A. Oro and Carmen Claver
Copyright © 2009 WILEY-VCH Verlag GmbH & Co. KGaA, Weinheim
ISBN: 978-3-527-31996-1

2.3	The Dihydrido Iridium Triisopropylphosphine Complex [IrH$_2$(NCMe)$_3$(P^{i}Pr$_3$)]BF$_4$ as Alkene Hydrogenation Catalysts	21
2.4	The Dihydrido Iridium Triisopropylphosphine Complex [IrH$_2$(NCMe)$_3$(P^{i}Pr$_3$)]BF$_4$ as Alkyne Hydrogenation Catalysts	26
2.5	Dihydrido Arene Iridium Triisopropylphosphine Complexes	29
2.6	Dihydrido Iridium Triisopropylphosphine Complexes as Imine Hydrogenation Catalysts	34
2.7	Conclusions	37
	Acknowledgments	37
	References	37
3	**Iridium N-Heterocyclic Carbene Complexes and Their Application as Homogeneous Catalysts**	**39**
	Eduardo Peris and Robert H. Crabtree	
3.1	Introduction	39
3.2	Types of Ir—NHC and Reactivity	40
3.2.1	Mono-NHCs and Intramolecular C—H Activation	40
3.2.2	Chelating bis-NHCs	43
3.2.3	Abnormal NHCs	46
3.3	Catalysis with Ir—NHCs	49
3.4	Conclusions	52
	References	52
4	**Iridium-Catalyzed C=O Hydrogenation**	**55**
	Claudio Bianchini, Luca Gonsalvi and Maurizio Peruzzini	
4.1	Introduction	55
4.2	Homogeneous C=O Hydrogenations	55
4.2.1	Chemoselective Hydrogenations	56
4.2.2	Enantioselective Hydrogenations	63
4.2.3	Transfer Hydrogenation (TH)	69
4.2.4	Asymmetric Transfer Hydrogenation (ATH)	81
4.3	Heterogeneous, Supported and Biocatalytic Hydrogenations	99
	References	103
5	**Catalytic Activity of Cp* Iridium Complexes in Hydrogen Transfer Reactions**	**107**
	Ken-ichi Fujita and Ryohei Yamaguchi	
5.1	Introduction	107
5.2	Hydrogen Transfer Oxidation of Alcohols (Oppenauer-Type Oxidation)	108
5.3	Transfer Hydrogenation of Unsaturated Compounds	112
5.3.1	Transfer Hydrogenation of Quinolines	112
5.3.2	Transfer Hydrogenation of Ketones and Imines	113
5.4	Asymmetric Synthesis Based on Hydrogen Transfer	113
5.4.1	Asymmetric Transfer Hydrogenation of Ketones	113
5.4.2	Dynamic Kinetic Resolution	118

5.5	Hydrogen Transfer Reactions in Aqueous Media	*119*
5.6	Carbon–Nitrogen Bond Formation Based on Hydrogen Transfer	*123*
5.6.1	N-Alkylation of Amines with Alcohols	*123*
5.6.2	Cyclization of Amino Alcohols	*126*
5.6.3	Cyclization of Primary Amines with Diols	*127*
5.6.4	Amidation of Alcohols with Hydroxylamine	*128*
5.7	Carbon–Carbon Bond Formation Based on Hydrogen Transfer	*130*
5.7.1	β-Alkylation of Secondary Alcohols	*130*
5.7.2	Alkylation of Active Methylene Compounds with Alcohols	*131*
5.8	Carbon–Oxygen Bond Formation Based on Hydrogen Transfer	*135*
5.8.1	Oxidative Lactonization of Diols	*135*
5.8.2	Inter- and Intra-Molecular Tishchenko Reactions	*137*
5.9	Dehydrogenative Oxidation of Alcohols	*138*
5.10	Conclusions	*140*
	References	*140*
6	**Iridium-Catalyzed Hydroamination**	**145**
	Romano Dorta	
6.1	Introduction	*145*
6.2	Iridium-Catalyzed Olefin Hydroamination (OHA)	*146*
6.2.1	The Ir(III)/Secondary Amines/Ethylene System	*146*
6.2.2	The Ir(I)/ZnCl$_2$/Aniline/Norbornene System	*146*
6.2.3	The Chiral Ir(I)/'Naked Fluoride'/Norbornene/Aniline System	*147*
6.2.4	The Chiral Ir(I)/Organic Base/Anilines/Olefins System	*150*
6.2.5	The Ir(I)/Piperidine/Methacrylonitrile System	*151*
6.3	Iridium-Catalyzed Alkyne Hydroamination (AHA)	*152*
6.3.1	Intromolecualar Aliphatic Systems	*152*
6.3.2	Indoles via Intramolecular AHA	*153*
6.3.3	Intermolecular Alkyne Hydroamination	*153*
6.4	Proposed Mechanisms	*156*
6.4.1	Olefin Hydroamination	*156*
6.4.2	Alkyne Hydroamination	*158*
6.5	Complexes and Reactions of Ir Relevant to Hydroamination	*160*
6.5.1	Ir(I)–Amine Complexes	*160*
6.5.2	Ir(I)–Anilido complexes	*161*
6.5.3	N—H Bond Activation Leading to Ir(III)–Amido-Hydrido Complexes	*162*
6.5.4	Alkyl–Amino-Hydrido Complexes of Ir(III)	*165*
6.5.5	Iridium–Fluoride Complexes	*168*
6.6	Conclusions	*169*
	References	*170*
7	**Iridium-Catalyzed Boron-Addition**	**173**
	Elena Fernández and Anna M. Segarra	
7.1	Introduction	*173*
7.2	Iridium–Boryl Complexes	*173*

7.3	Hydroboration 176
7.4	Diboration 184
7.5	Borylation 185
	References 191

8 Iridium-Catalyzed Methanol Carbonylation 195
Philippe Kalck and Philippe Serp
- 8.1 Introduction 195
- 8.2 Rhodium-Based Processes 197
- 8.2.1 The Monsanto Process 197
- 8.2.2 The Celanese Process 199
- 8.3 Iridium Reactivity in the Methanol Carbonylation Reaction 200
- 8.4 The Iridium-Based Cativa Process 204
- 8.5 The Iridium–Platinum-Based Process 206
- 8.6 The Iridium–Cocatalyst Mechanism, and Conclusions 207
- Acknowledgments 207
- References 208

9 Iridium-Catalyzed Asymmetric Allylic Substitutions 211
Günter Helmchen
- 9.1 Introduction 211
- 9.2 Ir-Catalyzed Allylic Substitutions: Fundamentals 212
- 9.2.1 Reactivity and Regioselectivity 212
- 9.2.2 Steric Course 214
- 9.2.3 Asymmetric Catalysis: The Beginnings with Phosphinooxazolines as Chiral Ligands 215
- 9.2.4 Phosphoramidites as Ligands for the Ir-Catalyzed Allylic Substitution 216
- 9.2.4.1 Survey 216
- 9.2.4.2 Catalyst Preparation, Reaction Conditions and Catalytic Cycle 217
- 9.2.4.3 Preparation of Phosphoramidites 219
- 9.2.4.4 Variation of the Phosphoramidite Ligands 219
- 9.2.4.5 Further Ligands Used in Ir-Catalyzed Allylic Substitutions 220
- 9.3 C-Nucleophiles 221
- 9.3.1 Stabilized Enolates as Nucleophiles 221
- 9.3.1.1 Malonates and Related Pronucleophiles 221
- 9.3.2 Aliphatic Nitro Compounds as Pronucleophiles 224
- 9.3.2.1 A Glycine Equivalent as Pronucleophile 225
- 9.3.3 Allylic Substitutions with Nonstabilized Enolates, Enamines and Organozinc Compounds 226
- 9.3.3.1 Ketone Enolates Derived from Silyl Enol Ethers as Nucleophiles 226
- 9.3.3.2 Allylation of Enamines 227
- 9.3.3.3 Decarboxylative Allylic Alkylation 227
- 9.3.3.4 Reactions with Aryl Zinc Compounds 228
- 9.4 N-Nucleophiles 229

9.4.1	Inter- and Intramolecular Reactions with Aliphatic Amines and Ammonia as Nucleophiles *229*	
9.4.2	Arylamines as Nucleophiles *232*	
9.4.3	Amination of Allylic Alcohols *232*	
9.4.4	Pronucleophiles Serving as Ammonia Surrogates: N,N-Diacylamines, Trifluoroacetamide and N-Sulfonylamines *233*	
9.4.5	Decarboxylative Allylic Amidation *236*	
9.4.6	Dihydropyrroles and γ-Lactams via Allylic Substitution and Ring-Closing Metathesis *237*	
9.4.7	Hydroxylamine Derivatives as N-Nucleophiles *238*	
9.5	O-Nucleophiles *239*	
9.5.1	Phenolates as Nucleophiles *239*	
9.5.2	Alkoxides as Nucleophiles *241*	
9.5.3	Hydroxylamine Derivatives as O-Nucleophiles *242*	
9.5.4	Silanolates as Nucleophiles *242*	
9.5.5	Dihydrofurans via Allylic Etherification in Combination with RCM *244*	
9.6	Synthesis of Biologically Active Compounds via Allylic Substitution *244*	
9.7	Conclusions *246*	
	Acknowledgments *247*	
	References *247*	

10   **Iridium-Catalyzed Coupling Reactions** *251*
Yasutaka Ishii, Yasushi Obora and Satoshi Sakaguchi

10.1	Introduction *251*	
10.2	Iridium-Catalyzed Dimerization and Cyclotrimerization of Alkynes *251*	
10.3	Iridium-Catalyzed, Three-Component Coupling Reactions of Aldehydes, Amines and Alkynes *253*	
10.4	Head-to-Tail Dimerization of Acrylates *256*	
10.5	A Novel Synthesis of Vinyl Ethers via an Unusual Exchange Reaction *258*	
10.6	Iridium-Catalyzed Allylic Substitution *260*	
10.7	Alkylation of Ketones with Alcohols *262*	
10.8	N-Alkylation of Amines *264*	
10.9	Oxidative Dimerization of Primary Alcohols to Esters *266*	
10.10	Iridium-Catalyzed Addition of Water and Alcohols to Terminal Alkynes *266*	
10.11	Iridium-Catalyzed Direct Arylation of Aromatic C—H Bonds *267*	
10.12	Iridium-Catalyzed Anti-Markovnikov Olefin Arylation *267*	
10.13	Iridium-Catalyzed Silylation and Borylation of Aromatic C—H Bonds *268*	
10.14	Miscellaneous Reactions Catalyzed by Iridium Complexes *269*	
	References *271*	

11	**Iridium-Catalyzed Cycloadditions** *277*
	Takanori Shibata
11.1	Introduction *277*
11.2	[2+2+2] Cycloaddition *278*
11.3	Enantioselective [2+2+2] Cycloaddition *281*
11.4	[2+2+1] Cycloaddition *283*
11.5	[4+2] and [5+1] Cycloadditions *288*
11.6	Cycloisomerization *289*
11.7	Ir(III)-Catalyzed Cyclizations *291*
11.8	Miscellaneous Cycloadditions *293*
11.9	Conclusions *295*
	References *296*

12	**Pincer-Type Iridium Complexes for Organic Transformations** *299*
	Martin Albrecht and David Morales-Morales
12.1	Introduction *299*
12.2	Iridium PCP-Catalyzed Activation of $C(sp^3)$—H Bonds in Unfunctionalized Alkanes *300*
12.2.1	Scope of the Reaction *300*
12.2.2	Mechanistic Considerations *302*
12.2.3	Catalyst Optimization *307*
12.2.4	Application of Alkane Functionalization *309*
12.2.4.1	Alkane Metathesis *309*
12.2.4.2	Polymer Functionalization *310*
12.3	Arene $C(sp^2)$—H and Alkyne $C(sp^1)$—H Bond Activation *311*
12.3.1	Activation of $C(sp^2)$—H Bonds *312*
12.3.2	Activation of $C(sp^1)$—H Bonds *315*
12.4	C—E Bond Activation *317*
12.4.1	Activation of Carbon–Halogen Bonds *317*
12.4.2	Activation of Carbon–Oxygen Bonds *318*
12.4.3	Activation of Carbon–Carbon Bonds *318*
12.5	Ammonia Borane Dehydrogenation *319*
12.6	Conclusions *321*
	Acknowledgments *321*
	References *321*

13	**Iridium-Mediated Alkane Dehydrogenation** *325*
	David Morales-Morales
13.1	Introduction *325*
13.1.1	The Beginning *326*
13.2	Alkane C—H Activation with Ir Derivatives *327*
13.3	Alkane Dehydrogenation with Ir Complexes *328*
13.4	Alkane Dehydrogenation Catalyzed by Ir Pincer Complexes *333*
13.4.1	Ir-PCP Pincer Compounds *333*
13.4.2	Ir-POCOP Pincer Compounds *336*

13.5	Final Remarks *342*	
	Acknowledgments *342*	
	References *342*	

14 Transformations of (Organo)silicon Compounds Catalyzed by Iridium Complexes *345*
Bogdan Marciniec and Ireneusz Kownacki

14.1	Introduction *345*	
14.2	Hydrosilylation and Dehydrogenative Silylation of Carbon–Carbon Multiple Bonds *346*	
14.2.1	Hydrosilylation and Dehydrogenative Silylation of Alkenes *346*	
14.2.2	Application of Hydrosilylation in Polymer Chemistry *348*	
14.2.3	Hydrosilylation and Dehydrogenative Silylation of Alkynes *349*	
14.3	Asymmetric Hydrosilylation of Ketones and Imines *352*	
14.4	Transformation of Organosilicon Compounds in the Presence of Carbon Monoxide *356*	
14.4.1	Hydroformylation of Vinylsilanes *356*	
14.4.2	Silylcarbonylation of Alkenes and Alkynes *357*	
14.5	Silylation of Aromatic Carbon–Hydrogen Bonds *359*	
14.6	Silylation of Alkenes with Vinylsilanes *360*	
14.7	Alcoholysis and Oxygenation of Hydrosilanes *361*	
14.8	Isomerization of Silyl Olefins *361*	
14.9	Addition of silylacetglenes ≡C—H Bonds into Imines *362*	
14.10	Conclusions *364*	
	References *365*	

15 Catalytic Properties of Soluble Iridium Nanoparticles *369*
Jackson D. Scholten and Jaïrton Dupont

15.1	Introduction *369*	
15.2	Synthesis of Soluble Iridium Nanoparticles *369*	
15.2.1	Polyoxoanions *369*	
15.2.2	Surfactants *370*	
15.2.3	Imidazolium Ionic Liquids *372*	
15.3	Kinetic Studies of Iridium Nanoparticle Formation: The Autocatalytic Mechanism *377*	
15.4	Catalytic Applications of Soluble Iridium Nanoparticles *380*	
15.5	Conclusions *387*	
	References *388*	

Index *391*

Preface

The impressive developments of organometallic chemistry during the past 50 years have allowed the preparation of a wide variety of soluble metal complexes useful for organic transformations under mild conditions. Among these metals, the noble triad of ruthenium, rhodium and palladium has played the major role, with rhodium being the most relevant when acquiring knowledge of the mechanisms of metal-mediated organic transformations. In fact, the discovery and application of Wilkinson's catalyst, $RhCl(PPh_3)_3$, proved to be a major milestone during the mid-1960s, mainly because it led to the initiation of many successful applications, including rhodium-catalyzed olefin hydroformylation, the Monsanto production of acetic acid, and industrial asymmetric hydrogenations. Details of some of the major applications of rhodium complexes were updated in a recent book, *Rhodium-Catalyzed Organic Reactions* (edited by P. A. Edwards, published by Wiley-VCH in 2006). In contrast, iridium complexes have been most often used as model compounds to acquire an understanding of the elementary steps of transition metal-catalyzed reactions. A good example of this was 'Vaska's compound', the details of which were first reported in 1965, and which has since provided valuable information on the oxidative addition reactions that serve as key steps in almost every catalytic cycle. Unfortunately, the $IrCl(PPh_3)_3$ complex – which can be seen as analogous to Wilkinson's catalyst – is not a good hydrogenation catalyst, due mainly to the inability of the $IrH_2Cl(PPh_3)_3$ to lose hydrogen. Nonetheless, a variety of cationic iridium complexes with a phosphine : iridium ratio of 1 to 2 have been used extensively as efficient catalysts for alkene hydrogenation, including the industrial enantioselective hydrogenation of imines. Although iridium complexes are less frequently used than their rhodium analogues, in some processes they may be more effective – the carbonylation of methanol being an excellent example in the arena of bulk chemistry. Perhaps the most important catalytic applications of iridium complexes, however, are in the manufacture of fine chemicals, most notably in areas of chemoselective and enantioselective hydrogenation. Iridium complexes have also been shown to play an important role in the enantioselective hydrogenation of C=N, which is used widely in the fine chemicals industry, and also of non-functionalized C=C bonds, with particular interest being centered on the creation of intermediates and building blocks in organic synthesis.

Iridium Complexes in Organic Synthesis. Edited by Luis A. Oro and Carmen Claver
Copyright © 2009 WILEY-VCH Verlag GmbH & Co. KGaA, Weinheim
ISBN: 978-3-527-31996-1

Today, although the value of iridium is becoming increasingly recognized in many organic transformations, it is still not used to any great degree by the chemicals industry. So, it is the aim of this book not only to evaluate the potential of the most promising reactions that involve iridium complexes, but also to provide an account of the role of these materials in important organic transformations such as hydrogenation, hydroamination, hydroboration, C—C bond formation, carbonylation and cycloadditions. The use of relevant ligands as carbenes or pincer ligands, as well as recent catalytic systems using iridium nanoparticles, are also described.

We, the editors, believe that organoiridium chemistry has not only a rich past, but that the application of iridium complexes in organic transformations also has a brilliant future!

October, 2008

Luis A. Oro, Zaragoza
Carmen Claver, Tarragona

List of Contributors

Martin Albrecht
University of Fribourg
Department of Chemistry
Chemin du Musée 9
1700 Fribourg
Switzerland

Claudio Bianchini
Institute of Chemistry of
  Organometallic Compounds
National Research Council
  (ICCOM-CNR)
Via Madonna del Piano 10
50019 Sesto Fiorentino
Firenze
Italy

Hans-Ulrich Blaser
Solvias AG
P.O. Box
4002 Basel
Switzerland

Carmen Claver
Universitat Rovira i Virgili
Dept. de Química Fisica i Jnorgànica
Facultat de Química
c/ Marcel.lí domingo, s/n
43007 Tarragona
Spain

Robert H. Crabtree
Yale University
Department of Chemistry
225 Prospect Street
New Haven, CT 06520-8107
USA

Romano Dorta
Universidad Simón Bolívar
Departamento de Quimica
Valle de Sartenejas
1080 A Baruta – Caracas
Venezuela

Jaïrton Dupont
Laboratory of Molecular Catalysis
Institute of Chemistry
UFRGSAv. Bento Gonçalves
9500 Porto Alegre 91501-970 RS
Brazil

Elena Fernández
Facultat de Químicas
Universitat Rovira i Virgili
Campus Sescelades
C. Marcel.lí Domingo
43007 Tarragona
Spain

Iridium Complexes in Organic Synthesis. Edited by Luis A. Oro and Carmen Claver
Copyright © 2009 WILEY-VCH Verlag GmbH & Co. KGaA, Weinheim
ISBN: 978-3-527-31996-1

Ken-ichi Fujita
Graduate School of Human and
  Environmental Studies
Kyoto University
Yoshida
Sakyo-ku
Kyoto 606-8501
Japan

Luca Gonsalvi
Institute of Chemistry of
  Organometallic Compounds
National Research Council
  (ICCOM-CNR)
Via Madonna del Piano 10
50019 Sesto Fiorentino
Firenze
Italy

Günter Helmchen
Universität Heidelberg
Organisch-Chemisches Institut
Im Neuenheimer Feld 270
69120 Heidelberg
Germany

Yasutaka Ishii
Kansai University
Department of Chemistry and
  Materials Engineering
Faculty of Chemistry
Materials and Bioengineering
Suita
Osaka 564-8680
Japan

Philippe Kalck
Université de Toulouse
Laboratoire de Chimie de
  Coordination du CNRS
UPR 8241
Equipe Catalyse et Chimie Fine
Composante INP-ENSIACET
118 route de Narbonne
31077 Toulouse
France

Ireneusz Kownacki
Adam Mickiewicz University
Department of Organometallic
  Chemistry
Faculty of Chemistry
Grunwaldzka 6
60-780 Poznan
Poland

Bogdan Marciniec
Adam Mickiewicz University
Department of Organometallic
  Chemistry
Faculty of Chemistry
Grunwaldzka 6
60-780 Poznan
Poland

David Morales-Morales
Universidad Nacional Autonoma de
  México
Instituto de Quimica
Circuito Exterior S/N
Ciudad Universitaria
Mexico D.F. Coyoacan
C. P. 04510
Mexico

Yasushi Obora
Kansai University
Department of Chemistry and
  Materials Engineering
Faculty of Chemistry
Materials and Bioengineering
Suita
Osaka 564-8680
Japan

Luis A. Oro
Instituto Universitario de Catálisis
  Homogénea
Department of Inorganic Chemistry
Faculty of Science-ICMA
University of Zaragoza-CSIC
50009 Zaragoza
Spain

Eduardo Peris
Universitat Jaume I
Department of Química Inorgánica y
 Orgánica
Avenue Vicente Sos Baynat s/n,
12080 Castellón
Spain

Maurizio Peruzzini
Institute of Chemistry of
 Organometallic Compounds
National Research Council
 (ICCOM-CNR)
Via Madonna del Piano 10
50019 Sesto Fiorentino
Firenze
Italy

Satoshi Sakaguchi
Kansai University
Department of Chemistry and
 Materials Engineering
Faculty of Chemistry
Materials and Bioengineering
Suita
Osaka 564-8680
Japan

Jackson D. Scholten
Laboratory of Molecular Catalysis
Institute of Chemistry
UFRGSAv. Bento Gonçalves
9500 Porto Alegre 91501-970 RS
Brazil

Anna M. Segarra
Escola Técnica Superior d'Enginyeria
 Quimica
Universitat Rovira i Virgili
Campus Sescelades
C. Marcel.lí Domingo
43007 Tarragona
Spain

Philippe Serp
Université de Toulouse
Laboratoire de Chimie de
 Coordination du CNRS
UPR 8241
Equipe Catalyse et Chimie Fine
Composante INP-ENSIACET
118 route de Narbonne
31077 Toulouse
France

Takanori Shibata
Waseda University
Department of Chemistry
School of Advanced Science &
 Engineering
Ohkubo
Shinjuku Tokyo 169-8555
Japan

Ryohei Yamaguchi
Graduate School of Human and
 Environmental Studies
Kyoto University
Yoshida
Sakyo-ku
Kyoto 606-8501
Japan

1
Application of Iridium Catalysts in the Fine Chemicals Industry
Hans-Ulrich Blaser

1.1
Introduction

Traditionally, catalysis is associated with the production of bulk chemicals, whereas fine and specialty chemicals are produced predominantly with noncatalytic organic synthesis. While these statements are basically still correct, there is a growing list of examples demonstrating that catalytic processes are also an opportunity for the production of more complex agrochemicals and pharmaceuticals [1]. However, our experience shows, that only relatively few catalytic transformations are actually applied on a larger scale and it is probably not surprising that the most important catalytic applications in fine chemicals manufacture are in the area of chemoselective and enantioselective hydrogenation [2, 3]. Not surprisingly, this is also the case for the – until now – rather scarce industrial applications of Ir-catalyzed reactions. For this reason, the enantioselective hydrogenation of C=N and of nonfunctionalized C=C bonds will be discussed in some detail, while miscellaneous catalytic transformations with industrial potential will be summarized only briefly.

1.2
Industrial Requirements for Applying Catalysts

In order to understand the challenges facing the application of catalysts in the fine chemicals industry, one has to understand not only the essential industrial requirements but also how process development is carried out, and which criteria determine the suitability of a catalyst.

1.2.1
Characteristics of the Manufacture of Enantiomerically Pure Products

The manufacture of chiral fine chemicals such as pharmaceuticals or agrochemicals can be characterized as follows (numbers given in parentheses reflect the experience of the authors):

Iridium Complexes in Organic Synthesis. Edited by Luis A. Oro and Carmen Claver
Copyright © 2009 WILEY-VCH Verlag GmbH & Co. KGaA, Weinheim
ISBN: 978-3-527-31996-1

- Multifunctional molecules produced via multistep syntheses (five to 10 or more steps for pharmaceuticals, and three to seven steps for agrochemicals) with short product lives (often <20 years).
- Relatively small-scale products (1–1000 t y^{-1} for pharmaceuticals, 500–10 000 t y^{-1} for agrochemicals), usually produced in multipurpose batch equipment [continuous processes are very rare and catalyst recycling is cumbersome, especially under a good manufacturing practice (GMP) regime].
- High-purity requirements (usually >99% and <10 ppm metal residue in pharmaceuticals).
- High added values and therefore tolerant to relative high process costs (especially for very effective, small-scale products).
- Short development time for the production process (less than a few months to 1–2 years), since the time to market affects the profitability of the product and, in addition, development costs for a specific compound must be kept low as process development starts at an early phase when the chance of the product to ever 'make it' is about 10%.
- At least in European companies, chemical development is carried out by all-round organic chemists, sometimes in collaboration with technology specialists. For this reason, catalysts must be available commercially as there is little expertise (and time) for making complex catalysts.

1.2.2
Process Development: Critical Factors for the Application of Catalysts

It is useful to divide the development of a manufacturing process into different phases:

Phase 1: Outlining and assessing possible synthetic routes on paper. Here, the decision is made whether to use chemocatalytic steps for making the desired product or to use classical organic synthesis. This decision will depend on a number of considerations, such as the goal of the development, the know-how of the investigators, the time frame, or the available manpower and equipment.

Phase 2: Demonstrating the chemical feasibility of the key step – which often is the catalytic reaction – and showing that the catalytic step fits into the overall synthetic scheme.

Phase 3: Optimizing the key catalytic reaction as well as the other steps. Showing the technical feasibility (catalyst separation, metal impurities, etc.)

In Phases 2 and 3, it is not only the results of the catalyst tests (selectivity, activity, productivity, catalyst costs, etc.) but also the total product costs that decide whether the catalytic route will be further developed, or abandoned.

Phase 4: Optimizing and scale-up of the catalytic step as well as of the overall process.

In the final analysis, the choice as to whether a synthesis with an catalytic step is selected depends very often on the answers to two questions:

- Can the costs for the overall manufacturing process compete with alternative routes?
- Can the catalytic step be developed in the given time frame?

1.2.3
Requirements for Practically Useful Catalysts

Based on the published results and our own experience, we consider that several requirements should be met in order to make catalysts feasible for industrial applications.

1.2.3.1 Preparation Methods
It is not really possible to predict what type of catalyst will be suitable for a given substrate and process. Therefore, flexible methods that allow different combinations of the most important catalyst components such as metal, ligand or support are vitally important.

1.2.3.2 Catalysts Cost
The catalyst costs will only be important later, when the costs of goods of the desired product are compared. For homogeneous catalysts, the (chiral) ligand often is the most expensive component (typical prices for the most important chiral phosphines are 100 to 500 \$ g^{-1} for laboratory quantities and 5000 to >40 000 \$ kg^{-1} on a larger scale).

1.2.3.3 Availability of the Catalysts
If a specific catalyst is not available at the right time, and in the appropriate quantity, it will not be applied due to time limitations for process development. Today, a sizable number of homogeneous catalysts and ligands (especially for hydrogenation) are available commercially in technical quantities.

1.2.3.4 Catalytic Performance
The performance concerning selectivity, activity, productivity and so on, must fulfill the requirements of a given product. As a rule of thumb, (enantio)selectivity should be >90–95%; turnover numbers (TON) should be >200 for small volume products and reactions with high added value and >10 000 for large-volume products; and reaction times should not exceed 5–10 h. Furthermore, functional group tolerance will often be important as most substrates will have other functional groups which can either be reduced or which can interfere with the catalyst via complexation.

1.2.3.5 Separation

Although the separation of homogeneous catalysts can be problematic, in many cases product distillation or crystallization and sometimes extraction is possible. Trace metals can also be problematic as the limit for metal contamination in biologically active compounds is in the ppm region. An additional purification step such as adsorption is sometimes required.

1.3 Enantioselective Hydrogenation of C=N Bonds

1.3.1 Catalysts and Scope

Even though the asymmetric hydrogenation of C=N has been investigated less systematically than that of C=C and C=O groups, the topic is of relevance here as chiral amines are important intermediates for biologically active compounds. The successful implementation of a large-scale process to produce the herbicide (S)-metolachlor, catalyzed by an Ir–josiphos catalyst (see below), has motivated increased efforts. As a consequence, in recent years a variety of catalytic systems has been described for the enantioselective hydrogenation of various C=N functions [4]. Ir-based catalysts have been shown to be very effective for the hydrogenation of N-aryl amines, selected endocyclic C=N bonds and of quinolines; a summary of the most important results reported up to September 2007 is provided in Table 1.1. It must be pointed out that, with very few exceptions, all results have been obtained with monofunctional model substrates (see Figure 1.1) and optimization was focused on enantioselectivity. As a consequence, the turnover number (TON) and turnover frequency (TOF) of most catalysts are (as yet) insufficient for industrial applications, and only time will tell whether real world substrates can be hydrogenated with the same high enantioselectivities.

Table 1.1 The state of the art for the Ir catalyzed reduction of various C=N groups.

Substrate	ee (%)[a]	TON[a]	TOF (h^{-1})[a]	Ligands[b]
N-Aryl imines	90->99	100–200	2–50	t-Bu-bis-P* [5], f-binaphane [6], 1[7], 2[8], 3[9], josiphos [10], 4[11]
Endocyclic imines	86–95	100–1000	2–300	binap [12, 13], bicp [14], josiphos [10], bcpm [15]
Quinolines	85–97	100–1000	5–80	5 [16], MeO-biphep [17], 6 [18], P-phos [19], H8-binapo [20], 7 [21]

a Typical range for suitable substrate (Figure 1.1) and optimized catalyst.
b For structures, see Figure 1.2.

1.3 Enantioselective Hydrogenation of C=N Bonds

N-aryl imines — **endocyclic imines** — **quinolines**

Figure 1.1 Structure of model substrates.

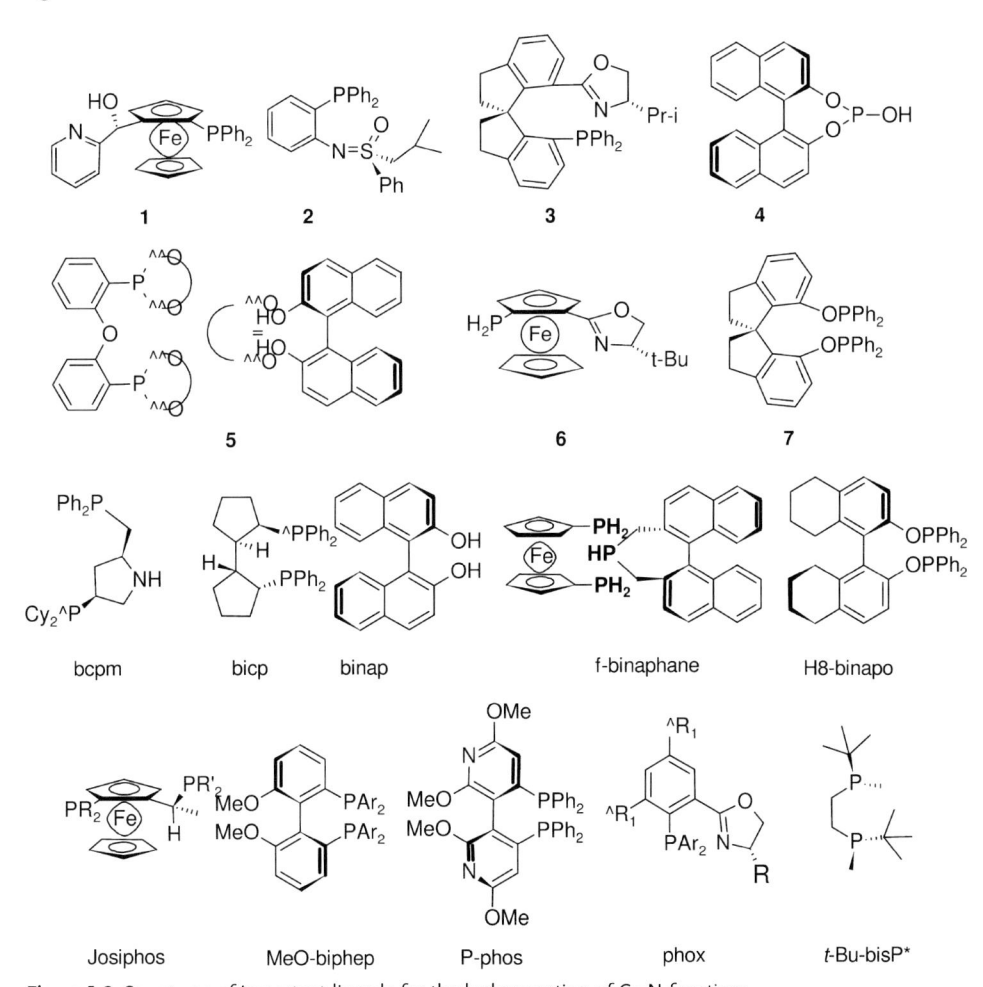

Figure 1.2 Structures of important ligands for the hydrogenation of C=N functions.

Here, a few comments should be made with regards to Table 1.1. It is clear that for all three substrate classes, enantioselectivities can range from good to very high, whereas the TON – and especially TOF – values are less impressive (and in most cases have not been optimized). Most catalyst systems require pressures of 20–100 bar ($2-10 \times 10^4$ hPa) to achieve realistic reaction times but, as a rule, the chemical yields are very high. In the presence of Ti(OiPr)$_4$, Ir/f-binaphane catalyzes the reductive amination of aromatic ketones with a variety of substituted anilines with high enantiomeric excess (ee) values but relatively low catalyst activity [6]. It was also shown that Ir–phox catalysts work well in supercritical carbon dioxide (sCO_2)-ionic liquid systems, allowing easy separation and recycling of the catalyst with good catalyst performance [22]. Although the hydrogenation of substituted heteroaromatic substrates provides access to a variety of cyclic amines, this is a very demanding reaction for any homogeneous catalyst, and very few substrate types have been hydrogenated successfully. A patent described the Ir–josiphos catalyzed hydrogenation of a pyrazine amide with up to 77% ee but very low catalyst activity (TON ~20, TOF $1 h^{-1}$) [23]. On the other hand, very impressive results were reported for the Ir-catalyzed hydrogenation of various substituted quinoline derivatives, again at high pressures.

1.3.2
Industrial Applications

Until now, only a small number of industrial applications has been reported. The metolachlor process was originally was developed in the Central Research Laboratories of Ciba-Geigy (now Solvias), but is now operated by Syngenta. With a volume of >10 000 ty^{-1} it is the largest known enantioselective catalytic production process [24–26].

Metolachlor is the active ingredient of Dual™, one of the most important grass herbicides for use in maize and a number of other crops. In 1997, after years of intensive research, Dual Magnum™, with a content of approximately 90% (1'S)-diastereomers and with the same biological effect at about 65% of the use rate, was introduced into the market. This 'chiral switch' was made possible by the new technical process that is briefly described below. The key step of this new synthesis is the enantioselective hydrogenation of the isolated MEA imine, as depicted in Figure 1.3.

The search for a commercially viable process took many years, during which time several approaches with Rh or Ir complexes using commercially available diphosphine ligands proved to be unsuccessful. However, a critical breakthrough was achieved when Ir complexes with a new class of ferrocenyl based ligands – now known as Solvias Josiphos – were used. Especially in the presence of acid and iodide ions, extremely active and productive catalysts were obtained, and conse-

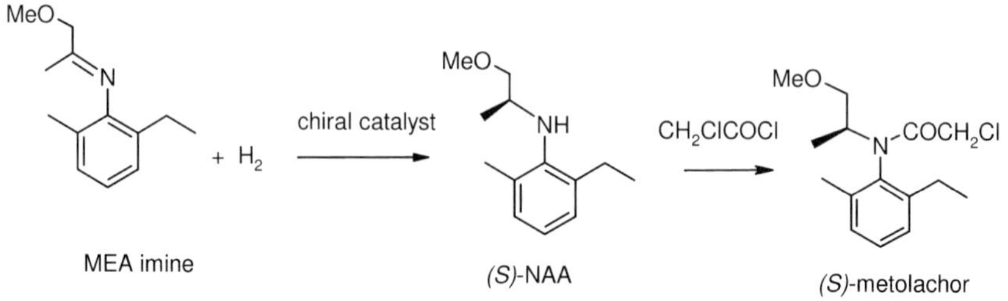

Figure 1.3 Enantioselective synthesis of (S)-metolachlor.

Table 1.2 The most successful Josiphos ligands for the hydrogenation of MEA imine (for ligand structures, see Figure 1.2).

R	R'	TON	TOF (h^{-1})	ee (%)	Comments
Ph	3,5-xylyl	1 000 000	>300 000	79	Production process
p-CF_3Ph	3,5-xylyl	800	400	82	Ligand screening
Ph	4-tBu-C_6H_4	5000	80	87	Low temperature
Ph	4-(n-Pr)$_2$N-3,5-xyl	100 000	28 000	83	Optimized conditions

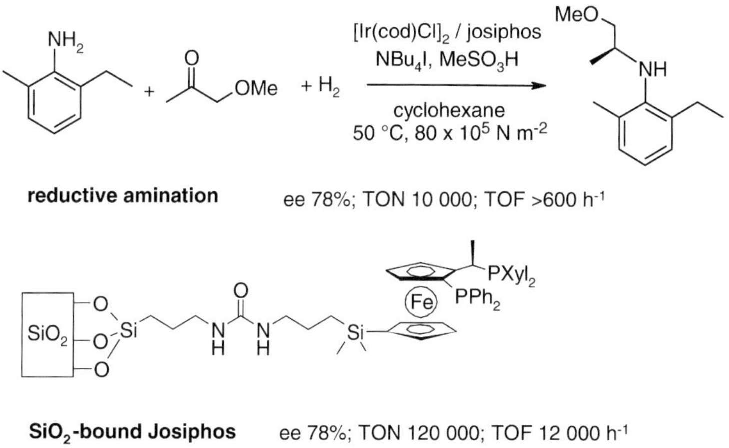

reductive amination   ee 78%; TON 10 000; TOF >600 h^{-1}

SiO$_2$-bound Josiphos   ee 78%; TON 120 000; TOF 12 000 h^{-1}

Figure 1.4 Alternative variants for the production of (S)-metolachlor.

quently several different Josiphos ligands were tested. A selection of the best results obtained is provided in Table 1.2.

The optimized process is carried out in a loop reactor at 80 bar (8×10^4 hPa) hydrogen and 50 °C with a catalyst generated *in situ* from [Ir(cod)Cl]$_2$ and the Josiphos ligand R = Ph, R' = Xyl at a substrate-to-catalyst ratio of >1 000 000 in the presence of trace amounts of HI. Complete conversion is reached within 3–4 h, the initial TOFs exceed 1 800 000 h^{-1}, and the ee-value is about 80%. The product (S)-NAA is distilled and the catalyst discarded. Today, this process is operated by Syngenta on a scale of >10 000 t y^{-1}.

Two alternatives to the homogeneous hydrogenation of the isolated MEA imine have been investigated (see Figure 1.4). The first method involved the direct amination of MEA with methoxyacetone in order to avoid isolation of the MEA imine [27]; the second method involved the use of immobilized josiphos ligands in order to avoid product distillation [28]. While both variants led to the desired product with similar ee-values as the homogeneous process, both the TON and TOF were insufficient for a technical application.

An Ir-catalyzed hydrogenation developed by Lonza for an intermediate of dextromethorphan was carried out on >100 kg scale [29]. Here, the important success factors were the ligand fine tuning and the use of a biphasic system. The

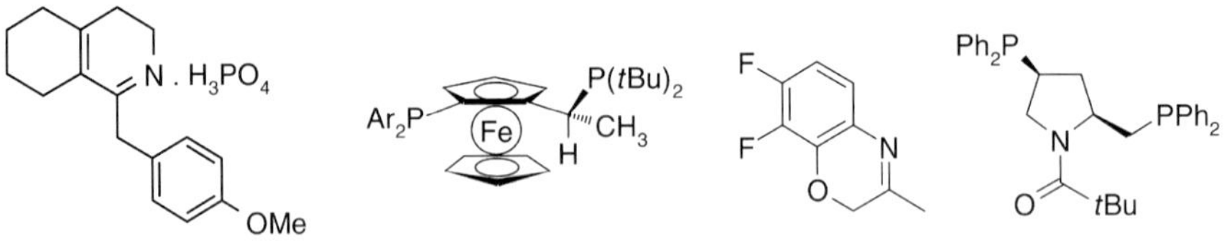

Figure 1.5 Industrial hydrogenation of various C=N bonds.

Ir / Josiphos
20 °C, 70 x 10^5 N m^{-2}
ee 90%; TON 1500; TOF n.a.
pilot process, Solvias/Lonza

Josiphos
Ar = 3,5-Me$_2$-4-OMe-Ph

Ir / bppm / BiI$_3$
-10 °C, 40 x 10^5 N m^{-2}
ee 90%; TON 100; TOF ~30 h^{-1}
feasibility study, Daiichi

bppm

chemoselectivity was high but catalyst productivity rather low for an economical technical application. Satoh *et al.* reported up to 90% ee for the hydrogenation of an intermediate of the antibiotic levofloxacin using Ir–diphosphine complexes. Best results were obtained with bppm and modified diop ligands in the presence of bismuth iodide at low temperature [30] (Figure 1.5).

1.4
Enantioselective Hydrogenation of C=C Bonds

1.4.1
Catalysts and Scope

Generally, the enantioselective hydrogenation of C=C bonds is the domain of Rh and Ru catalysts [3, 31]. Whilst this is due to the high effectiveness of these catalysts, it is still somewhat surprising as one of the most active catalysts for hindered C=C bonds is Crabtree's [Ir(cod)(PCy$_3$(py)]PF$_6$ catalyst [32]. The very remarkable exception is the hydrogenation of alkenes without 'privileged' functional groups, which is much more difficult. Here, Ir catalysts are clearly superior, especially in presence of a variety of P,N ligands such as the above-mentioned phox ligands, originally developed by Pfaltz [33, 34] which, for the first time, gave high enantioselectivities for unfunctionalized C=C (see Figure 1.6) as well as for cyclic analogues. Later, Burgess reported a carbene/N ligand **9** that was similarly effective [35]. Today, a range of ligands exists which with can hydrogenate what Burgess refers to as 'largely unfunctionalized alkenes' with enantioselectivities >99%. Most catalysts require pressures of 5–50 bar (0.5–5 × 10^4 hPa) and, with few exceptions, the TON (typically 50–200) and TOF (typically 10–100 h^{-1}) of most catalysts are relatively low. Pfaltz has shown however that, under optimized conditions, a TON of up to 5000 can be realized and that the reactions are initially very rapid, although catalyst deactivation may be problematic. One special feature of these Ir/PN catalysts is their high sensitivity to anions (by far the best results are obtained with [Ir(ligand)(cod)]BARF complexes) and to protic and polar solvents. The very high activity of the selected Ir/PN system was confirmed with the effective hydrogenation of tetrasubstituted C=C bonds as well as (benzo)furanes (Figure 1.6), with

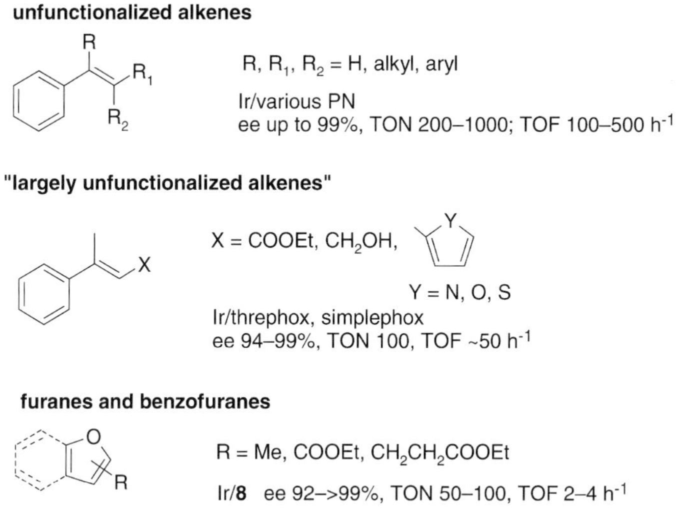

1.4 Enantioselective Hydrogenation of C=C Bonds

unfunctionalized alkenes

R, R$_1$, R$_2$ = H, alkyl, aryl

Ir/various PN
ee up to 99%, TON 200–1000; TOF 100–500 h^{-1}

"largely unfunctionalized alkenes"

X = COOEt, CH$_2$OH,

Y = N, O, S

Ir/threphox, simplephox
ee 94–99%, TON 100, TOF ~50 h^{-1}

furanes and benzofuranes

R = Me, COOEt, CH$_2$CH$_2$COOEt

Ir/**8** ee 92–>99%, TON 50–100, TOF 2–4 h^{-1}

Figure 1.6 Hydrogenation of olefins without privileged substitution and of (benzo)furanes.

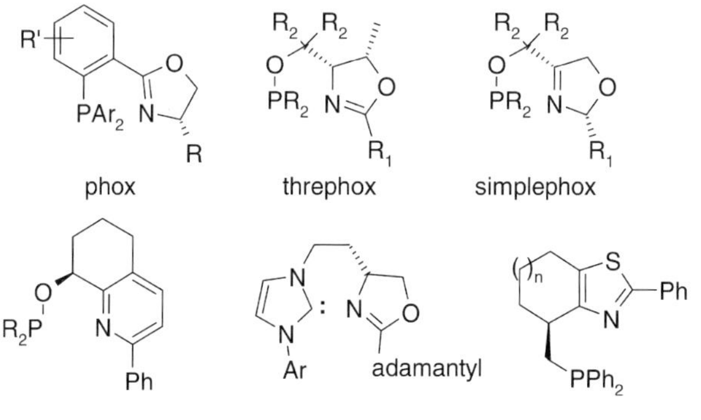

phox   threphox   simplephox

8   **9**   **10**

Figure 1.7 Selected PN ligands.

very high enantioselectivities and decent TONs. Although purely alkyl-substituted olefins are still difficult substrates, Ir catalysts with selected PN ligands (e.g. **8**) have recently been identified that perform well in this case. As this would be beyond the scope of this chapter, at this point we simply depict some of the most effective ligands (see Figure 1.7) and refer the reader to reviews by Pfaltz [34] (phox, threphox, simplephox, **8**), Burgess [35] (ligand **9**) and Andersson [36] (ligand **10**).

1.4.2
Industrial Applications

Although no large-scale Ir-catalyzed C=C hydrogenation processes have yet been reported, two interesting cases indicate their industrial potential. Yue and Nugent

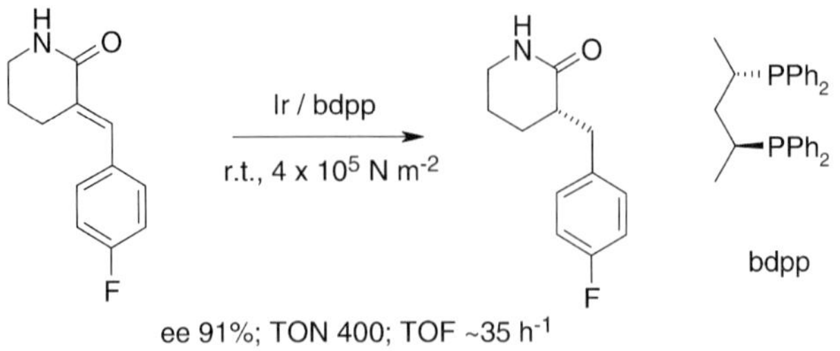

Figure 1.8 Hydrogenation of an α,β-unsaturated lactam.

[37], at Bristol-Meyers Squibb, described the hydrogenation of an exocyclic α,β-unsaturated lactam which could be carried out with an Ir/bdpp catalyst with up to 91% ee (Figure 1.8). The catalyst—which was selected after an intensive screening—involving 32 chiral phosphines and eight Ru, Rh and Ir precursor complexes. The process was carried out on a 20-kg scale but, presumably, the TON and TOF were too low for consideration as a manufacturing process.

The power of the Ir/PN catalysts was demonstrated in a feasibility study carried out by Pfaltz in collaboration with DSM for the hydrogenation of γ-tocotrienyl acetate (Figure 1.9) [38]. This reaction involves the reduction of three C=C bonds and creates two new stereocenters in a single step. Because the two prochiral C=C units are both *E*-configured, the sense of asymmetric induction at the two reaction sites is expected to be the same. The best stereoselectivity was achieved with Ir/8 (R = 2-Tol), which gave almost exclusively the natural (*RRR*)-isomer of γ-tocopheryl acetate, generating two stereogenic centers in a single step with virtually perfect stereocontrol. Because these catalysts do not require the presence of any particular directing substituents near the C=C bond, they should be applicable to a much wider range of olefins than Rh or Ru catalysts. This was actually confirmed by the hydrogenation of the four stereoisomers of farnesol, where all four stereoisomers of the product were accessible with ≥99% enantioselectivity using the same catalyst but starting from the corresponding *E* or *Z* C=C bond leading to *R* and *S* configuration, respectively [34].

1.5
Miscellaneous Catalytic Applications with Industrial Potential

Besides the reaction discussed above, several Ir-catalyzed reactions are available which probably have some potential to be applied in industrial syntheses.

Asymmetric allylic substitutions are widely applied in organic synthesis, using various metal complexes, chiral ligands, nucleophiles and allyl systems [39]. Although Pd is often the metal of choice, this is not the case for monosubstituted allylic substrates, where most Pd catalysts predominantly produce the achiral linear product. In contrast, Mo, W and Ir catalysts preferentially give rise to the desired branched products and, in recent years, a number of very effective Ir catalysts for various substrates have been developed [40]. Since, to the best of our

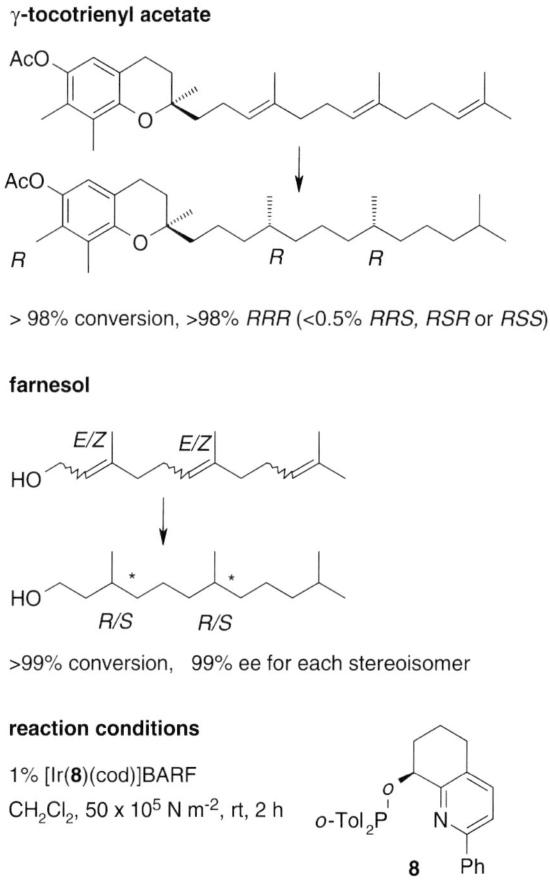
γ-tocotrienyl acetate
AcO
O
AcO
O
R
R
R
> 98% conversion, >98% RRR (<0.5% RRS, RSR or RSS)
farnesol
E/Z
E/Z
HO
HO
R/S
R/S
>99% conversion, 99% ee for each stereoisomer
reaction conditions
1% [Ir(8)(cod)]BARF
CH2Cl2, 50 x 10^5 N m^-2, rt, 2 h
O
o-Tol2P
N
8
Ph

Figure 1.9 Hydrogenation of γ-tocotrienyl acetate and farnesol.

knowledge, no industrial application has yet been described, we provide only a brief summary at this point (see Figure 1.10). While the first results have been obtained with Ir/phox catalysts, phosphorous amidites are the ligands of choice for reactions with malonates (and related compounds), aliphatic nitro compounds, ketone enolates, various N-nucleophiles (amines, sulfonyl- and acylamines), phenolates and Cu-alcoholates. Reactions are usually carried out in the presence of $[Ir(cod)Cl]_2$ and 2 equivalents of ligand (Ir/L = 1) at about room temperature and with rather low substrate-to-catalyst ratios of between 5 and 50. The yields are generally good to high, with branched products often >90% and ee-values of typically 90–97%. The TOFs vary between 5 and 10 h^{-1}. Pybox (8–12%), under similar conditions, is preferred for the N-allylation with hydroxylamines, and O-allylation with hydroxylamines and oximes. Yields vary between 70 and 95%, typically with 70–95% branched and ee-values of 80–95%. The high enantioselectivities obtained with various model compounds indicate considerable synthetic potential, even though the TON and TOF are still rather low.

effective ligands

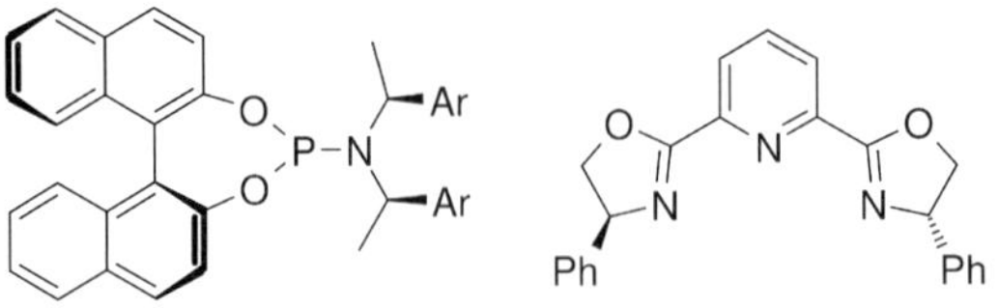

phosphorous amidites   pybox

allylic substrates

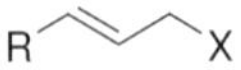

R⎯⎯⎯X

R = Ar, Alk   X = OAc, OCO$_2$R, OP(O)(OR)$_2$

Figure 1.10 Ligands and allylic substrates for Ir-catalyzed allylic substitution reactions.

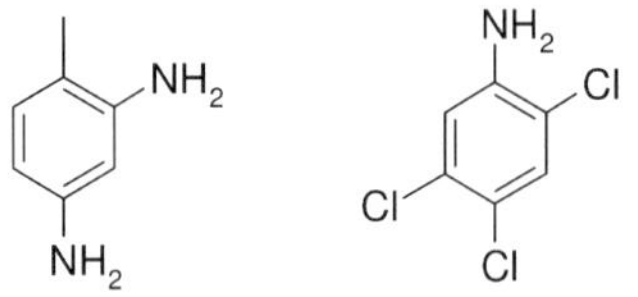

Catalyst	Yield (%)	Yield (%)
Pt/C	n.a.	96.0
Pd/C	98.2	n.a.
Ir/C	n.a.	>98.5
Ir-Fe-Cu/C	n.a.	>99.3
Ir-Fe-Co/C	99.4	n.a.

Figure 1.11 Hydrogenation of nitroarenes with Ir/C catalysts.

Ir–diphosphine complexes have been shown to catalyze a number of potentially interesting C—C coupling reactions [41]. Special mention should be made of the various transformations of alkynes (cycloaddition, isomerization, dimerization, alkynylation), selected aldol and carbonylation reactions, even though the synthetic potential of some of these reactions has not been explored.

Although the focus of this chapter is on homogeneous Ir catalysis, it would be advantageous at this point briefly to describe the unusual selectivity of a number of heterogeneous Ir/C catalysts developed by Degussa for the chemoselective hydrogenation of chloro nitroarenes [42] and dinitrotoluene [43] (Figure 1.11). The results obtained for 2,4-dinitrotoluene and the rather demanding substrate 2,4,5-trichloronitrobenzene, show that especially Ir (5%) catalysts doped with 0.15% of Fe and Cu or Co are inherently more selective than the Pd and Pt catalysts classically used for these substrates. Both, the selectivity with respect to hydrodechlorination and the formation of byproducts (such as tars in the case of dinitrotoluene) were significantly decreased. These catalysts are commercially available from Degussa and have been applied to unspecified industrial problems on a development scale (D. Ostgard, Degussa, personal communication).

Besides preparative applications, Ir complexes such as Ir(cod)(acac) or [cp*Ir(PMe$_3$)H$_3$]OTf have successfully been applied for hydrogen/deuterium (H/D) exchange reactions, which are important for producing labeled compounds for kinetic and metabolic studies in the pharmaceutical industries [44]. While we do not know of any 'real world' applications, it has been shown that both selective as well as extensive deuteration are possible, depending on the Ir catalyst and the deuterating agent.

1.6
Conclusions and Outlook

To date, only a few iridium catalysts have been applied to industrially relevant targets, especially on the larger scale. It is likely that several types of Ir catalyst are, in principle, feasible for technical applications in the pharmaceutical and agrochemical industries. At present, the most important problems are the relatively low catalytic activities of many highly selective systems and the fact, that relatively few catalysts have been applied to multifunctional substrates. For this reason, the scope and limitations of most catalysts known today have not yet been explored. For those in academic research, the lesson might be to employ new catalysts not only with monofunctional model compounds but also to test functional group tolerance and – as has already been done in some cases – to apply the catalysts to the total synthesis of relevant target molecules.

References

1 For an overview on the significance of catalysis in fine chemicals production, see Blaser, H.U. and Studer, M. (1999) *Applied Catalysis A: General*, **189**, 191.

2 For an overview of the application of enantioselective catalysis in fine chemicals production, see (a) Blaser, H.U., Spindler, F. and Studer, M. (2002) *Applied Catalysis A: General*, **221**, 119; (b) Blaser, H.U. and Schmidt, E. (eds) (2003) *Large-Scale Asymmetric Catalysis*, Wiley-VCH Verlag GmbH, Weinheim, p. 1.

3 For an assessment of enantioselective catalysis from an industrial point of view, see Blaser, H.U., Pugin, B. and Spindler, F. (2005). *Journal of Molecular Catalysis A – Chemical*, **231**, 1.

4 For a recent review, see Blaser, H.U. and Spindler, F. (2007) *Handbook of Homogeneous Hydrogenation* (eds J.G. de Vries and C.J. Elsevier), Wiley-VCH Verlag GmbH, p. 1193.

5 Imamoto, T., Iwadate, N. and Yoshida, K. (2006) *Organic Letters*, **8**, 2289.

6 Xiao, D. and Zhang, X. (2001) *Angewandte Chemie – International Edition*, **40**, 3425.

7 Cheemala, M.N. and Knochel, P. (2007) *Organic Letters*, **9**, 3089.

8 Moessner, C. and Bolm, C. (2005) *Angewandte Chemie – International Edition*, **44**, 7564.

9 Zhu, S.-F., Xie, J.-B., Zhang, Y.-Z., Li, S. and Zhou, Q.-L. (2006) *Journal of the American Chemical Society*, **128**, 12886.

10 Blaser, H.U., Buser, H.P., Häusel, R., Jalett, H.P. and Spindler, F. (2001) *Journal of Organometallic Chemistry*, **621**, 34.

11 Reetz, M.T. and Bondarev, O. (2007) *Angewandte Chemie – International Edition*, **46**, 4523.

12 Yamagata, T., Tadaoka, H., Nagata, M., Hirao, T., Kataoka, Y., Ratovelomana, V., Genêt, J.P. and Mashima, K. (2006) *Organometallics*, **25**, 2505.
13 Morimoto, T., Suzuki, N. and Achiwa, K. (1998) *Tetrahedron: Asymmetry*, **9**, 183.
14 Zhu, G. and Zhang, X. (1998) *Tetrahedron: Asymmetry*, **9**, 2415.
15 Morimoto, T., Suzuki, N. and Achiwa, K. (1996) *Heterocycles*, **43**, 2557.
16 Reetz, M.T. and Li, X. (2006) *Chemical Communications*, 2159.
17 Wang, W.-B., Lu, S.-M., Yang, P.-Y., Han, X.-W. and Zhou, Y.-G. (2003) *Journal of the American Chemical Society*, **125**, 10536.
18 Lu, S.-M., Han, X.-W. and Zhou, Y.-G. (2004) *Advanced Synthesis Catalysis*, **346**, 905.
19 Xu, L., Lam, K.H., Ji, J., Wu, J., Fan, Q.-H., Lo, W.-H. and Chan, A.S.C. (2005) *Chemical Communications*, 1390.
20 Lam, K.H., Xu, L., Feng, L., Fan, Q.-H., Lam, F.L., Lo, W.-G. and Chan, A.S.C. (2005) *Advanced Synthesis Catalysis*, **347**, 1755.
21 Tang, W.-J., Zhu, S.-F., Xu, L.-J., Zhou, Q.-L., Fan, Q.-H., Fan, Q.-H., Zhou, H.-F., Lam, K. and Chan, A.S.C. (2007) *Chemical Communications*, 613.
22 Solinas, M., Pfaltz, A., Cozzi, P.G. and Leitner, W. (2004) *Journal of the American Chemical Society*, **126**, 16142.
23 Fuchs, R. (1997) E P 0803502 A2, assigned to Lonza AG.
24 Blaser, H.U., Buser, H.P., Coers, K., Hanreich, R., Jalett, H.P., Jelsch, E., Pugin, B., Schneider, H.D., Spindler, F. and Wegmann, A. (1999) *Chimia*, **53**, 275.
25 Blaser, H.U. (2002) *Advanced Synthesis Catalysis*, **344**, 17.
26 Hofer, R. (2005) *Chimia*, **59**, 10.
27 Blaser, H.U., Buser, H.P., Jalett, H.P., Pugin, B. and Spindler, F. (1999) *Synlett*, 867.
28 Pugin, B., Landert, H., Spindler, F. and Blaser, H.U. (2002) *Advanced Synthesis Catalysis*, **344**, 974.
29 Imwinkelried, R. (1997) *Chimia*, **51**, 300.
30 Satoh, K., Inenaga, M. and Kanai, K. (1998) *Tetrahedron: Asymmetry*, **9**, 2657.
31 de Vries, J.G. and Elsevier, C.J. (2007) *Handbook of Homogeneous Hydrogenation*, Wiley-VCH Verlag GmbH, p. 1586.
32 (a) Crabtree, R.H., Felkin, H., Fillebeen-Khan, T. and Morris, G.E. (1979) *Journal of Organometallic Chemistry*, **168**, 183; (b) Crabtree, R.H. (1979) *Accounts of Chemical Research*, **12**, 331.
33 Lightfoot, A., Schnider, P. and Pfaltz, A. (1998) *Angewandte Chemie – International Edition*, **37**, 2897.
34 For a recent update, see Roseblade, S.J. and Pfaltz A. (2007) *Accounts of Chemical Research*, **40**, 1402–11.
35 Cui, X. and Burgess, K. (2005) *Chemical Reviews*, **105**, 3272.
36 Källström, K., Munslow, I. and Andersson, P.G. (2006) *Chemistry – A European Journal*, **12**, 3194.
37 Yue, T.-Y. and Nugent, W.A. (2002) *Journal of the American Chemical Society*, **124**, 13692.
38 Bell, S., Wüstenberg, B., Kaiser, S., Menges, F., Netscher, T. and Pfaltz, A. (2006) *Science*, **311**, 642.
39 Trost, B.M. and Lee, C. (2000) *Catalytic Asymmetric Synthesis*, 2nd edn (ed. I. Ojima), Wiley-VCH Verlag GmbH, New York, p. 593.
40 For a recent overview, see Helmchen, G., Dahnz, A., Dübon, P., Schelwies, M. and Weihofen, R. (2007) *Chemical Communications*, 675.
41 For an update, see Takeuchi, R., Kezuka, S. (2006) *Synthesis*, 3349.
42 Auer, E., Gross, M., Panster, P. and Takemoto, K. (2001) *Catalysis Today*, **65**, 31.
43 Auer, E., Freund, A., Gross, M., Hartung, R. and Panster, P. (1998) *Catalysis of Organic Reactions (Dekker)*, **75**, 551.
44 Atzrodt, J., Derdau, V., Fey, T. and Zimmermann, J. (2007) *Angewandte Chemie – International Edition*, **46**, 7744.

2
Dihydrido Iridium Triisopropylphosphine Complexes: From Organometallic Chemistry to Catalysis
Luis A. Oro

2.1
Introduction

The impressive developments of organometallic chemistry have allowed the preparation of a wide variety of soluble metal complexes useful for organic transformations under mild conditions. There is no question that much of the general knowledge on homogeneous catalysis has stemmed from studies of organometallic chemistry and homogeneous hydrogenation. Following the discovery of Wilkinson's hydrogenation catalyst, $RhCl(PPh_3)_3$, in 1964 [1] – which is by far the most-studied homogeneous catalyst – Schrock and Osborn reported the catalytic potential of $[Rh(diene)L_2]^+$ complexes as hydrogenation catalysts. Under hydrogen, the diene is hydrogenated to generate the reactive $[RhH_2S_2(PR_3)_2]^+$ species which, in some cases, can be isolated relatively easily from coordinating solvents (S) such as acetone or ethanol [2]. In contrast to Wilkinson's catalyst, a large number of donor ligands can be used, while the use of a $PR_3:Rh$ ratio of 2:1 avoids the need for phosphine dissociation which is required with Wilkinson's catalyst.

Interestingly, cationic diene iridium complexes of general formula $[Ir(COD)(PR_3)_2]^+$ (COD = 1,5-cyclo-octadiene), mainly developed by Crabtree, may be even more active than the cationic diene rhodium complexes, particularly when dichloromethane (DCM) is used as solvent. The solvent is less coordinating and must be easily displaced by the substrates. Furthermore, its polarity favors the solubility of the ionic catalyst. As in the case of rhodium systems, $[IrH_2S_2(PR_3)_2]^+$ species can be observed by hydrogenation of the diene [3, 4]. Another important discovery was the very active mixed ligand complex $[Ir(COD)(py)(PCy_3)]^+$ (py = pyridine), usually known as Crabtree's catalyst, that in DCM as solvent effectively hydrogenates tetrasubstituted amido-alkenes [3]. The related benzonitrile (bzn) derivatives $[Ir(COD)(bzn)(PR_3)]^+$ ($PR_3 = PCy_3$ or the chiral phosphines, neomenthyldiphenylphosphine or phenyl-(o-methoxyphenyl)methylphosphine) are also very active catalysts for the hydrogenation of prochiral didehydro amino acid derivatives [5], although poor enantiomeric excess (ee) values are obtained with chiral monodentate phosphines. The effectiveness of the $[Ir(COD)(L)(PR_3)]^+$ cations for

Iridium Complexes in Organic Synthesis. Edited by Luis A. Oro and Carmen Claver
Copyright © 2009 WILEY-VCH Verlag GmbH & Co. KGaA, Weinheim
ISBN: 978-3-527-31996-1

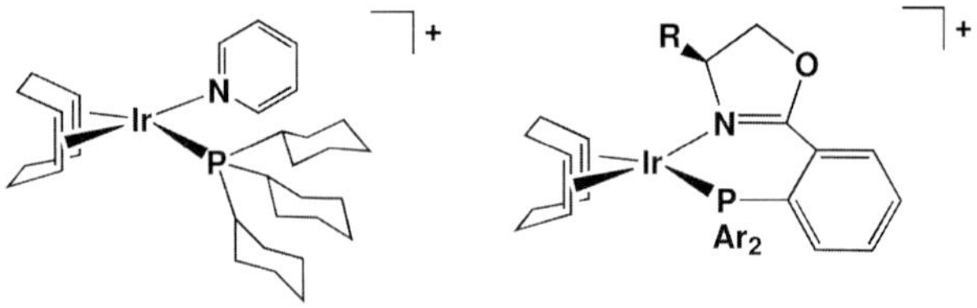

Figure 2.1 Crabtree's catalyst and Pfalz's catalyst.

the hydrogenation of hindered alkenes has been boosted towards remarkable enantioselectivities by replacing the PR_3/N-donor ligand pair with a chiral P,N bidentate ligand such as phosphinooxazolidine (PHOX) in Pfalz's catalyst [Ir(COD)(PHOX)]$^+$. In this particular case the nature of the anion has proved to be crucial in avoiding catalyst deactivation and reducing the moisture-sensitivity of the reactions (Figure 2.1) [6, 7].

Crabtree's catalyst shows relevant directing effects when the catalyst binds to directing groups such as ketone, ether, ester or OH groups, due to the low PR_3:Rh ratio of 1. The coordination of those groups on an alkene substrate directs the attack of the hydrogenation catalyst from the face of the molecule containing the directing group [8]. One limitation of the above-mentioned iridium catalysts, in DCM, is their decomposition to inactive cluster hydrido compounds when the substrate has been hydrogenated. In general, although weakly coordinating solvents provide optimum activity for Crabtree's catalyst, the use of more-coordinating solvents such as nitriles or pyridine should be avoided.

2.2
[Ir(COD)(NCMe)(PR$_3$)]BF$_4$ (PR$_3$ = P^{i}Pr$_3$, PMe$_3$) and Related Complexes as Catalyst Precursors: Is 1,5-Cyclo-Octadiene an Innocent and Removable Ligand?

The complexes [Ir(COD)(NCMe)(PR$_3$)]BF$_4$ (PR$_3$ = P^{i}Pr$_3$, PMe$_3$), which are closely related to Crabtree's catalyst, also contain mixed ligands, very basic phosphines, P^{i}Pr$_3$ or PMe$_3$, and a nitrogen donor ligand, acetonitrile. These mixed-ligand pair combinations have been shown to be very useful for spectroscopic observations and have provided detailed information on the mechanism of reactions performed by those complexes [9, 10].

Scheme 2.1 indicates a likely general scheme for activation of the [Ir(COD)(L)(PR$_3$)]$^+$ catalyst precursors, where PR$_3$ are basic phosphines such as P^{i}Pr$_3$, PMe$_3$ or PCy$_3$, and L is acetonitrile or pyridine.

Under hydrogen, these catalyst precursors undergo partial or total hydrogenation of the cyclo-octadiene ligand to yield the catalytically active species. Thus, the treatment of [Ir(COD)(NCMe)(PMe$_3$)]BF$_4$ with dihydrogen in DCM at 253 K afforded the dihydrido complex [IrH$_2$(1,2,5,6-η-C$_8$H$_{12}$)(NCMe)(PMe$_3$)]BF$_4$ that, upon the addition of 1 equivalent of acetonitrile readily led to the formation of the

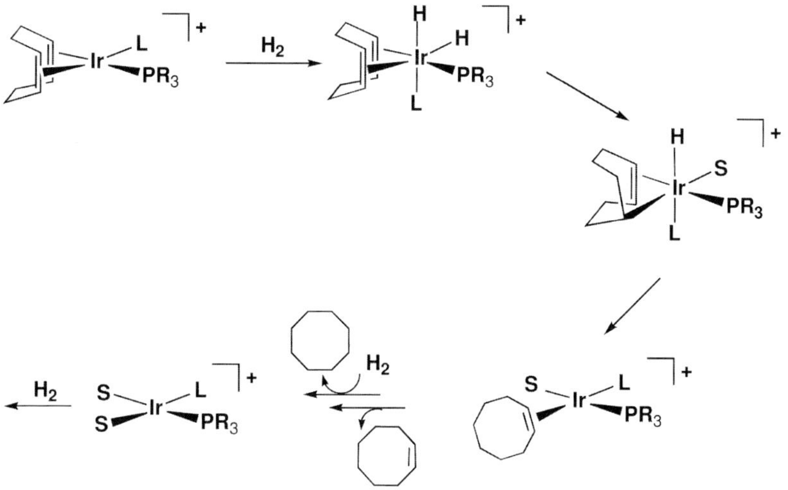

2.2 [Ir(COD)(NCMe)(PR₃)]BF₄ (PR₃ =PⁱPr₃, PMe₃) and Related Complexes as Catalyst Precursors

Scheme 2.1

monohydrido compound [IrH(1-k-4,5-η-C₈H₁₃)(NCMe)₂(PMe₃)]BF₄; this can be isolated as a pale yellow solid by precipitation in diethylether at room temperature (Scheme 2.2) [10].

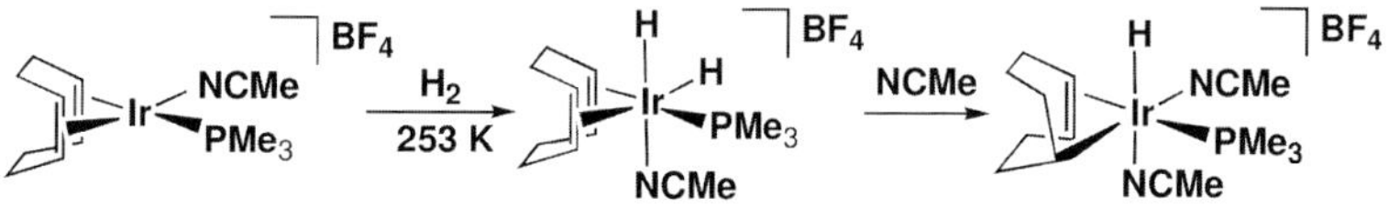

Scheme 2.2

A *cis*-reductive elimination will give rise to the formation of cyclo-octene and finally cyclo-octane, as suggested in Scheme 2.1. In these hydrogenation processes catalyzed by cationic diene rhodium and iridium complexes with phosphine ligands, the diene is considered to be removed from the coordination sphere of the metal by partial or total hydrogenation. On the same lines, diene dihydrido intermediates of formula [IrH₂(COD)(PHOX)]BArF, where BArF is the weakly coordinating tetrakis(3,5-bis(trifluoromethyl)phenyl)borate (BArF), have been observed by reacting Pfalz's catalyst with hydrogen at low temperature, followed by the formation of cyclo-octane [7].

In contrast, 1,5-cyclo-octadiene remains coordinated during the catalytic cycle of hydrogenation of phenylacetylene to styrene, catalyzed by the related iridium complex [Ir(COD)(ⁱPr₂PCH₂CH₂OMe)]BF₄. This complex, which contains an ether-phosphine-chelated ligand, catalyzes the selective hydrogenation reaction via a dihydrido-cyclo-octadiene intermediate. The reaction is first order in each of catalyst, phenylacetylene and hydrogen [11]; the proposed catalytic cycle is shown in Scheme 2.3.

The [Ir(COD)(ⁱPr₂PCH₂CH₂OMe)]BF₄ complex also catalyzes the hydrosilylation of phenylacetylene with triethylsilane, to yield – along with the normal

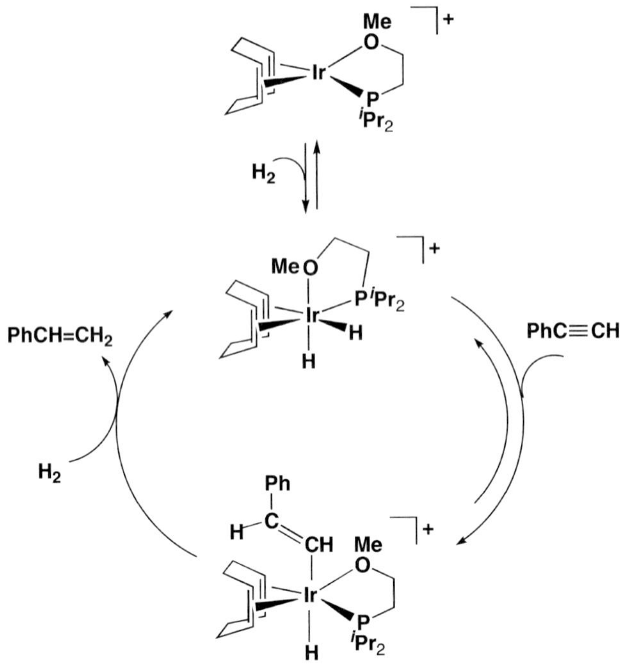

Scheme 2.3

hydrosilylation products – dehydrogenative silylation products [12]. Scheme 2.4 shows the suggested catalytic cycle for the formation of cis-PhCH=CH(SiEt$_3$), showing that 1,5-cyclo-octadiene ligand remains coordinated during the catalytic cycle.

Formation of the dehydrogenative silylation products, PhC≡CSiEt$_3$ and PhC=CH$_2$, is favored by increasing the alkyne:silane ratio, highlighting that the resultant products are formed via the hydridoalkenyl intermediate (Scheme 2.5), that it is the main compound under catalytic conditions. The oxidative addition of triethylsilane to the hydridoalkenyl intermediate could give rise to an iridium(V) intermediate which, after the subsequent reductive elimination of PhC≡CSiEt$_3$, may yield the dihydrido [Ir H$_2$(COD)(^{i}Pr$_2$PCH$_2$CH$_2$OMe)]BF$_4$ which is responsible for the hydrogenation of PhC≡CH to PhCH=CH$_2$ [12]. In these reactions the hybrid hemilabile ether-phosphine ligand can provide a dynamic 'on and off' chelating effect during catalysis, allowing coordination sites for the reactants.

It is likely that the diene also remains coordinated during the hydrosilylation reactions of phenylacetylene with HSiEt$_3$, catalyzed by [Ir(diene)(NCMe)(PR$_3$)]BF$_4$ complexes [PR$_3$ = P^{i}Pr$_3$, PMe$_3$; diene = 1,5-cyclo-octadiene, tetraflorobenzobarrelene (TFB)]. However, detailed studies on the [Ir(COD)(NCMe)(PMe$_3$)]BF$_4$ complex show that cyclo-octadiene cannot be considered, in this case, as an ideal innocent ligand because it transform into complexes containing cyclo-octadiene, cyclo-octadienyl or cyclo-octenyl ligands in a variety of coordination modes, due to an

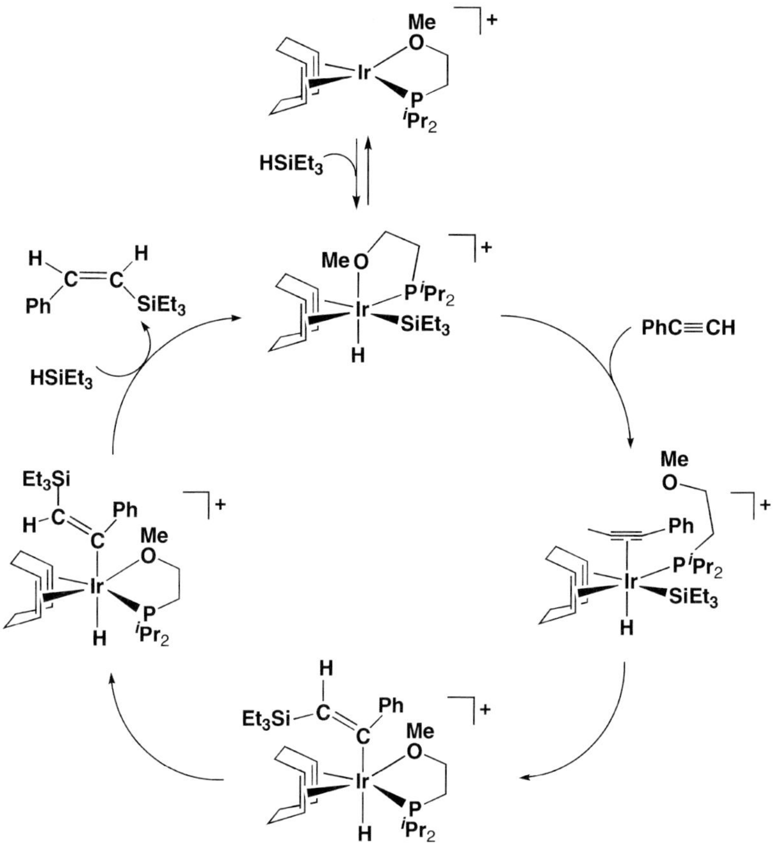

2.2 [Ir(COD)(NCMe)(PR₃)]BF₄ (PR₃ =PⁱPr₃, PMe₃) and Related Complexes as Catalyst Precursors | 19

Scheme 2.4

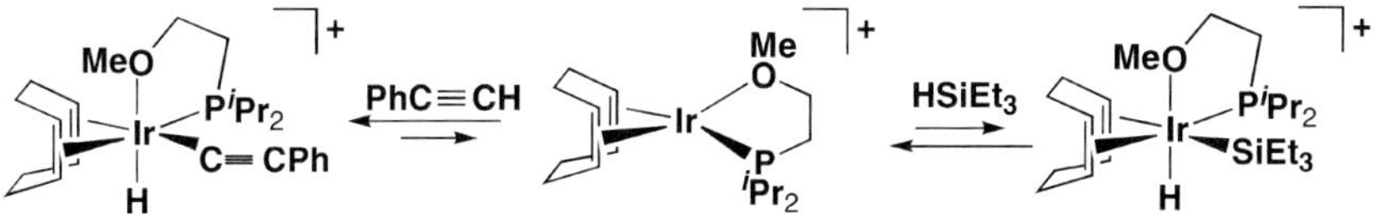

Scheme 2.5

initial intramolecular C—H activation. Thus, the catalytic hydrosilylation process catalyzed by [Ir(COD)(NCMe)(PMe₃)]BF₄ affords *cis*-alkenylsilane as the major product. However, on the addition of acetonitrile to the catalytic reaction the selectivity decreases and a formation of disilane is observed. This change in selectivity has been attributed to the appearance of new species promoted by the addition of acetonitrile (Scheme 2.6).

In contrast, the analogous [Ir(TFB)(NCMe)(PⁱPr₃)]BF₄ complex where a C—H activation is less favorable does not show any dependence upon the addition

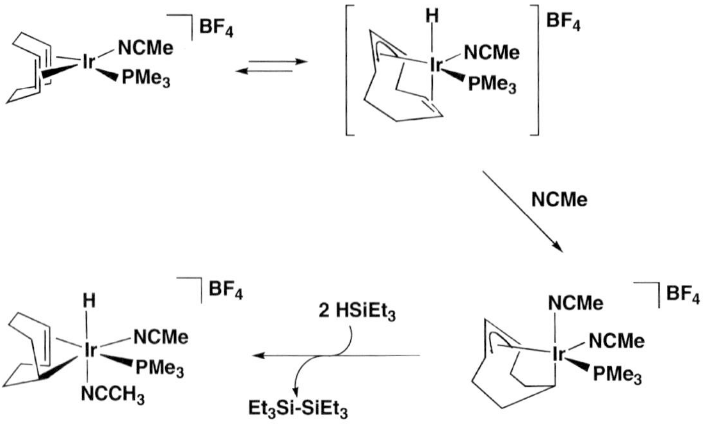

Scheme 2.6

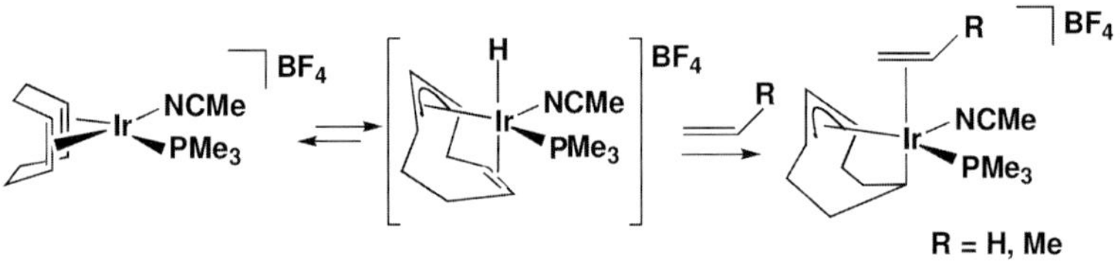

Scheme 2.7

of acetonitrile [10]. Up to eight different coordination modes involving cyclo-octadiene, cyclo-octadienyl or cyclo-octenyl ligands have been observed by studying the reactivity of the [Ir(COD)(NCMe)(PMe$_3$)]BF$_4$ complex, showing the extreme versatility of the cyclo-octadiene ligand and its dependence on the presence of different reactants or solvents and experimental conditions [10]. Thus, simple alkenes such as ethene or propene react with [Ir(COD)(NCMe)(PMe$_3$)]BF$_4$ to yield [IrH(1-k-4,5,6-η-C$_8$H$_{12}$)(NCMe)(alkene)(PMe$_3$)]BF$_4$ (alkene = C$_2$H$_4$, C$_3$H$_6$) complexes via a sequence of C—H activation and insertion steps (Scheme 2.7) [13]. This shows that, in some cases, simple substrates or donor solvents can promote isomerization of the cyclo-octadiene ligands at the iridium center from an Ir(I)(1,2,5,6-η-C$_8$H$_{12}$) coordination to an Ir(III)(1-k-4,5,6-η-C$_8$H$_{12}$) coordination via an Ir(III)H(1,2,4,5,6-η-C$_8$H$_{11}$) intermediate.

These results show the relevant versatility of the cyclo-octadiene ligand, which may encompass different coordination modes and bonding hapticities, depending on the electronic and steric properties of the iridium center. Different coordination modes of the cyclo-octadiene ligand have been previously proposed in order to explain the diastereomeric interconversion and isotopomer exchange for the [IrD$_2$(COD)(Me-DuPHOS)]BF$_4$ complex [14]. The possibility of using the observed cyclo-octadiene versatility in some iridium catalytic systems to achieve more selective and more versatile catalysts for some particular reactions should be not excluded.

2.3
The Dihydrido Iridium Triisopropylphosphine Complex [IrH$_2$(NCMe)$_3$(P^{i}Pr$_3$)]BF$_4$ as Alkene Hydrogenation Catalysts

The reaction of the cationic complex [Ir(COD)(NCMe)(P^{i}Pr$_3$)]$^+$ with hydrogen in coordinating solvents, gives rise to the formation of tris-solvato complexes of formula [IrH$_2$(S)$_3$(P^{i}Pr$_3$)]$^+$. In particular, in the presence of acetonitrile the white dihydrido tris-acetonitrile derivative [IrH$_2$(NCMe)$_3$(P^{i}Pr$_3$)]BF$_4$ can be isolated. These tris-solvato complexes contain only one phosphine per iridium and the three labile coordinated solvent molecules in *fac*-disposition, facilitating access of the incoming substrates. An alternative and more general procedure for the synthesis of cationic dihydrido iridium complexes of formula [IrH$_2$(S)$_3$(PR$_3$)]$^+$, consists of a reaction of the dimer [Ir(μ-OMe)(COD)]$_2$ [15] with phosphonium salts in the presence of hydrogen and an appropriate solvent. Thus, [IrH$_2$(NCMe)$_3$(P^{i}Pr$_3$)]BF$_4$ was prepared in high yield by the reaction of [Ir(μ-OMe)(COD)]$_2$ with [HP^{i}Pr$_3$]BF$_4$ in the presence of dihydrogen and acetonitrile (Scheme 2.8) [16].

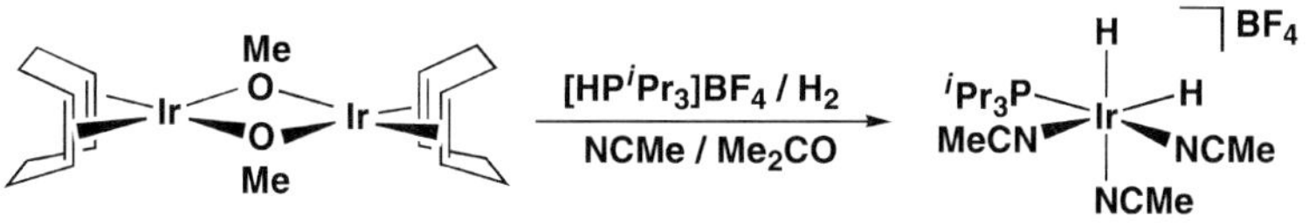

Scheme 2.8

The three acetonitrile ligands of complex [IrH$_2$(NCMe)$_3$(P^{i}Pr$_3$)]BF$_4$ are labile, being readily replaced by acetonitrile-d_3 in solution at room temperature. A detailed kinetic study of these substitution reactions revealed that any of the acetonitrile ligands coordinated *trans* to hydride can be easily dissociated to give a five-coordinate species. This unsaturated species is fluxional, allowing the formation of an intermediate with the coordination vacancy *trans* to phosphine. The life-time of this latter intermediate is long enough to allow substitution reactions at the position *trans* to phosphine (Scheme 2.9).

The results of reactivity studies on the [IrH$_2$(NCMe)$_3$(P^{i}Pr$_3$)]BF$_4$ complex have suggested that small incoming ligands (e.g. carbon monoxide or ethene) prefer coordination at the most labile position, *trans* to hydride, whereas bulkier ligands substitute at the less-hindered position, *trans* to phosphine [16].

The [IrH$_2$(NCMe)$_3$(P^{i}Pr$_3$)]BF$_4$ complex is an active catalyst for alkene hydrogenation, in acetone solution, at 298 K and 1 atm of hydrogen. Thus, under these conditions, cyclohexene was hydrogenated with a TOF of 1.1 min^{-1}. This complex has also been very useful in the mechanistic investigation of alkene hydrogenation.

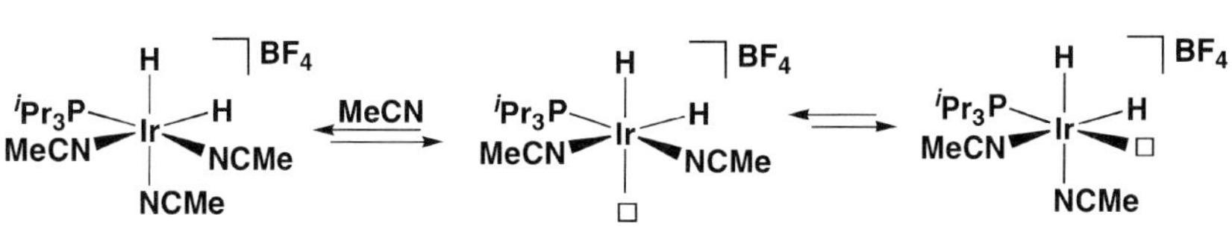

Scheme 2.9

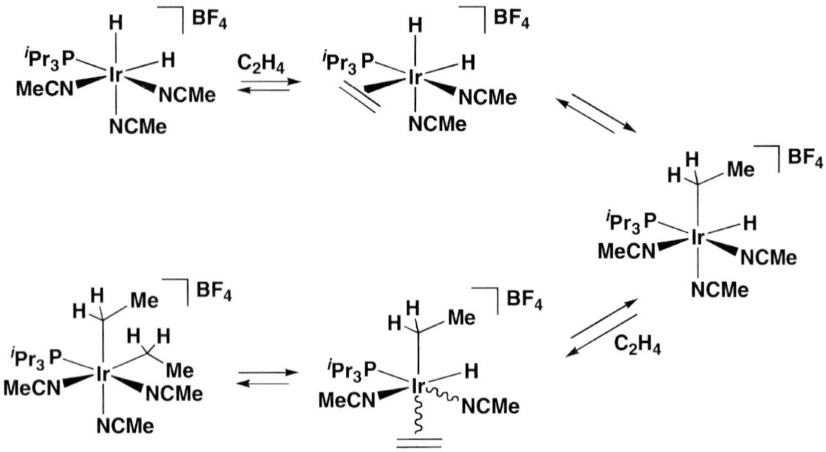

Scheme 2.10

Accordingly, spectroscopic observation of the reaction of complex [IrH$_2$(NCMe)$_3$(P^{i}Pr$_3$)]BF$_4$ with ethene reveals the consecutive formation of several products, as shown in Scheme 2.10. The complex [IrH$_2$(η^2-C$_2$H$_4$)(NCMe)$_2$(P^{i}Pr$_3$)]BF$_4$ is observed after a slow stream of ethene is passed through a solution of [IrH$_2$(NCMe)$_3$(P^{i}Pr$_3$)]BF$_4$ in deuterochloroform at 233 K. On raising the temperature to 273 K, and in the presence of an excess of dissolved ethene, the stable diethyl complex [Ir(Et)$_2$(NCMe)$_3$(P^{i}Pr$_3$)]BF$_4$ is formed, as a result of two ethene insertions into the Ir—H bonds. The spectroscopic control of this reaction reveals the appearance, and subsequent disappearance, of two monoethyl intermediates, which can be identified as [IrH(Et)(NCMe)$_3$(P^{i}Pr$_3$)]BF$_4$ and [IrH(Et)(η^2-C$_2$H$_4$)(NCMe)$_2$(P^{i}Pr$_3$)]BF$_4$. Interestingly, if argon is bubbled through a solution of the diethyl complex at 273 K, a quantitative formation of the starting [IrH$_2$(NCMe)$_3$(P^{i}Pr$_3$)]BF$_4$ complex is observed, thus demonstrating the reversibility of the process.

However, if solutions of [Ir(Et)$_2$(NCMe)$_3$(P^{i}Pr$_3$)]BF$_4$ are warmed to room temperature, the diethyl-complex disappears and an elimination of ethane is observed. No traces of butane were detected, which shows that the potential C—C reductive coupling of the mutually *cis*-disposed ethyl ligands is not a favorable process in comparison with the hydrogen β-elimination that seems to be kinetically favored over the reductive elimination of butane. This unfavorable relationship between β-elimination and C—C reductive elimination rates can be overcome by using 1-alkynes as substrates (see Section 2.3). The reductive elimination of ethane should give rise to iridium(I) species. In the presence of an excess of ethene, the nuclear magnetic resonance (NMR) spectra of the white solution resulting from the reductive elimination of ethane display several broad signals. However, after removal of the ethene excess the color of the solution changes to orange, and the observed NMR spectra indicate that the solution contains a mixture of the iridium(I) complexes [Ir(NCMe)$_3$(P^{i}Pr$_3$)]BF$_4$ and [Ir(η^2-C$_2$H$_4$)(NCMe)$_2$(P^{i}Pr$_3$)]BF$_4$, in a ratio which

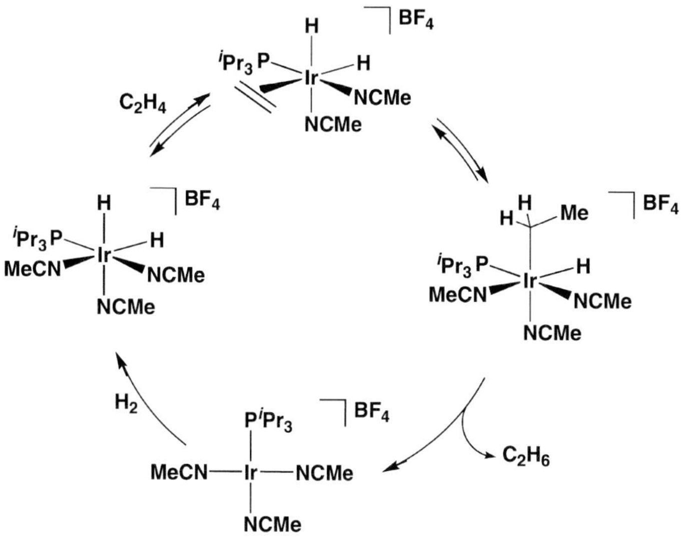

2.3 The Dihydrido Iridium Triisopropylphosphine Complex [IrH₂(NCMe)₃(PⁱPr₃)]BF₄

Scheme 2.11

is dependent on the time taken for the removal of ethene. These species react readily with hydrogen to regenerate the starting dihydrido complex, thereby closing a cycle for ethene hydrogenation in which most of the participating species have been spectroscopically characterized (Scheme 2.11). The rate-determining step of this catalytic cycle is the reductive elimination of ethane, which is the only reaction that requires temperatures above 273 K to proceed.

Scheme 2.11 illustrates the stepwise nature of the ethene hydrogenation process, where the sequence observed might be considered as a model for the hydrogenation of alkenes by [IrH₂(S)₂(L)(PR₃)]⁺ catalysts, derived from [Ir(COD)(L)(PR₃)]⁺ catalyst precursors. Here, the different steps involving iridium species can be clearly distinguished spectroscopically as the temperature is raised. Although iridium(V) species have not been detected, this cannot be excluded. Indeed, it is likely that under catalytic conditions an iridium(III)–iridium(V) cycle could be more favorable than an iridium(I)–iridium(III) cycle, despite the fact that the iridium(I) complex [Ir(NCMe)₃(PⁱPr₃)]BF₄ has been detected in the conditions described above. On the basis of kinetic and density functional theory (DFT) studies, an iridium(III)–iridium(V) cycle has been proposed [17] for the hydrogenation of unfunctionalized alkenes by the Pfalz catalyst [Ir(COD)(PHOX)]BAr_F, containing the chiral P,N bidentate ligand, phosphinooxazolidine (PHOX). The proposed ethene hydrogenation cycle (Scheme 2.12) suggests that a dihydrogen co-ligand undergoes oxidative addition to the iridium during the migratory insertion step.

Nevertheless, controversy persists regarding the oxidation state of the intermediates in alkene hydrogenation by the Pfalz's catalyst, and a conventional iridium(I)–iridium(III) cycle has also been proposed [18]. It has been suggested that both cycles are energetically very similar [7].

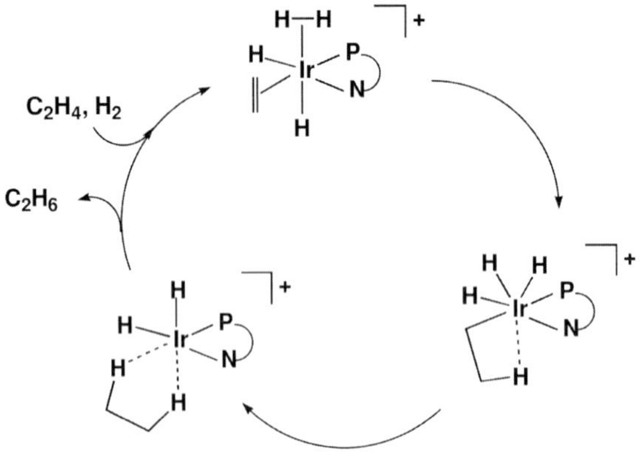

Scheme 2.12

The complex [IrH$_2$(NCMe)$_3$(P^{i}Pr$_3$)]BF$_4$ also reacts with other alkenes, such as propene, though some significant differences are apparent compared to ethene. Consequently, the reaction of [IrH$_2$(NCMe)$_3$(P^{i}Pr$_3$)]BF$_4$ with propene at 233 K shows formation of the complex [IrH$_2$(η^3-C$_3$H$_6$)(NCMe)$_2$(P^{i}Pr$_3$)]BF$_4$, in which propene coordinates *trans* to the phosphine, in contrast to the *cis* mutual position of the analogous ethene complex, most probably for steric reasons. On raising the temperature to 273 K, the propene ligand inserts into one of the Ir—H bonds to yield the complex [IrH(^{n}Pr)(NCMe)$_3$(P^{i}Pr$_3$)]BF$_4$ which does not undergo any observable reaction at this temperature. It appears that the regioselectivity of this insertion is very high, as a species containing the isomeric 2-propenyl ligand cannot be detected. On warming to room temperature, propane is reductively eliminated, but the expected iridium(I) species resulting from this elimination cannot be observed. Instead, a mixture of two allyl-hydrido complexes, [IrH(η^3-C$_3$H$_5$)(NCMe)$_2$(P^{i}Pr$_3$)]BF$_4$ and [IrH(η^3-C$_3$H$_5$)(η^2-C$_3$H$_6$)(NCMe)(P^{i}Pr$_3$)]BF$_4$, is obtained (Scheme 2.13). The [IrH(η^3-C$_3$H$_5$)(NCMe)$_2$(P^{i}Pr$_3$)]BF$_4$ compound is formed by C—H activation of propene at the iridium(I) nonobserved intermediate, whilst the [IrH(η^3-C$_3$H$_5$)(η^2-C$_3$H$_6$)(NCMe)(P^{i}Pr$_3$)]BF$_4$ complex is the result of the replacement of one acetonitrile ligand by propene from [IrH(η^3-C$_3$H$_5$)(NCMe)$_2$(P^{i}Pr$_3$)]BF$_4$ when the reaction is made under a propene atmosphere. Both allyl-hydrido complexes have been isolated and fully characterized [16].

A cycle for the hydrogenation of propene can be closed by reaction of the [IrH(η^3-C$_3$H$_5$)(NCMe)$_2$(P^{i}Pr$_3$)]BF$_4$ complex with dihydrogen in the presence of 1 equivalent of acetonitrile. At room temperature and 1 atm of dihydrogen, formation of the starting [IrH$_2$(NCMe)$_3$(P^{i}Pr$_3$)]BF$_4$ complex is completed within a few minutes. The NMR spectra of the reaction mixtures formed with substoichiometric amounts of H$_2$ indicate that the formation of [IrH$_2$(NCMe)$_3$(P^{i}Pr$_3$)]BF$_4$ occurs with the simultaneous release of propene (not propane). Under the same experimental conditions, reaction with deuterium leads to the formation of [IrHD(NCMe)$_3$(P^{i}Pr$_3$)]BF$_4$ as the major product, together with very small quantities of the [IrD$_2$(NCMe)$_3$(P^{i}Pr$_3$)]BF$_4$ complex, the latter being deuterated at both hydride positions. With

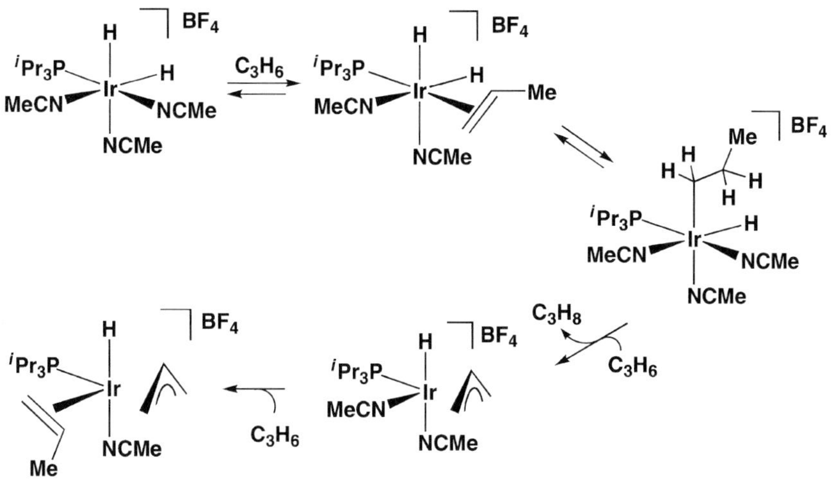

2.3 The Dihydrido Iridium Triisopropylphosphine Complex [IrH₂(NCMe)₃(PⁱPr₃)]BF₄

Scheme 2.13

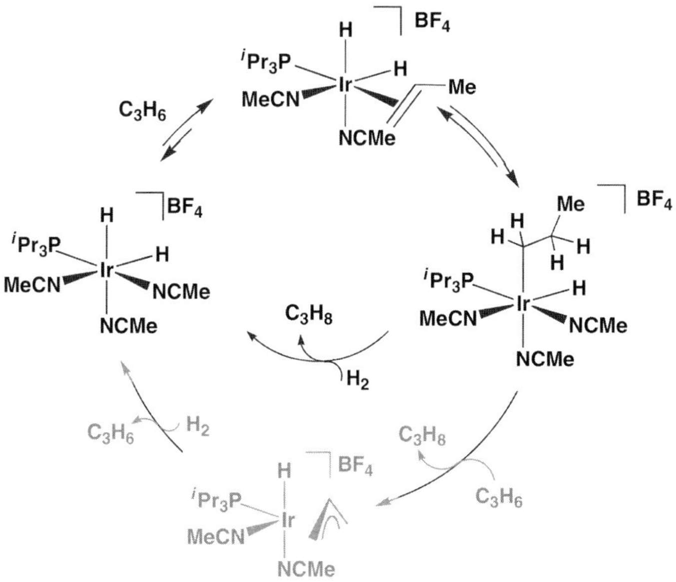

Scheme 2.14

regards to this lack of H/D scrambling, it appears likely that the elimination of propene from the allyl-hydrido [IrH(η³-C₃H₅)(NCMe)₂(PⁱPr₃)]BF₄ complex results from a σ-bond metathesis process in the transient dihydrogen complex [IrH(η²-H₂)(η³-C₃H₅)(NCMe)(PⁱPr₃)]BF₄. However, under catalytic conditions it is likely that the formation of propane takes place by a direct reaction of the [IrH(ⁿPr)(NCMe)₃(PⁱPr₃)]BF₄ complex with hydrogen, being the allyl-hydrido [IrH(η³-C₃H₅)(NCMe)₂(PⁱPr₃)]BF₄ complex in a resting state (Scheme 2.14).

2.4
The Dihydrido Iridium Triisopropylphosphine Complex [IrH$_2$(NCMe)$_3$(P^{i}Pr$_3$)]BF$_4$ as Alkyne Hydrogenation Catalysts

The reaction of [IrH$_2$(NCMe)$_3$(P^{i}Pr$_3$)]BF$_4$ with excess of phenylacetylene, at 273 K, gives rise to the formation of a 7:1 mixture of disubstituted butadiene complexes of formula [Ir(η^4-(1,3-Ph$_2$C$_4$H$_4$)](NCMe)$_2$(P^{i}Pr$_3$)]BF$_4$ [Ir(η^4-(1,4-Ph$_2$C$_4$H$_4$))(NCMe)$_2$(P^{i}Pr$_3$)]BF$_4$. However, when the reaction with 1 equiv of phenylacetylene was made at 253 K an equimolar mixture of two isomeric alkenyl-hydrido complexes, [IrH(C(Ph)=CH$_2$)(NCMe)$_3$(P^{i}Pr$_3$)]BF$_4$ and [IrH(Z-CH=CHPh)(NCMe)$_3$(P^{i}Pr$_3$)]BF$_4$, containing α- and β-(Z)-alkenyl ligands is observed by NMR [19] (Scheme 2.15). Both isomers are stable towards the reductive elimination of styrene at room temperature.

At this point it is worthy of mention that solutions of these alkenyl-hydrido isomers react with hydrogen, at room temperature, to yield styrene and the starting [IrH$_2$(NCMe)$_3$(P^{i}Pr$_3$)]BF$_4$ complex. Deuterium treatment of the alkenyl-hydrido isomers shows an easy H/D hydride exchange, which suggests that the reaction with hydrogen is more favorable than C—H reductive elimination. Therefore, the hydrogenation is dominated by an iridium(III) species, and most probably iridium(I) species are not involved under catalytic conditions.

The [Ir(η^4-(Ph$_2$C$_4$H$_4$))(NCMe)$_2$(P^{i}Pr$_3$)]BF$_4$ complexes should be formed by reaction of the isomeric alkenyl-hydrido complexes with a second equivalent of phenylacetylene, suggesting a double insertion/C—C-coupling reaction sequence. This behavior indicates that for 1-alkynes, the C—C-coupling reaction is more favorable than the reductive elimination of styrene, whilst for alkenes there is an unfavorable relationship between β-elimination and C—C formation. These observations agree with the previous reports regarding a more easy coupling of two sp^2 carbons, in comparison with sp^3 carbons [20], as well as the lack of reports on β-hydrogen elimination from alkenyl ligands. It is interesting to point out the different reactivity of these isomeric alkenyl-hydrido complexes towards the second equivalent of the phenylacetylene. Thus, [IrH(Z-CH=CHPh)(NCMe)$_3$(P^{i}Pr$_3$)]BF$_4$ readily reacts

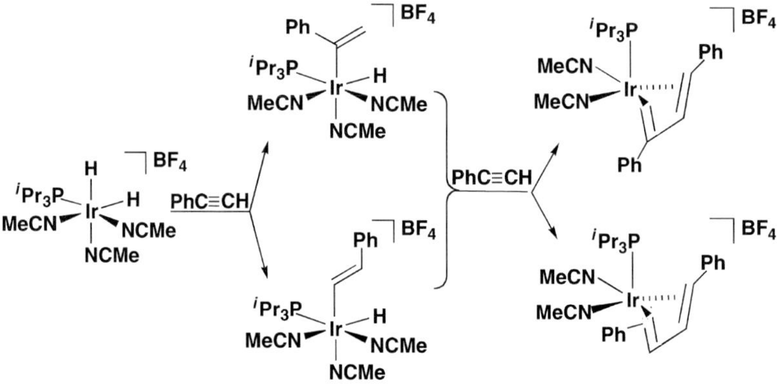

Scheme 2.15

2.4 The Dihydrido Iridium Triisopropylphosphine Complex [IrH₂(NCMe)₃(P^iPr₃)]BF₄

with an excess of the alkyne already at 253 K giving [Ir(η⁴-(Ph₂C₄H₄)(NCMe)₂(P^iPr₃)]BF₄, whereas the disappearance of the α-alkenyl-hydrido [IrH(C(Ph)=CH₂)(NCMe)₃(P^iPr₃)]BF₄ is slow even at temperatures above 273 K. In both cases, the spectroscopic observations do not allow the detection of any additional intermediates of this second reaction step. This suggests that, once the second insertion takes place, the subsequent C—C reductive coupling is rapid. The isomeric distribution resulting from evolution of the alkenyl-hydrido intermediates suggests that a sequence of two α-insertions is not favored. Moreover, after an initial β-(Z)-insertion, a subsequent insertion of α-stereochemistry seems to be preferred.

The reaction of [IrH₂(NCMe)₃(P^iPr₃)]BF₄ with an excess of *tert*-butylacetylene is more selective, and only the [Ir(η⁴-1,3-^tBu₂C₄H₄)](NCMe)₂(P^iPr₃)]BF₄ complex is obtained and the intermediate [IrH(Z-CH=CH^tBu)(NCMe)₃(P^iPr₃)]BF₄ species can be observed (Scheme 2.16).

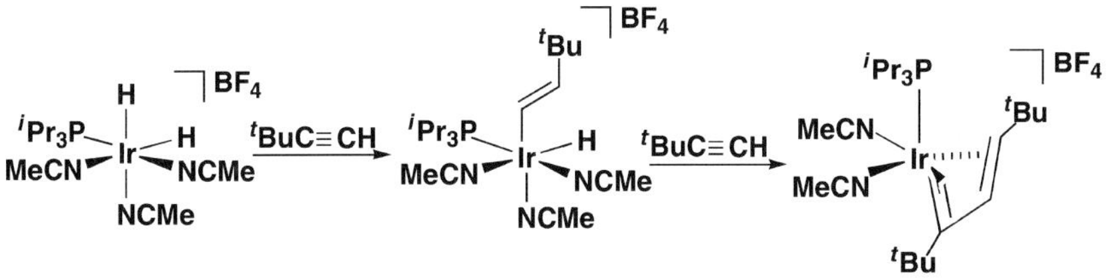

Scheme 2.16

The [Ir(η⁴-1,3-^tBu₂C₄H₄)(NCMe)₂(P^iPr₃)]BF₄ complex reacts with hydrogen, in the presence of an excess of acetonitrile, to give alkenes and regenerate the starting [IrH₂(P^iPr₃)(NCMe)₃]BF₄ complex (Scheme 2.17).

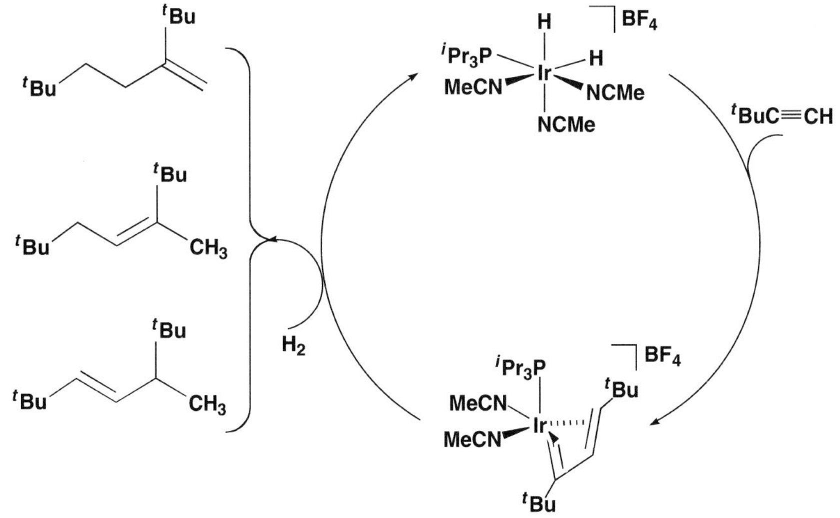

Scheme 2.17

This reaction suggests an unusual catalytic cycle for the hydrogenative dimerization of terminal alkynes at room temperature and an atmospheric pressure of hydrogen. Thus, when *tert*-butylacetylene and hydrogen are treated with catalytic amounts of the [IrH$_2$(NCMe)$_3$(P^{i}Pr$_3$)]BF$_4$ complex in 1,2-dichloroethane, the alkenes shown in Scheme 2.17 are the main products formed. In contrast, under the same catalytic conditions, phenylacetylene affords (almost exclusively) the simple hydrogenation products of styrene and ethylbenzene. The reason for this different behavior, and dependence of selectivity on the alkyne, may arise from the competition of substrates for the alkenyl-hydrido reaction intermediates. Thus, in the case of *tert*-butylacetylene, the second alkyne insertion in the [IrH(Z-CH=CH^{t}Bu)(NCMe)$_3$(P^{i}Pr$_3$)]BF$_4$ intermediate seems to be kinetically favored over its reaction with dihydrogen, whereas this latter reaction seems to be the fastest alternative for the [IrH(C(Ph)=CH$_2$)(NCMe)$_3$(P^{i}Pr$_3$)]BF$_4$ and [IrH(Z-CH=CHPh)(NCMe)$_3$(P^{i}Pr$_3$)]BF$_4$ intermediates. In agreement with such rationalization based on substrate competition, it can be observed that reaction selectivity depends upon the relative concentration of the reactants. Thus, the proportion of ^{t}BuCH=CH$_2$ and ^{t}BuCH$_2$CH$_3$ hydrogenation products has been found to increase at low alkyne concentrations. Under the latter conditions, the lower substrate:catalyst ratio leads to a faster isomerization of the kinetic terminal alkene into its internal isomers, as well as a faster hydrogenation of ^{t}BuCH=CH$_2$ to ^{t}BuCH$_2$CH$_3$. The observed competitive processes are summarized in Scheme 2.18.

As expected, the decrease in the initial alkyne concentration results in less dimerization and higher initial hydrogenation rates. A similar substrate-inhibition effect has also been observed with PhC≡CH as substrate, revealing a complex dependence of the hydrogenation rate upon the alkyne concentration. To the best

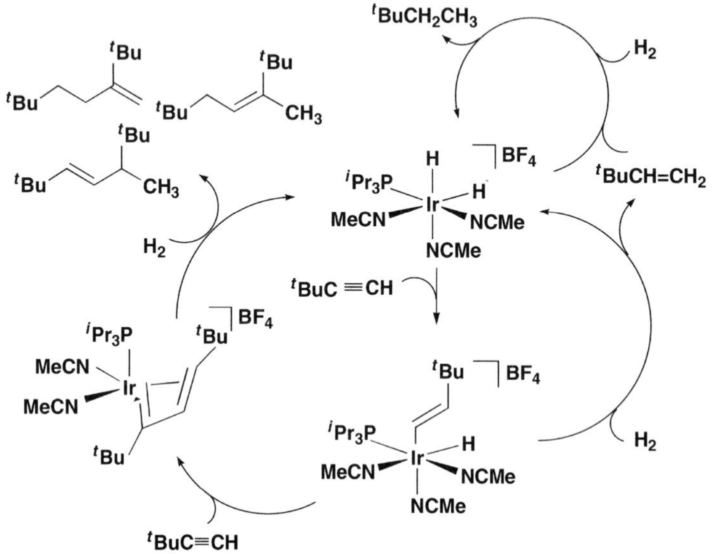

Scheme 2.18

of our knowledge, such a catalytic hydrogenative dimerization of alkynes has not been previously reported.

In conclusion, the [IrH$_2$(NCMe)$_3$(P^{i}Pr$_3$)]BF$_4$ complex has been found to be a good hydrogenation catalyst, and has demonstrated a high value for mechanistic investigation. Thus, the spectroscopic studies of this compound in the presence of alkenes, alkynes and hydrogen have allowed the observation and characterization of those reaction intermediates likely involved in homogeneous hydrogenation, as well as the identification of other feasible side reactions which compete with those leading to hydrogenation. This has allowed a direct comparison to be made among most elementary organometallic reactions, and the conception of new catalytic transformations under hydrogenation conditions.

2.5
Dihydrido Arene Iridium Triisopropylphosphine Complexes

Dihydrido arene iridium complexes of formula [(η^6-C$_6$H$_6$)IrH$_2$(PR$_3$)]BF$_4$ (PR$_3$ = P^{i}Pr$_3$, PCy$_3$) can be prepared by treatment of the dimer [Ir(μ-OMe)(COD)]$_2$ [15] with phosphonium salts in acetone/benzene, followed by reaction of hydrogen. Following this procedure, the [(η^6-C$_6$H$_6$)IrH$_2$(P^{i}Pr$_3$)]BF$_4$ complex was obtained as a white solid according to Scheme 2.19 [21, 22].

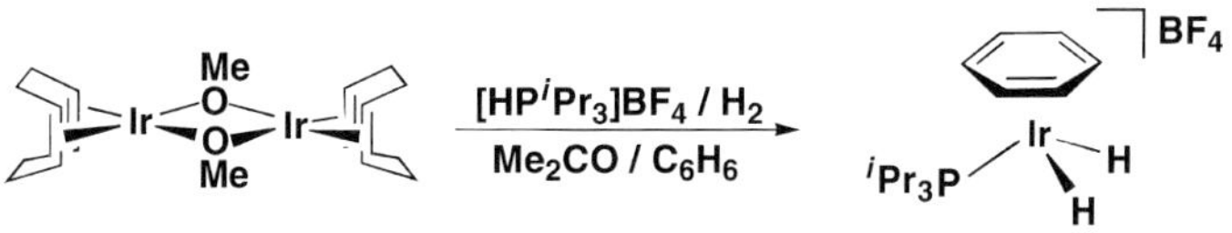

Scheme 2.19

NMR studies have shown the expected signals of coordinated benzene, phosphine and the two hydride ligands, when the spectrum was registered in CDCl$_3$, but when in coordinating solvents (e.g. (CD$_3$)$_2$CO) the labile benzene ligand is partially substituted by the three acetone ligands (Scheme 2.20). The benzene ligand can be effectively displaced by arene ligands with more coordinating capability than benzene, such as toluene, 1,3,5-benzene, hexamethylbenzene, 1-methylstyrene or phenol. The spectroscopic data for these arene complexes are consistent with the rapid rotation around the iridium–arene axis.

Synthesis of the above-mentioned arene complexes of formula [(η^6-arene)IrH$_2$(P^{i}Pr$_3$)]BF$_4$ was carried out by the treatment of acetone solutions of [(η^6-C$_6$H$_6$)IrH$_2$(P^{i}Pr$_3$)]BF$_4$ with an excess of the arene. However, in the case of aniline

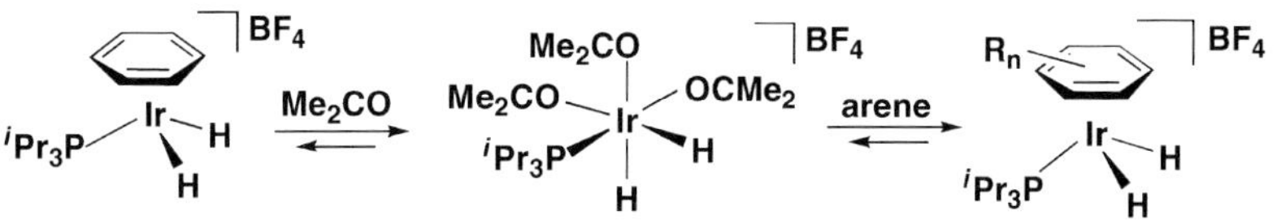

Scheme 2.20

the synthesis of [(η⁶-C₆H₅NH₂)IrH₂(P^iPr₃)]BF₄ requires the use of stoichiometric amounts of aniline because, in the presence of an excess, the η⁶-aniline is replaced by three N-bonded aniline ligands (Scheme 2.21).

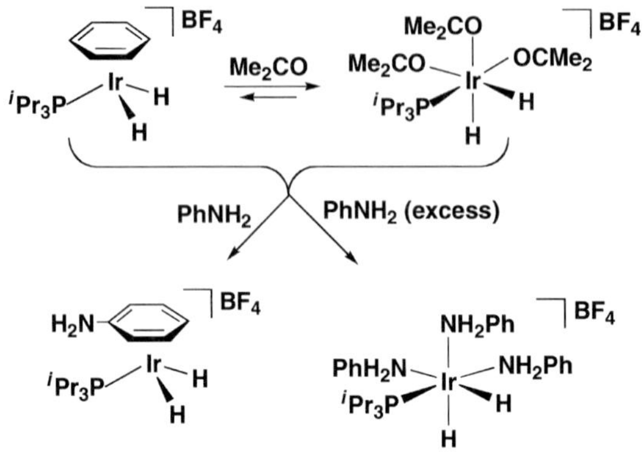

Scheme 2.21

An estimation of the relative coordinating capabilities of the different arenes has been obtained by measuring the equilibrium constants for arene replacement by acetone-d_6 (Scheme 2.20). The values of the equilibrium constants show the lack of acetone-d_6 replacement in the case of hexamethylbenzene, and the higher coordination capability of aniline in comparison with trimethylbenzene or toluene [22].

It is interesting to mention that, whilst 1-methylstyrene gives rise to the arene coordination according to Scheme 2.20, under the same conditions styrene did not yield the arene-substituted product. Thus, the styrene was hydrogenated to ethylbenzene to yield the [(η⁶-C₆H₅Et)Ir(η²-CH₂=CHPh)]BF₄ complex, where ethylbenzene remained coordinated to the iridium center. The suggested hydrogenation/coordination pathway is shown in Scheme 2.22.

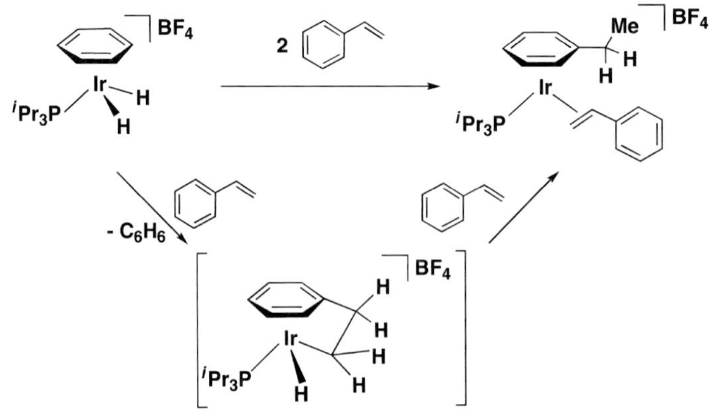

Scheme 2.22

2.5 Dihydrido Arene Iridium Triisopropylphosphine Complexes

This reaction highlights the possibility of preparing arene iridium(I) complexes by the treatment of dihydrido arene iridium(III) complexes with alkenes that, initially, will act as hydrogen acceptors to yield the corresponding alkanes. This proposal has been proved by treating a variety of dihydrido arene iridium(III) complexes with ethene (Scheme 2.23). The NMR spectra for the [(η^6-arene)Ir(η^2-C$_2$H$_4$)(P^{i}Pr$_3$)]BF$_4$ complexes (arene = benzene, toluene, 1,3,5-benzene) indicate rotation of the arene ligands around the iridium–arene axis, whilst the ethene ligand is not rotating.

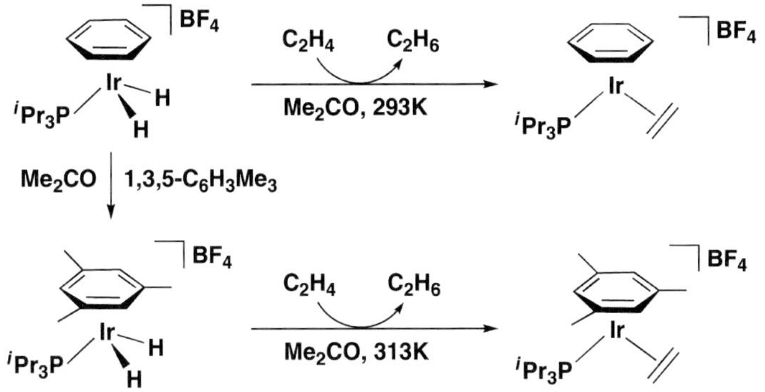

Scheme 2.23

The behavior shown in Schemes 2.22 and 2.23 suggests that these hydrido arene–iridium complexes might act as hydrogenation catalysts. Thus, the [(η^6-C$_6$H$_6$)IrH$_2$(P^{i}Pr$_3$)]BF$_4$ complex has been shown to be an efficient hydrogenation catalysts for alkenes and alkynes, as well as some carbonyl groups [22].

A recent gas-phase study [18] on the hydrogenation of styrene by the Pfalz's catalyst [Ir(COD)(PHOX)]$^+$, suggests an iridium(I)–iridium(III) cycle, as shown in Scheme 2.24. This cycle would imply substitution of the ethylbenzene ligand by

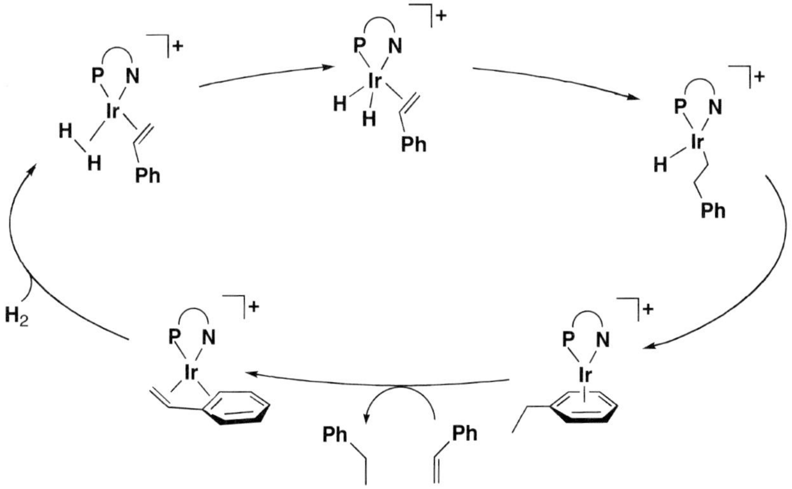

Scheme 2.24

styrene on the iridium(I) center. However, the evidences found in the related [(η^6-arene)Ir(η^2-C$_2$H$_4$)(P^{i}Pr$_3$)]BF$_4$ complexes indicate that the cationic arene iridium(I) species are less labile than the corresponding dihydrido arene iridium(III) species, and that substitution of the coordinated η^6-ethylbenzene from [(η^6-C$_6$H$_5$Et)Ir(η^2-alkene)(P^{i}Pr$_3$)]BF$_4$ did not take place in the presence of styrene. Thus, it should also be considered that, under catalytic conditions, alkene hydrogenations can be performed without the participation of iridium(I) intermediates.

Electron-rich iridium(I) complexes can perform C—H activation reactions under mild conditions [13]. In this line, acetone-d_6 solutions of the [(η^6-1,3,5-C$_6$H$_3$Me$_3$))Ir(η^2-C$_2$H$_4$)(P^{i}Pr$_3$)]BF$_4$ complex, at room temperature, show deuterium incorporation to the ethane ligand, most likely due to the participation of hydrido vinyl iridium(III) species, formed by the C—H activation of ethane, according to Scheme 2.25 [21].

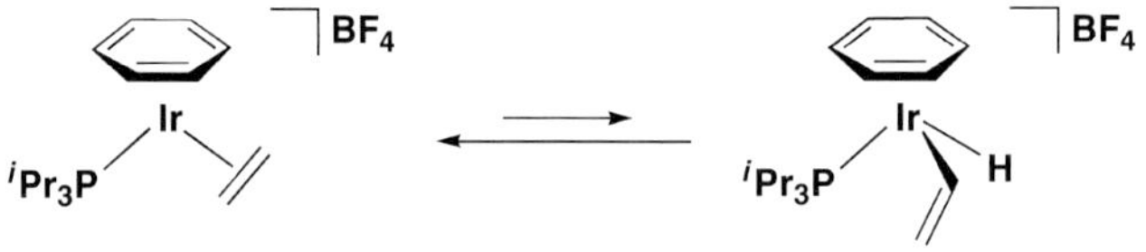

Scheme 2.25

The iridium(III) complex [(η^6-C$_6$H$_6$)IrH$_2$(P^{i}Pr$_3$)]BF$_4$ also undergoes H/D scrambling, in acetone-d_6 solution, initially at the hydride ligands and extended after a longer time to the benzene moiety and phosphine ligand. Such behavior may involve either oxidative addition to iridium(V) intermediates or a σ-bond metathesis process [21]. It has been reported that, for iridium(III) complexes, there is a preponderance of evidence supporting the participation of iridium(V) intermediates [23].

A related dihydrido iridium complex of formula [(η^6-C$_6$H$_5$CH$_2$CH$_2$P^{i}Pr$_2$-κ-P)IrH$_2$]BF$_4$ [24] can be prepared by treatment of the dimer [Ir(μ-OMe)(COD)]$_2$ [15] with the phosphonium salt, [HP^{i}Pr$_2$CH$_2$CH$_2$C$_6$H$_5$]BF$_4$, in acetone followed by the reaction of hydrogen (Scheme 2.26).

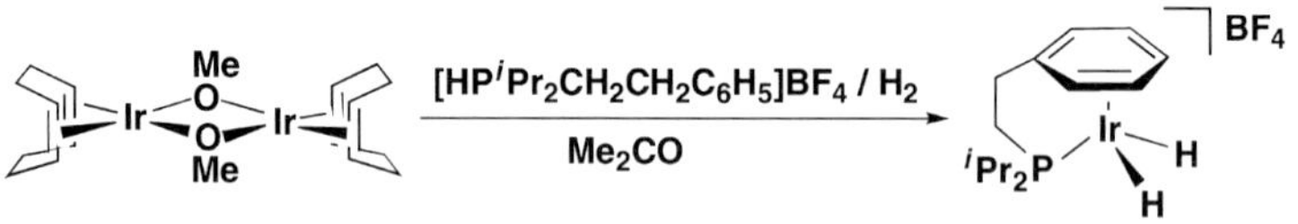

Scheme 2.26

The functionalized phosphine coordinates to the iridium center as a 6+2 electron donor ligand. This dihydrido iridium(III) complex reacts with ethene or propene to yield ethane or propane and formation of the iridium(I) complexes [(η^6-C$_6$H$_5$CH$_2$CH$_2$P^{i}Pr$_2$-κ-P)Ir(CH$_2$=CHR)]BF$_4$ (R=H, Me) (Scheme 2.27).

The complex [(η^6-C$_6$H$_5$CH$_2$CH$_2$P^{i}Pr$_2$-κ-P)IrH$_2$]BF$_4$ also reacts with diphenylacetylene, that acts as a hydrogen acceptor, to yield the stilbene complex [(η^6-C$_6$H$_5$CH$_2$CH$_2$P^{i}Pr$_2$-κ-P)Ir(Z-PhCH=CHPh)]BF$_4$. The latter compound can also be

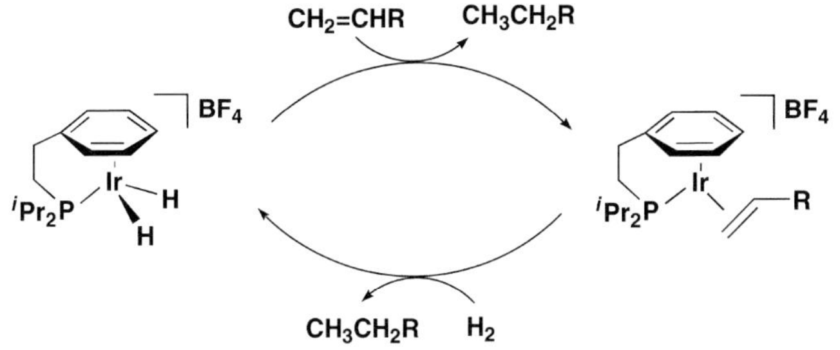

2.5 Dihydrido Arene Iridium Triisopropylphosphine Complexes

Scheme 2.27

obtained by replacement of the propene analogue with Z-stilbene. If diphenylacetylene were to be added to the propene complex, $[(\eta^6\text{-}C_6H_5CH_2CH_2P^iPr_2\text{-}\kappa\text{-}P)Ir(CH_2=CHMe)]BF_4$, the alkene would be replaced by the alkyne to yield $[(\eta^6\text{-}C_6H_5CH_2CH_2P^iPr_2\text{-}\kappa\text{-}P)Ir(PhC\equiv CPh)]BF_4$ (Scheme 2.28).

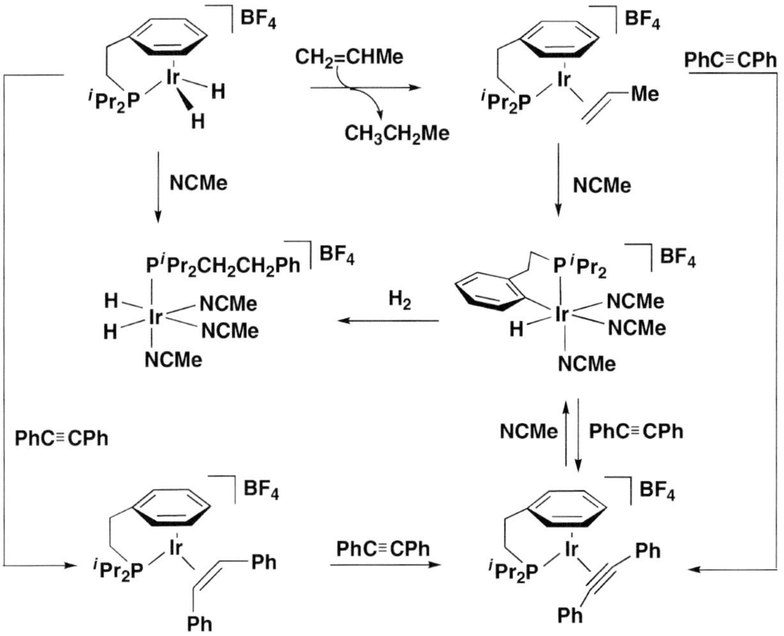

Scheme 2.28

The $[(\eta^6\text{-}C_6H_5CH_2CH_2P^iPr_2\text{-}\kappa\text{-}P)Ir(CH_2=CHMe)]BF_4$ complex reacts with acetonitrile to yield the arylhydrido complex $[IrH(C_6H_4CH_2CH_2P^iPr_2\text{-}\kappa^2\text{-}C,P)(NCMe)_3]BF_4$ as a consequence of an intramolecular C—H metalation. This arylhydrido complex can also be obtained by treatment of the π-alkyne complex $[(\eta^6\text{-}C_6H_5CH_2CH_2P^iPr_2\text{-}\kappa\text{-}P)Ir(PhC\equiv CPh)]BF_4$ with excess of acetonitrile in acetone [22]. These observations support the concept that the insertion of a metal into one of the arene C—H bonds is reversible, as was previously observed in related complexes [25]. The reaction of the arylhydrido complex $[IrH(C_6H_4CH_2CH_2P^iPr_2\text{-}\kappa^2\text{-}C,P)(NCMe)_3]BF_4$

with hydrogen, in acetone, gives rise to formation of the dihydrido iridium complex [IrH$_2$(NCMe)$_3$(C$_6$H$_5$CH$_2$CH$_2$P^{i}Pr$_2$-κ-P)]BF$_4$, in which the ligand is only coordinated by the P-atom. This dihydrido complex is also obtained by treatment of the [(η^6-C$_6$H$_5$CH$_2$CH$_2$P^{i}Pr$_2$-κ-P)IrH$_2$]BF$_4$ complex with excess of acetonitrile in acetone (Scheme 2.28). A kinetic study of the latter reaction suggests a η^6–η^4 phenyl ring slippage, as shown in Scheme 2.29. Slippage from η^6- to η^4-arene coordination has been considered previously to be a key step in ligand exchange processes and catalytic reactions [26].

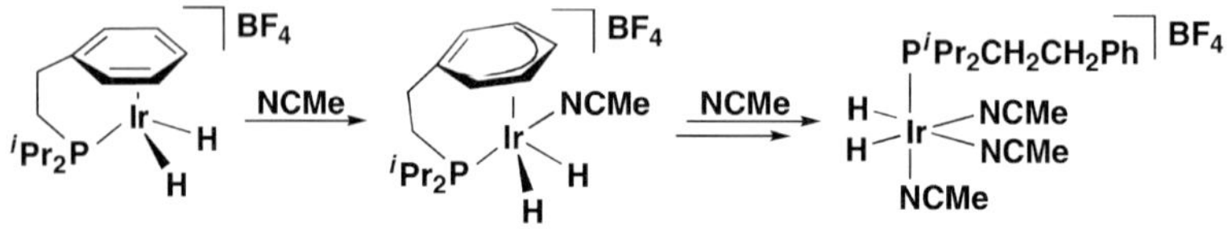

Scheme 2.29

2.6
Dihydrido Iridium Triisopropylphosphine Complexes as Imine Hydrogenation Catalysts

The complexes [IrH$_2$(NCMe)$_3$(P^{i}Pr$_3$)]BF$_4$ and [(η^6-C$_6$H$_6$)IrH$_2$(P^{i}Pr$_3$)]BF$_4$ have shown to be efficient hydrogenation catalysts for alkenes, and spectroscopic studies on these compounds have allowed the observation and characterization of several reaction intermediates [16, 22]. Whilst the hydrogenation of C=C functionalities is easily performed by a large number of transition-metal complexes [27], the homogeneous catalytic hydrogenation of C=N bonds is generally considered to be a difficult process [28]. Interestingly, the triisopropylphosphine complexes [IrH$_2$(NCMe)$_3$(P^{i}Pr$_3$)]BF$_4$ and [(η^6-C$_6$H$_6$)IrH$_2$(P^{i}Pr$_3$)]BF$_4$ are also capable of hydrogenating of N-benzylideneaniline under mild conditions, although at very different reaction rates (the arene complex being the more efficient). NMR observations under conditions similar to those of catalysis showed the formation of new monohydrido or dihydrido iridium resting states. Thus, the solutions obtained from the more active catalyst precursor [(η^6-C$_6$H$_6$)IrH$_2$(P^{i}Pr$_3$)]BF$_4$ have been found to contain a sole observable iridium complex, [IrH$_2$(η^5-{(C$_6$H$_5$)NHCH$_2$Ph}(P^{i}Pr$_3$)]BF$_4$, in which the hydrogenation product coordinates the metal through a phenyl ring. In the case of the [IrH$_2$(NCMe)$_3$(P^{i}Pr$_3$)]BF$_4$ precursor the formation of cyclometalated complexes of formula [IrH{PhN=CH(C$_6$H$_4$)-κ-N,C}(NCMe)$_2$(P^{i}Pr$_3$)]BF$_4$ and [IrH{PhN=CH(C$_6$H$_4$)-κ-N,C}(NCMe)(NH$_2$Ph)(P^{i}Pr$_3$)]BF$_4$ have been observed (Scheme 2.30).

Both monohydrido complexes contain a cyclometalated molecule derived from the N-benzylideneaniline substrate, and the presence of the aniline ligand in the later complex seems to be due to some imine hydrolysis caused by adventitious water. Accordingly, this aniline complex practically disappears when carefully dried substrates and solvents are used [29].

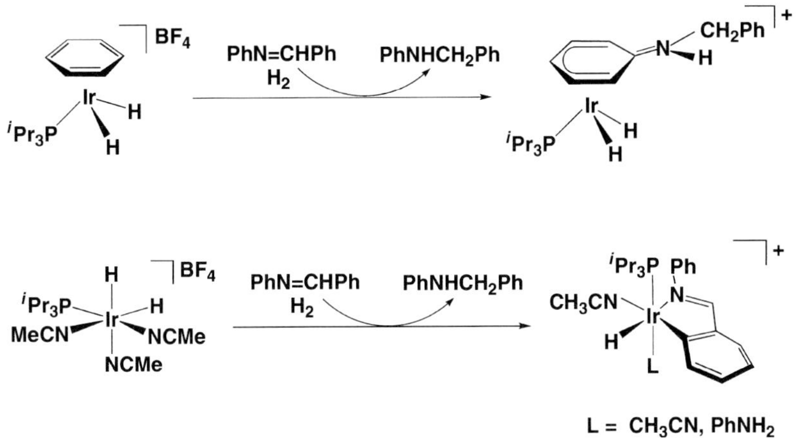

2.6 Dihydrido Iridium Triisopropylphosphine Complexes as Imine Hydrogenation Catalysts

Scheme 2.30

Scheme 2.31

The complex [IrH$_2${η^5-(C$_6$H$_5$)NHCH$_2$Ph}(P^{i}Pr$_3$)]BF$_4$ can be obtained directly by substituting the benzene ligand in [(η^6-C$_6$H$_6$)IrH$_2$(P^{i}Pr$_3$)]BF$_4$ by N-benzylaniline, according to Scheme 2.31.

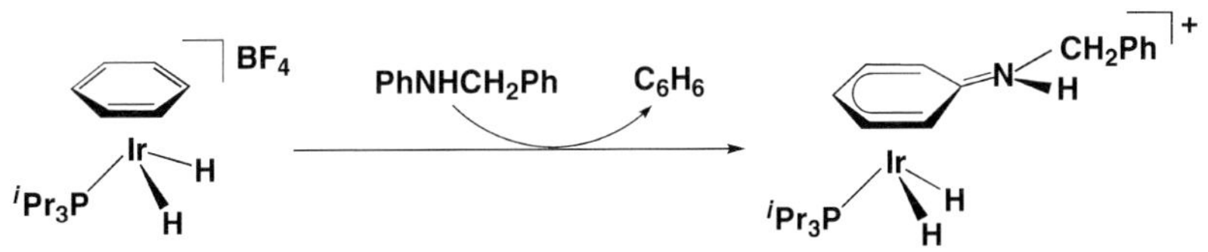

The X-ray structure of the complex confirms coordination of the amine as an η^5-ligand containing a C=N double bond. Thus, five of the six Ir—C distances are within the range 2.24–2.35 Å, whereas one Ir—C(1) distance is clearly longer, at 2.500(6) Å, whilst the N—C(1) distance of 1.335(9) Å confirms the proposed C=N double bond. The ^{1}H NMR spectra of the compound also agree with such a bonding description, revealing a hindered rotation around the C=N bond at low temperature. It is of interest to note that the reactions between [IrH$_2$(η^5-(C$_6$H$_5$)NHCH$_2$Ph)(P^{i}Pr$_3$)]BF$_4$ and deuterium reveal a selective sequence of deuterium incorporation into the complex which is accelerated by the amine product.

Under catalytic conditions the more efficient catalyst precursor [(η^6-C$_6$H$_6$)IrH$_2$(P^{i}Pr$_3$)]BF$_4$ transforms into the resting state [IrH$_2$(η^5-(C$_6$H$_5$)NHCH$_2$Ph)(P^{i}Pr$_3$)]BF$_4$ in which the hydrogenation product coordinates the metal through a phenyl ring. The catalytic hydrogenation process is strongly dependent upon concentrations, temperature and solvents, with methanol being the solvent that leads to the fastest reactions. Although some activation periods have been observed, the addition of amine at the start of the catalytic reactions shortened the induction periods. Such behavior suggests that, besides its contribution to form the resting state [IrH$_2$(η^5-(C$_6$H$_5$)NHCH$_2$Ph)(P^{i}Pr$_3$)]BF$_4$, the amine product may play a significant

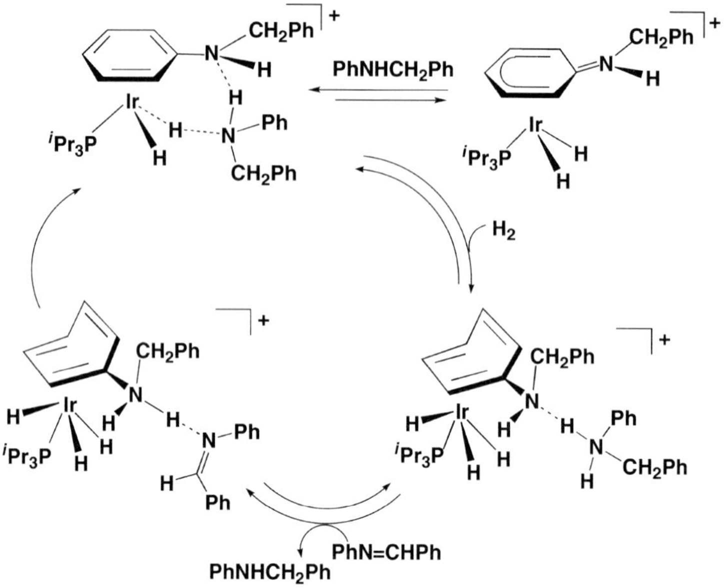

Scheme 2.32

role in the hydrogenation mechanism. Indeed, the results of kinetic studies have indicated a first-order dependence of the reaction rate on the catalyst, imine and dihydrogen concentrations.

All of these experimental observations point to a remarkable hydrogen-bonding chemistry among the components of the catalytic reaction. In addition, MP2 calculations performed on model compounds allow the formulation of an ionic, outer sphere, bifunctional hydrogenation mechanism, as shown in Scheme 2.32.

The amine adduct that initiates the process could be formally described as an iridium(I) complex containing an ammonium moiety connected to both the metal and the arene ligand nitrogen through hydrogen bonds. This adduct implies a formal deprotonation of the metal atom by the amine product which allows for a subsequent dihydrogen oxidative addition that eventually forces the proton to migrate to the nitrogen. Such a process is feasible also due to the coordination versatility of the arene ligand, which can change its coordination mode from η^6 to η^4. The result of the oxidative addition is the formation of an iridium(III) intermediate containing a hydrogen-bonded amine that, in a subsequent step, is replaced by a hydrogen-bonded imine [29]. This exchange is followed by a bifunctional hydrogenation of the imine which, in accordance with theoretical calculations, presents a transition state similar to those previously described in C=O hydrogenations and transfer hydrogenations by hydrido(diamine)ruthenium species, where a hydride on Ru and a proton of the NH_2 ligand are simultaneously transferred to the C=O function via a six-membered, pericyclic transition state [30]. Thus, the last step involves the simultaneous transfer of proton and hydride from the catalyst to the imine.

2.7
Conclusions

Dihydrido iridium cationic complexes with basic isopropylphosphine ligands are effective catalysts for a variety of reactions involving alkenes and alkynes as substrates. Interestingly, these complexes have allowed the identification of several reaction intermediates derived from the initial coordination of alkenes or alkynes to the iridium center, as well as the detailed observation of various elementary reactions involved in catalytic hydrogenation processes, that most likely follow an iridium(III)–iridium(V) cycle. In contrast, observations on the mechanism for imine hydrogenation indicate that C=N bonds are hydrogenated by an ionic outer-sphere bifunctional mechanism similar to those found in C=O hydrogenation, and transfer hydrogenation by Noyori-type ruthenium catalysts [30]. However, a special feature of the above-mentioned imine hydrogenation iridium catalyst is that protons are transferred to the substrate from the NH moiety of a coordinated molecule of the amine hydrogenation product, so that ancillary ligands bearing NH functions are not required.

Acknowledgments

It is a pleasure to acknowledge the experimental and intellectual contributions made to the various aspects of these studies by the coworkers whose names appear in the references, and without whose contributions this chapter could not have been written. In particular, I would like to thank Dr. Eduardo Sola and Prof. Helmut Werner for their stimulating discussions and collaborations on this subject.

References

1 (a) Osborn, J.A., Wilkinson, G. and Young, J.F. (1965) *Chemical Communications*, 17; (b) Osborn, J.A., Jardine, F.H., Young, J.F. and Wilkinson, G. (1966) *Journal of the Chemical Society A*, 1711.
2 (a) Schrock, R.R. and Osborn, J.A. (1976) *Journal of the American Chemical Society*, **98**, 2134; (b) Schrock, R.R. and Osborn, J.A. (1976) *Journal of the American Chemical Society*, **98**, 2143; (c) Schrock, R.R. and Osborn, J.A. (1976) *Journal of the American Chemical Society*, **98**, 4450.
3 Crabtree, R.H. (1979) *Accounts of Chemical Research*, **12**, 331.
4 Usón, R., Oro, L.A. and Fernández, M.J. (1980) *Journal of Organometallic Chemistry*, **193**, 127.
5 (a) Oro, L.A., Cabeza, J.A., Cativiela, C., Díaz de Villegas, M.D. and Meléndez, E. (1983) *Journal of the Chemical Society, Chemical Communications*, 1383; (b) Cabeza, J.A., Cativiela, C., Díaz de Villegas, M.D. and Oro, L.A. (1988) *Journal of the Chemical Society, Perkin Transactions I*, 1881.
6 Lightfoot, A., Schnider, P. and Pfaltz, A. (1998) *Angewandte Chemie – International Edition*, **37**, 2897.
7 Roseblade, S.J. and Pfaltz, A. (2007) *Comptes Rendus Chimie*, **10**, 178.

8 Crabtree, R.H. and Davis, M.W. (1986) *Journal of Organic Chemistry*, **51**, 2655.
9 Sola, E., Bakhmutov, V.I., Torres, F., Elduque, A., López, J.A., Lahoz, F.J., Werner, H. and Oro, L.A. (1998) *Organometallics*, **17**, 683.
10 Martín, M., Sola, E., Torres, O., Plou, P. and Oro, L.A. (2003) *Organometallics*, **22**, 5406.
11 Esteruelas, M.A., Oro, A.M., López, L.A., Pérez, A., Schulz, M. and Werner, H. (1993) *Organometallics*, **12**, 1823.
12 Esteruelas, M.A., Oliván, M., Oro, L.A. and Tolosa, J.I. (1995) *Journal of Organometallic Chemistry*, **487**, 143.
13 Martín, M., Torres, O., Oñate, E., Sola, E. and Oro, L.A. (2005) *Journal of the American Chemical Society*, **127**, 18074.
14 Kimmich, B.F.M., Somsook, E. and Landis, C.R. (1998) *Journal of the American Chemical Society*, **120**, 10115.
15 Usón, R., Oro, L.A. and Cabeza, J.A. (1985) *Inorganic Synthesis*, **23**, 126.
16 Sola, E., Navarro, J., López, J.A., Lahoz, F.J., Oro, L.A. and Werner, H. (1999) *Organometallics*, **18**, 3534.
17 Brandt, P., Hedberg, C. and Andersson, P.G. (2003) *Chemistry–A European Journal*, **9**, 339.
18 Dietiker, R. and Chen, P. (2004) *Angewandte Chemie–International Edition*, **43**, 5513.
19 Navarro, J., Sági, M., Sola, E., Lahoz, F.J., Dobrinovitch, I.T., Katho, A., Joó, F. and Oro, L.A. (2003) *Advanced Synthesis Catalysis*, **345**, 280–8.
20 (a) Evitt, E.R. and Bergman, R.G. (1980) *Journal of the American Chemical Society*, **102**, 7003;
(b) Thompson, J.S. and Atwood, J.D. (1991) *Organometallics*, **10**, 3525.
21 Torres, F., Sola, E., Martín, M., López, J.A., Lahoz, F.J. and Oro, L.A. (1999) *Journal of the American Chemical Society*, **121**, 10632.
22 Torres, F., Sola, E., Martín, M., Ochs, C., Picazo, G., López, J.A., Lahoz, F.J. and Oro, L.A. (2001) *Organometallics*, **20**, 2716.
23 Alaimo, P.J. and Bergman, R.G. (1999) *Organometallics*, **18**, 2707.
24 Canepa, G., Sola, E., Martín, M., Lahoz, F.J., Oro, L.A. and Werner, H. (2003) *Organometallics*, **22**, 2151.
25 Canepa, G., Brandt, C.D. and Werner, H. (2001) *Organometallics*, **20**, 604.
26 (a) Basolo, F. (1994) *New Journal of Chemistry*, **18**, 19;
(b) Bowyer, W.J. and Geiger, W.E. (1985) *Journal of the American Chemical Society*, **107**, 5657.
27 Oro, L.A. (2003) *Encyclopedia of Catalysis*, Vol. 4 (eds I.V. Horváth, E. Iglesia, M.T. Klein, J.A. Lercher, A.J. Rusell and E.I. Stiefel), John Wiley & Sons, Inc., New York, pp. 55–107.
28 (a) James, B.R. (1997) *Catalysis Today*, **37**, 209;
(b) Tang, W. and Zhang, X. (2003) *Chemical Reviews*, **103**, 3029;
(c) Kobayashi, S. and Ishitani, H. (1999) *Chemical Reviews*, **99**, 1069; (d) Tang, W. and Zhang, X. (2003) *Chemical Reviews*, **103**, 3029.
29 Martín, M., Sola, E., Tejero, S., Andrés, J.L. and Oro, L.A. (2006) *Chemistry–A European Journal*, **12**, 4043.
30 (a) Sandoval, C.A., Ohkuma, T., Muñiz, K. and Noyori, R. (2003) *Journal of the American Chemical Society*, **125**, 13490;
(b) Noyori, R., Yamakawa, M. and Hashiguchi, S. (2001) *Journal of Organic Chemistry*, **66**, 7931;
(c) Abdur-Rashid, K., Clapham, S.E., Hadzovic, A., Harvey, J.N., Lough, A.J. and Morris, R.H. (2002) *Journal of the American Chemical Society*, **124**, 15104.

3
Iridium N-Heterocyclic Carbene Complexes and Their Application as Homogeneous Catalysts

Eduardo Peris and Robert H. Crabtree

3.1
Introduction

N-Heterocyclic carbene (NHC) ligands, which were first reported as ligands by Öfele [1] and Wanzlick and Schonher [2] and later isolated in the free state by Arduengo and coworkers [3], have emerged as an extremely useful class of ligands for transition-metal catalysis. It is well recognized that the replacement of phosphines by N-heterocyclic carbenes can provide complexes with enhanced catalytic performances and greater stability [4]. This interest in the design of new NHCs and their coordination to metal complexes can be exemplified by the continuous increase in the number of articles on NHC-compounds, as shown in Figure 3.1. Although most of the early NHC-complexes were Pd-based and were mainly used in C—C-coupling reactions, the ligands soon proved to be highly versatile and capable of coordinating to many other transition metals. A noteworthy example is the application of NHCs in Ru-complexes as the later-generation catalysts for alkene metathesis reactions [5]. After palladium and ruthenium, iridium is most likely one of the metals for which the most NHCs have been described, and indeed the number articles concerning Ir—NHCs has also shown a dramatic increase during the past ten years (see Figure 3.1). Besides the obvious catalytic interest raised by the introduction of a new family of ligands for iridium complexes, Ir—NHCs have provided the first information on a new coordination form for NHCs (the so-called 'abnormal' NHC), as well as mechanistic aspects concerning the metallation processes of the precursor imidazolium salts.

Although not a comprehensive treatise on Ir—NHC complexes, this chapter first describes the different types of Ir—NHC described in the literature, with particular attention to their reactivity and special chemical features. Later in the chapter the most relevant catalytic applications of this type of complex are described, mainly in the field of C—H activation. The general atom-numbering scheme for NHCs, which is used throughout the text, is shown in Figure 3.2.

Iridium Complexes in Organic Synthesis. Edited by Luis A. Oro and Carmen Claver
Copyright © 2009 WILEY-VCH Verlag GmbH & Co. KGaA, Weinheim
ISBN: 978-3-527-31996-1

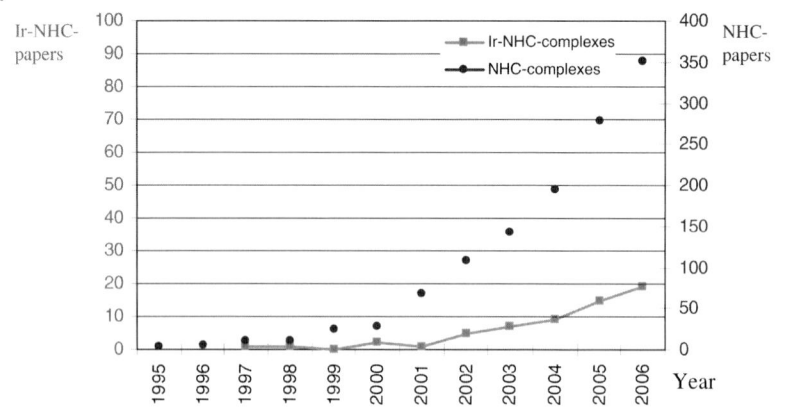

Figure 3.1 Time course showing numbers of published articles on NHC complexes (•) and Ir—NHCs (■).

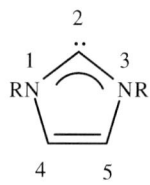

Figure 3.2 The general atom-numbering system in NHCs.

3.2
Types of Ir—NHC and Reactivity

The wide diversity of topologies that can today be found for NHC-precursors, together with the different efficient metallation strategies [6], have provided a large set of Ir—NHC complexes among which monodentate, bis-chelate and chiral species are abundant.

3.2.1
Mono-NHCs and Intramolecular C—H Activation

The preparation of carbonyl-Ir—NHC complexes (Scheme 3.1) and the study of their average CO-stretching frequencies [7], have provided some of the earliest experimental information on the electron-donor power of NHCs, quantified in terms of Tolman's electronic parameter [8]. The same method was later used to assess the electronic effects in a family of sterically demanding and rigid N-heterocyclic carbenes derived from bis-oxazolines [9]. The high electron-donor power of NHCs should favor oxidative addition involving the C—H bonds of their N-substituents, particularly because these substituents project towards the metal rather than away, as in phosphines. Indeed, NHCs have produced a number of unusual cyclometallation processes, some of which have led to electron-deficient

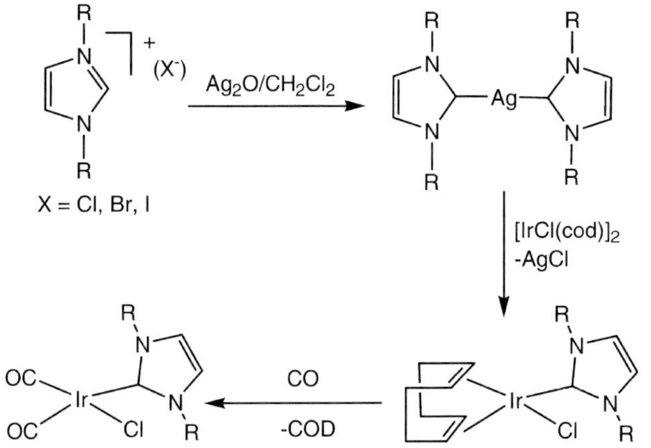

Scheme 3.1

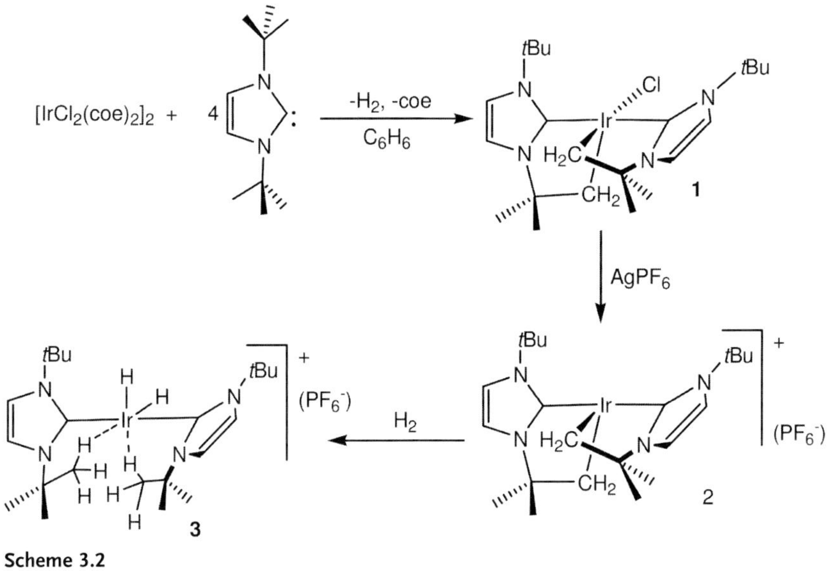

Scheme 3.2

species that had never been observed with other types of ligand. For example, Nolan recently described the interaction of N,N'-di(t-butyl)imidazol-2-ylidene (ItBu) with [IrCl$_2$(coe)$_2$]$_2$ (coe = cyclo-octene) leading to the formation of a doubly cyclometallated product (**1**, Scheme 3.2). Abstraction of the halide with AgPF$_6$ allowed the preparation of an unusually stable 14-electron species, **2** [10]. The stability of this electron-deficient complex is undoubtedly due to the high electron-donor power of the NHC ligands. The reaction of **2** with H$_2$ leads to a remarkable electron-deficient dihydride complex [IrH$_2$(ItBu)$_2$](PF$_6$), **3**, in which two C—H agostic interactions are observed [11].

In a recent investigation, Yamaguchi and coworkers reported the intramolecular alkyl C—H activation of a series of *iso*-propyl functionalized imidazolylidenes coordinated to the Cp*Ir fragment [12]. The ability of Cp*Ir(NHC) complexes to undergo intramolecular C—H activation had been reported by Herrmann and coworkers, who described the cyclometallation of the complex Cp*Ir(ICy)(Me)$_2$ (ICy = 1,3-dicylohexylimidazol-2-ylidene) [13]. For the analogous phosphine complexes of the 'Cp*Ir(PR$_3$)' fragment, Bergman and coworkers have extensively studied a number of inter- and intramolecular C—H activations [14]. For this type of complex, both aromatic and aliphatic C—H activations have been described by us, both being highly favorable processes [15–18]. In the cases where both aliphatic and aromatic C—H activation is possible – as in the case depicted in Scheme 3.3 – steric factors seem to determine the selectivity of the reaction [16].

For the reaction outcome of this reaction, it seems that the steric hindrance of the reaction products determines the selectivity of the process. For the carbene containing an *i*Pr group, the reaction proceeds through an aromatic C—H activation pathway because the Ir—C$_{aromatic}$ bond is stronger than its aliphatic analogue. In addition, the orientation of the *i*Pr group allows the hydrogen atom to point towards the Cp* ring, thus providing steric relief in the final product (**A**, Scheme 3.4). For the *t*Bu-substituted NHC, metallation via the aromatic ring would provide a highly strained structure, with one of the methyl groups of the *t*Bu pointing towards the Cp* ring (**B**). For this case, metallation through one of the methyl groups affords a more favorable situation, as the benzyl group can orient such that it points away from the coordination sphere of the Ir atom (**C**).

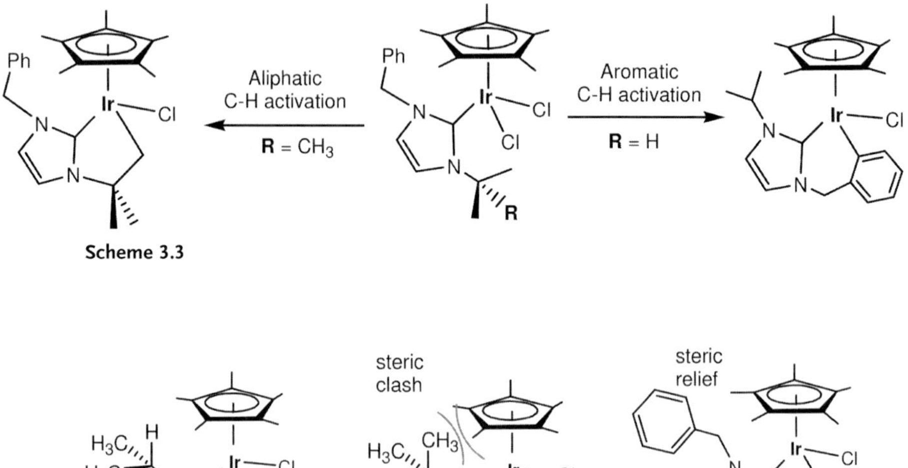

Scheme 3.3

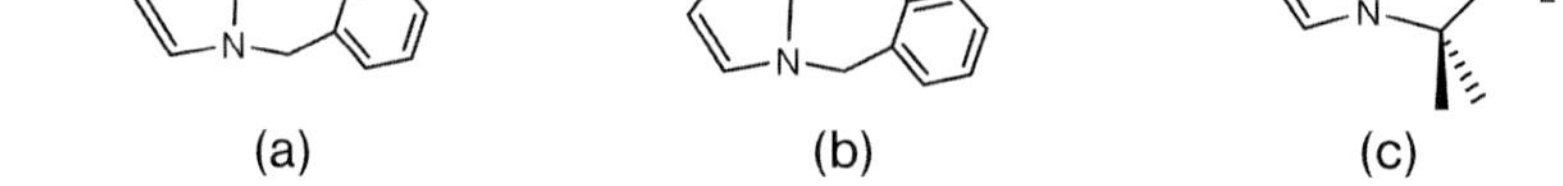

Scheme 3.4

3.2.2
Chelating bis-NHCs

Polydentate NHC ligands have provided numerous new compounds [19]. Chelate-N-heterocyclic ligands also allow the fine-tuning of topological properties such as steric hindrance, bite angles, chirality and fluxional behavior. In the case of Ir—chelate—NHC compounds, most of the data that have been reported were obtained as an extension of investigations into their rhodium analogues [19, 20].

Crabtree and coworkers prepared a series of complexes with a methylene linker and different wingtips. Changing the bulk of the wingtip groups has a strong influence on whether a chelating conformation is achieved. For example, in the case of the *tert*-butyl group a monocarbene complex was formed (Scheme 3.5) instead of the chelate complex [21].

Some mechanistic studies on the formation of bis-chelating NHCs from imidazolium precursors have been performed, indicating that C—H oxidative addition of the imidazolium salt may play a decisive role in the overall process, at least for Ir(I). The reaction of a ferrocenyl-bis-imidazolium salt with [IrCl(cod)]$_2$ in the presence of NEt$_3$ provided the first evidence of the intermediacy of a stable NHC—Ir^{III}—H complex by direct oxidative addition of the imidazolium salt [22]. Although initially it was proposed that the ferrocenyl fragment might be sterically protecting the M—H from subsequent reductive elimination, it was later shown that this fragment was not necessary in order to obtain the intermediate NHC—Ir^{III}—H complexes (Scheme 3.6) [23]. The oxidative addition of the C$_2$—H bond of the imidazolium salt may be followed by a reductive elimination of HX initiated by a weak base, and this could explain why NHC—M—H complexes are so scarce (Scheme 3.7). A combined experimental and theoretical approach provided a

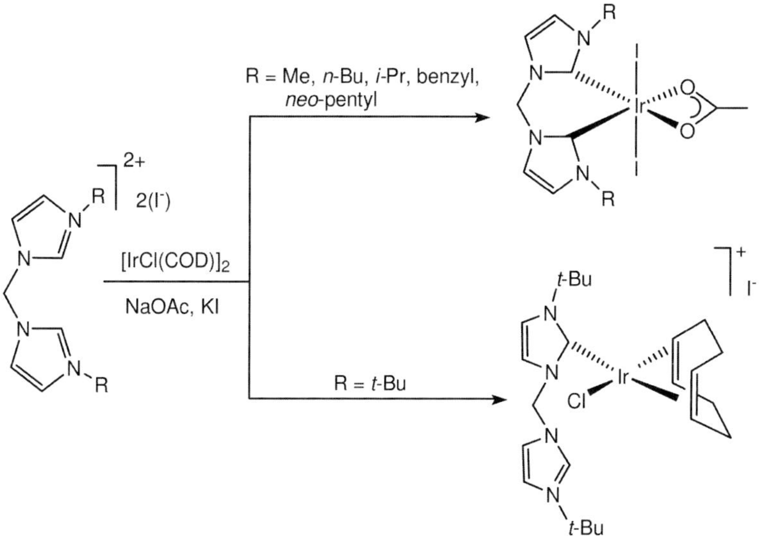

Scheme 3.5

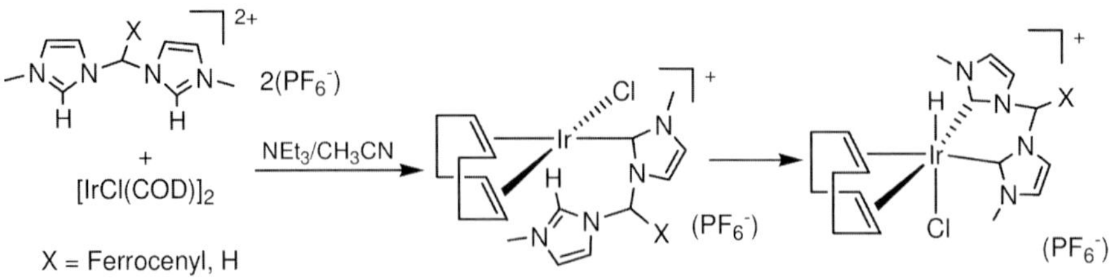

Scheme 3.6

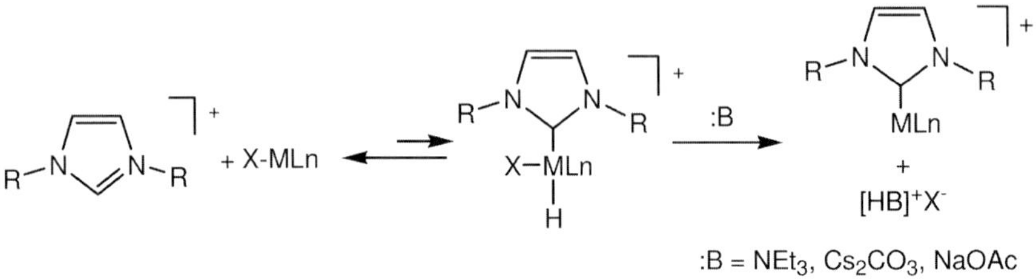

Scheme 3.7

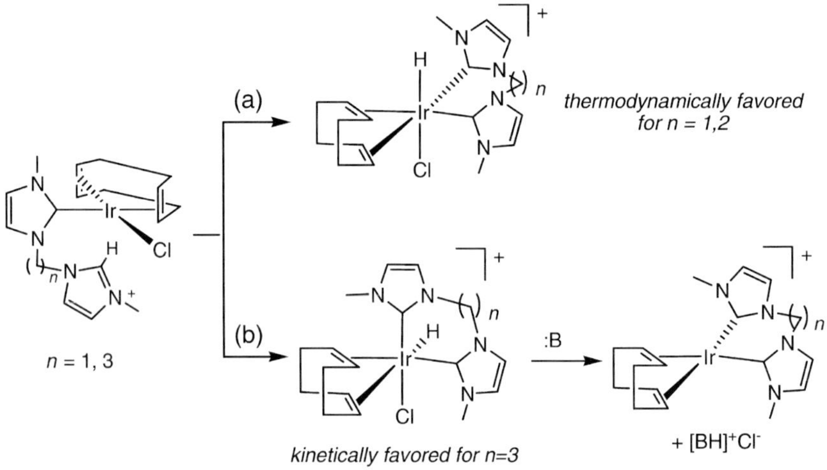

:B = NEt$_3$, Cs$_2$CO$_3$, NaOAc

Scheme 3.8

unified mechanism for the metallation of a series of bis-imidazolium salts with linkers of different lengths between the azolium rings [23]. It was concluded that metallation of the second imidazolium ring proceeds by C$_2$—H oxidative addition. The final formation of the bis-NHC—Ir^{III}—H (short linker) or bis-NHC—Ir^I (long linker) depends on whether the oxidative addition product yields the *trans* [short linker, $n = 1,2$, mechanism (a)] or *cis* [long linker, $n = 3$, mechanism (b)] products (Scheme 3.8). The *trans* products are the thermodynamically favored complexes, but in the case of the ligands with long linker lengths, the *cis* complexes are kinetically favored, thus providing the bis-NHC—Ir^I reductive elimination products [23].

3.2 Types of Ir—NHC and Reactivity

The chelate effect also favors oxidative addition of the C_2—H bonds of imidazolium salts because it provides stabilized complexes. The reaction of a pyridine–imidazolium salt with [IrCl(cod)]$_2$ yields the oxidative addition product, even in the absence of a base (Scheme 3.9), thus confirming that the oxidative addition of an imidazolium salt should be considered as a valid process for the preparation of NHC—M—H complexes [24].

In an attempt to obtain tripodal tris-NHCs of Rh and Ir, the ligand 1,1,1-[tris(3-alkylimidazole-2-ylidene)methyl]ethane (TIME) [25], was used. Depending on the metallation strategy, bis-NHC mononuclear or tris-NHC-dinuclear species were formed, as depicted in Scheme 3.10. The N-anchored analogue of TIME is the ligand tris[2-(3-alkylmethylimidazole-2-ylidene)ethyl]amine (TIMEN^R); this can coordinate to Ir, affording the trinuclear species shown in Figure 3.3.

Scheme 3.9

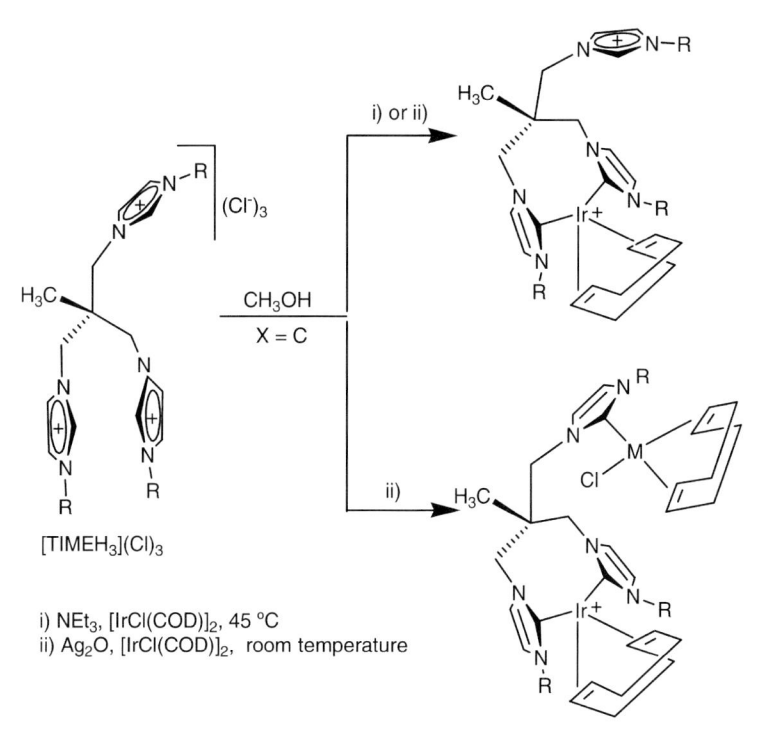

[TIMEH$_3$](Cl)$_3$

i) NEt$_3$, [IrCl(COD)]$_2$, 45 °C
ii) Ag$_2$O, [IrCl(COD)]$_2$, room temperature

Scheme 3.10

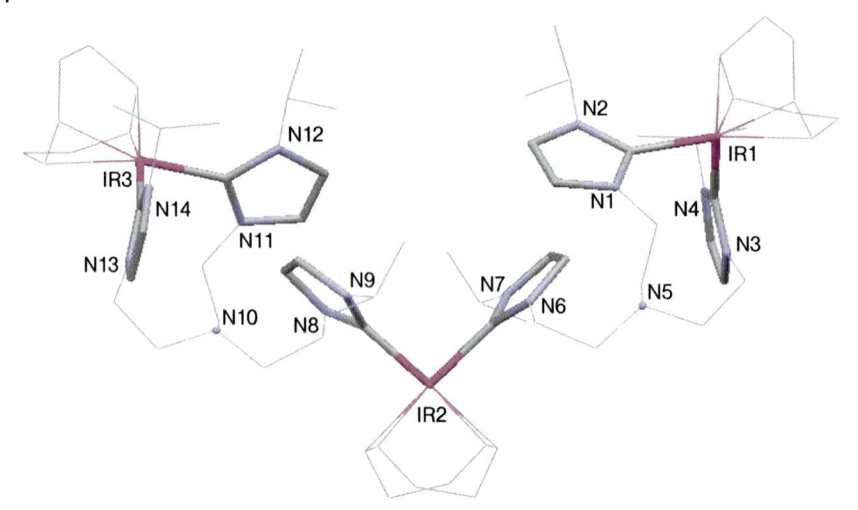

Figure 3.3 Molecular diagram of [(TIMEN^{iPr})$_3$Ir(COD)$_3$](PF$_6$)$_3$. The hydrogen atoms and counterions (PF$_6$) have been omitted for clarity.

Scheme 3.11

3.2.3
Abnormal NHCs

Iridium complexes were the first to show an interesting type of NHC binding, in which the metallation occurred in the 'wrong way' – that is, with the metal being bound not to the activated C$_2$ position but rather to the C$_5$ position [26, 27]. This mode of NHC binding – which today is generally referred to as 'abnormal' NHC (aNHC) binding – was first observed for the reaction of an N-isopropyl-substituted, methylene-linked pyridine–imidazolium salt and IrH$_5$(PPh$_3$)$_2$ (Scheme 3.11). Formation of the abnormal NHC is favored by the lower steric strain at the metal center. On moving to the smaller wingtip Me group, a mixture of the normal and abnormal carbenes is obtained (Scheme 3.12).

Theoretical calculations suggested that the normal binding is thermodynamically favored when the anion-free complexes are considered, but ion-pairing has

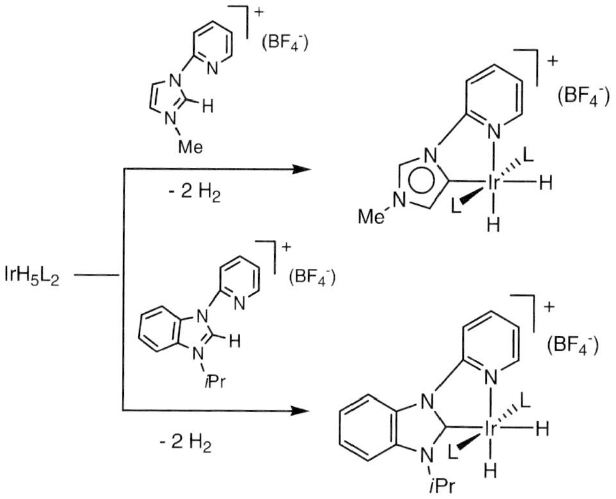

Scheme 3.12

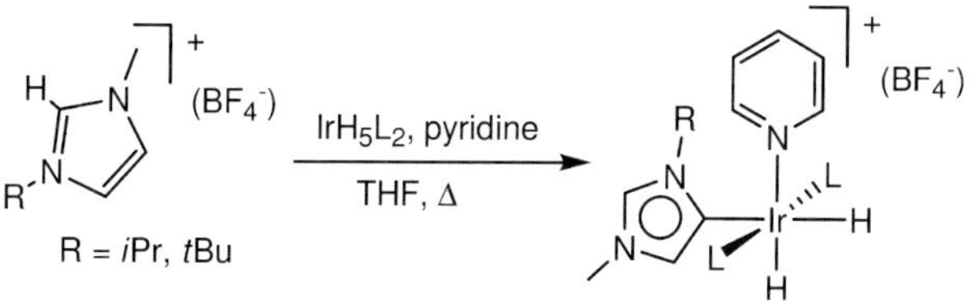

Scheme 3.13

a significant effect in lowering the energy of the abnormal-binding species, most likely due to the energy differences in the C_2/C_5 C—H hydrogen bonding to the anions. The counteranions BF_4, PF_6 and SbF_6 give predominantly an abnormal product, whereas Br favors the normal C_2 bond. This happens because the C_5—H oxidative addition is anion-insensitive, whereas the C_2—H metallation occurs by proton transfer and hence is highly sensitive to the anion. The better ion pair former, bromide, accompanies the proton during its migration and hence favors formation of the normal NHC [28].

The 'abnormal' metallation is also favored when the conjugated pyridine–imidazolylidene precursor (with a smaller bite angle) is used. For these precursors, an abnormal binding is produced even with a small wingtip group, as shown in Scheme 3.12. Under the same conditions, the pyridine–benzimidazolium analogue afforded the C_2 carbene complex (Scheme 3.12) [27].

Chelation is not necessary to promote the abnormal metallation. When imidazolium salts with bulky substituents (e.g. *i*Pr, *t*Bu) are refluxed with pyridine and $IrH_5(PPh_3)_2$ in tetrahydrofuran (THF), aNHC complexes are obtained in good yields, with the least sterically hindered of the three imidazolium carbons selectively bound to the metal (Scheme 3.13) [29]. Infrared spectroscopy on carbonyl derivatives indicated that the aNHCs are much stronger electron donors than their C_2 analogues [29].

A very convenient method for the selective preparation of aNHC complexes is to block the imidazolium ligand precursors by substitution at C_2 with alkyl or aryl groups [29, 30]. Following this procedure, the coordination of 1,2,3-trimethylimidazolium iodide to [Cp*IrCl$_2$]$_2$ afforded Cp*Ir(aNHC), as shown in Scheme 3.14 [31].

The use of methylene- and ethylene-bridged bisimidazolium salts with C_2-Me gave unexpected results. Depending on the length of the linker, and on the nature of the bisimidazolium salt used, the corresponding aNHC complexes or the products resulting from the activation of the C—H bond in the C—Me group were obtained. As depicted in Scheme 3.15, the reaction of 1,1′-ethylene-2,3,3′-trimethylbis(1H-imidazolium) dibromide with [Cp*IrCl$_2$]$_2$ in the presence of NaOAc in refluxing acetonitrile allows the preparation of a compound, in which the chelating-biscarbene ligand is coordinated by both the abnormal and normal modes.

The reaction gave completely different products when doubly C_2-Me-substituted bisimidazolium salts were used, for which the length of the linker clearly determined the reaction outcome. For the methylene-bridged ligand, the major product obtained contained a bischelating ligand coordinated by an aNHC, and a metallated methylene group. For the ethylene-bridged bisimidazolium, the major product was the expected chelating bis-aNHC, together with the chelating C_2-Me-activated compound and a new neutral species with a 1,2-dimethylimidazole ligand (Scheme 3.16) [31].

Formation of the reaction products, in the case of the methylene linker, was rationalized by means of density functional theory (DFT) calculations with the inclusion of a solvent effects polarized continuous model (PCM). The calculations

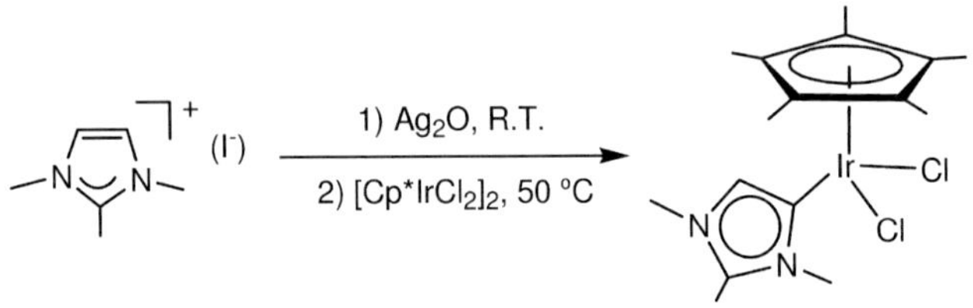

Scheme 3.14

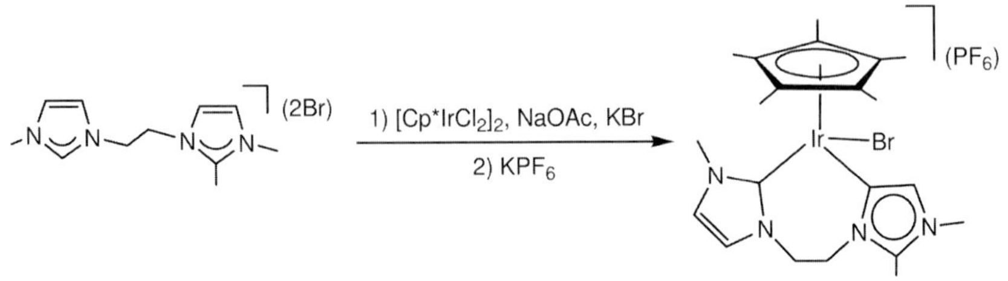

Scheme 3.15

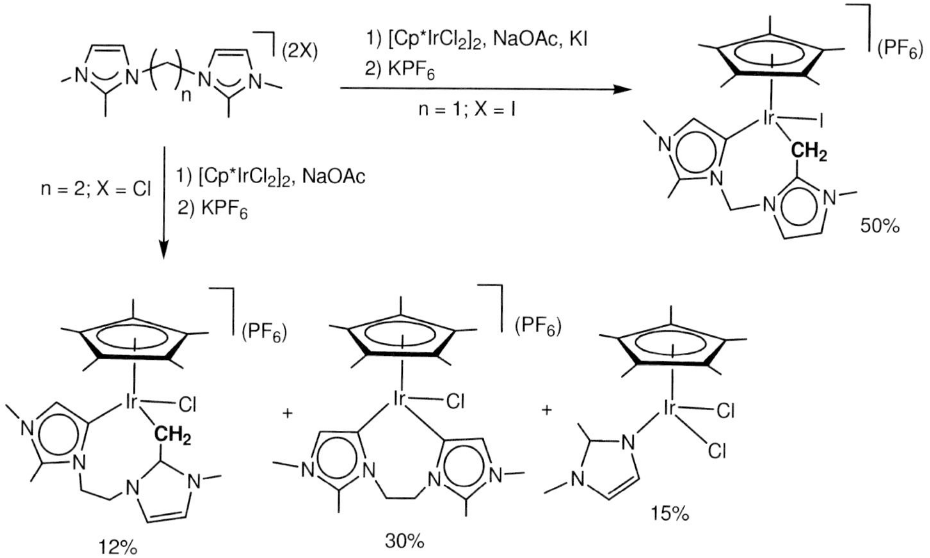

Scheme 3.16

could not identify whether the first metallation occurs by direct deprotonation of the ligand by the base, or metallation through C—H activation at Ir. However, both cases point to a kinetic preference for first metallation at the C_2-Me group. The second metallation process is the result of kinetically preferred C—H activation at the C_5 position [31].

3.3
Catalysis with Ir—NHCs

The wide application of NHCs to catalysis places them along with phosphines and cyclopentadienyls in their utility for organometallic catalysis. Being much more donor in nature than phosphines, the NHCs occupy a distinct region of the Tolman ligand map [32]. A number of pincer NHC derivatives are catalytically active, but are not discussed here because they have been fully described elsewhere [33]. Some other important recent reviews discuss catalysis by NHC complexes in general [10, 34, 35].

A large number of reports have concerned transfer hydrogenation using isopropanol as donor, with imines, carbonyls – and occasionally alkenes – as substrate (Scheme 3.17). In some early studies conducted by Nolan and coworkers [36], NHC analogues of Crabtree catalysts, [Ir(cod)(py)(L)]PF$_6$ (L = Imes, Ipr, Icy) all proved to be active. The series of chelating iridium(III) carbene complexes shown in Scheme 3.5 (upper structure) proved to be accessible via a simple synthesis and catalytically active for hydrogen transfer from alcohols to ketones and imines. Unexpectedly, iridium was more active than the corresponding Rh complexes, but

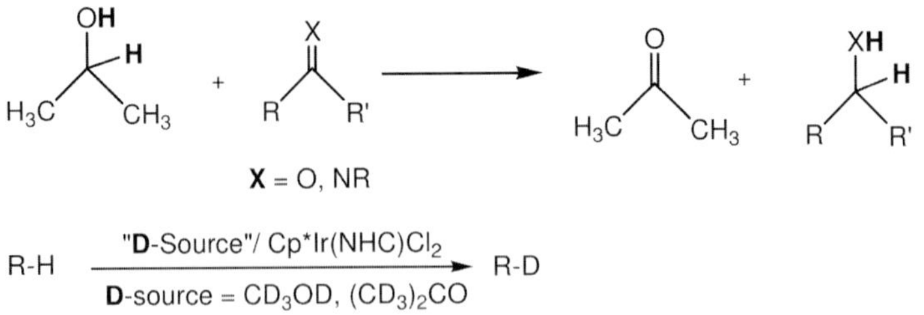

Scheme 3.17

neopentyl substituents at the azole nitrogen were required for high activity [21]. Even aldehydes were successfully reduced with this system [34]. Aldehydes are normally problematic substrates because of their tendency to undergo metal-mediated decarbonylation and base-mediated aldol condensation.

A series of new iridium(I) triazole-based NHC complexes [(cod)Ir(NHC)L]BF$_4$ (L = PPh$_3$, pyridine) were later prepared and showed excellent activity for the transfer hydrogenation of C=O, C=N and C=C double bonds in refluxing 2-propanol with K$_2$CO$_3$ as mild base. The phosphine series was shown to be more active than the pyridine series in the case of imine transfer hydrogenation. A neopentyl wingtip substituent on the NHC once again gave the best catalytic activity with the following competitive order: aldehyde > ketone > imine. In a substrate containing both aldehyde and ketone functionalities, only the aldehyde was reduced. A rare extension of transfer hydrogenation to polarized and even nonpolarized C=C bonds also proved possible. In a useful organic synthetic application, a direct, one-pot catalytic reductive amination of RCHO with R'NH$_2$ to give RCH$_2$NHR' was demonstrated for a variety of cases [37]. NHC–P hybrids were also effective for iridium-catalyzed asymmetric transfer hydrogenation [38].

In a very recent study, trialkylated 1,2,4-triazoles were shown to produce the novel bridging diylidynes with metals coordinated at C$_3$ and C$_5$. For example, with (cod)M (M = Rh and Ir) as the metal fragment they proved active in transfer hydrogenation, although it is possible that their activity in the cyclization of alkynoic acids may be more relevant [39].

The catalytic activity of Cp*Ir(III) complexes in the Oppenauer-type oxidation of alcohols was considerably enhanced by the introduction of N-heterocyclic carbene ligands. Here, high turnover numbers (TONs) of up to 950 were achieved in the oxidation of secondary alcohols [40].

Interestingly, Cp*Ir(NHC) complexes were also shown to be efficient catalysts for the deuteration of organic molecules using CD$_3$OH or (CD$_3$)$_2$CO as deuterium sources (Scheme 3.17). A wide set of organic molecules (including ketones, alcohols, olefins and ethers) were deuterated in high yields [15].

Cp*Ir(NHC) complexes are a very versatile type of catalyst, with a wide range of applications. In a chemoenzymatic application, Cp*Ir complexes activated by fluorinated and nonfluorinated NHC ligands were shown to be catalysts for racemization in the one-pot chemoenzymic dynamic kinetic resolution (DKR) of secondary

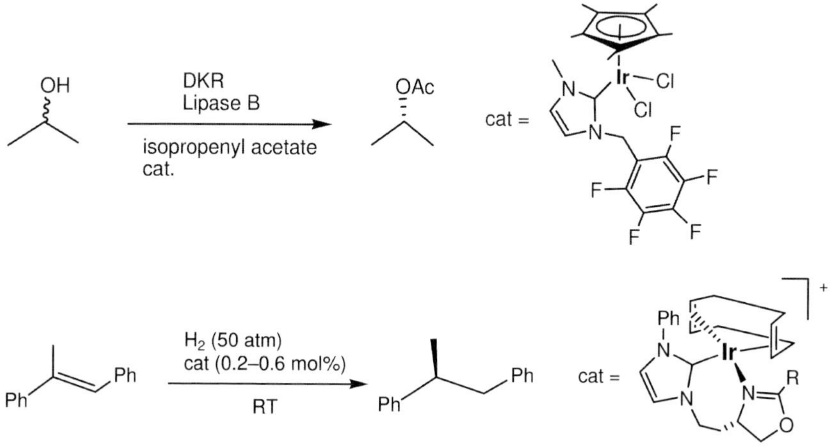

3.3 Catalysis with Ir–NHCs | 51

Scheme 3.18

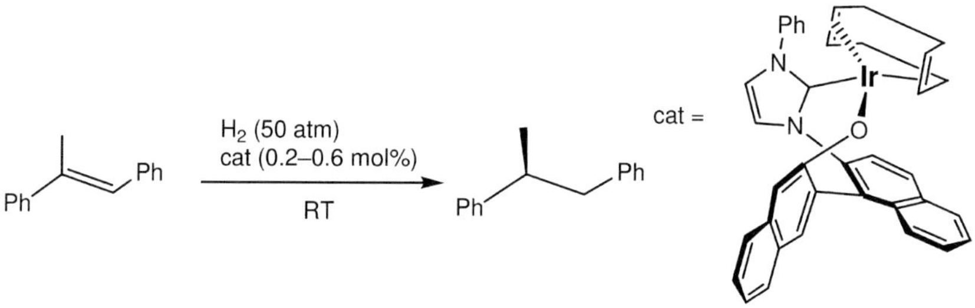

Scheme 3.19

alcohols. Excellent conversions and good enantioselectivities were observed for alkyl aryl and dialkyl secondary alcohols (Scheme 3.18) [41].

Five-coordinate complexes of the type shown below have the attractive feature of containing four alkene ligands which, in principle, are capable of being removed by hydrogenation to reveal an Ir(NHC) cation. Although these, too, are active for transfer hydrogenation, in related complexes the allyl substituents do not seem to be hydrogenated [42].

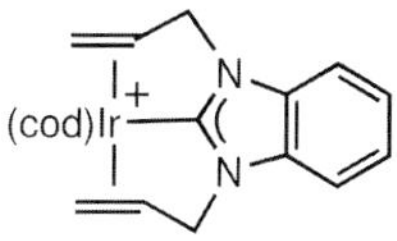

Despite the very low number of examples describing effective iridium-based asymmetric catalysts, there are some which deserve to be mentioned at this point. The preparation of hybrid NHC–oxazoline ligands, allowed Burgess and coworkers to prepare Ir(I) complexes for the asymmetric hydrogenation of aryl alkenes (Scheme 3.19) [43]. These catalysts are among the most efficient in terms of yield and asymmetric inductions for this reaction.

In asymmetric ketone hydrosilylation, axially chiral bidentate N-heterocyclic carbene ligands derived from BINAM proved to be moderately effective when bound to iridium, but less so with rhodium (Scheme 3.19) [44].

New N-heterocyclic carbene rhodium and iridium complexes derived from 2,2′-diaminobiphenyl were successfully synthesized and their structures unambiguously characterized by X-ray diffraction (XRD) analysis. These are catalytically active for the hydrosilylation of ketones with diphenylsilane, although an NHC—rhodium complex was found to be the best among those investigated [45].

The reaction of (S,S)-1,3-di(methylbenzyl)imidazolium chloride with [Cp*IrCl$_2$]$_2$ in the presence of NaOAc affords the diastereoselective formation of a Cp*Ir(NHC) complex with a chelating ligand coordinated through the carbene and the *ortho*-position of one of the phenyl groups. The complex was used in the catalytic diboration of olefins, providing high efficiencies and chemoselectivities for the organodiboronate products [17].

Perhaps most dramatically of all, for the first time, bis(carbene)-substituted iridium complexes, such as [Ir(cod)(NHC)$_2$]$^+$ (NHC = 1,3-dimethyl- or 1,3-dicyclohexylimidazolin-2-ylidene] were successfully used by Herrmann and coworkers as C—H-activation catalysts in the synthesis of arylboronic acids starting from pinacolborane and arene derivatives [46].

3.4
Conclusions

Despite many studies having been reported regarding the reactivity of Ir—NHC complexes, it seems that the catalytic properties of these materials have not yet been fully explored. Studies on the reactivity, and the access to new molecular architectures in which the metal in highly electron-rich, envisage a wide set of applications in homogeneous catalysis. It is noteworthy to point out that Ir—NHCs offer a clear advantage over other metal complexes in processes implying C—H activations, as observed by the number of intramolecular versions of this process that have been reported and fully studied.

Despite Ir—NHCs having demonstrated their worth as catalysts, there are clearly many more reports of transfer hydrogenation. Consequently, many further investigations must be conducted in other areas of catalysis, where there is every reason to feel that the Ir—NHCs will be equally useful.

References

1 Ofele, K. (1968) *Journal of Organometallic Chemistry*, **12**, 42.
2 Wanzlick, H.W. and Schonher, H.J. (1968) *Angewandte Chemie – International Edition*, **7**, 141.
3 Arduengo, A.J., Harlow, R.L. and Kline, M. (1991) *Journal of the American Chemical Society*, **113**, 361.
4 (a) Herrmann, W.A. (2002) *Angewandte Chemie – International Edition*, **41**, 1291;

(b) Herrmann, W.A., Elison, M., Fischer, J., Kocher, C. and Artus, G.R.J. (1995) *Angewandte Chemie – International Edition in English*, **34**, 2371.

5 (a) Trnka, T.M. and Grubbs, R.H. (2001) *Accounts of Chemical Research*, **34**, 18;
(b) Colacino, E., Martinez, J., Lamaty, F. (2007) *Coordination Chemistry Reviews*, **251**, 726;
(c) Grubbs, R.H. (2006) *Angewandte Chemie – International Edition*, **45**, 3760;
(d) Grubbs, R.H. (2004) *Tetrahedron*, **60**, 7117;
(e) Furstner, A. (2000) *Angewandte Chemie – International Edition*, **39**, 3013;
(f) Furstner, A. (1997) *Topics in Catalysis*, **4**, 285.

6 Peris, E. (2007) *Topics in Organometallic Chemistry*, **21**, 83.

7 Chianese, A.R., Li, X.W., Janzen, M.C., Faller, J.W. and Crabtree, R.H. (2003) *Organometallics*, **22**, 1663.

8 Tolman, C.A. (1977) *Chemical Reviews*, **77**, 313.

9 Altenhoff, G., Goddard, R., Lehmann, C.W. and Glorius, F. (2004) *Journal of the American Chemical Society*, **126**, 15195.

10 Scott, N.M. and Nolan, S.P. (2005) *European Journal of Inorganic Chemistry*, 1815.

11 Scott, N.M., Pons, V., Stevens, E.D., Heinekey, D.M. and Nolan, S.P. (2005) *Angewandte Chemie – International Edition*, **44**, 2512.

12 Hanasaka, F., Tanabe, Y., Fujita, K. and Yamaguchi, R. (2006) *Organometallics*, **25**, 826.

13 Prinz, M., Grosche, M., Herdtweck, E. and Herrmann, W.A. (2000) *Organometallics*, **19**, 1692.

14 (a) Arndtsen, B.A. and Bergman, R.G. (1995) *Science*, **270**, 1970;
(b) Tellers, D.M. and Bergman, R.G. (2000) *Journal of the American Chemical Society*, **122**, 954;
(c) Golden, J.T., Andersen, R.A. and Bergman, R.G. (2001) *Journal of the American Chemical Society*, **123**, 5837;
(d) Peterson, T.H., Golden, J.T. and Bergman, R.G. (2001) *Journal of the American Chemical Society*, **123**, 455;
(e) Klei, S.R., Golden, J.T., Burger, P. and Bergman, R.G. (2002) *Journal of Molecular Catalysis A – Chemical*, **189**, 79;
(f) Tellers, D.M., Yung, C.M., Arndtsen, B.A., Adamson, D.R. and Bergman, R.G. (2002) *Journal of the American Chemical Society*, **124**, 1400;
(g) Klei, S.R., Golden, J.T., Tilley, T.D. and Bergman, R.G. (2002) *Journal of the American Chemical Society*, **124**, 2092;
(h) Skaddan, M.B., Yung, C.M. and Bergman, R.G. (2004) *Organic Letters*, **6**, 11;
(i) Yung, C.M., Skaddan, M.B. and Bergman, R.G. (2004) *Journal of the American Chemical Society*, **126**, 13033.

15 Corberan, R., Sanau, M. and Peris, E. (2006) *Journal of the American Chemical Society*, **128**, 3974.

16 Corberan, R., Sanau, M. and Peris, E. (2006) *Organometallics*, **25**, 4002.

17 Corberan, R., Lillo, V., Mata, J.A., Fernandez, E. and Peris, E. (2007) *Organometallics*, **26**, 4350.

18 Corberan, R., Sanau, M. and Peris, E. (2007) *Organometallics*, **26**, 3492.

19 Mata, J.A., Poyatos, M. and Peris, E. (2007) *Coordination Chemistry Reviews*, **251**, 841.

20 Poyatos, M., Uriz, P., Mata, J.A., Claver, C., Fernandez, E. and Peris, E. (2003) *Organometallics*, **22**, 440.

21 Albrecht, M., Miecznikowski, J.R., Samuel, A., Faller, J.W. and Crabtree, R.H. (2002) *Organometallics*, **21**, 3596.

22 Viciano, M., Mas-Marza, E., Poyatos, M., Sanau, M., Crabtree, R.H. and Peris, E. (2005) *Angewandte Chemie – International Edition*, **44**, 444.

23 Viciano, M., Poyatos, M., Sanau, M., Peris, E., Rossin, A., Ujaque, G. and Lledos, A. (2006) *Organometallics*, **25**, 1120.

24 Mas-Marza, E., Sanau, M. and Peris, E. (2005) *Inorganic Chemistry*, **44**, 9961.

25 Mas-Marza, E., Peris, E., Castro-Rodriguez, I. and Meyer, K. (2005) *Organometallics*, **24**, 3158.

26 Grundemann, S., Kovacevic, A., Albrecht, M., Faller, J.W. and Crabtree, R.H. (2001) *Chemical Communications*, 2274.

27 Grundemann, S., Kovacevic, A., Albrecht, M., Faller, J.W. and Crabtree, R.H. (2002) *Journal of the American Chemical Society*, **124**, 10473.

28 (a) Kovacevic, A., Grundemann, S., Miecznikowski, J.R., Clot, E., Eisenstein, O. and Crabtree, R.H. (2002) *Chemical Communications*, 2580;

(b) Appelhans, L.N., Zuccaccia, D., Kovacevic, A., Chianese, A.R., Miecznikowski, J.R., Macchioni, A., Clot, E., Eisenstein, O. and Crabtree, R.H. (2005) *Journal of the American Chemical Society*, **127**, 16299.

29 Chianese, A.R., Kovacevic, A., Zeglis, B.M., Faller, J.W. and Crabtree, R.H. (2004) *Organometallics*, **23**, 2461.

30 (a) Chianese, A.R., Zeglis, B.M. and Crabtree, R.H. (2004) *Chemical Communications*, 2176;
(b) Bacciu, D., Cavell, K.J., Fallis, I.A. and Ooi, L.L. (2005) *Angewandte Chemie–International Edition*, **44**, 5282;
(c) Alcarazo, M., Roseblade, S.J., Cowley, A.R., Fernandez, R., Brown, J.M. and Lassaletta, J.M. (2005) *Journal of the American Chemical Society*, **127**, 3290.

31 Viciano, M., Feliz, M., Corberan, R., Mata, J.A., Clot, E. and Peris, E. (2007) *Organometallics*, **26**, 5304–14.

32 Crabtree, R.H. (2005) *Journal of Organometallic Chemistry*, **690**, 5451.

33 Peris, E. and Crabtree, R.H. (2004) *Coordination Chemistry Reviews*, **248**, 2239.

34 Miecznikowski, J.R. and Crabtree, R.H. (2004) *Polyhedron*, **23**, 2857.

35 (a) Crudden, C.M. and Allen, D.P. (2004) *Coordination Chemistry Reviews*, **248**, 2247;
(b) Cesar, V., Bellemin-Laponnaz, S., Gade, L.H. (2004) *Chemical Society Reviews*, **33**, 619.

36 Hillier, A.C., Lee, H.M., Stevens, E.D. and Nolan, S.P. (2001) *Organometallics*, **20**, 4246.

37 Gnanamgari, D., Moores, A., Rajaseelan, E. and Crabtree, R.H. (2007) *Organometallics*, **26**, 1226–30.

38 Hodgson, R. and Douthwaite, R.E. (2005) *Journal of Organometallic Chemistry*, **690**, 5822.

39 Mas-Marza, E., Mata, J.A. and Peris, E. (2007) *Angewandte Chemie–International Edition*, **46**, 3729.

40 Hanasaka, F., Fujita, K.I. and Yamaguchi, R. (2004) *Organometallics*, **23**, 1490.

41 Marr, A.C., Pollock, C.L. and Saunders, G.C. (2007) *Organometallics*, **26**, 3283.

42 (a) Hahn, F.E., Heidrich, B., Pape, T., Hepp, A., Martin, M., Sola, E. and Oro, L.A. (2006) *Inorganica Chimica Acta*, **359**, 4840;
(b) Hahn, F.E., Holtgrewe, C., Pape, T., Martin, M., Sola, E., Oro, L.A. (2005) *Organometallics*, **24**, 2203.

43 Powell, M.T., Hou, D.R., Perry, M.C., Cui, X.H. and Burgess, K. (2001) *Journal of the American Chemical Society*, **123**, 8878.

44 Chianese, A.R. and Crabtree, R.H. (2005) *Organometallics*, **24**, 4432.

45 Chen, T., Liu, X.G. and Shi, M. (2007) *Tetrahedron*, **63**, 4874.

46 Frey, G.D., Rentzsch, C.F., von Preysing, D., Scherg, T., Muhlhofer, M., Herdtweck, E. and Herrmann, W.A. (2006) *Journal of Organometallic Chemistry*, **691**, 5725.

4
Iridium-Catalyzed C=O Hydrogenation
Claudio Bianchini, Luca Gonsalvi and Maurizio Peruzzini

4.1
Introduction

The homogeneous hydrogenation of carbonyl groups from organic compounds is a valuable tool for the catalytic synthesis of alcohols, many of which bear other functionalities such as C=C double bonds. Thus, early reports have focused on α,β-unsaturated carbonyl compounds as test substrates for novel chemoselective and regioselective catalysts. The hydrogenation protocols of choice have used either hydrogen gas or transfer hydrogenation conditions as terminal reducing agents, and the literature review will be therefore divided into these two topics in the following sections. Metals other than iridium have been successfully used to catalyze the chemoselective, regioselective and stereoselective reduction of C=O double bonds [1]. In this chapter, for the sake of completeness, one section is dedicated to ketone and aldehydes hydrogenations in heterogeneous phase by catalysts comprising either iridium metal particles or iridium single-site catalysts tethered to various support materials. Among these, ruthenium and rhodium play a major role, with the former reported as being generally more active. Although, as a general trend, iridium catalysts exhibit a lower activity as compared to ruthenium and rhodium catalysts, they are often more selective, especially in terms of chemoselectivity and regioselectivity. Both, experimental and theoretical studies have been conducted to address the different activities of iridium complexes versus ruthenium analogues [2]. It is generally agreed that the different binding of the supporting ligands controls the energy of the transition state, thus ultimately determining the reduction kinetics [1, 2].

4.2
Homogeneous C=O Hydrogenations

The selective reduction of the C=O functional group of organic substrates has been reported using hydrogen gas and various iridium complexes as precatalysts.

Iridium Complexes in Organic Synthesis. Edited by Luis A. Oro and Carmen Claver
Copyright © 2009 WILEY-VCH Verlag GmbH & Co. KGaA, Weinheim
ISBN: 978-3-527-31996-1

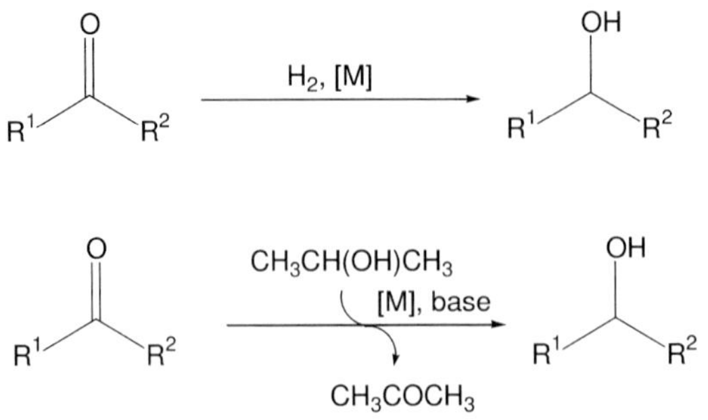

Scheme 4.1

In the following sections, the most relevant reports dealing with Ir-catalyzed hydrogenations using a homogeneous phase (molecular single-site catalysts soluble in the reaction media) using hydrogen are summarized, and divided into chemoselective and enantioselective reactions, respectively. Under these conditions, the actual catalytic species have been invariably associated with metal hydrides (Scheme 4.1).

4.2.1
Chemoselective Hydrogenations

Chemoselectivity is often a major issue in the reduction of multifunctional organic substrates such as substituted conjugated enones. The corresponding unsaturated alcohols have found use as building blocks for pharmaceutically active molecules; for example β-amino-α-phenylethanol is used for the synthesis of β-blockers which are the active molecules for controlling hypertension and other cardiac disorders.

One of the first accounts on Ir-catalyzed CO hydrogenation dates back to 1967 by Coffey, in which a 5×10^{-3} M solution of [IrH$_3$(PPh$_3$)$_3$] (**1**) in AcOH catalyzed the hydrogenation of 0.56 M PrCHO quantitatively to BuOH at 50 °C and 1 atm of H$_2$ in 4 h. Whilst octenes were not reduced under similar conditions, activated olefins, CH$_2$=CHCO$_2$R (R=H or Me) were hydrogenated to EtCO$_2$R (R = H or Me). A mixture of Ir hydrido-acetates formed by the reaction of **1** with AcOH was indicated as the actual hydrogenation catalyst [3]. A patent was filed in the same year [4] which claimed that aldehydes or ketones were hydrogenated to alcohols in the presence of a carboxylic acid and an iridium compound. As an example, IrH$_3$(PPh$_3$)$_3$ was dissolved in AcOH to give IrH(OAc)(PPh$_3$)$_3$ (**2**) and IrH$_2$(OAc)(PPh$_3$)$_3$ (**3**). The solution obtained was stirred for 20 min under H$_2$ at 50 °C, PrCHO was added, and the mixture stirred for 3 h to give an 88% conversion to BuOH and BuOAc.

The same precatalyst was used by Strohmeier *et al.*, who prepared the homogeneous catalyst *in situ* by the reaction of IrH$_3$(PPh$_3$)$_3$ with HOAc, and then used it to hydrogenate saturated and unsaturated aldehydes at 10 atm of H$_2$, without solvent. High conversions [turnover number (TON) up to 8000] could be obtained.

However, the hydrogenation of unsaturated aldehydes was found not to be selective, and ketones were not hydrogenated at pressures up to 10 atm [5]. Similar results were obtained by Kalinkin et al., using **1** and $(Ph_3P)IrH_5$ (**4**) in CF_3CO_2H. These authors proposed that an ionic mechanism, including H-transfer from the Ir complex to the protonated ketone, was active in these systems. It was confirmed that C=C and C=O groups of α,β-unsaturated ketones were hydrogenated non-selectively [6].

A comparative study on the hydrogenation of α,β-unsaturated ketones using Ir and Ru (carbonyl) complexes such as $RuCl_2(CO)_2(PPh_3)_2$, $RuCl_2(CO)_2(PCy_3)_2$, $RuHCl(CO)(PPh_3)_2$, $RuCl_2(PPh_3)_4$, $Ir(CO)Cl(PCy_3)_2$ (**5**) and $Ir(CO)Cl(PPh_3)_2$ [Vaska's compound, (**6**)] was carried out by Strohmeier et al. [7]. The study results showed that all these complexes were selective catalysts for the reduction of C=C double bond instead of the C=O bond. The TONs ranged from 6 to 196, depending on the substrate, catalyst and reaction conditions. The highest selectivity to saturated ketone in the hydrogenation of mesityl oxide was achieved using $RuCl_2(PPh_3)_4$ or **5** at 160 °C and 15 bar (1.5×10^4 hPa) of H_2.

By replacement of the chloride ligand with perchlorate ClO_4^- in **6**, $Ir(ClO_4)(CO)(PPh_3)_2$ (**7**) was synthesized [8] and further reacted with unsaturated aldehydes to give the k^1-O coordinated $[Ir(trans\text{-}RCH_2=CHCHO)(CO)(PPh_3)_2](ClO_4)$ (R = Me, **8a**; Ph, **8b**), in analogy with the ligand-substitution reactions carried out earlier on $[IrL(CO)(PPh_3)_2](ClO_4)$ (L = unsaturated nitriles). Under an atmosphere of H_2 at room temperature, the precatalyst was converted into $[Ir(H)_2(ClO_4)(CO)(PPh_3)_2]$ (**9**), which catalyzed the C=C bond hydrogenation of cinnamaldehyde and acrolein to the saturated aldehydes. Complex **7** (typically 0.1 mmol) was then used as the precatalyst for ketone (3.0 mmol) hydrogenations in $CHCl_3$ at 70 °C under 3 atm of H_2, to give conversions which ranged from 6% (acetophenone) to 100% (cyclohexanone) [9]. Under the same conditions, saturated aldehydes were hydrogenated to the corresponding alcohols [10]. Further investigations of the reaction conditions showed that, by raising the pressure to 9 atm and lowering the temperature to 50 °C, complete selectivity to cinnamol was achieved after 6 h using $[Ir(CO)(PPh_3)_3](ClO_4)$ (**10**), albeit with lower yields [11]. Replacement of the CO ligand led to similarly active complexes $[Ir(cod)(PPh_3)L](ClO_4)$ (**11**) [12].

A clearer understanding of the type of metal hydride required to achieve selective C=O bond hydrogenation in α,β-unsaturated substrates was provided by Graziani and coworkers. The selective hydrogenation of cinnamaldehyde and benzylidene acetone to the corresponding unsaturated alcohols (Scheme 4.2) was achieved in the presence of $[H_2Ir(phosphine)_4]^+$ complexes in toluene; the introduction of a chiral phosphine such as $(S)\text{-}(+)\text{-}PPh_2(CH_2CHMeCH_2Me)$ gives a 7.4% enantiomeric excess (ee) of $(S)\text{-}(-)\text{-}1$-phenylbut-1-en-3-ol [13].

As a general trend a large excess (up to 10:1 to Ir) of ligand is required to increase selectivity to C=O bond hydrogenation, which is in parallel accompanied by a decrease in the catalyst activity and the need for longer reaction times (Table 4.1).

In a subsequent detailed study, other iridium phosphine systems prepared *in situ* were used. Depending on the steric properties of the phosphine employed and

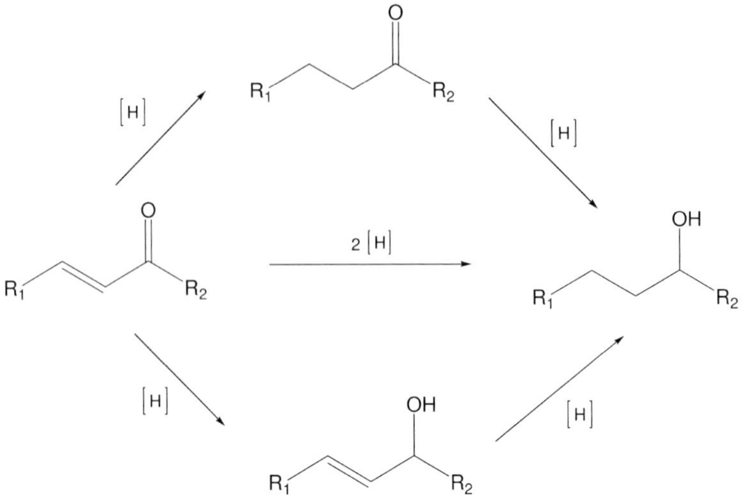

4 Iridium-Catalyzed C=O Hydrogenation

Scheme 4.2

Table 4.1 Hydrogenation of α,β-unsaturated carbonyl compounds [13].

Catalyst	P:Ir ratio	% Conversion (h)	% Unsaturated alcohol[a]	% Saturated aldehyde/ketone[a]	% Saturated alcohol
[Ir(cod)(OMe)]₂ + PEt₂Ph	2	93 (5)[b]	0	83	10
	10	97 (48)[b]	90	5	2
[Ir(PEt₂Ph)₄]⁺ (12)	–	96 (70)[b]	91	4	1
cis-[H₂Ir(PEt₂Ph)₄]⁺(13)	–	92 (28)[b]	81	7	4
[Ir(cod)(OMe)]₂ + PEtPh₂	10	99 (10)[b]	96	2	1
[Ir(cod)(OMe)]₂ + PEt₂Ph	10	98 (7)[c]	97	0	1
12	–	99 (28)[c]	98	0	1
13	–	95 (22)[c]	94	0	1

a Conditions: [Ir] = 4 × 10⁻⁴ M; substrate:Ir ratio = 500; toluene 75 ml, H₂ 30 atm; 100 °C.
b Substrate: benzylidene acetone.
c Substrate: cinnamaldehyde.

the P:Ir ratio, different species are formed in solution, as evidenced by NMR spectroscopy. [IrH₅(PR₃)₂] (R = alkyl or aryl), formed in the presence of bulkier phosphines under an atmosphere of H₂, was found to be a catalyst for the hydrogenation of the C=C double bond, whereas mer-[IrH₃(PR₃)₃] catalyzes the reduction of the carbonyl group with a selectivity up to 100% (PR₃=PEt₂Ph, **14a**; PMePh₂, **14b**) [14].

Spogliarich et al. published the details of a study [15] on the electronic effect implicated in reactions catalyzed by iridium/phosphine systems, and found there to be a slight dependence on the charge distribution around the carbonyl group of the substrates, while electron-withdrawing groups enhance the reduction rate of the C=O function and selectivity of the reaction. In the presence of **14** generated *in situ* under a pressure of H_2 (generally 20 atm) at 100 °C starting from the [Ir(cod)(OMe)]$_2$ precursor, enones such as CH_2=CHC(O)Et, Me—CH=CHC(O)Et, Me$_2$—C=CH—C(O)Me, PhCH=CHC(O)Me (benzylidene acetone) and PhCH=CHC(O)Me (chalcone) were hydrogenated to various activities (24–100% conversion) and selectivities (0–92% to the unsaturated alcohol). These authors concluded that, for this system, the steric hindrance of the catalytic species was necessary to achieve chemoselectivity in the reduction of conjugated enones, and for chalcone electronic parameters to either increase or decrease the C=O reduction rate. Moreover, the enantioselectivity towards chiral alcohol is also related to the steric hindrance of the substrate (see Section 4.2.4). For example, when using [Ir(S,S-DIOP)$_2$]$^+$ (**15**; S,S-DIOP = (+)-2,3-O-isopropylidene-2,3-dihydroxy-1,4-bis(diphenylphosphino)butane) in the reduction of 4-substituted chalcones, a maximum ee-value of 25% was obtained (Scheme 4.3). It was concluded that both steric and electronic effects were active in ruling the stereoselectivity and chemoselectivity of the reaction. In a subsequent, more detailed, study Tolman's cone angle of the phosphine ligands was correlated to the selectivity towards unsaturated alcohol for hydrogenation of substrates such as cinnamaldehyde, which was found to be close to 100%, with cone angles greater than 140°; this was in contrast to what was observed for transfer hydrogenation with the same complexes [16].

The use of bidentate P,N ligands for Ir-catalyzed hydrogenations of saturated and α,β-unsaturated substrates such as crotonaldehyde, ethylvinyl ketone, 2-cyclohexen-1-one, benzylidene acetone and chalcone was reported by Dahlenburg *et al.* [17]. [Ir(CO)(PPh$_3$)(2-Ph$_2$PC$_6$H$_4$NR-κN,κP)] (R = H, **16a**; R = Me, **16b**), [Rh(CO)(PPh$_3$)(2-Ph$_2$PC$_6$H$_4$NR)] (R = H, Me), [Rh(PPh$_3$)$_2$(2-Ph$_2$PC$_6$H$_4$NH)] and [Ir(CO)(PPh$_3$)(2-Ph$_2$PC$_6$H$_4$N{C(O)camph-(1S)}-κN,κP)] (**17**, C(O)camph-(1S) = (1S)-camphanoyl) were tested. Only **16a** showed significant activity for the reduction of the C=O group. The catalytically active species was found to be *cis*-[IrH$_2$(CO)(PPh$_3$)(2-Ph$_2$PC$_6$H$_4$NH)] (**18**), which was observed by ^{1}H NMR under 45 bar (4.5×10^4 hPa) of H_2 (Scheme 4.4). The homogeneous hydrogenation proceeded with chemoselectivities for the corresponding allylic alcohols with different performances, varying between zero for EtC(O)CH=CH$_2$ and 78% for MeCH=CHCHO. Complex **16a** was characterized by X-ray structural analysis; the molecular structure is shown in Figure 4.1.

Later, the same group [18] reported more detailed studies on the hydrogenation of prochiral ketones in the presence of P,N-chelated Ir complexes and a strong base under hydrogen pressure. A general scheme summarizing the cationic β-aminophosphine complexes used is shown in Scheme 4.5.

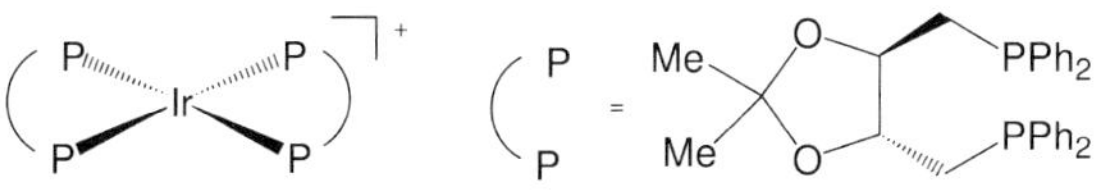

15   (+)-DIOP

Scheme 4.3

4 Iridium-Catalyzed C=O Hydrogenation

Scheme 4.4

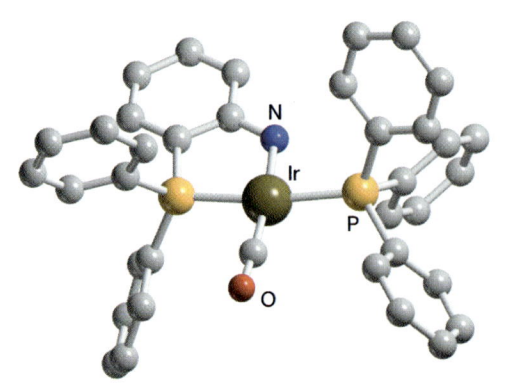

Figure 4.1 The molecular structure of [Ir(CO)(PPh$_3$)(2-Ph$_2$PC$_6$H$_4$NH-κN,κP)] (**16a**).

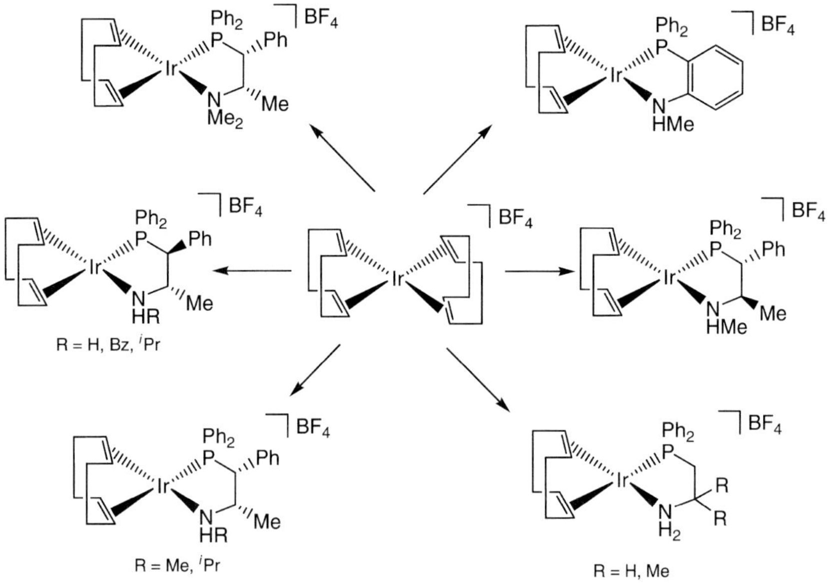

Scheme 4.5

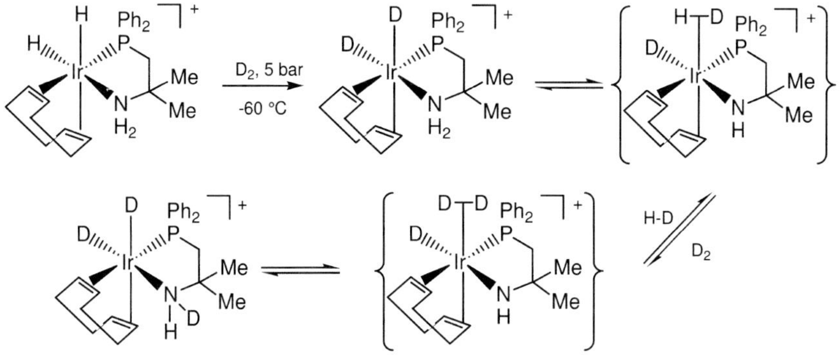

Scheme 4.6

If combined with an alkaline or amine base (1–5 equiv.) in MeOH under 10–50 bar (1–5 × 10^4 hPa) of H$_2$ and temperatures ranging from 25 to 50 °C, all complexes catalyze the reduction of aryl alkyl ketones to the corresponding 1-arylalkanols. The role of the base was described by the authors as 'metal-assisted direct transfer of H$^{\delta-}$/H$^{\delta+}$ to the C=O function', and higher rates were obtained in the presence of stronger bases such as LiOH, K$_2$CO$_3$ and KOH. Mechanistic studies on hydrogen/deuterium (H/D) exchange highlighted the role of the amine in assisting heterolytic hydrogen splitting, and a series of *cis*-dihydrides derived from oxidative addition of H$_2$ was observed (Scheme 4.6).

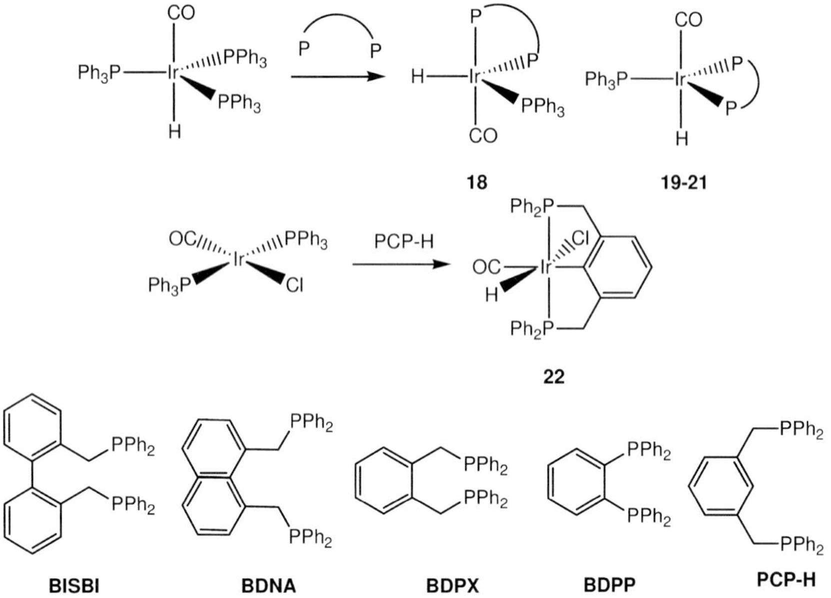

Scheme 4.7

Five iridium hydrido carbonyl complexes containing bidentate phosphine ligands, IrH(CO)(PPh$_3$)(L) [L = BPPB, 1,2-bis(diphenylphosphino)benzene, **18**; BISBI, 2,2′-bis(diphenylphosphinomethyl)-1,1′-biphenyl, **19**; BDNA, 1,8-bis(diphenylphosphinomethyl)naphthalene, **20**; BDPX = 1,2-bis(diphenylphosphinomethyl)benzene, **21**; PCP-H, 1,3-bis(diphenylphosphino methyl)benzene, **22**] were synthesized by Li et al. [19] (Scheme 4.7). The catalytic hydrogenation activities and selectivities for citral and cinnamaldehyde reductions were studied. Complexes **18–21** showed high selectivity for the hydrogenation of C=O group in citral. In the case of cinnamaldehyde using complex **21**, high selectivity to C=O bond hydrogenation could be obtained in the presence of excess of ligand (BDPX).

The influence of ligand structure on the biphasic hydrogenation of aldehydes was investigated by Gulyas et al. Rh and Ir complexes of trisulfonated triarylphosphines, TPPTS (tris(3-sulfonatophenyl)phosphine), T(p-A)PTS (tris(3-sulfonato-4-methoxyphenyl)phosphine), T(2,4-X)TS (tris(2,4-dimethyl-5-sulfonatophenyl)phosphine; Scheme 4.8), were synthesized starting from the precursors [M(cod)Cl]$_2$ (M = Rh, Ir) and tested in the biphasic (H$_2$O/EtOAc) hydrogenation of benzaldehyde and caproaldehyde at 65 °C and 20 bar (2×10^4 hPa) of H$_2$. T(2,4-X)TS could not stabilize the Rh complex under the applied conditions, due to the large cone angle (196°, 210°). Interestingly, the T(2,4-X)TS/Ir catalyst is stable and effective in the hydrogenation of benzaldehyde and caproaldehyde (Table 4.2). The reason for this behavior was explained by the authors as a dependency on the size of the metal center, which results is a less-crowded region, even in the presence of bulky phosphines for Ir (with a larger atom radius) [20].

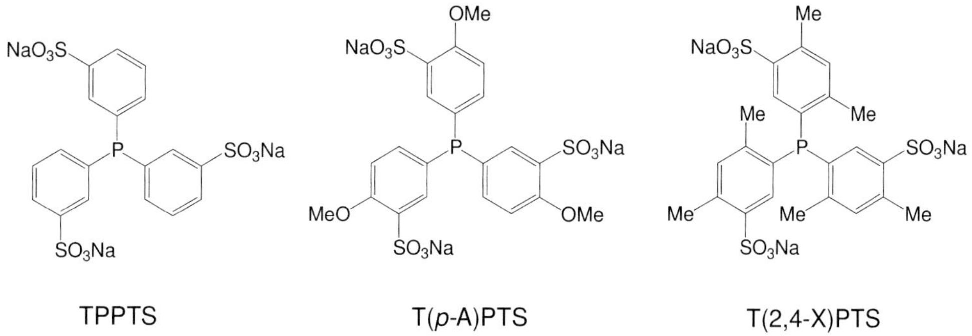

Scheme 4.8

TPPTS T(*p*-A)PTS T(2,4-X)PTS

Table 4.2 Biphasic hydrogenation of aldehydes using sulfonated phosphines [20].

Ligand	Substrate	Conversion (%)
TPPTS	Benzaldehyde	7
T(*p*-A)PTS	Benzaldehyde	6
T(2,4-X)PTS	Benzaldehyde	99
TPPTS	Caproaldehyde	4
T(*p*-A)PTS	Caproaldehyde	4
T(2,4-X)PTS	Caproaldehyde	93

Conditions: $[Ir(cod)Cl]_2$ = 0.01 mmol; substrate, 4 mmol; ligand = 0.1 mmol, buffer (pH 7):EtOAc 3 ml : 3 ml, H_2 20 atm; 65 °C.

4.2.2
Enantioselective Hydrogenations

The asymmetric reduction of ketones to form optically active secondary alcohols attracted the attention of research groups at Monsanto during the late 1970s, at which time a patent was filed involving the use of trisubstituted monodentate phosphines or arsines $ER_1R_2R_3$ (E = P, As; R_1, R_2, R_3 = H, aryl, alkyl) and Rh(I), Ir(I) precursors [21]. Under pressures ranging from 5 to 50 atm of H_2 and temperatures ranging from −20 °C to 100 °C, alcohols with percentage optical purities (defined as observed optical activity of the mixture × 100/optical activity of pure enantiomer) from 5.9 to 19.9 were obtained using complexes such as $[Ir(cod)L_2]BF_4$ (L = cyclohexyl (*o*-methoxyphenyl)methylphosphine, **23**) (Scheme 4.9).

The asymmetric hydrogenation of α-amino-acetophenones such as $PhCOCH_2NRR^1$ [R = CH_2Ph, R^1 = Me, CH_2Ph; RR^1 = $(CH_2)_5$] to β-amino-α-phenylethanol derivatives was described by Graziani and coworkers using the cationic Ir(I) complexes [Ir(cod)(prolophos)]BPh$_4$ (**24**) and [Ir(prolophos)$_2$]BPh$_4$ (**25**) (prolophos = (*S*)-(−)-*N*-(diphenylphosphino)-2-diphenylphosphinoxymethylpyrrolidine) (Scheme 4.10). All reactions were carried out using 2 mol% of catalysts at 60 °C and 70 atm of H_2, as no activity was observed at lower temperatures and pressures.

Scheme 4.9

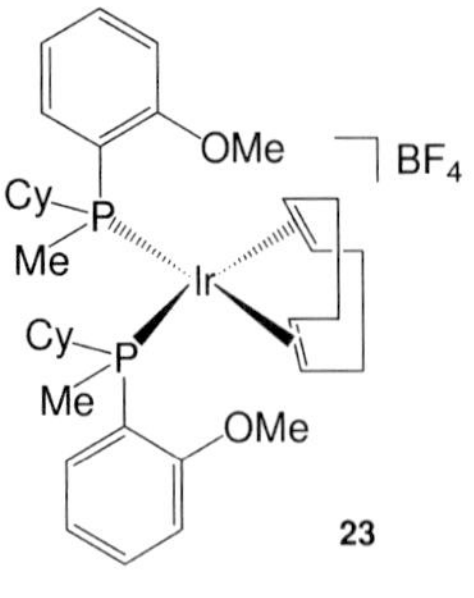

23

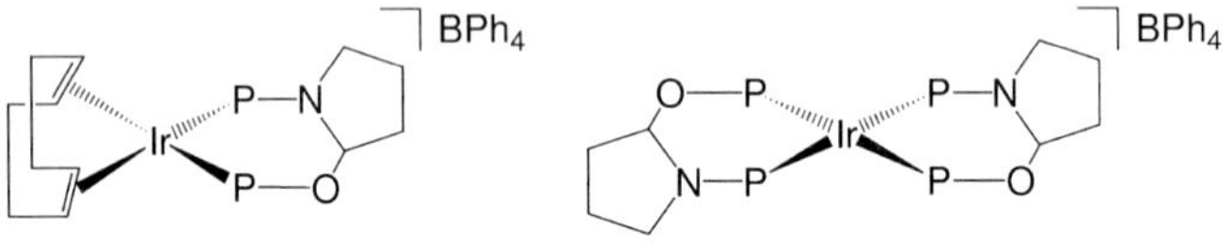

Scheme 4.10

24   **25**

Interestingly, whereas the highest ee-value in the product was found using **24** (30%), the configuration of the chiral center was observed to be strongly solvent-dependent in the presence of **25** – that is, the (S)-(+) was obtained in benzene and the (S)-(−) amino alcohol in toluene (only the (S)-(+) form was produced by **24** irrespective of the solvent). These authors suggested that this effect might be due to different hydride catalytic species being generated in the different solvent media [22]. The same catalysts were applied to the hydrogenation of α,β-unsaturated ketones such as benzylidene acetone. 1-Phenyl-1-buten-3-ol was selectively produced in high yields using **25** and [Ir(S,S-DIOP)$_2$]$^+$ (**15**; S,S-DIOP = (+)-2,3-O-isopropylidene-2,3-dihydroxy-1,4-bis(diphenylphosphino)butane) and as catalyst precursor or high P:Ir ratios in the systems prepared *in situ* [23]. The use of **15** in the reduction of cyclic α,β-unsaturated ketones such as 2-cyclopenten-1-one, 3-substituted-2-cyclohexen-1-ones and R-(−)-carvone (5R-isoprenyl-2-methylcyclohexen-1-one) was tested under a pressure of 20 bar (2 × 10^4 hPa) of H$_2$, 100 °C, substrate:catalyst ratio = 500, in toluene and gave the highest ee-value for the latter (28%) after a reaction time of 144 h and at 10% conversion [24].

The enantioselective hydrogenation of acyclic alkyl aryl ketones PhCOR (R = alkyl or cycloalkyl group) has been accomplished using the catalytic systems derived from [Ir[(S)-BINAP](cod)]BF$_4$ (**26**) or [Ir[(S)-H8-BINAP](cod)]BF$_4$ [**27**, H8-BINAP = 2,2′-bis(diphenylphosphino)-5,5′,6,6′,7,7′,8,8′-octahydro-1,1′-BINAPhthyl] and bis(o-dimethylaminophenyl)phenylphosphine in dioxane:MeOH (5:1) under hydrogen pressures varying from 54 to 61 atm [25] (Scheme 4.11). The ee-values, as well as the absolute configurations of the corresponding alcohols, were found to depend heavily on the steric size of the alkyl or cycloalkyl groups, and the highest ee-values were obtained with R = ^{i}Pr (84%, R) and C$_6$H$_{11}$ (80%, R). Other related active catalysts were described as complexes [H$_2$IrL$_1$L$_2$]Y (**28**, L$_1$ = optically active phosphine, e.g. BINAP; L$_2$ = tertiary phosphine, e.g. o-

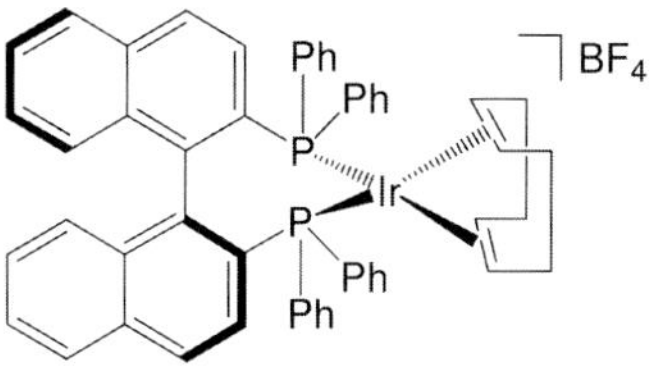

4.2 Homogeneous C=O Hydrogenations

26

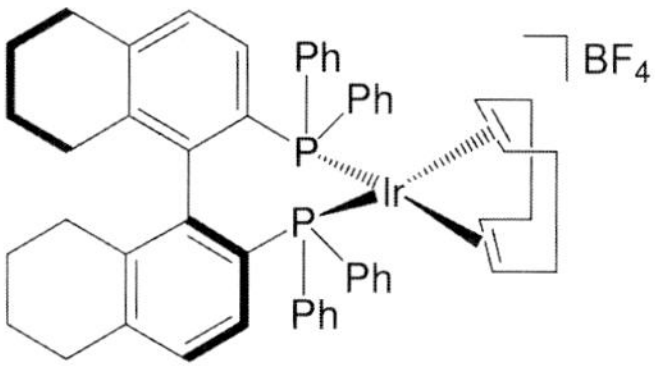

27

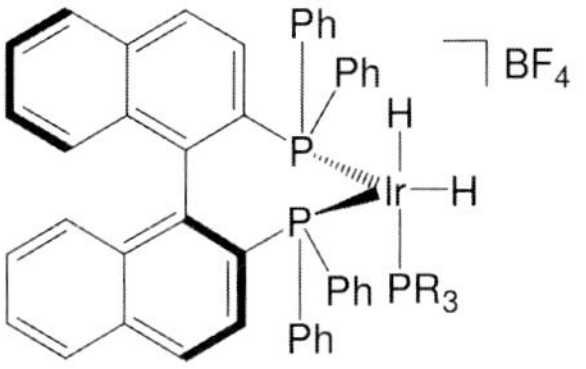

28

PR$_3$ = (o-NMe$_2$-C$_6$H$_4$)PPh$_2$
Scheme 4.11

dimethylaminodiphenylphosphine; Y = BF$_4$, PF$_6$ which were prepared by stirring [Ir(cod)[(+)-BINAP]]BF$_4$ with L$_2$ in tetrahydrofuran (THF) under H$_2$ for 18 h. The hydrogenation of 3-oxotetrahydrothiophene (214.5 g) in the presence of **28** gave (+)-3-hydroxytetrahydrothiophene (63 g, 63.2% ee) [26].

Other phosphines based on the same principle of C$_2$-symmetry and the binaphthyl framework were reported for the synthesis of transition-metal complexes (including Ir, but mainly Ru-based) and application as catalysts for the enantioselective hydrogenation of C=O. (R)-(+)-CM-BIPHEMP was used as a chiral ancillary ligand to hydrogenate substrates such as HOCH$_2$COMe in MeOH at 65 °C under 10 atm of H$_2$ for 16 h to 97.8% (R)-HOCH$_2$CHMeOH [(R)-(+)-CM-BIPHEMP = (R)-(+)-2,2'-bis(diphenylphosphino)-5,5'-dichloro-4,4',6,6'-tetramethyl-1,1'-biphenyl] [27]. The use of chiral diphosphines such as (R)- or (S)-2,2'-diphosphino-6,6'-bis(acyloxy)-1,1'-biphenyls, and in particular (R)-2,2'-bis(diphenylphosphino)-6,6'-diacetoxy-1,1'-biphenyl, as optically active ligands for transition-metal complexes (e.g. M$_x$H$_y$X$_z$L$_2$(Sv)$_p$; M = Ru, Rh, Ir; X = Cl, Br, F, I; Sv = tertiary amine, ketone, ether; L = chiral ligand; y = 0, 1; x = 1, 2; z = 1, 4; p = 0, 1) and asymmetric catalytic reduction methods was also patented [28].

Asymmetric C=O hydrogenations in water were also reported by Lemaire et al. This catalytic system is based on Ir(cod)L* complexes, where L* is a hydrophilic chiral C$_2$-symmetric diamine ligand such as p-substituted (1R 2R)-(+)-1,2-diphenylethylenediamine derivatives (**29a–e**; Scheme 4.12). The use of such ligands allowed catalyst recovery without loss of activity and enantioselectivity in at least four acetophenone hydrogenation cycles [29]. The ee-values observed in the reduction of phenyl glyoxylate in the water phase were, however, lower than were found when running the tests in THF (Table 4.3), when the substituents were H and Me, and about the same with OH, OMe and O-(C$_2$H$_4$O)$_3$Me.

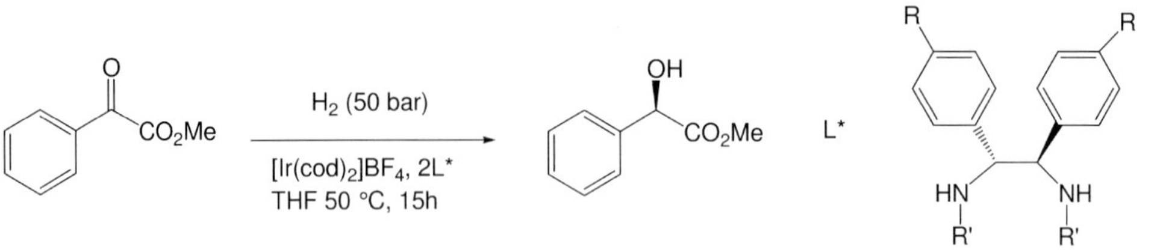

4 Iridium-Catalyzed C=O Hydrogenation

Scheme 4.12

29a, R = H, R' = H
29b, R = H, R' = Me
29c, R = OMe, R' = Me
29d, R = OH, R' = Me
29e, R = O-(C$_2$H$_4$O)$_3$CH$_3$, R' = M

Table 4.3 Phenylglyoxylate methyl ester reduction in the presence of Ir(I) and **29a–e**.

Ligand	Solvent	Yield (%)	ee (%) (R)
29a	THF	99	31
29a	H$_2$O	100	15
29b	THF	100	80
29b	H$_2$O	73	56
29c	THF	100	45
29c	H$_2$O	96	50
29d	THF	100	49
29d	H$_2$O	100	42
29e	THF	98	43
29e	H$_2$O	100	54

Conditions: [Ir(cod)$_2$]BF$_4$ = 5 mol%; substrate, 0.5 mol l^{-1}; H$_2$ 50 atm; 50 °C, 15 h.

By introducing phosphonic acid groups at the *para* position of the aromatic rings, the same authors [30] developed another chiral N,N ligand and tested the corresponding iridium complex for asymmetric hydrogenation of ketones under MeOH/water conditions. The system was shown to perform as well as the homogeneous organic phase counterpart, but with noticeably higher reaction rates. For the hydrogenation of acetophenone in MeOH at 45 atm H$_2$, total conversion and ee-values varying between 34% and 55% were obtained in the function of metal : ligand : substrate ratio. Other, different ketones were tested and, under the same conditions, a maximum ee of 72% was obtained for the asymmetric hydrogenation of phenyl *t*-butyl ketone (Scheme 4.13).

As part of the above-mentioned study on chemoselective hydrogenation performed by Dahlenburg, 18 chiral versions of P,N-chelating ligands were synthesized and used in the asymmetric hydrogenation of aryl alkyl ketones under a pressure of hydrogen. Ir(I) complexes **31–33** catalyzed the direct hydrogenation of alkyl phenyl ketones to the corresponding 1-phenylalkanols, combined with an

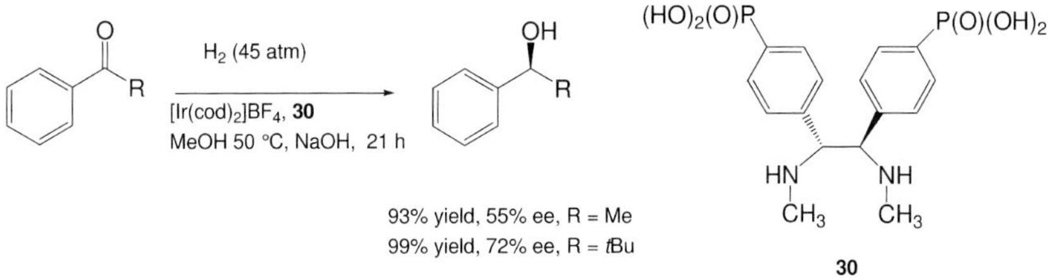

4.2 Homogeneous C=O Hydrogenations

Scheme 4.13

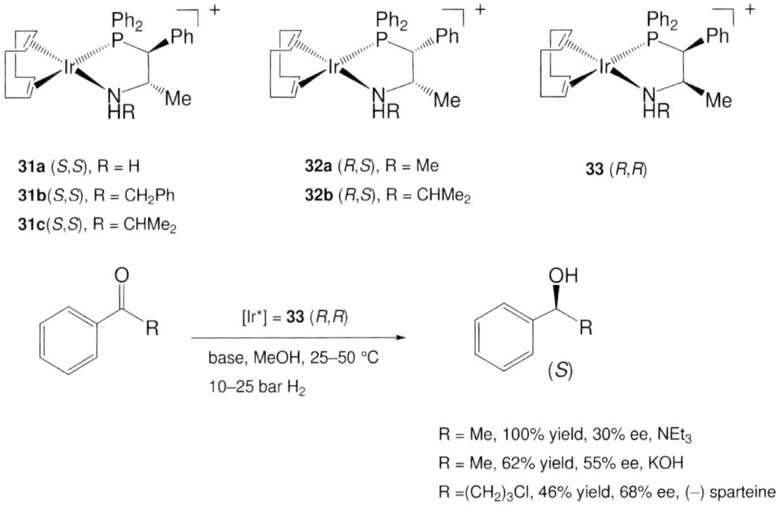

PhC(O)R + H$_2$ (45 atm) → PhCH(OH)R
[Ir(cod)$_2$]BF$_4$, **30**
MeOH 50 °C, NaOH, 21 h

93% yield, 55% ee, R = Me
99% yield, 72% ee, R = tBu

Ligand **30**: (HO)$_2$(O)P- and -P(O)(OH)$_2$ substituted diphenyl ethylenediamine with N-CH$_3$ groups.

31a (S,S), R = H
31b (S,S), R = CH$_2$Ph
31c (S,S), R = CHMe$_2$

32a (R,S), R = Me
32b (R,S), R = CHMe$_2$

33 (R,R)

Scheme 4.14

PhC(O)R + H$_2$ → (S)-PhCH(OH)R
[Ir*] = **33** (R,R)
base, MeOH, 25–50 °C
10–25 bar H$_2$

R = Me, 100% yield, 30% ee, NEt$_3$
R = Me, 62% yield, 55% ee, KOH
R = (CH$_2$)$_3$Cl, 46% yield, 68% ee, (−) sparteine

alkaline base or an amine in methanol under H$_2$ (10–50 bar; 1–5 × 10^4 hPa) and between 25 and 50 °C (Scheme 4.14). In the case of acetophenone, modest to moderate ee-values (30–55%) were observed using **33** (R,R) and different bases. When using KOH or (−)sparteine as a base the results showed, in all cases, that enantioselectivities obtained in the presence of KOH were higher than with (−)sparteine (19–55% and 4–47%, respectively). Other alkyl aryl ketones were hydrogenated with **33** (R,R) as catalyst, and a maximum of 68% ee was obtained in the case of phenyl 3-chloropropyl ketone.

A series of Ir(III) complexes bearing a chelating chiral diamine and mono- or bidentate (chiral) phosphines was described very recently by Dahlenburg and coworkers [31]. Reactions of [IrH$_2$(OCMe$_2$)$_2$(PPh$_3$)$_2$]BF$_4$ or [IrH$_5$(PR$_3$)$_2$]/HBF$_4$ (R = ^{i}Pr, cy) with chelating diamines H$_2$NNH$_2$ gave the corresponding cis-dihydridoiridium(III) chelate complexes [IrH$_2$(H$_2$NNH$_2$)(PR$_3$)$_2$]BF$_4$ [PR$_3$ = PPh$_3$, H$_2$NNH$_2$ = 1,2-(H$_2$N)$_2$C$_6$H$_4$ (**34a**); (1R,2R)-(H$_2$N)$_2$C$_6$H$_{10}$ {(R,R)-dach} (**34b**); (R)-2,2′-diamino-1,1′-binaphthyl {(R)-dabin} (**34c**); PR$_3$ = P^{i}Pr$_3$, H$_2$NNH$_2$ = (R,R)-dach (**35a**), (R)-dabin (**35b**); PR$_3$ = PCy$_3$, H$_2$NNH$_2$ = (R)-dabin (**36**)]. By oxidative addition of HCl to Ir[(cod){(S,S)-bdpcp}]BF$_4$ [(S,S)-bdpcp = (1S,2S)-(Ph$_2$P)$_2$C$_5$H$_8$], the mononuclear adduct [Ir(cod)(H)(Cl){(S,S)-bdpcp}]BF$_4$ (**37**) was obtained. In

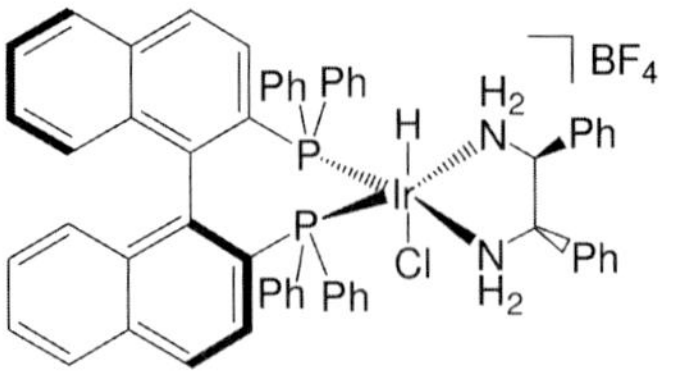

Scheme 4.15

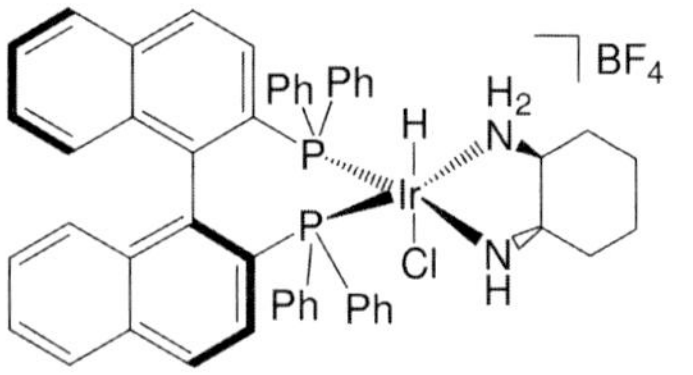

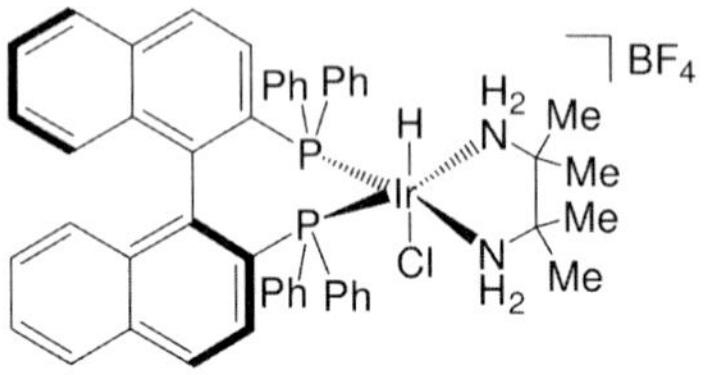

contrast, the reaction of [Ir(cod){(R)-binap}]BF$_4$ with HCl gave the triply chlorido-bridged diiridium complex [{(R)-binap}$_2$Ir$_2$H$_2$(μ-Cl)$_3$]BF$_4$ (**38**). Subsequent reaction with chelating diamines yielded complexes [Ir(H)(Cl)(H$_2$NNH$_2$){(R)-binap}]BF$_4$ [H$_2$NNH$_2$ = (1R,2R)-H$_2$NCH(Ph)CH(Ph)NH$_2$ {(R,R)-dpen} (**39a**), (R,R)-dach (**39b**) and H$_2$NCMe$_2$CMe$_2$NH$_2$ {tmen} (**39c**), as shown in Scheme 4.15. It was shown that, for the enantioselective hydrogenation of acetophenone, the dihydrides [IrH$_2${(R,R)-dach}(PR$_3$)$_2$]BF$_4$ and [IrH$_2${(R)-dabin}(PR$_3$)$_2$]BF$_4$ (R = ^{i}Pr, Ph) performed poorly, while complexes **39a–c** catalyzed the formation of 1-phenylethanol in good enantiomeric excess [ee$_{max}$ = 82–84% (S)] in the presence of KOH at 40–50 °C under a pressure of 20 bar (2 × 10^4 hPa) of H$_2$.

A highly efficient method for the asymmetric hydrogenation of substituted acetophenones catalyzed by Ir complexes with chiral planar ferrocenyl phosphinothioether ligands was disclosed by Le Roux *et al.* [32]. Optimization of the conditions led to a highly active catalytic system with TONs up to 915 and a turnover frequency (TOF) of up to 250 h^{-1}. The reactions were run in isopropanol at 30 bar (3 × 10^4 hPa) H$_2$, in the presence of strong bases such as NaOMe, ^{t}BuOK or KOH. At room temperature the asymmetric hydrogenation of acetophenone gave 43–77% ee, the best value being obtained with the complex having benzyl as substituent on the sulfur atom (Scheme 4.16). When the reaction was carried out at 10 °C, much higher enantioselectivities (up to >99% for 4-fluoroacetophenone) were obtained while the catalytic activities remained invariably high. Whereas, the need for an alcoholic medium is not required as the reaction could be carried out also in toluene, the presence of both base and molecular H$_2$ is mandatory in order to obtain substrate conversion (Table 4.4).

Finally, the asymmetric hydrogenation of a series of α-hydroxy aromatic ketones in methanol catalyzed by Cp*Ir(OTf)(MsDPEN) (**42**, MsDPEN = N-(methanesul-

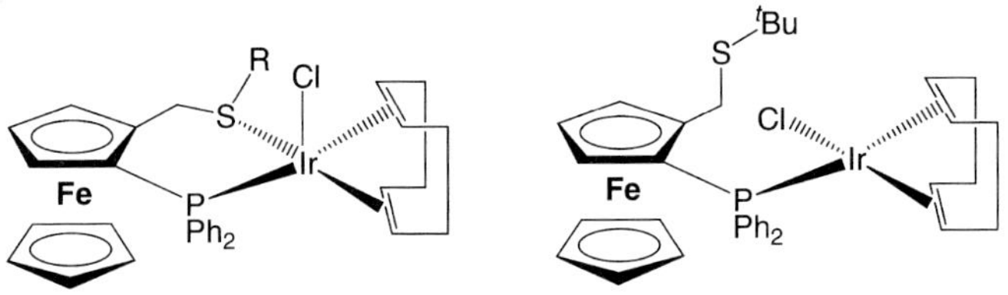

Scheme 4.16

40a, R = Ph
40b, R = Et
40c, R = Bz

41

Table 4.4 Asymmetric hydrogenation of acetophenones using Ir complexes bearing planar chiral ferrocenyl phosphino thioethers.[a]

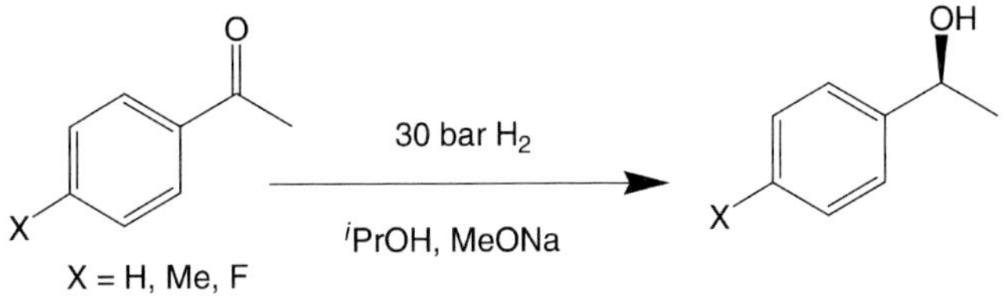

X = H, Me, F; 30 bar H_2, ^{i}PrOH, MeONa

Catalyst	X	Conversion (%)[b]	ee (%)[b]
40b	H	>99	78
40c	H	99	87
40b	Me	86	93
40c	Me	84	93
40b	F	>99	>99
40c	F	96	>99

a Reaction conditions: catalyst, 6.4×10^{-3} mmol; NaOMe, 3.2×10^{-2} mmol; substrate, 3.2 mmol (1/5/500), 10 °C, ^{i}PrOH, 30 bar (3×10^4 hPa) H_2, 8 h.
b Conversion and ee-value determined using chiral gas chromatography.

fonyl)-1,2-diphenylethylenediamine) affords the 1-aryl-1,2-ethanediols in up to 99% ee [33] (Scheme 4.17). The reactions could be carried out with a substrate: catalyst molar ratio as high as 6000 at 60 °C under 10 atm of H_2. 1-Hydroxy-2-propanone was also hydrogenated with high enantioselectivity (80% ee) to (R)-1,2-propanediol.

4.2.3
Transfer Hydrogenation (TH)

Reduction methods based on the use of a sacrificial reagent such as an alcohol (usually isopropanol) together with a metal or transition metal, or other

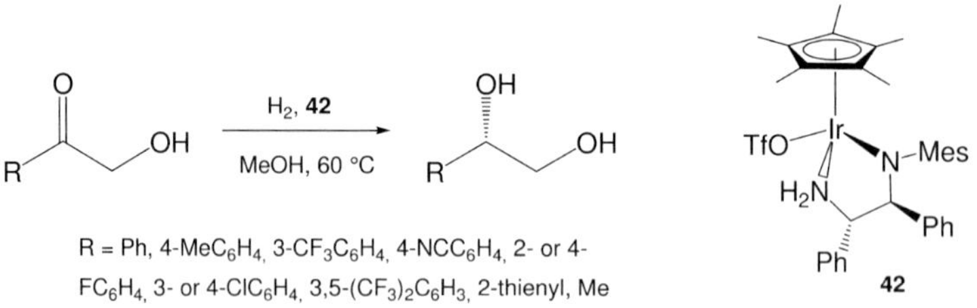

Scheme 4.17

R = Ph, 4-MeC$_6$H$_4$, 3-CF$_3$C$_6$H$_4$, 4-NCC$_6$H$_4$, 2- or 4-FC$_6$H$_4$, 3- or 4-ClC$_6$H$_4$, 3,5-(CF$_3$)$_2$C$_6$H$_3$, 2-thienyl, Me

compounds able to transfer hydride ligands to the catalysts (NaOMe, HCO$_2$H, etc.) have received much attention during recent years, and are generally known as transfer hydrogenation (TH) reactions. On the topic of selective C=O reduction, ruthenium-based catalysts represent the vast majority of protocols studied to date, and apply to many diverse multifunctional substrates, based largely on the seminal studies of the groups of Noyori [34] and Zhang [35]. Although Ir-based TH systems are known to a lesser extent (and will be reviewed here), this subject is best covered by dividing it into those reports which deal with standard TH reactions, and those that deal with enantioselective versions (asymmetric transfer hydrogenation; ATH).

The mechanism of reversible β-hydrogen elimination from square planar Ir(I) alkoxide complexes with labile dative ligands, followed by associative displacement of the coordinated ketone or aldehyde by incoming phosphine, which can be implied in TH reactions, was proposed by Hartwig and coworkers [36].

The biphasic TH of aldehydes catalyzed by water-soluble phosphine TPPMS complexes of Ru(II), Rh(I) and Ir(I) (e.g. Ir(CO)Cl(PR$_2$R$_s$), R = Ph, R$_s$ = m-SO$_3$NaPh, TPPMS) was reported by Benyei and Joo [37]. Aromatic and aliphatic aldehydes can be reduced to the corresponding alcohols by hydrogen transfer from formate, catalyzed by water-soluble complexes bearing the monosulfonated triphenylphosphine TPPMS, in aqueous/organic biphasic systems without the need for phase-transfer catalysts. Whereas, a complete conversion of benzaldehyde (5 ml, 0.5 M solution in PhCl) was obtained after 20 min in the presence of 0.01 mM RuCl$_2$(TPPMS)$_2$ and an excess of free ligand (0.1 mM) using 3 ml of 5 M solution of NaHCO$_2$ at 80 °C, and 80% conversion after 1 h using HRuCl(CO)(TPPMS)$_3$, only 11% conversion was achieved using IrCl(CO)(TPPMS)$_2$ (43).

The hydrogenation of ketones and olefins via hydrogen transfer catalyzed by rhodium and iridium phosphine complexes in organic solvents was described by Graziani and coworkers in 1982 [38]. It was shown that a combination of [Rh(coe)$_2$Cl]$_2$ and [M(cod)Cl]$_2$ (M = Rh, Ir) in the presence of tertiary phosphines (PR$_3$, R = Ph, p-MePh, p-MeOPh, o-MePh, o-MeOPh, Cy, ^{t}Bu, Cy$_2$Ph, CyPh$_2$, MePh$_2$, Me$_2$Ph) or bidentate phosphines R$_2$PCH$_2$CH$_2$PR$_2$ (R = Ph, dppe; Et, depe; Me, dmpe) would catalyze H transfer from isopropanol to 4-t-butylcyclohexanone with almost complete conversion using KOH as a base, at 50 °C. The catalytic activity was found to depend on the electronic and steric properties of the tertiary phosphine ligands and on the activation time. A 1:1 mixture of ketone and olefin was seen to slow the reaction, without affecting chemoselectivity. The reduction of unsaturated ketones such as MeCOCH$_2$CH$_2$CMe=CH$_2$ was also carried out under the same conditions (Ir/PPh$_3$), but no conversion was observed. Later, the study was expanded by testing specific Ir complexes such as [Ir(cod)(PPh$_3$)$_2$]$^+$ (44) for the TH of cyclic ketones such as 4-*t*-butylcyclohexanone in ^{i}PrOH and KOH (initial TON 175 000, after 15 s) which are nominally rivalling with horse liver alcohol dehydrogenase (HLAD) for performance. With the substrate as above, complete conversion was reached after 4 min at a substrate:catalyst ratio of 30 000, running the reaction at 83 °C [39]. The catalytic activity was seen to depend on the basicity of the coordinated phosphine, in the order PPh$_3$ > P(*p*-anisyl)$_3$ >> PMePh$_2$ > dppe, in parallel with classic hydrogenation reactions, where a nonoxidative addition of hydrogen was proposed.

An alternative system which was developed during the same period and by the same group [40] used a combination of Ir(III) and Sn(II) for the TH of cyclohexanones in isopropanol under acidic conditions (HCl). Both, IrCl$_3$ hydrate and H$_3$IrCl$_6$ could be used as precatalysts and showed essentially the same activity. The best results (93% yield in the alcohol) were obtained for the reduction of 4-*t*-butylcyclohexanone using an SnCl$_2$ × 2H$_2$O/IrCl$_3$ × 3H$_2$O ratio of 3, at a substrate:catalyst ratio of 150, for 5 h at 83 °C. Both, an increase in the Sn:Ir ratio and the presence of excess of chloride decreased the activity of the catalytic system.

Complexes of the type [Ir(NN)Hd]ClO$_4$ (45) and [Ir(NN)HdX], (46, NN = bipy, phen and substituted derivatives; Hd = 1,5-hexadiene; X = Cl, Br, I) in the presence of a mineral base ([KOH]:[cat] = 1) are described as very active catalyst precursors for the selective reduction of unhindered cyclohexanones to the equatorial alcohols by TH in isopropanol. The catalytic activity depends on the [KOH]:[cat] ratio and the coordinating power of the ligands, and is higher at high base:Ir ratios. The stereoselectivity reached its highest value with the 3,4,7,8-Me$_4$phen derivative and was independent of the KOH concentration, with high activities and selectivities with substrate:catalyst ratios of 20 000 [41].

The system formed *in situ* from [Ir(cod)Cl]$_2$ and P(*o*-MeOPh)$_3$ catalyzed the selective TH of the CO group in 5-hexen-2-one [42]. The maximum yield of 91% of unsaturated alcohol was obtained after 1 h reflux in ^{i}PrOH in the presence of KOH using a P:Ir ratio of 100. This effect was explained by the presence of an intramolecular interaction between the substrate and the MeO groups in the *ortho*-position of the phosphine. The same system, when tested with benzylidene acetone, showed a lower selectivity under the same conditions, thus reflecting the higher reactivity of the C=C bond due to activation by the C=O group. 2-Hexanone was also reduced to the alcohol, while the sterically crowded Me$_2$C=CHCH$_2$CH$_2$COMe also gave (selectively) the unsaturated alcohol upon reduction. It was established that the possibility of chelation affects the reactivity of the molecule as compared to the simple ketone–olefin mixture. The same group also reported cinnamaldehyde and crotonaldehyde selective C=O hydrogenation, using the same system [43].

The anchimeric assistance by the methoxy group of P(*o*-MeOPh)$_3$ in determining the selectivity of C=O hydrogenation, with OMe coordination to Ir and an increase in electron density, was further verified by replacing the oxygen atom with a nitrogen donor as in the ligand (P)-NMe$_2$ (o-diphenylphosphino)-N,N-dimethyl aniline to form the iridium complex {HIr(cod)[(P)-NMe$_2$]} (**47**). Both, NMR and X-ray structural data indicated that a six-membered chelate ring is formed upon oxidative addition of the N-methyl C—H bond, when **47** is placed at reflux under 1 atm H$_2$. The product {H$_2$Ir[(P)-NCH$_2$Me][(P)-NMe$_2$]} (**48**) thus obtained was used to reduce a series of α,β-unsaturated ketones at a substrate:Ir ratio of 500 in isopropanol at reflux (Scheme 4.18). In the case of benzylidene acetone, the unsaturated alcohol was obtained with 93% selectivity after 1 h of reaction [44].

Following this trend, the development of Ir complexes for TH reactions has involved the use of polydentate aminophosphine ligands. Bianchini et al. [45] reported on the use of electron-rich Ir complexes with mixed-donor polydentate ligands such as ^{n}PrN(CH$_2$CH$_2$PPh$_2$)$_2$ (PNP) and Et$_2$NCH$_2$CH$_2$N(CH$_2$CH$_2$PPh$_2$)$_2$ (P$_2$N$_2$) for the chemoselective reduction of benzylidene acetone to the corresponding allylic alcohol (C=O bond reduction). An interesting coordination isomerism was observed between the σ,η^2-cyclo-octenyl complexes [(PNP)Ir(σ,η^2-C$_8$H$_{13}$)] (**49**) and the η^4-cod isomer [(PNP)IrH(η^4-C$_8$H$_{13}$)] (**50**) via the β-elimination/hydride migration pathway, driven by the central nitrogen atom which plays a hemilabile 'arm-off' role in driving the equilibrium (Scheme 4.19). A similar behavior was observed for the P$_2$N$_2$ analogues, namely [(P$_2$N$_2$)Ir(σ,η^2-C$_8$H$_{13}$)] (**51**) and [(P$_2$N$_2$)IrH(η^4-C$_8$H$_{13}$)] (**52**). Both, preformed **49** and **51** and the precatalysts formed in situ from [Ir(cod)(OMe)]$_2$ + PNP/P$_2$N$_2$ catalyzed the reduction of benzylidene acetone with about 90% conversion. However, the P$_2$N$_2$-based system gave a higher

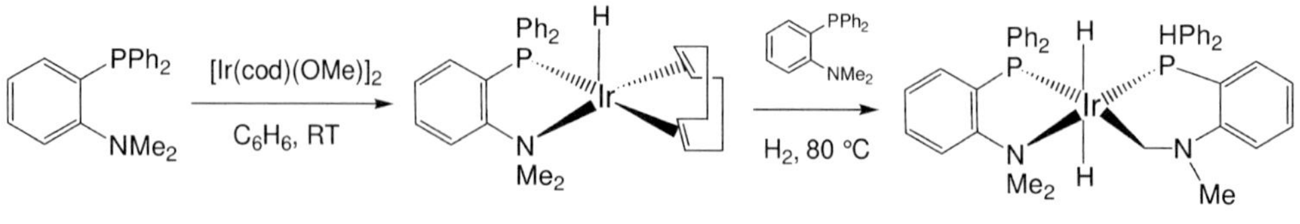

Scheme 4.18

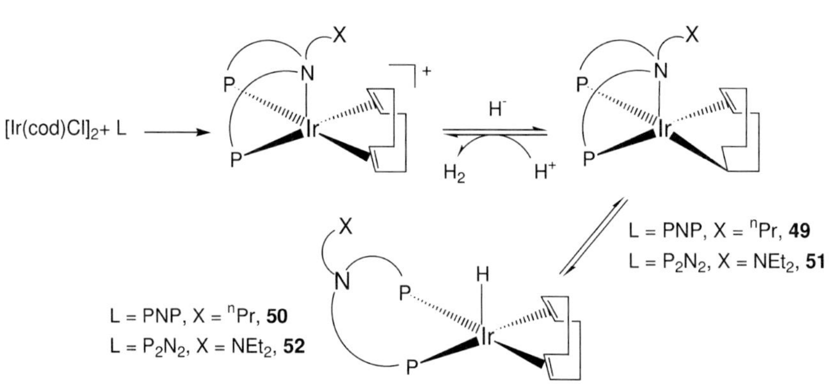

Scheme 4.19

activity than the PNP analogue as the reaction time could be halved (120 min versus 240 min for 90% conversion, 83 °C in ^{i}PrOH) and it also gave the best performance overall (93% conversion, 2 min, 140 °C in cyclopentanol). The major factor determining the selectivity in such reductions remains the degree of nucleophilicity of the coordinated hydride.

Later, this study was extended to other tripodal tetraphosphines such as PP$_3$ [PP$_3$ = P(CH$_2$CH$_2$PPh$_2$)$_3$] and other transition metals (Fe, Ru, Os), as Ir was observed to be inactive in the presence of such ligands [46]. In the case of the latter system, a vacant site for substrate coordination was easily generated by the loss of a weakly bound ligand from the metal precursor, that is [(PP$_3$)MH(H$_2$)]$^+$.

A TH protocol based on the common Ir(I) precursor [Ir(cod)Cl]$_2$ together with dppp [1,3-bis(diphenylphosphino)propane] and Cs$_2$CO$_3$ in isopropanol, was used by Ishii and coworkers to hydrogenate cyclic and acyclic ketones at 80 °C, with a 4 h reaction time. Cyclohexanone, octanone and hexanal were smoothly converted into the corresponding alcohols, with selectivities varying from 94 to 98%. Subsequently, it was found that the same protocol could be applied to alkene hydrogenation, and interestingly competition experiments, with the hydrogenation of α,β-unsaturated ketones or aldehydes giving a selective reduction of the C=C double bond [47].

Cationic iridium complexes bearing imidazol-2-ylidene ligands were applied as TH catalysts by Hillier et al. [48]. Here, [Ir(cod)(py)(L)]PF$_6$ (L = IMes, 1,3-bis(2,4,6-trimethylphenyl)imidazol-2-ylidene (**53a**), IPr, 1,3-bis(2,6-diisopropylphenyl)imidazol-2-ylidene (**53b**) and ICy, 1,3-bis(cyclohexyl)imidazol-2-ylidene (**53c**) were employed as catalysts for TH from 2-propanol to various unsaturated substrates, and compared to [Ir(cod)(py)(PCy$_3$)]PF$_6$ and complexes formed *in situ* from [Ir(cod)(py)$_2$]PF$_6$ and diazabutadienes (RN=CHCH=NR, DAB-R; R = cyclohexyl, DAB-Cy; 2,4,6-trimethylphenyl, DAB-Mes; adamantyl, DAB-Ad; 2,4,6-trimethoxyphenyl, DAB-trimethoxyphenyl). All complexes were active catalysts for the TH of acyclic ketones, with complex **53c** being the most active, showing 100% conversion of ^{t}BuC(O)Me and PhC(O)Ph in 25 and 10 min, respectively, using 0.025 mol% of catalyst and KOH in ^{i}PrOH at reflux (Scheme 4.20).

In being considered as phosphine analogues, Arduengo-type carbenes have naturally attracted much attention for use as ancillary ligands in catalytic applications. In 2002, Faller, Crabtree and coworkers published the details of an

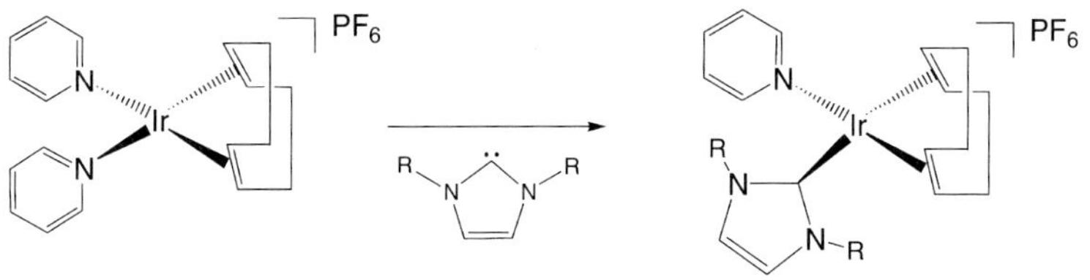

53a, R = 2,4,6-Me$_3$C$_6$H$_2$
53b, R = 2,6-^{i}Pr$_2$C$_6$H$_3$
53c, R = C$_6$H$_{11}$

Scheme 4.20

interesting study on the effect of chelated Ir(III) bis-carbene complexes as air-stable catalysts for the TH of substituted acetophenone and other ketones (Scheme 4.21) [49].

The primary advantage of these ligands is the higher stability of the complexes, which can be handled in air and are not moisture-sensitive. Furthermore, high TOFs (h^{-1}) were obtained for the reduction of benzophenone, with the highest at 50 000 using **54d** (0.1 mol%, ^{i}PrOH, reflux, KOH), acetophenone (TON 5300, **54d** 0.01 mol%) and substituted acetophenones, for which a high compatibility with the functional groups was observed. By contrast, benzylidene acetone reduction was shown to be selective mainly to the saturated ketone and alcohol. A second generation of NHC carbenes was developed which included a modification of the imidazolium ring, while the neopentyl-substituted bis-N-heterocyclic carbene iridium acetates obtained were found to catalyze the reduction of (enolizable) aldehydes under TH conditions using alkali metal carbonates as bases.

The iridium-bis-NHC complexes **54d, 55a,b** (Scheme 4.22) were obtained by the reaction of [Ir(cod)Cl]$_2$ or [Ir(coe)$_2$Cl]$_2$ with the corresponding bis-imidazolium or bis-triazolium salts in the presence of NaOAc, and used as catalysts for the TH of 4-substituted benzaldehydes and other aldehydes RCHO (R = PhCH$_2$, PhCHMe, PhCH$_2$CH$_2$, 2-naphthyl, Me$_2$C=CH(CH$_2$)$_2$CHMeCH$_2$) by 2-propanol in the presence of alkali metal carbonate or KOH; as a consequence, high yields of ArCH$_2$OH and R$_2$CH$_2$OH were afforded, even in the case of enolizable aldehydes [50]. The higher TOF was measured for R = 2-naphthyl in the presence of **54d** (0.1 mol%; 3000 h^{-1}), and the most efficient base was found to be K$_2$CO$_3$. The effect of the

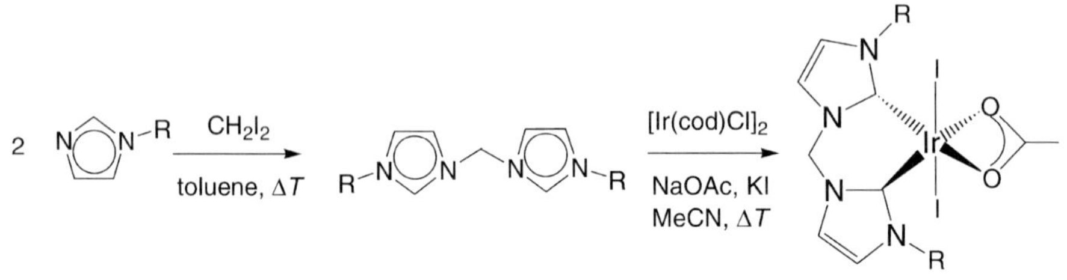

54a, R = Me   **54d**, R = neo-pentyl
54b, R = nBu   **54e**, R = Bn
54c, R = iPr   **54f**, R = tBu

Scheme 4.21

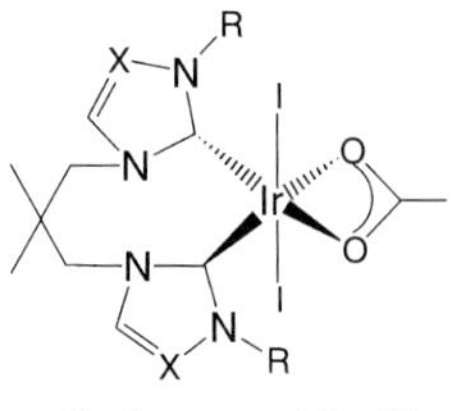

55a, R = neo-pentyl, X = CH
55b, R = neo-pentyl, X = N

Scheme 4.22

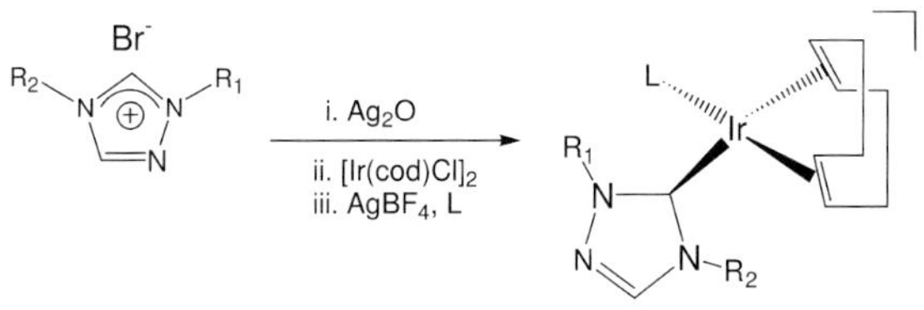

4.2 Homogeneous C=O Hydrogenations

Scheme 4.23

56a, R_1 = nBu, R_2 = PhCH$_2$
56b, R_1 = nPn, R_2 = nBu
56c, R_1 = nPn, R_2 = PhCH$_2$
L = PPh$_3$, py

number of carbon atoms in the linker was also examined for the test reduction of acetophenone. The bis-imidazoles generally exhibited higher catalytic activities than the bis-triazoles, with CH$_2$ being the best spacer for imidazoles and neopentyl for the triazoles [51].

Following on the triazole motif, a series of new iridium(I) 1,2,4-triazole-3-ylidene NHC complexes [Ir(cod)(NHC)L]BF$_4$ (**56a–c**, L = PPh$_3$, pyridine; Scheme 4.23) were synthesized and tested and good results found for TH on C=O, C=N and C=C double bonds in 2-propanol with K$_2$CO$_3$ [52].

Complexes bearing PPh$_3$ showed a higher activity than those with pyridine for imine TH. A neopentyl wingtip substituent on the NHC gave the best catalytic activity in the order: aldehyde > ketone > imine, and in competitive experiments with mixtures containing both aldehyde and ketone functionalities (i.e. 3-acetyl-bezaldehyde) only the aldehyde was reduced with selectivity greater than 95%. The direct, one-pot reductive amination of RCHO with R'NH$_2$ gave secondary amines RCH$_2$NHR' in the presence of such catalysts (**56b**, L = PPh$_3$), via hydrogenation of the imine formed by a room-temperature reaction in ^{i}PrOH/K$_2$CO$_3$. These complexes are among the most active homogeneous catalysts for imine hydrogenation (TOF 333 h^{-1}).

An interesting variation in the imidazole-based ligand design is the presence of N-allyl substituents, which can behave as hemilabile chelating ligands and stabilize free coordination sites at catalytically active metal centers. Oro and coworkers [53] described the synthesis of mono- and bis-N-allyl-substituted benzimidazol-2-ylidene NHCn ligands (NHC$_1$ = 1-methyl-3-(2-propenyl)benzimidazolium iodide; NHC$_2$ = 1,3-di-(2-propenyl)benzimidazolium bromide, respectively) and use of the corresponding Ir(I) catalysts [IrX(cod)(NHCn)] (n = 1, X = I, **57a**; n = 2, X = Br, **57b**) formed by reaction with [Ir(cod)(OMe)]$_2$ in the catalytic TH of cyclohexanone in ^{i}PrOH. The cationic derivative of **57b**, namely [Ir(cod)(η^2:η^2-C-NHC$_2$)] (**57c**) obtained by reaction with AgBF$_4$ (Scheme 4.24), features an Ir center which is exclusively coordinated by sp^2 carbon atoms, alkene, allyl and carbene type, as shown by the X-ray crystal structure obtained (Figure 4.2).

Reaction of the mono-hapto bis-allyl complex **57b** with NaOEt in EtOH gave allyl group hydrogenation yielding [Ir(cod)Br(NHC$_3$)] (**55d**, NHC$_3$ = 1,3-di(propyl)benzimidazol-2-ylidene). Complexes **57a, b** and **d** were tested for the TH of cyclohexanone (4 mM) using an equimolar amount of KOH to the catalyst (0.2 mM) in ^{i}PrOH at 80 °C. Complex **57d** showed the highest activity, reaching a TOF of 6000 h^{-1}, whereas **57a** and **57b** achieved TOFs of 70 and 50 h^{-1}, respectively, as might be expected from the presence of additional η^2–allyl coordination which

Scheme 4.24

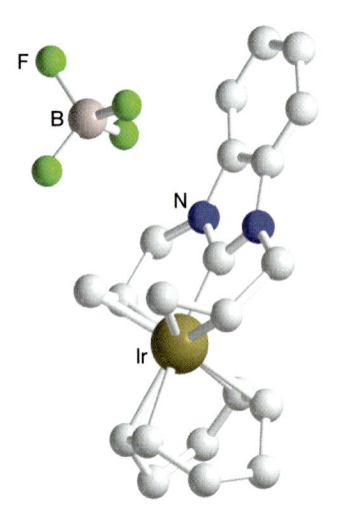

R = allyl, X = Br, **57b**
R = Me, X = I, **57a**

57c

57d

Figure 4.2 The molecular structure of **57c**.

is likely to reduce availability of the vacant coordination site required for substrate activation.

Pyridine-functionalized N-heterocyclic carbene Rh and Ir complexes have also been described as active precatalysts for C=O bond TH. For example, Peris and coworkers observed the formation of metal hydrides by C—H oxidative addition of a pyridine-N-substituted imidazolium salt such as N-^{n}Bu-N-(2-pyridylmethyl-imidazolium) hexafluorophosphate in the reaction leading to M-pyNHC complexes, that is [Ir(cod)H(pyNHC)Cl] (**58**) [54]. Transmetallation from silver carbene

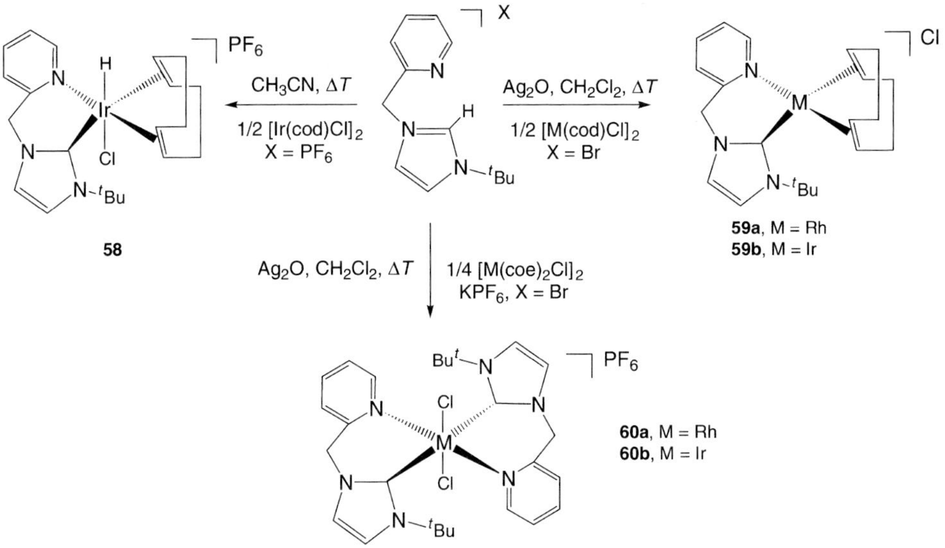

4.2 Homogeneous C=O Hydrogenations

Scheme 4.25

yielded the M(I) corresponding N,C-chelates [M(cod)(pyNHC)]Cl (**59a,b**) or M(III) bischelates [M(pyNHC)$_2$Cl$_2$]PF$_6$ (**60a,b**), depending on the starting precursor and reaction conditions (Scheme 4.25). Complexes **60a,b** were tested as catalysts for the TH of benzophenone and acetophenone, using KOH and ^{t}BuOK in ^{i}PrOH under reflux conditions. Both complexes were seen to provide complete conversion to the corresponding alcohols, with **60b** (M = Ir) performing better for both substrates and with a catalyst loading as low as 0.1 mol%.

Iridium 2-pyridinylmethyl imidazolylidene C,N-chelates were obtained by transmetallation of the silver carbene complexes and tested for catalytic activity in the TH of benzophenone and nitroarenes by isopropanol [55]. The neutral monodentate complexes [(L-κC)Ir(COD)Cl] [**61a,b**; L = 1-methyl-3-(6-mesityl-2-pyridinylmethyl)-2-imidazolylidene, 1-mesityl-3-(6-mesityl-2-pyridinylmethyl)-2-imidazolylidene] were converted into cationic chelates [(L-κC,κNPy)Ir(COD)][BF$_4$] (**62a,b**) by chloride abstraction. Complexes **62a,b**, once exposed to CO, yielded the carbonyl complexes [(L-κC,κNPy)Ir(CO)$_2$][BF$_4$] (**63a,b**). The pyridine nitrogen donor is weakly bonded to the metal, as evidenced by NMR by reversible replacement of the MeCN ligand in acetonitrile solution of such complexes. Complexes **61–63** showed catalytic activity in the TH of benzophenone, at 0.1 mol% catalyst loading, with TOFs ranging from 42 (**62a**) to 78 h^{-1} (**61b**) (Scheme 4.26).

Another category of multidentate molecules which has received interest for use in homogeneous reductions by TH are 'pincer ligands'. The air-stable complex IrH$_2$Cl[(^{i}Pr$_2$PC$_2$H$_4$)$_2$NH] (**64**) was synthesized by Gusev et al. [56] from the reaction of [IrCl(coe)$_2$]$_2$ with the P,N,P pincer ligand (^{i}Pr$_2$PC$_2$H$_4$)$_2$NH in isopropanol at reflux. The reaction of **64** with ^{t}BuOK in THF resulted in the formation of the air-sensitive amidodihydride IrH$_2$[(^{i}Pr$_2$PC$_2$H$_4$)$_2$N] (**65**), which in isopropanol gives the

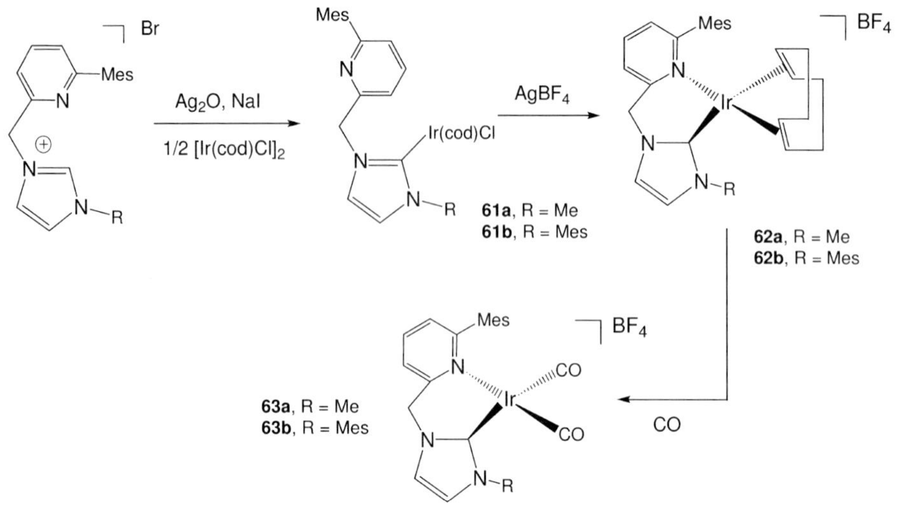

Scheme 4.26

moderately air-stable trihydride complex $IrH_3[(^iPr_2PC_2H_4)2NH]$ (**66**). Complexes **65** and **66** were found to be exceptionally active catalysts for the TH of ketones such as acetophenone, benzylidene acetone, benzophenone, cyclohexanone and linear alkyl ketones, in the absence of a base (complete conversions of acetophenone at catalyst load as low as 0.0001 mol%), whereas **64** was inactive under similar reaction conditions. ^tBuOK in isopropanol was required to activate **64** and to obtain complete reduction of acetophenone under comparable reaction conditions. A bifunctional TH mechanism was proposed by these authors for the catalytic process (Scheme 4.27).

The same group followed on this research line by synthesizing the P,N,N pincer analogue $^tBu_2PC_2H_4NHC_2H_4NEt_2$ (PNHN) and using to coordinate Ir from the readily available precursor $[Ir(coe)_2Cl]_2$. Under an atmosphere of H_2 the dihydride cis-$[IrH_2Cl(\kappa^3$-PNHN$)]$ (**67**) was formed; this was then reacted with ^tBuOK to produce the 16-electron amido complex cis-$[IrH_2Cl(\kappa^3$-PNN$)]$ (**68**). The subsequent exposure of **68** to H_2 led to production of the trihydrido complex mer-$[IrH_3Cl(\kappa^3$-PNHN$)]$ (**69**), which was identified and isolated as the dimer $[IrH_2Cl(\kappa^2$-PNHN$)]_2(\mu$-H$)_2$ (**70**) [57] bearing an Ir=Ir double bond. X-ray crystallographic studies also showed a hemilabile behavior on the NEt_2 arm (Figure 4.3). Complexes **68–70** were tested for the TH of acetophenone and butanone at catalyst load of 0.1 mol% and 0.12 mol% for cyclohexanone, reaching initial TOFs of 1500, 1850 and 1600 h^{-1}, respectively.

Bis(imino)aryl iridium N,C,N pincer complexes (Scheme 4.28) were obtained through the oxidative addition of 2-bromoisophthalaldimines 2-$BrC_6H_3(CH=NR)_2$ to $[Ir(cod)Cl]_2$ in the presence of NaBr to form $\kappa N,\kappa C,\kappa N$-$[IrBr_2(MeCN)(C_6H_3(CH=NR)_2]$ (R = Ph (**71a**), Mes (**71b**), Mes = 2,4,6-mesityl), ^iPr (**71c**), Me (**71d**)). The X-ray crystal structure of **71a** is shown in Figure 4.4 [58].

Scheme 4.27

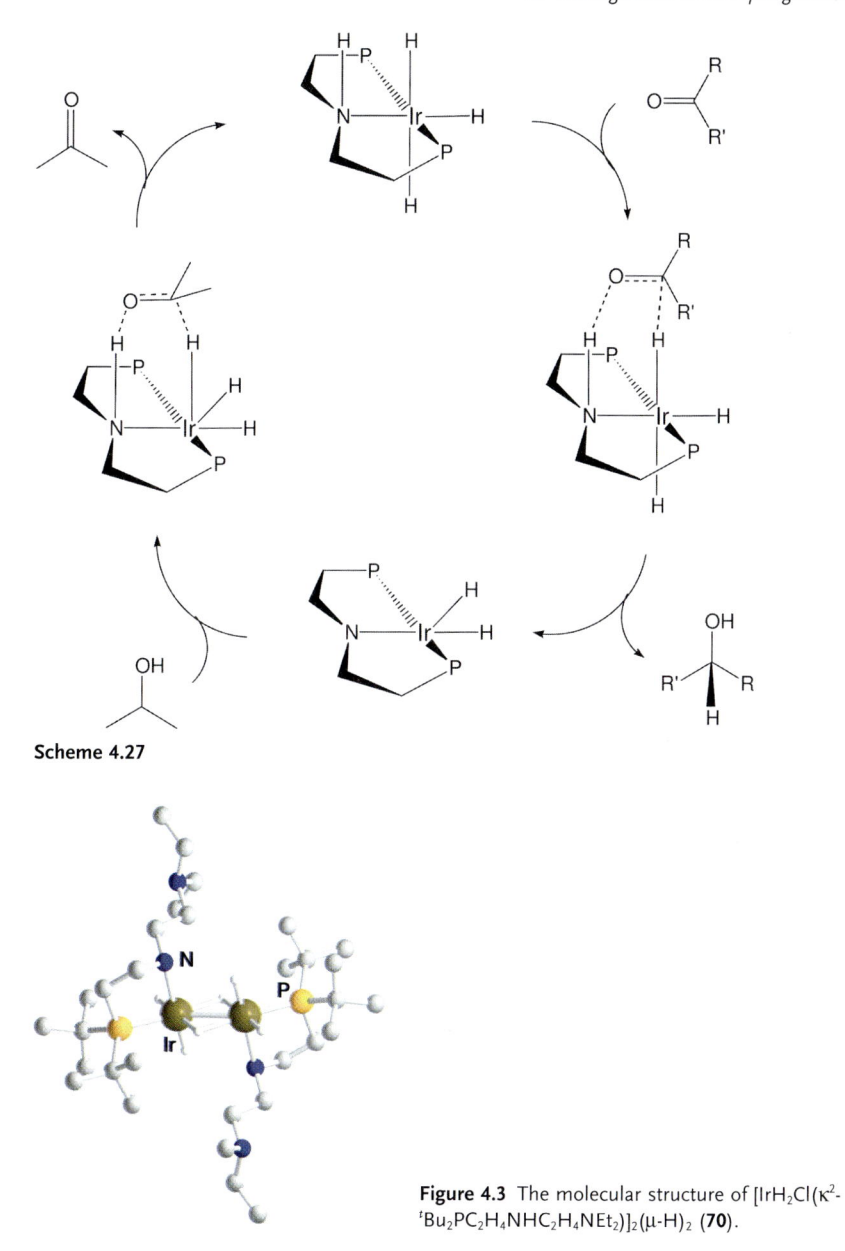

Figure 4.3 The molecular structure of [IrH$_2$Cl(κ^2-^{t}Bu$_2$PC$_2$H$_4$NHC$_2$H$_4$NEt$_2$)]$_2$(μ-H)$_2$ (**70**).

Complex **71a** (0.5 mol%) was used as the catalyst for the TH of acetophenone and benzophenone at room temperature in ^{i}PrOH in the presence of KOH. In the case of acetophenone, 50% conversion was reached after 5 h with a KOH : catalyst ratio of 2, whereas for the reduction of benzophenone, a strong excess of base was required (25 equiv. to the catalyst) in order to obtain a meaningful conversion (92% after 3.5 h).

Scheme 4.28

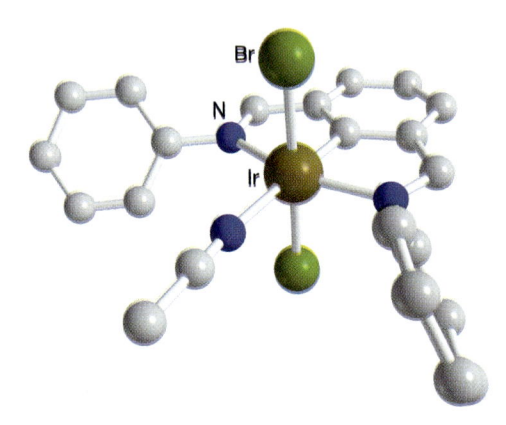

71a, R = Ph
71b, R = Mes
71c, R = iPr
71d, R = Me

Reagents: 1/2 [Ir(cod)Cl]$_2$; i. NaBr, ethoxyethanol, 110 °C, 15 h, N$_2$; ii. MeCN/H$_2$O

Figure 4.4 The molecular structure of iridium(III) κN,C,N-bis(phenyl)isophthalaldimin-2-yl dibromide·MeCN (0.5 H$_2$O) (**71a**).

Some reports have appeared describing the use of water-soluble iridium catalysts for TH in either aqueous phase or biphasic systems. Ajjou and Pinet [59] reported the use of water-soluble ligands such as 2,2′-bisquinoline-4,4′-dicarboxylic acid potassium salt (BQC) and m-trisulfonated triphenylphosphine (TPPTS), together with [Ir(cod)Cl]$_2$ and various reducing agents (^{i}PrOH, HCO$_2$Na, H$_2$) for the TH reduction of benzaldehyde and acetophenone. All of the systems tested performed rather poorly when compared to the corresponding Rh analogues, and a maximum yield of 19% benzyl alcohol was obtained in H$_2$O/^{i}PrOH/Na$_2$CO$_3$.

A robust and highly active catalyst for water-phase, acid-catalyzed THs of carbonyl compounds at pH 2.0–3.0 at 70 °C was disclosed by Ogo and coworkers [60]. The water-soluble hydride complex [Cp*Ir(bipy)H]$^+$ (**72**, Cp* = η^5-C$_5$Me$_5$, bipy = 2,2′-bipyridine) was synthesized from the reaction of [Cp*Ir(bipy)(H$_2$O)]$^{2+}$ (**73**) with HCOOX (X = H or Na) in H$_2$O under controlled pH conditions (2.0 < pH < 6.0, 25 °C). The pH control is pivotal in avoiding protonation of the hydrido ligand of **72** below pH ca. 1.0 and deprotonation of the aquo ligand of **73** above pH ca. 6.0. The rate of the reaction is heavily dependent on the pH of the solution, the reaction temperature, and the concentration of HCOOH. High TOFs of the acid-catalyzed transfer hydrogenations at pH 2.0–3.0, ranging from 150 to 525 h^{-1}, were observed for a variety of linear and cyclic ketones, as summarized in Table 4.5.

Table 4.5 Transfer hydrogenation of various ketones in water, catalyzed by **73** using HCOOH at pH 2.0.

Substrate	TOF (h^{-1})	Time (h)	Yield (%)
Cyclohexanone	376	1	99
Acetophenone	343	1	97
Trifluoromethylacetophenone	525	1	99
Ethyl methyl ketone	150	4	99
Pyruvic acid	481	1	98
p-SO$_3$Na-acetophenone	419	1	99
α-Tetralone	203	3	98

Conditions: **73** : substrate : HCOOH = 1 : 200 : 1000; substrate, 0.32 mol; H$_2$O 3 ml; 70 °C.

Finally, rapid, selective and high-yielding THs of a wide range of aldehydes are achieved using Ir(III) catalysts with simple monotosylated ethylenediamine ligands. This procedure is suitable for aldehydes with a wide range of functional groups. [(Cp*IrCl$_2$)$_2$] and N-tosylethylenediamine in water catalyzed the TH of PhCHO in the presence of HCO$_2$Na to give 98% PhCH$_2$OH [61]. The introduction of a CF$_3$ withdrawing group in the *para* position of the tosylate end of the diamine increased the TOF for the TH of benzaldehyde to 132 000 h^{-1}, with a substrate : catalyst ratio as low as 1 : 50 000. Another advantage of this system is that the reactions can be carried out in air, without any observable loss of either activity or selectivity. Until now, this system has been considered as the benchmark for chemoselective homogeneous TH reduction of aldehydes using Ir catalysts.

4.2.4
Asymmetric Transfer Hydrogenation (ATH)

Although ruthenium and rhodium phosphine or amine complexes predominate, some reports have been made which deal with Ir-based ATH reactions. These have been published in the peer-reviewed and patent literature during the past 20 years, and the subject has also been reviewed [62]. During the early 1980s, investigations into the enantioselective synthesis of secondary optically pure alcohols from the catalyzed reduction of ketones via ATH protocols focused mainly on the use of simple chiral phosphines as ancillary ligands. Graziani and coworkers used [Ir(cod)Cl]$_2$ and [Ir(cod)(py)$_2$]PF$_6$ precursor and chiral phosphines NMDPP (neomenthyldiphenylphosphine), (R)-PROPHOS (R-(+)-bis(1,2-diphenylphosphino) propane), (S,S)-CHIRAPHOS [(−)-(2S,3S)-bis(diphenylphosphino) butane] (Scheme 4.29) and (+)-DIOP [(+)-2,3-o-isopropylidene-2,3-dihydroxy-1,4-bis (diphenylphosphino)butane] (see Scheme 4.3) as catalysts for the asymmetric reduction of PhCOR (R = Me, Et, Pr), 2-octanone and cyclohexylmethylketone with ^{i}PrOH to the corresponding alcohols [63]. Rapid conversions (4 h) and high yields (97%) were observed using NMDPP (20 : 1 excess to Ir) for the reduction of acetophenone, although the optical yield was found to be about 18% at best, when running

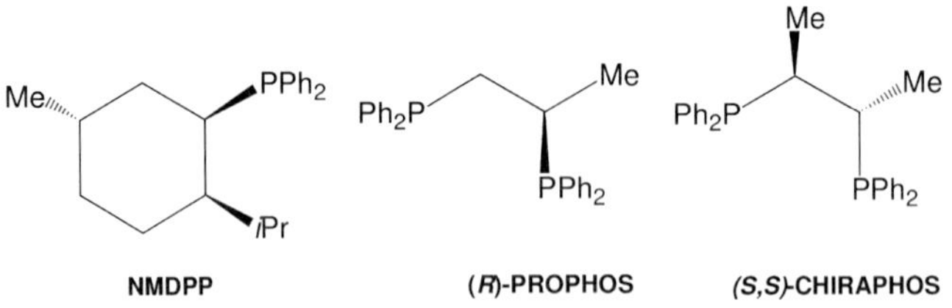

Scheme 4.29 NMDPP, (R)-PROPHOS, (S,S)-CHIRAPHOS

the tests in ⁱPrOH at reflux in the presence of KOH. In the case of (R)-PROPHOS and (S,S)-CHIRAPHOS, longer reaction times and lower yields were accompanied by a slight increase in ee-value (30 and 28%, respectively) or no increase for (+)-DIOP (14%).

This initial study was followed by a broader substrate scope and more detailed investigations on the effect of reaction conditions and the catalysts compared to the Rh analogues [64]. It was noted that not only the presence of a small amount of KOH but also activation of the precatalyst was necessary for higher catalytic activity. [Rh(nbd)L$_2$]$^+$ (L$_2$ = (S,S)-CHIRAPHOS), when activated for 30 min in refluxing degassed isopropanol, to which aqueous KOH was added, gave (R)-(+)-PhCHEtOH (34.3% ee) from PhCOEt with a conversion of 59% after 4 h. The Ir counterparts [Ir(cod)(CHIRAPHOS)]$^+$ (**74**) and [Ir(cod)(PROPHOS)]$^+$ (**75**) required a longer activation time (up to 15 h reflux, then 5.5 h in the presence of KOH) but gave a higher conversion (71%) and ee-value (58%, S-phenylethanol, using **75**).

The use of P-based ligands other than phosphines was reported in the literature. Whereas, the use of [IrHClX(dmso)$_3$] (X = Cl, RCO$_2^-$) including chiral systems proved to be inactive in homogeneous hydrogenations either under H$_2$ pressure or using TH conditions, the replacement of dimethylsulfoxide (DMSO) with phosphites gave efficient precatalysts, formed from [IrCl(coe)$_2$]$_2$, a carboxylic acid, and P(OR$_2$)$_3$ (R$_2$ = ⁿBu, Ph, Me), for the TH of cyclohexanone and ATH of acetophenone in isopropanol under Ar [65]. In the case of TH of cyclohexanone, the best carboxylic acids were BzOH, AcOH, MeCH=CHCO$_2$H, PhCH=CHCO$_2$H and the chiral acids (R)-(−)-PhCH(OH)CO$_2$H and (R)-(+)-HO$_2$CCH$_2$CH(OH)CO$_2$H. The addition of Et$_3$N increased conversion to 78% when the reaction mixture contained BzOH and P(OMe)$_3$. The *cis/trans* ratio of cyclohexanol increased as the P(OMe)$_3$-Ir ratio increased to approximately 4. Acetophenone was converted to (S)-PhCH(OH)Me in 75% yield, but with only 1.0% ee in the presence of NaOMe as base. (R)-(−)-PhCH(OAc)CO$_2$H gave a 12% ee of the S-isomer, with 20% conversion.

Phosphinite–Ir complexes were also tested in the ATH of ketones. Ligands such as BDPODP, (1R,3R)-(−)-1,3-bis(diphenylphosphinoxy)propane, and BDPOP, (2R,4R)-(−)-2,4-bis(diphenylphosphinoxy)pentane (Scheme 4.30) were used together with both [Ir(coe)$_2$Cl]$_2$ and [Ir(cod)Cl]$_2$ precursors and a strong base such as NaOMe, KOH, ᵗBuOK, Et$_3$N in ⁱPrOH. The highest optical yield observed was 17.9% (R-isomer) at 85% conversion after 9 h using [Ir(cod)Cl]$_2$, BDPOP, NaOMe (1:2:10 ratios). The authors suggested that the role of the strong base in the reac-

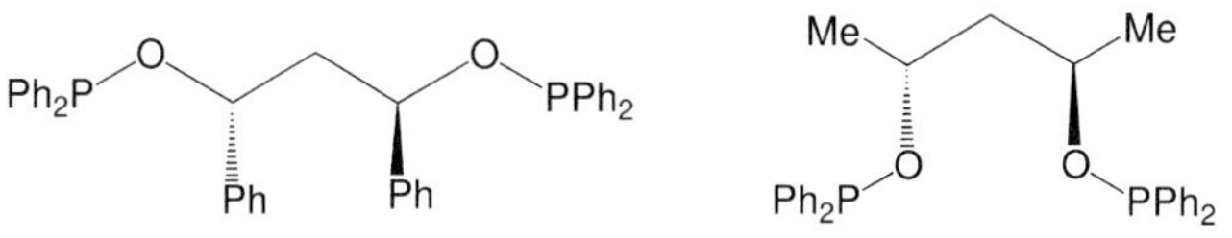

4.2 Homogeneous C=O Hydrogenations

Scheme 4.30

BDPODP

BDPOP

Scheme 4.31

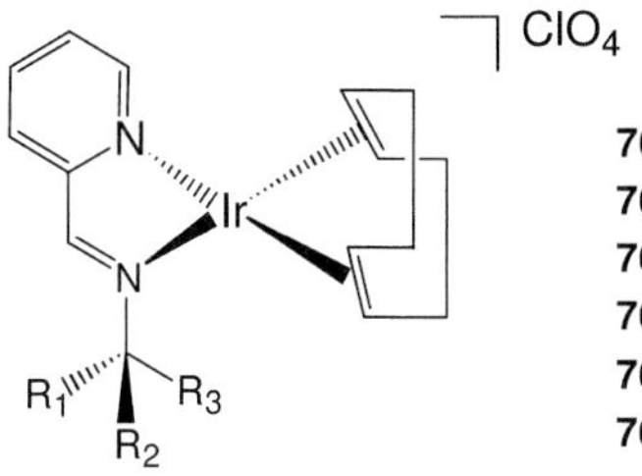

76a, R$_1$ = H, R$_2$ = H, R$_3$ = H, **PMI**
76b, R$_1$ = H, R$_2$ = Me, R$_3$ = Et, **PMEI**
76c, R$_1$ = H, R$_2$ = Me, R$_3$ = Ph, **PMPI**
76d, R$_1$ = H, R$_2$ = Et, R$_3$ = Ph, **PPEI**
76e, R$_1$ = H, R$_2$ = Bz, R$_3$ = Ph, **PPBI**
76f, R$_1$ = H, R$_2$ = Me, R$_3$ = α-C$_{10}$H$_7$, **PNEI**

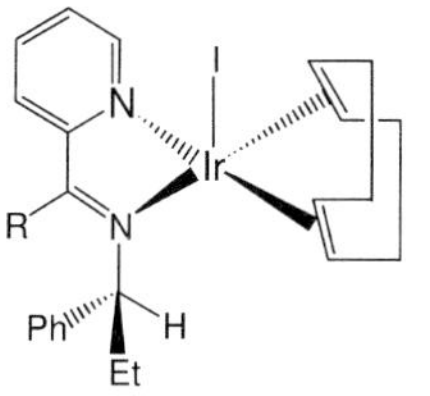

R = H, **77a**
R = Me, **77b**

Scheme 4.32

tions with [Ir(cod)Cl]$_2$ was to retard the substitution of cod and open reaction pathways, leading to higher enantioselectivities.

Another class of ligands for ATH is represented by multidentate Schiff bases and their derivatives. Zassinovich and Mestroni reported on the effective reduction of alkyl aryl ketones catalyzed by a series of Ir(I) complexes with chiral bidentate pyridylaldimines, of the form [Ir(cod)(NNR*)]ClO$_4$ (**76a–f**; see Scheme 4.31). It was observed that both the activity and selectivity depended heavily on the nature of the substituents at the chiral center of the ligand, and also at the prochiral center of the substrate. Optical yields of up to 50% (R-isomer) at 100% conversion were obtained in the ATH of ^{t}BuC(O)Ph and PhCH$_2$C(O)Ph using [Ir(cod)(PPEI)]ClO$_4$ as the precatalyst (0.1% mol, 83 °C, ^{i}PrOH, KOH) [66].

Based on these findings, new pentacoordinated Ir complexes [Ir(cod){(S)-(+)-(NNR*)}I] were synthesized and characterized also by X-ray crystal structure determination by the same authors (NNR* = 2-pyridinal-1-phenyhethylimine, PPEI, **77a**; 2-acetylpyridine-1-phenylethylimine, APPEI, **77b**; Scheme 4.32). The complexes were tested in the ATH of ketones in isopropanol and the data compared with results obtained using the corresponding square planar complexes [Ir(cod){(R)-(−)PPEI}]ClO$_4$ (**76d**) [67].

Both, the activities and selectivities were observed to depend heavily on many variables, including KOH concentration, the presence of water, and the nature of

Table 4.6 Asymmetric transfer hydrogenation of ^{t}BuC(O)Ph in isopropanol, catalyzed by **76d** and **77a**.

Catalyst	KOH:Ir ratio	H$_2$O (%, v/v)	Conversion (%)	Time (min)	ee (%)
77a	1.5	0	98	120	66.0 S-(−)
77a	2.0	0	95	105	66.0 S-(−)
77a	1.5	0.5	94	105	73.5 S-(−)
77a	1.5	1.0	94	120	78.0 S-(−)
77a	1.5	1.5	94	120	80.0 S-(−)
76d	1.5	0	91	120	50.0 R-(+)
76d	2.0	0	99	75	51.5 R-(+)

Conditions: [cat] = 1.6 × 10^{-4} M; substrate:catalyst ratio = 1000; ^{i}PrOH 125 ml; 83 °C.

Table 4.7 Asymmetric transfer hydrogenation of prochiral ketones in isopropanol, catalyzed by **77a** and **77b**.

Catalyst	Substrate	Conversion (%)	TOF (h^{-1})	ee (%)
77a	MeC(O)Ph	88	160	36.5 S-(−)
77a	^{i}PrC(O)Ph	91	183	52.0 S-(−)
77a	^{t}BuC(O)Ph	94	472	79.5 S-(−)
77b	MeC(O)Ph	92	153	19.5 S-(−)
77b	^{i}PrC(O)Ph	92	154	27.0 S-(−)
77b	^{t}BuC(O)Ph	96	382	42.0 S-(−)

Conditions: [cat] = 1.6 × 10^{-4} M; KOH:Ir ratio = 2; H$_2$O (vol.%) = 2.0; substrate:catalyst ratio = 1000; ^{i}PrOH 125 ml; 83 °C.

the chiral ligand. In general, the pentacoordinate species proved to be less active but more selective catalysts (66% versus 50% ee) for the reduction of ^{t}BuC(O)Ph. The asymmetric induction was observed to depend more heavily on the water content (Table 4.6). An increase in KOH concentration had a positive effect on the activity of **76d**; for example, in the case of **77a** a higher substrate:catalyst ratio caused a sharp decrease in optical induction, whereas for **76d** increases in both activity and selectivity were noted (TOF 1660 h^{-1}, 63.5% ee, substrate:catalyst ratio = 3000).

A substrate screening was also carried out and reported [67]. For both penta- and tetracoordinate complexes the reaction rate and selectivity increased on going from methyl to *t*-butyl ketone, thus following the steric hindrance and electrophilicity of the substrate. In the comparison between PPEI and APPEI, the former were markedly more selective in both the neutral (Table 4.7) and cationic complexes. An increase in the iodide:Ir ratio also had an improving effect on the asymmetric induction for **77a**, reaching a maximum 84% ee at [I$^-$]:[Ir] = 3.

Industrial applications of this type of Ir complex can be found in the patent literature. Optically active Ir complexes of 6-alkyl substituted pyridylimine Ir(I) complexes **78** [R = Me, R₁ = chiral carbon hydrocarbyl group; R = H, R₁ = 1,2-diphenylethyl, five- to six-membered-heteroalicyclic group containing two chiral carbon atoms and one Ph or alkyl group at α-position; X = basic (in)organic acid anion; Y, Z = ethylene or YZ = cyclic dienes], were used as enantioselective homogeneous catalysts for the TH of prochiral ketones. As an example, it was reported that 1.43 g of butyrophenone was treated with 36 ml ^{i}PrOH in the presence of catalyst (1.33×10^{-4} M) at 60 °C for 21 h to give 69.7% of 1-phenyl-1-butanol with a 49.2% enantiomer excess of the (S)-isomer [68].

Chirality was also achieved by using cyclic (chiral) substituents on the imino nitrogen. Complexes **79** and **80** (the latter obtained by reduction of the C=N double bond) were tested for the ATH of butyrophenone in isopropanol at 60 °C with NaOH. The activity (26% versus 68% conversion for **79a,b**) and stereoselectivity (64% versus 3% for **79a,b**) were sharply reduced when a methyl group was introduced into the 6-position of the pyridine ring. Tetracoordinated cationic complexes and pentacoordinated neutral species displayed similar efficiencies (64 versus 66% ee for **79a** and **80a**, respectively).

Variations on the theme of pyridylimines as ligands for ATH reactions included use of the C=N double bond reduced versions – that is, substituted pyridylamino ligands such as DHPPEI (**81**; Scheme 4.33; R = H, R₁ = H, R₂ = Me, R₃ = Ph). Among simple alkyl phenyl ketones, an increase in the steric bulk of the alkyl substituents affected only slightly the enantioselectivity and the relative rates. However, a marked increase in stereoselectivity was obtained by introducing a

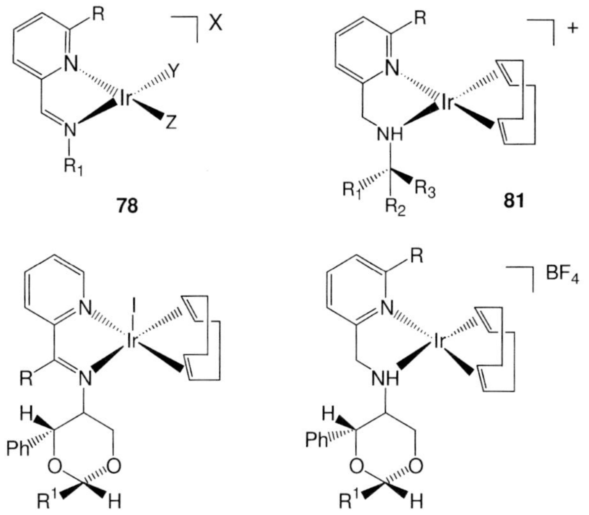

79a, R = H, R¹ = Ph
79b, R = Me, R¹ = Ph

80a, R = H, R¹ = Ph
80b, R = Me, R¹ = Ph
80c, R = H, R¹ = 4-styryl

Scheme 4.33

phenyl group into the alkyl chain – that is, phenyl 3-phenylpropyl ketone was reduced with more than 90% ee under similar ATH conditions [62, 69].

Among other examples of Schiff base Ir complexes applied to ATH reactions, worthy of mention are the chiral N,N,N-terdentate bis-pyridylimino ligands PDPBI and the N,N-bidentate (1R,2R)- and (1S,2S)-diaminocyclohexyl arylimines (Scheme 4.34). Whereas, poor selectivity was observed for the ATH of acetophenone using a catalyst generated *in situ* by the reaction of [Ir(coe)$_2$Cl]$_2$ and PDPBI, 4-phenyl-3-buten-2-(S)-ol could be obtained in 95% yield and 67% ee by ATH reduction of the corresponding α,β-unsaturated ketone [70]. In the case of **82**, the corresponding Ir(cod)Cl complexes catalyzed the ATH of alkyl aryl ketones ArC(O)R (Ar = Ph, 1-naphthyl, 2-naphthyl, 2-fluorenyl, R = Me, Et, CHMe$_2$) in ^{i}PrOH at room temperature to give the corresponding chiral secondary alcohols with enantioselectivities up to 61% ee at 88% yield (**82b**) [71].

Mixed donor chiral tetradentate PNNP diaminodiphosphines **83** (Scheme 4.35) have found application in the ATH of ketones, catalyzed by iridium. The results showed that, in the presence of [IrCl-(R,R)-**83**], chiral alcohols could be obtained with high activity (up to 99.4% yield) and excellent enantioselectivities (up to 99.0% ee) under mild conditions using a ketone:catalyst ratio of 5000 with KOH in ^{i}PrOH. In the case of propiophenone, the TON reached 4780 mol product per mole iridium, while the TOF was as high as 1593 h^{-1} at 55 °C [72].

The same authors further developed the catalytic system described above, wherein the chiral Ir catalytic system generated *in situ* from the iridium hydride complex [IrH(cod)Cl$_2$]$_2$ and chiral diaminodiphosphine ligand **83** was employed in the ATH of aromatic ketones PhC(O)R (R = Me, Et, ^{i}Pr, cy, ^{t}Bu, etc.) to produce

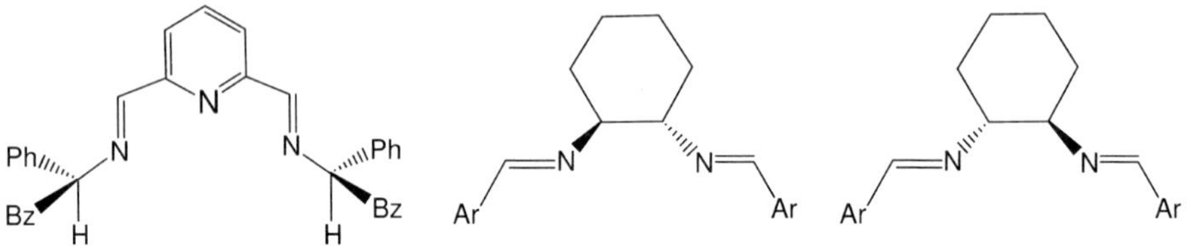

(R,R)-(+)-PDPBI   (S,S)-**82**   (R,R)-**82**

Scheme 4.34

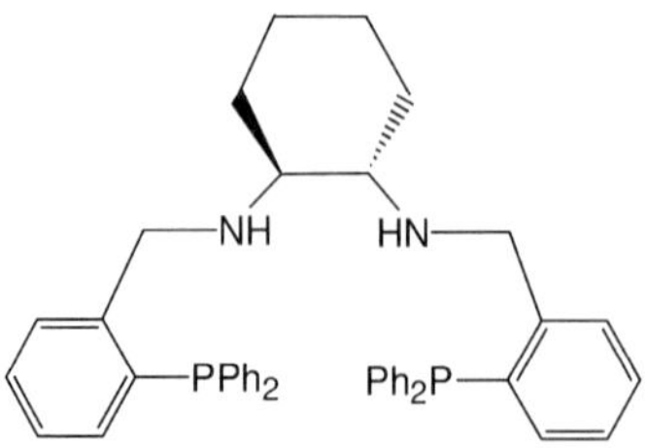

(R,R)-**83**

Scheme 4.35

Table 4.8 Asymmetric transfer hydrogenation of aromatic ketones, catalyzed by [IrH(CO)(PPh$_3$)$_3$]/**83** under base-free conditions.

Substrate[a]	Temperature (°C)	Time (h)	Yield (%)	ee (%)
PhCOEt	75	0.5	97	90
PhCOEt[b]	82	2	98	84
PhCO^{i}Bu	75	6	90	93
PhCO^{t}Bu	75	4	99	80
o-NO$_2$C$_6$H$_4$COMe	70	1	99	92
m-OMeC$_6$H$_4$COMe	75	1	91	70
1,1-Ph$_2$-COMe	75	8.5	99	97
PhCOCy	75	3	96	94

a   Conditions: substrate:catalyst ratio = 200; solvent ^{i}PrOH.
b   Substrate:catalyst ratio = 2000.

chiral alcohols with up to 99% ee in quantitative yield for substrates such as 1,1-diphenylacetone. The same ketone could be reduced using KOH as base, with a substrate:catalyst ratio of up to 10 000:1, although a reaction time of 63 h was required to reach 89% conversion [73]. Base-free conditions could be also applied by using a precursor such as [IrH(CO)(PPh$_3$)$_3$] [74]. In this way, it was possible to reduce PhC(O)Et to the corresponding chiral alcohol at a substrate:catalyst ratio of 200, in ^{i}PrOH at 75 °C, reaching 97% yield and 90% ee after 30 min. A broader catalyst scope was demonstrated, the results of which are summarized in Table 4.8.

Later, chiral PNNP ligand **83** together with precursor [IrH(cod)Cl$_2$]$_2$ were applied for the first time in the ATH of aromatic ketones with HCOONa in water using 5% mol of a phase-transfer catalyst, giving the corresponding alcohols in high yield and excellent enantioselectivity (for α-tetralone up to 99% ee). The reduction of propiophenone was carried out at a substrate:catalyst molar ratio as high as 8000, while still maintaining a high enantioselectivity (85% ee at 88% yield after 101 h, at 60 °C [75]. Finally, the water-soluble analogue of **83**, namely the PNNP tetradentate diaminodiphosphine ligand [(R,R)-C$_6$P$_2$(NH)$_2$(SO$_3$Na)$_4$] (**84**), was prepared by the sulfonation of **83** with 50% SO$_3$ oleum. The water-soluble iridium catalyst obtained in situ from [Ir(cod)(PPh$_3$)Cl] and **84** was tested for the ATH of various aromatic ketones in ^{i}PrOH:H$_2$O (2:1) solvent mixture using KOH as base, with good results for both yields and enantioselectivities. Even those ketones which had great bulk within the alkyl group (e.g. isobutyrophenone, phenyl cyclohexyl ketone, 1,1-diphenylacetone) were smoothly converted to optically active alcohols with ee-values of 98–99% and yields ranging from 89 to 97% [76].

A tetradentate macrocylic ONNO-type ligand was used recently by the same authors for Ir-catalyzed ATH reactions. The reaction of 1,3-bis(2-formylphenoxy)-2-propanol with chiral 1,2-diaminocyclohexanes gave *(S,S)*- or *(R,R)*-macrocyclic C$_6$O$_2$N$_2$ ligands (**85**) (Scheme 4.36), while the catalysts generated in situ in the

Scheme 4.36

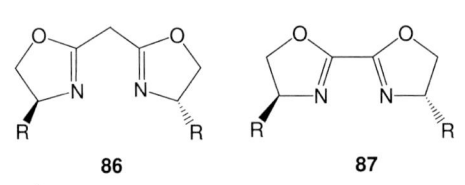

(R,R)- or (S,S)-**85**

Scheme 4.37

86   **87**

presence of the iridium hydride complex [IrH(cod)Cl$_2$]$_2$ were used in the ATH reduction of aromatic ketones PhC(O)R (R = Me, Et, ^{i}Pr, cy, ^{t}Bu, etc.), using 2-propanol as a source of hydrogen. The yields of the chiral alcohols ranged from 41 to 99%, depending on the substitution on the phenyl ring, and showed ee-values of 50–91%, when using 1% of catalyst and running the reactions at 40 °C. The addition of water to the system caused a decrease in activity for the reduction of propiobenzophenone, without affecting the enantioselectivity (85% ee). One advantage of this system is that the reactions can be performed in air, without affecting conversions and selectivities, as compared to tests run under a rigorously inert atmosphere [77].

Another class of NN ligand used in enantioselective catalysis is represented by bis-oxazoles and their derivatives. C_2-symmetric 4,4′,5,5′-tetrahydro-2,2′-methylenebis[oxazoles] (**86**) and 4,4′,5,5′-tetrahydrobis(oxazoles) (**87**) were synthesized by Pfaltz and coworkers and applied as chiral ligands for a series of reactions, including Ir-catalyzed ATH of ketones using isopropanol [78] (Scheme 4.37). Alkyl aryl ketones were readily reduced in the presence of **87** and [Ir(cod)Cl]$_2$, affording the corresponding alcohols in 47–91% ee, whereas dialkyl ketones were less reactive and gave low yields of the racemic products. The best result was obtained with the isopropylsubstituted ligand and isopropyl phenyl ketone as the substrate (91% ee at 70% conversion).

Despite these promising results, bis(oxazoline) ligands seem to perform better in the ATH of ketones when associated with Ru precursors. Pinel et al. [79] described a comparative study on the effect of the metal in ATH of acetophenone using the hydroxymethyl-substituted bis(oxazoline) **88** (Scheme 4.38). Whereas, Ru(II) complexes gave 89% ee at 50% conversion, only 22% conversion was

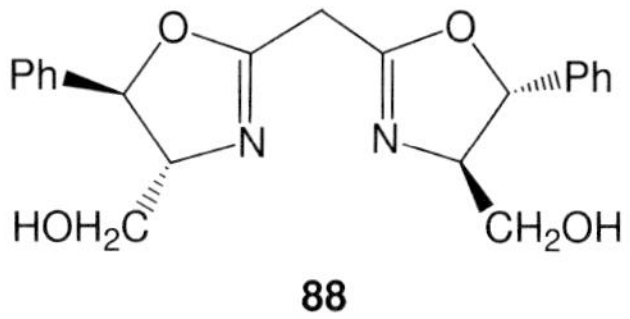

Scheme 4.38

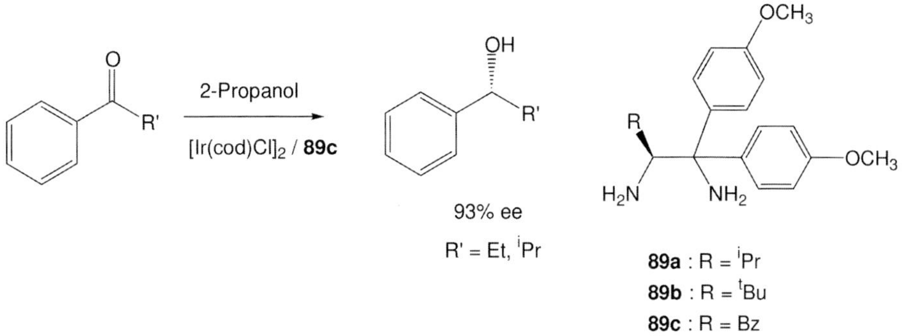

Scheme 4.39

observed for the Rh and Ir(cod)Cl complexes, giving ee-values of 16% and 20%, respectively.

Chiral diamines constitute another well-represented class of bidentate NN ligands which has been developed for application in ATHs. Noyori and coworkers [80] reported the development of an Ir catalyst made *in situ* from [Ir(cod)Cl]$_2$ and chiral-substituted 1,1-di(*p*-anisyl)ethylenediamines **89a–c**, which were found to catalyze the ATH of acetophenone in isopropanol/KOH at room temperature. Depending on the nature of the substituent, the yields and ee-values varied from 68 to 96% and from 53 to 78%, respectively (Scheme 4.39). When using benzyl-substituted ligand **89c**, a series of alkyl aryl ketones were hydrogenated smoothly to almost complete conversion and excellent ee-values (93% for PhCOEt after 12 h at room temperature, substrate:catalyst ratio 500). The reactivities were decreased by the introduction of an electron-donating group on the ketone (Me, OMe), in contrast to the presence of an electron-withdrawing substituent in the *ortho* or *meta* positions, which accelerated the reaction rates. Of relevance here, the Ir-based systems were observed to perform as well as the Ru analogues.

Among examples of complexes bearing diamine ligands, pentamethycyclopentadienyl Rh and Ir chlorides bearing Noyori's monotosylated diamine TsDPEN (1*S*,2*S*)-*N*-(*p*-toluenesulfonyl)-1,2-diphenylethylenediamine, namely [Cp*MCl(TsDPEN)] (R = Rh, **90a**; Ir, **90b**) (Scheme 4.40) were found to be active catalysts for the ATH of acetophenone, 2-acetonaphthone, 1-tetralone and 1-indanone, giving the *S*-alcohols in 85–99% ee (1-indanol using **90a**) in ^{i}PrOH/KOH using 1% of catalyst at room temperature after 48 h [81]. For Ir, it was observed that doubling the amount of KOH (2:1 ratio to Ir) caused a decrease in conversion, without affecting the enantioselectivity of the process. Complex **90a** was generally found to perform better than the Ir analogue **90b**.

4 Iridium-Catalyzed C=O Hydrogenation

Scheme 4.40

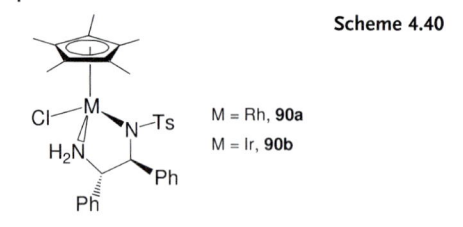

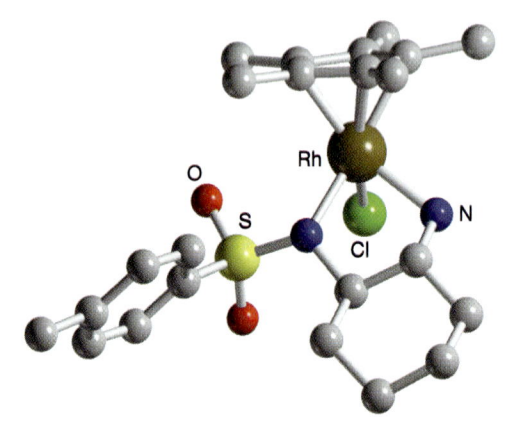

M = Rh, **90a**
M = Ir, **90b**

Figure 4.5 The molecular structure of [Cp*Rh(Ts-CYDN)Cl] (**91a**).

At about the same time as the studies in the previous report were being conducted, [Cp*M(Ts-DPEN)Cl] and the analogues [Cp*M(Ts-CYDN)Cl] [M = Rh, **91a**, Ir, **91b**; TsCYDN = (1R,2R)-N-(p-toluensulfonyl)-1,2-cyclohexanediamine] were used by Murata et al. for the ATH of substituted aryl alkyl ketones in the presence of ^{t}BuOK/^{i}PrOH at 30 °C [82]. The X-ray crystal structure of **91a** shows a λ-configured five-membered ring caused by the R,R stereochemistry of the ligands (Figure 4.5). The mechanistic details were highlighted by reacting **91b** with NaOH in ^{i}PrOH to yield [Cp*IrH(Ts-CYDN)] (**92**) as a single stereoisomer. Further reaction with acetone gave the amide complex **93**, where the amine is deprotonated. All of the complexes catalyzed the ATH reductions of ketones, at a substrate: catalyst ratio of 200. Complex **91a** gave better rates and enantioselectivities – even higher than the benchmark Ru complex [Ru(p-cymene)(TsDPEN)Cl] under these conditions (85% conversion and 97% ee, 12 h). The Ir analogue **91b** performed best with m-CF$_3$-substituted acetophenone (99% conversion, 94% ee), although longer reaction times were required compared to **91a** (24 versus 12 h). Worthy of note here is the observation that hydride **92** catalyzed the ATH of m-CF$_3$-substituted acetophenone with the same activity and selectivity as the chloride parents, with the reactions being run without the need of any added base.

Use of the water-soluble analogues of complexes **90** and **91** were later reported by Williams and coworkers [83]. In this study, water solubility was achieved by replacing the chiral diamines TsDPEN and TsCYDN with the p-sulfonated derivatives TsDPEN-SO$_3$Na and TsCYDN-SO$_3$Na (Scheme 4.41). The ATH of aryl alkyl

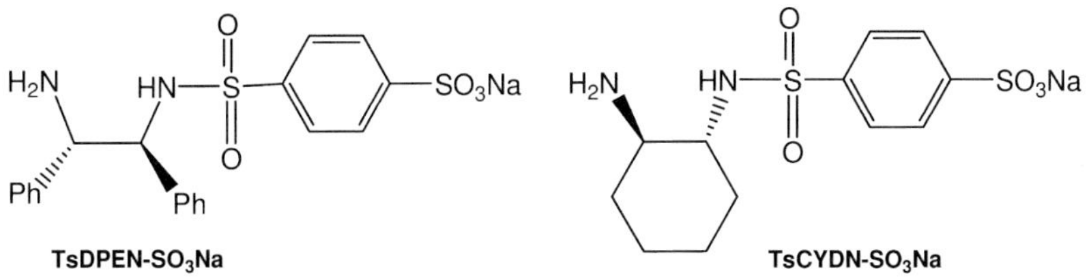

TsDPEN-SO₃Na **TsCYDN-SO₃Na**

Scheme 4.41

Table 4.9 Asymmetric transfer hydrogenation of aryl alkyl ketones in ⁱPrOH/H₂O, catalyzed by **93a,b** and **94a,b**.[a]

Catalyst	Substrate	Time (h)	Conversion (%)	ee (%)
93a	PhC(O)Me	24	92	97
94a	PhC(O)Me	18	94	95
93b	PhC(O)Me	140	90	82
94b	PhC(O)Me	43	95	86
93b	m-CF$_3$C$_6$H$_3$C(O)Me	43	95	86
94b	m-CF$_3$C$_6$H$_3$C(O)Me	4	98	93
93b	m-FC$_6$H$_3$C(O)Me	51	83	85
93b[b]	m-FC$_6$H$_3$C(O)Me	22	74	92
93b[c]	m-FC$_6$H$_3$C(O)Me	22	90	92
94b	m-FC$_6$H$_3$C(O)Me	26	99	94
94b[b]	m-FC$_6$H$_3$C(O)Me	2.5	82	94
94b[c]	m-FC$_6$H$_3$C(O)Me	2.5	94	93
93b	p-MeOC$_6$H$_3$C(O)Me	150	22	78
93b[b]	p-MeOC$_6$H$_3$C(O)Me	115	20	91
93b[c]	p-MeOC$_6$H$_3$C(O)Me	115	33	92
94b	p-MeOC$_6$H$_3$C(O)Me	141	80	95
94b[b]	p-MeOC$_6$H$_3$C(O)Me	116	76	92
94b[c]	p-MeOC$_6$H$_3$C(O)Me	116	89	87
93b	Naphthylacetone	139	77	73
94b	Naphthylacetone	45	96	96
93b	Indanone	139	41	91
94b	Indanone	45	55	97

a Conditions: substrate, 2 mmol; [cat] = 0.01 mmol; ᵗBuOK, 0.20 mmol; ⁱPrOH/H₂O; 22 °C, [H₂O] 15% vol.
b [H₂O] 34% vol.
c [H₂O] 51% vol.

ketones such as (p- or m-substituted) acetophenones, 1-naphthyl acetone and indanone were carried out using ᵗBuOK as base in a mixture of ⁱPrOH/H₂O at increasing water contents (15, 34 and 51%) in the presence of catalysts [Cp*MCl(wsNN)] (ws NN = TsDPEN-SO₃Na, M = Rh, **93a**; M = Ir, **93b**; ws NN = TsCYDN-SO₃Na, M = Rh, **94a**; M = Ir, **94b**). The selected catalytic data, as summarized in Table 4.9, show that both the choice of metal and ligand affected the rate and enantioselectivity of the reaction. Rh catalysts gave higher activity and enantioselectivity in the ⁱPrOHl-rich system, and electron-poor ketones were reduced at a higher rate. Between the two Ir-based systems, **94b** outperformed **94a** in most cases. In contrast to what was expected, the increase in water content led to a significant increase in both conversions and ee-values.

The use of terminal reducing agents other than ᵗBuOK and KOH was reported for the water-phase ATH of aromatic ketones, without the need for an inert atmosphere. Ikariya, Xiao and coworkers [84] showed that HCO_2Na could be used efficiently, and in air, to obtain chiral secondary alcohols in good yields and ee-values. These authors used complexes **91a,b** to hydrogenate acetophenone in the presence of 5 equiv. of HCO_2Na, and obtained complete conversion to (R)-1-phenylethanol in 95% ee with **91a** in water at 40 °C after 15 min. Complex **91b** was slightly less active, with a conversion of 99% and 93% ee after 1 h. The use of an $HCOOH/NEt_3$ azeotrope gave longer reaction times. A range of substituted acetophenones and heteroaryl ketones were successfully converted in the presence of **91a** with substrate:catalyst ratios as high as 1000, giving TOFs of up to $3500 h^{-1}$. Further modifications of the chiral diamine ligand led to the synthesis and use in ATH reactions of Cp*MCl complexes of CsDPEN [(R,R,R)- or (S,S,S)-N-camphorsulfonyl-1,2diphenylethylendiamine] (Scheme 4.42), with M = Rh (**95a**) and Ir (**95b**), of aryl ketones by formate in neat water. Worthy of note here is the Ir-(R,R,R)-CsDPEN catalyst, which catalyzes the ATH of a wide range of ketones with almost full conversions within a few hours, at a substrate:catalyst ratio of 1000 at 40 °C in most cases, and with enantioselectivities up to 98% ee [85]. Unsaturated ketones such as (E)-chalcone were fully hydrogenated to the saturated 1,3-diphenylpropan-1-ol. Gas chromatographic monitoring of the reaction showed that the C=C double bond was reduced first.

A series of interesting N-based ligands derived from naturally occurring alkaloids was derived from the *Cinchona* analogues, quinine and quinidine, and tested together in the catalytic ATH of aromatic ketones in ⁱPrOH/ⁱPrOK. Ru, Rh and Ir precursors were tested and the best activities observed with $[Ir(cod)Cl]_2$. A wide range of aromatic ketones was reduced with excellent conversions and good enantioselectivities using the reduced forms (H-derivative) of quincorine (QCI)-amine and quincoridine(QCD)-amine (Scheme 4.43). For acetophenone, the enantioselectivities obtained were seen to decrease from 80% ee (94% conversion, 1 h) to 76% ee (97% conversion, 10 h) using HQCI-amine, whereas HQCD-amine gave a slightly higher ee (85% at 76% conversion, 1 h). For *p*- and *o*-substituted acetophenones, complete conversions and ee-values between 70% and 95% were

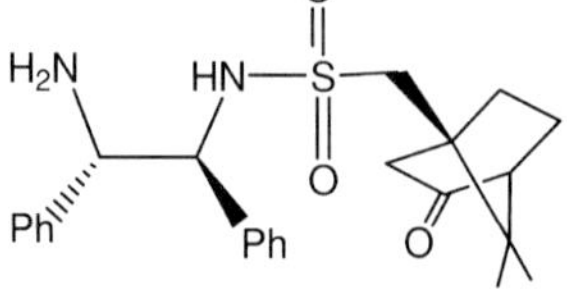

CsDPEN

Scheme 4.42

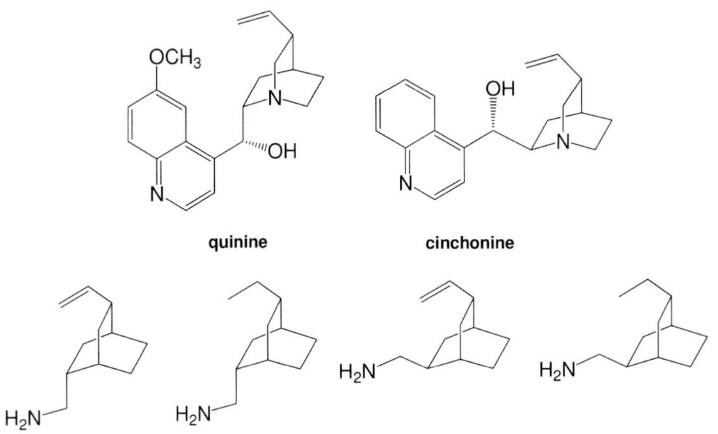

quinine cinchonine

QCI-amine HQCI-amine QCD-amine HQCD-amine

Scheme 4.43

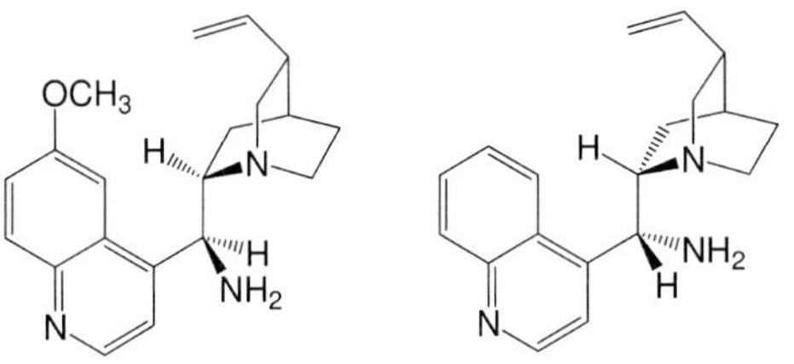

96a 96b

Scheme 4.44

obtained, the worse value with the electron-withdrawing *o*-NO_2 substituent. By making the aliphatic group of the ketone larger than methyl, higher selectivities (95% ee) were possible, albeit at lower yields (35%, 1 h, ^{i}PrC(O)Ph) [86].

He *et al.* elaborated on the concept of cinchona derivatives and produced the ligands 9-amino(9-deoxy)epiquinine (**96a**) and epicinchonine (**96b**) (Scheme 4.44) [87]. The amines were tested with both Rh and Ir precursors for the ATH of ketones in ^{i}PrOH/KOH. The best results were achieved using **96b** as the ligand and $[Ir(cod)Cl]_2$ as the metal precursor, and for isobutyrophenone the conversion and enantioselectivity were obtained in 90 and 97% ee, respectively. Later it was shown that the Ir complex of 9-amino(9-deoxy)epicinchonine **96b** could be recovered in high yields with dilute HCl. The yields (90–94%) and enantioselectivities of 1-phenylethanol (93–95% ee) were maintained with small variations after six cycles [88].

A series of chiral N,S-chelates was synthesized as ligands for the iridium(I)-catalyzed reduction of ketones using either $HCOOH/NEt_3$ or isopropanol as hydrogen sources. The ligands were obtained by sulfoxidation of an (*R*)-cysteine-based aminosulfide, providing a diastereomeric ligand family containing a chiral sulfur atom [89]. In parallel, a series of aminosulfides, each bearing two asymmetric carbon atoms in the backbone and derived from 1,2-disubstituted amino alcohols, were obtained. Both, the sulfoxide-based β-amino alcohols and the aminosulfides gave rise to high reaction rates when tested in the ATH of acetophenone with formic acid in presence of triethylamine and [Ir(cod)Cl]$_2$. The aminosulfide ligands **97a–k** gave, in some cases, a good conversion (98% after 1 h at 60 °C with **97d**), but the enantioselectivities in (S)-1-phenylethanol were very poor (5–18%) when HCOOH/NEt$_3$ was used (Scheme 4.45). A clear effect of chiral cooperativity was observed between the sulfoxide functionality and the α position of the amino alcohol with **98b**, which gave 65% ee and almost complete conversion after 30 min. For the ligands **99a–f** bearing a secondary amine group, the stereoselectivities were moderate (23–65%, with **99c** being the best-performing) and a total conversion was obtained after 3 h for **99b** and **99c** with HCOOH/NEt$_3$, running the tests at 60 °C. Introduction of the sulfoxide group in this class of ligands (**100a** and **100b**) considerably reduced the conversion and stereoselectivity of the reaction (Scheme 4.45). By screening the effect of the nature of the catalyst precursor, it was observed that the presence of a diene ligand such as 1,5-cyclo-octadiene (cod) is necessary for moderate to good activities, although the stereoselectivity of the reaction is not affected by the choice of iridium precursor. A strong temperature effect was noted, as decreasing the reaction temperature to 20 °C resulted in an increase in enantio-

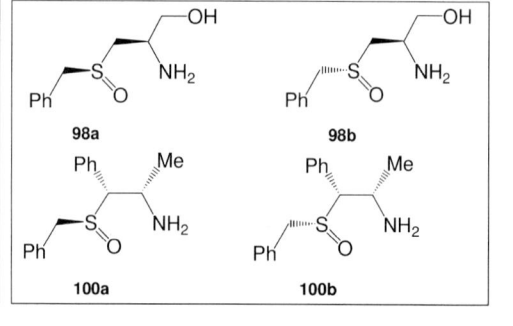

Scheme 4.45

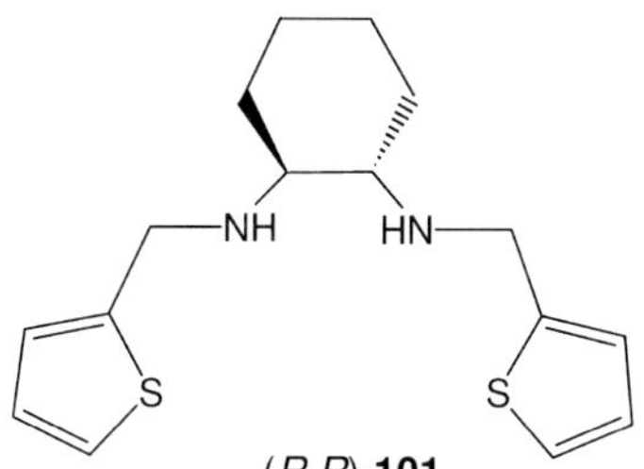

(R,R)-**101**

Scheme 4.46

selectivity in all cases, with a maximum of 80% ee when using **98b** (57% conversion, 5 h). Various prochiral ketones were tested, and the enantioselectivity of the reaction was markedly influenced by the electronic and steric properties of the substrate – that is, a larger steric bulk and the presence of an electron-withdrawing substituent in the substrate resulted in a higher enantioselectivities. When using 2-propanol as hydrogen donor instead of HCOOH/NEt$_3$, the maximum ee was obtained (97%) with 1-naphthyl-methyl-ketone at 99% conversion after 1 h at 20 °C.

More recently, Zhang et al. [90] showed that the combination of [Ir(cod)(PPh$_3$)Cl] and chiral SNNS-type ligands **101** (Scheme 4.46) derived from thiophene cyclohexyldiamines catalyzed the ATH of prochiral ketones under mild conditions and in the air. Enantioselectivities up to 96% were observed for the ATH of PhC(O)Cy after 22 h at 97% conversion, giving the R-alcohol using **101** in the (S,S)-form, running the reaction at 25 °C in an ^{i}PrOH/KOH reducing system. In all cases, inversion of the conformation from ligand to product was observed. The presence of p-, m- or o-substituents on the phenyl ring of the ketone allowed for faster rates, albeit with lower ee-values (91% ee, 96% conversion, 9 h for o-methyl acetophenone; 82% ee, 94% conversion, 7 h for m-methyl acetophenone; 71% ee, 99% conversion, 5 h for m-chloro acetophenone; 58% ee, 94% conversion, 5 h for p-chloro acetophenone). The major advantage of these systems resides in their ease of manipulation and stability of the catalysts (in air), as well as the wide applicability and functional group tolerance.

The combination of chirality and hemilability in the ancillary ligand of a transition-metal complex is often beneficial for catalytic applications. The use of p-bonded olefins or dienes (coe or cod) is ubiquitous for Rh and Ir complexes. In some cases, p-coordination can be achieved by having olefin pendant arms in the chiral ligand, which in turn can either stabilize active species [53] or allow for hemilabile complex behavior. The chiral tetrachelating amino-olefins (R,R)-N,N'-bis(5H-dibenzo[a,d]cyclohepten-5-yl)-1,2-diaminocyclohexane ((R,R)-trop$_2$dach) and (S,S)-N,N'-bis(5H-dibenzo[a,d]cyclohepten-5-yl)-1,2-diphenyl-1,2-ethylenediamine ((S,S)-trop$_2$dpen) were prepared and used as ligands [91] in the complexes (R,R)-[Rh(trop$_2$dach)]OTf and (S,S)-[Rh(trop$_2$dpen)]OTf OTf$^-$ = CF$_3$SO$_3^-$. Reactions of (S,S)-trop$_2$dpen with [Ir(cod)Cl]$_2$ in the presence of TlPF$_6$ gave the NNC$_2$-coordinated complex (S,S)-[Ir(cod)(trop$_2$dpen)]PF$_6$ (**102**; Scheme 4.47 and Figure 4.6). The reaction of (S,S)-trop$_2$dpen with [Ir(cod)Cl]$_2$, when run under an

4 Iridium-Catalyzed C=O Hydrogenation

Scheme 4.47

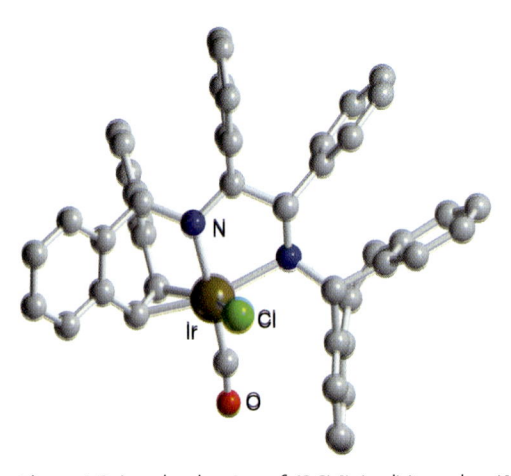

Figure 4.6 A molecular view of (S,S)-[Ir(cod)(trop$_2$dpen)]PF$_6$ (**103**).

atmosphere of CO, gave the distorted-trigonal-bipyramidal iridium complex (S,S)-[IrCl(CO)(trop$_2$dpen)] (**103**). The rhodium complexes and **102** are inactive for the transfer and direct hydrogenation of ketones, whereas in contrast **103** is a precursor to an active phosphine-free chiral catalyst in the ATH of acetophenone with 2-propanol, and the R isomer of 1-phenylethanol was obtained in 82% ee (98% conversion at substrate: ^{t}BuOK:**103** ratios 100:10:1, 80 °C, 1 h).

Amino acidate N,O-coordinated chiral complexes of Ru, Rh and Ir have found applications in the catalytic hydrogenations of ketones and unsaturated aldehydes

[92]. The mononuclear chlorides [(η-ring)M(Aa)Cl] ((η-ring)-M = Cp*Rh, Cp*Ir, (p-cymene)Ru; Aa = α-amino acidate) were reacted with AgBF$_4$ to yield the corresponding new chiral trimers [{(η-ring)M(Aa)}$_3$](BF$_4$)$_3$. The amino acidates include alaninate (Ala), 2-aminobutyrate (Abu), valinate (Val), terleucinate (Tle) phenylalaninate (Phe), L-prolinate (L-Pro), N-methyl-L-prolinate (Me-Pro), 4-hydroxy-L-prolinate (Hyp). The trimers [(Cp*Ir(Ala)$_3$](BF$_4$)$_3$ (**104**) and [{Ru(p-cymene)(L-Pro)}$_3$](BF$_4$)$_3$ were characterized by X-ray diffraction (XRD). Trimerization occurs by chiral self-recognition: the trimers RMRMRM (ρ isomer) or SMSMSM (σ isomer), which have equal configuration at the metal center, were the only diastereomers detected. The chiral trimers [(Cp*Rh(L-Pro)$_3$](BF$_4$)$_3$ and [{Ru(p-cymene)(L-Pro)}$_3$](BF$_4$)$_3$ catalyze the chemoselective reduction of unsaturated aldehydes (e.g. citral, neral, geranial) to the corresponding unsaturated alcohols by TH from sodium formate. When the mononuclear complexes [Ru(p-cymene)(Aa)Cl] (Aa = L-Pro, Phe), [Cp*Rh(D-Pro)Cl], [Cp*Ir(Aa)Cl] (Aa = L-Pro, **105a**; Me-Pro, **105b**) were tested for the ATH of acetophenone using 2-propanol and sodium formate (2 equiv.) as base, the performances compared well to those obtained in the presence of the corresponding cationic trimers. In the case of iridium, [(Cp*Ir(L-Pro)$_3$](BF$_4$)$_3$ (**106**) gave 15.6% (R)-(+)-1-phenylethanol with 59% ee. When 1 equiv. of base was used, however, a slight increase was observed for both conversion (20.6%) and ee (64%). The mononuclear analogue **105a** was tested in the presence of 2 equiv. of HCOONa and gave similar results to **106** (17% conversion, 58% ee), whereas **105b** was almost inactive (1.8% conversion, 2% ee).

Sinou and coworkers evaluated a range of enantiopure amino alcohols derived from tartaric acid for the ATH reduction of prochiral ketones. Various (2R,3R)-3-amino- and (alkylamino)-1,4-bis(benzyloxy)butan-2-ol were obtained from readily available (+)-diethyl tartrate. These enantiopure amino alcohols have been used with Ru(p-cymene)Cl$_2$ or Ir(I) precursors as ligands in the hydrogen transfer reduction of various aryl alkyl ketones; ee-values of up to 80% have been obtained using the ruthenium complex [93]. Using (2R,3R)-3-amino-1,4-bis(benzyloxy)butan-2-ol and (2R,3R)-3-(benzylamino)-1,4-bis(benzyloxy)butan-2-ol with [Ir(cod)Cl]$_2$ as precursor, the ATH of acetophenone resulted in a maximum yield of 72%, 30% ee, 3 h, 25 °C in ^{i}PrOH/KOH with the former, and 88% yield, 28% ee, 120 h with the latter.

Chiral β-amino alcohols such as (1R,2S)-(+)-cis-1-amino-2-indanol, (1S,2R)-(+)-2-amino-1,2-diphenylethanol, (1S,2R)-(+)-norephedrine and (1R,2S)-(−)-ephedrine (Scheme 4.48) were used as ligands for ruthenium-, rhodium- and iridium-catalyzed ATH of acetophenone and derivatives in water at 40 °C, using either sodium formate or HCOOH/NEt$_3$ azeotrope as hydrogen sources [94]. A strong effect of the formic acid to triethylamine ratio was observed on activities and enantioselectivities for all metal complexes, and this was correlated to the pH values of the water phase, with a higher pH favoring higher rates and better enantioselectivity. The best results using [Cp*IrCl$_2$]$_2$ as precursors were obtained in the presence of (1S,2R)-(+)-2-amino-1,2-diphenylethanol, at a HCCOH:NEt$_3$ ratio of 1:1.7, 1.5 h giving complete conversion and 55% ee.

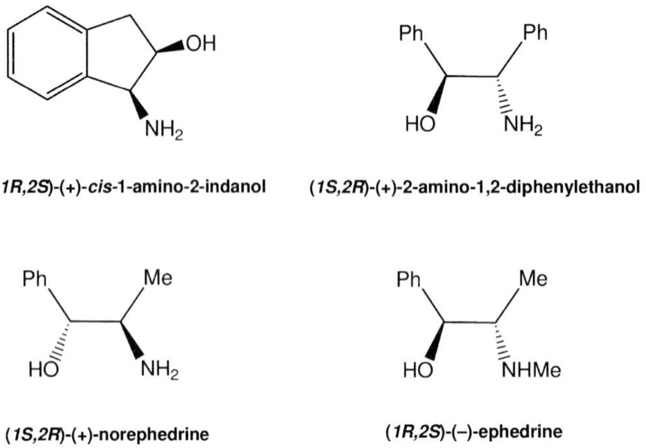
OH
NH2
Ph
Ph
HO
NH2
(1R,2S)-(+)-cis-1-amino-2-indanol
(1S,2R)-(+)-2-amino-1,2-diphenylethanol
Ph
Me
HO
NH2
Ph
Me
HO
NHMe
(1S,2R)-(+)-norephedrine
(1R,2S)-(−)-ephedrine

Scheme 4.48

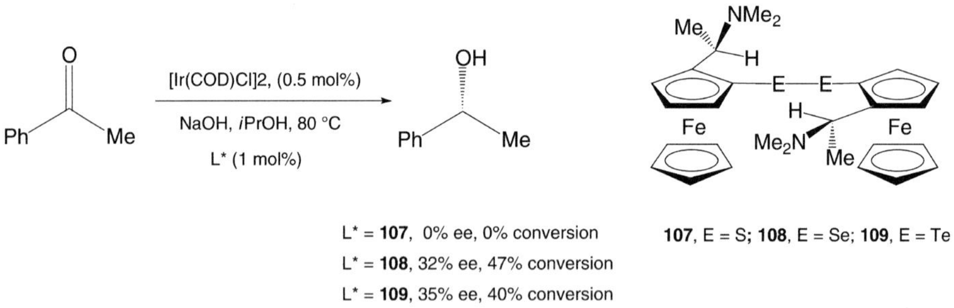
O
Ph
Me
[Ir(COD)Cl]2, (0.5 mol%)
NaOH, iPrOH, 80 °C
L* (1 mol%)
OH
Ph
Me
Me
NMe2
H
E
E
Fe
H
Me2N
Me
Fe
L* = 107, 0% ee, 0% conversion
L* = 108, 32% ee, 47% conversion
L* = 109, 35% ee, 40% conversion
107, E = S; 108, E = Se; 109, E = Te

Scheme 4.49

Asymmetric ferrocenyl ligands bearing different donor atoms connected through a chiral carbon atom have been used by many groups for different enantioselective reactions. Diferrocenyl dichalcogenides **107–109** were synthesized by Uemura *et al.* and tested in the ATH of acetophenone in the presence of $[Ir(cod)Cl]_2$ in ^{i}PrOH/NaOH to give *(R)*-phenylethan-1-ol at moderate conversions and enantioselectivities (Scheme 4.49) [95].

The combination of chiral ferrocenyl moieties and imidazolium salts was described by Chung *et al.* [96]. Whereas, the asymmetry in the ligand is derived from the chiral substituents on the upper ring of the ferrocene, the metal is coordinated by the carbenoid atom of the imidazolium salt. Different combinations of the two moieties (ligands **110–114**) were obtained through an accurate choice of experimental conditions (Scheme 4.50). ATH of 4′-methylacetophenone was carried out with 52.6% ee using complex [IrCl(cod){1,3-bis[(*R*)-1-ferrocenylethyl] benzymidazol-2-ylidene}] (**115**), the X-ray crystal structure of which is shown in Figure 4.7.

110a, R = Me
110b, R = Ph
110c, R = 2,6-diisopropylphenyl
110d, R = (R)-1-ferrocenylethyl

111a, X = Cl
111b, X = I

112

113

114

Scheme 4.50

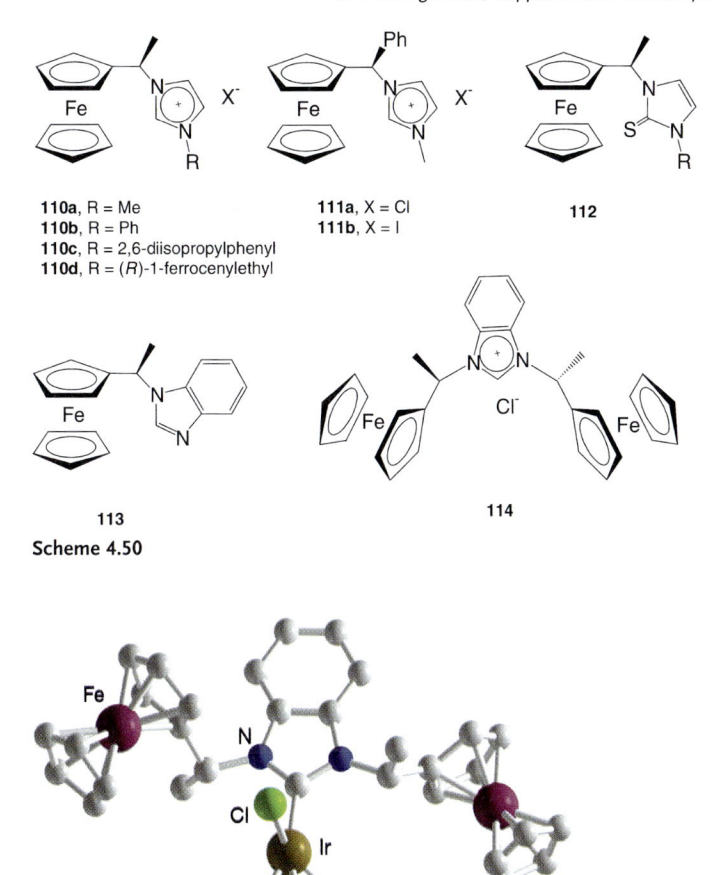

Figure 4.7 The molecular structure of complex [IrCl(cod){1,3-bis[(R)-1-ferrocenylethyl]benzymidazol-2-ylidene}] (**115**).

4.3
Heterogeneous, Supported and Biocatalytic Hydrogenations

The use of heterogeneous catalysis for the selective reduction of C=O groups using Ir has been reported in the literature. In a first report made by USSR scientists during the early 1970s it was claimed that, in the presence of an Ir/C catalyst, unsaturated aldehydes such as CH_2=CHCHO, MeCH=CHCHO and PhCH=CHCHO were hydrogenated to unsaturated alcohols with 70–100% yields at normal pressure of H_2 and at room temperature. As an example, 3 mmol of cinnamaldehyde in 10 ml of EtOH with 0.5 g Ir-C (5% load) gave cinnamol selectively as the only product. Over-reduction of the C=C double bond was not observed and the catalyst could be reused repeatedly [97].

Later, a Japanese patent disclosed the use of heterogeneous Ir, Fe and Rh catalysts for acetaldehyde reduction. Silica gel impregnated with a solution of $IrCl_4 \cdot H_2O$ and $FeCl_2 \cdot 4H_2O$ in EtOH was dried and reduced under hydrogen. A reactor containing the catalyst was fed with a mixture of CO (40 l h^{-1}), H_2 (80 l h^{-1}), and MeCHO (33.0 mmol h^{-1}) at 276 °C and a total pressure of 50 kg cm^{-2} gave 97.5% EtOH at 93.6% conversion based on acetaldehyde [98].

The reduction of crotonaldehyde, 4-methyl-3-penten-2-one and 5-hexen-2-one was carried out in a pulse reactor by initially reducing the catalysts $IrCl_3/TiO_2$, $IrCl_3/Al_2O_3$ (1.5% load) at 200 °C with a hydrogen flow of 20 ml min^{-1}, then cooling the reactor to 50–100 °C and injecting 0.1–1.0 μl of organic reagent into the reactor. A maximum selectivity to the unsaturated alcohol of 60% was measured at a conversion of 18% when using Ir/TiO_2 for crotonaldehyde at 100 °C [99].

Within the patent literature, a process described findings that monoaromatic ketones (e.g. acetophenone) are selectively hydrogenated at 20–70 °C and 20–80 bar (2–8 × 10^4 hPa) to the corresponding aryl-substituted carbinol (e.g. 1-phenylethanol) in high yield in the presence of a carrier catalysts comprising Ir, Ni, Pt, Rh and a Group 14 metal (Sn, Ge, Pb) on SiO_2, Al_2O_3 or aluminosilicates [100].

More recently, an interesting application of the Ir-based heterogeneous reduction of C=O bond was reported by Jacobs and coworkers. Starting from an $Ir(acac)_3$ precursor, an acid H-β zeolite (surface area 740 m^2 g^{-1}, average crystallite size 0.2 mm, Si:Al ratio = 12) was impregnated (2 wt%) and used as a support and promotor, to give an effective catalyst for the chemoselective hydrogenation of α,β-unsaturated aldehydes and ketones such as cinnamaldehyde, a-Me-cinnamaldehyde and citral, to the corresponding allylic alcohols (Table 4.10). The H-β-zeolite-supported iridium catalyst (1 wt %) allowed for the 100% stereoselective hydrogenation of testosterone to androst-4-ene-3β,17β-diol in ^{i}AmOH at room temperature and 30% conversion after 9 h and 20 bar (2 × 10^4 hPa) of H_2, on the basis of the structural rigidity of the cyclic enone, which decreased the reactivity of the sterically hindered C=C double bond (Scheme 4.51). Other substrates such as cholestenone and isolongifolenone were also reduced [101].

Table 4.10 Hydrogenation of α,β-unsaturated aldehydes using Ir/H-β zeolite.

Catalyst	Substrate	Pressure (bar)^a	Time (h)	Conversion (%)	Chemoselectivity (%)^b
Ir/H-β (2%)	Cinnamaldehyde	30 (3 × 10^4)	18	71	82
Ru/H-β (2%)	Cinnamaldehyde	30 (3 × 10^4)	35	81	52
Ir/H-β (2%)	α-Me-cinnamaldehyde	30 (3 × 10^4)	3	72	89
Ir/H-β (2%)	Citral	24 (2.4 × 10^4)	2.25	>98	90

Conditions: substrate, 50 mg; solvent, ^{i}PrOH 6.5 g; catalyst 25 mg (calcined at 300 °C and reduced at 450 °C).
a Values in parentheses are pressures in hPa.
b Chemoselectivity to the C=O bond.

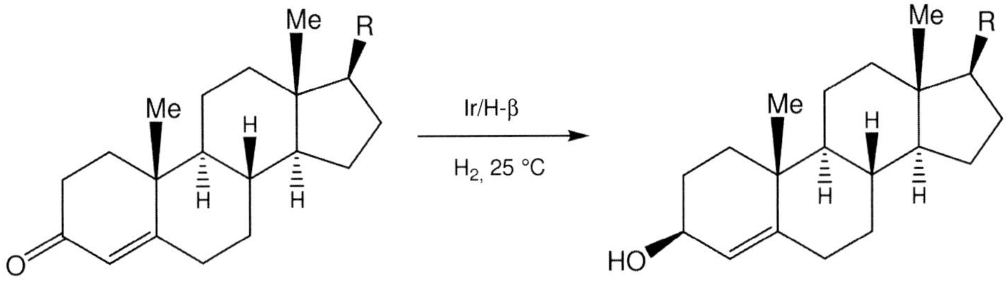

Scheme 4.51

Iridium nanoparticles generated in 1-*n*-butyl-3-methylimidazolium (BMI)-based ionic liquids were found to be excellent recyclable catalytic systems for the hydrogenation of a variety of substrates, including ketones such as simple ketones. The Ir nanoparticles were prepared by simple reduction of [Ir(cod)Cl]$_2$ dispersed in BMI·PF$_6$ at 75 °C under 4 atm of H$_2$. Benzaldehyde, cyclopentanone, methyl butanone and derivatives were hydrogenated with almost complete conversion, with TOFs ranging from 17 to 96 h^{-1} under solventless conditions (substrate : Ir ratio = 250, 75 °C, 4 atm H$_2$) [102].

Rhodium and iridium nanoparticles entrapped in aluminum oxyhydroxide nanofibers were shown by Park *et al.* to be suitable catalysts for the hydrogenation of arenes and ketones at room temperature, with hydrogen at ambient pressure [103]. Rhodium in aluminum oxyhydroxide [Rh/AlO(OH)] and iridium in aluminum oxyhydroxide [Ir/AlO(OH)], were simply prepared from readily available reagents such as RhCl$_3$ and IrCl$_3$ hydrates, 2-butanol and Al(O-*sec*-Bu) at 100 °C. Substrates such as cyclopentanone, 2-heptanone, ethyl pyruvate, acetone and 2,6-dimethyl-4-heptanone were reduced to the corresponding alcohols either in *n*-hexane at room temperature (maximum TOF 99 h^{-1} for ethyl pyruvate) or in solventless conditions at 75 °C using 4 atm of H$_2$ (maximum TOF 660 h^{-1} for acetone, 330 for 2-heptanone).

A few reports concerning various ways to support homogeneous catalysts for selective C=O hydrogenation have appeared in the literature. After Blum and coworkers [104] had reported the activity of polystyrene-bound Vaska's complex IrCl(CO)(PPh$_3$)$_2$ (6) in the C=C bond-transfer hydrogenation of α,β-unsaturated ketones using HCO$_2$H as reducing agent, scientists at Ciba-Geigy [69] successfully used a copolymer-bound Ir(I) catalyst prepared from [Ir(coe)$_2$Cl]$_2$ with optically active styrene derivatives. AIBN-initiated polymerization of *N*-(6-methyl-2-pyridylmethylene)-1-(4-vinylphenyl)ethylamine with styrene at 70 °C gave the copolymer ligand with [α]$_D$23 +0.44° and ligand capacity 0.499 mmol g^{-1}. The catalyst obtained was tested in the TH of 1-phenylbutanone (60 °C, 19 h) in ^{i}PrOH, yielding 12.4% 1-phenylbutanol with a 42.2% ee of the *R*-isomer.

The selective hydrogenation of α,β-unsaturated aldehydes into allylic alcohols catalyzed by supported aqueous-phase and organic-phase catalysts was demonstrated by Fache *et al.* [105]. Among the Ir-based catalysts tested, worthy of notice was the catalyst obtained by dissolving IrCl$_3$ in EtOH and treating it with PhP[Me$_2$Si(OMe)$_3$]$_2$ and Si(OMe)$_4$ at reflux in the presence of N[Me$_3$Si(OEt)$_3$]$_3$ and water. The gel formed was further treated with octan-1-ol, water and polyvinyl alcohol at reflux, to yield the catalyst as yellow spheres. All-*trans* retinal hydroquinone (RHQ) was hydrogenated to all-*trans* retinol hydroquinone (a precursor of vitamin A) at 25–50 °C under 100 bar (10×10^4 hPa) after 3.5 h, with a maximum selectivity of 97.5%.

Phosphonate-based supported catalysts for the hydrogenation of ketones under hydrogen pressure were obtained by Bujoli and coworkers [106]. Supported rhodium– and iridium–2,2′-bipyridine complexes were used for the hydrogenation of aromatic ketones under hydrogen pressure. The immobilization was achieved via functionalization of the 2,2′-bipyridine unit with two phosphonic acid moieties, and subsequent covalent grafting of the complex onto TiO_2 particles generated *in situ*. The resultant materials showed comparable activity to the homogeneous counterparts, which suggested that the major part of the catalytic sites were readily accessible. The catalyst was reused, and no significant metal leaching was observed. At 40 bar (4×10^4 hPa) of H_2, in $MeOH/H_2O$ at room temperature for 21 h, using 5% mol of catalyst, *p*-OMe-acetophenone was completely converted into *p*-OMe-2-phenylethanol.

Two styryl complexes **116a** and **116b** derived from aminopyridines such as DHPPEI (**81**; see Scheme 4.33, R = H, R_1 = H, R_2 = Me, R_3 = Ph) have been synthesized and used as homogeneous catalysts for the ATH of ketones (Scheme 4.52) [62]. The single-site catalysts were anchored via copolymerization of the monomers with 2-ethylhexyl methacrylate in the presence of di-2-butylbenzene as crosslinking agent, and resulted in more active catalysts (**117a** and **117b**) for the ATH of propiophenone. The reaction time could be decreased from 15 h (**116a**) to 6 h (**117a**), with an increase in ee-value from 46.6 to 83.5%; a similar situation was apparent in the case of **116b** and **117b** (from 10 to 5 h, from 51.9% ee to 84.3%, respectively).

An example of biocatalytic C=O bond reduction has also been reported in the literature. The asymmetric reduction of ketones via whole-cell bioconversions and TH was tested by van Leeuwen *et al.* as complementary approaches to asymmetric

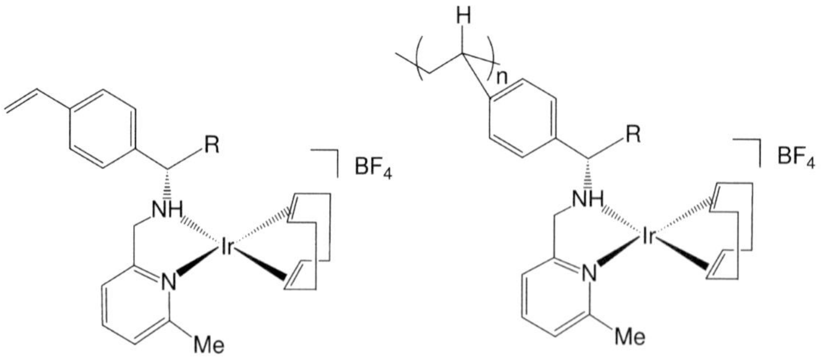

116a, R = Me
116b, R = Bz

117a, R = Me
117b, R = Bz

Scheme 4.52

synthetic reactions. Prochiral aryl and dialkyl ketones were enantioselectively reduced to the corresponding alcohols using whole cells of the white-rot fungus *Merulius tremellosus ono991* as a biocatalytic reduction system, with complexes such as ruthenium(II)-amino alcohol [Ru(*p*-cymene)Cl(N,O)]$^+$ (N,O = N,H-benzyl (1*R*,2*S*)-norephedrine) and iridium(I)-amino sulfides [Ir(cod)(N,S)]$^+$ (N,S = (1*R*,2*S*)-2-amino-1-phenyl-1-benzylthio-propane or ethane) [89] as metal catalysts in ATH reductions [107]. In the case of alkyl aryl ketones ArC(O)R, the choice of substrate did not affect the outcome of the reactions, and generally Ru catalysts gave the best results (acetophenone, 91% conversion, 95%-*R* ee, 2 h), whereas the biocatalytic system gave 61% yield and 95% ee (*S*) and the Ir complex yielded 82% of *R*-alcohol in 80% ee. For α,β-unsaturated ketones such as benzylidene acetone, the white rot fungus was unable to catalyze a conversion to the unsaturated alcohol, although unexpectedly a mixture of saturated ketone (55%) and hydroxy-substituted ketone derived from a Michael addition reaction (45%) was obtained. By contrast, both Ru (79% yield, 35% ee, 5 h) and Ir (40% yield, 25% ee, 5 h) were able to provide the corresponding chiral allylic alcohol. The biocatalytic approach must be preferred in the presence of substrates such as functionalized ketones (i.e. 2-chloroacetophenone and 3-chloropropiophenone), with conversions of up to 95% and ee-values of 88% versus no activity for the metal-catalyzed processes.

References

1 De Vries, J.-G. and Elsevier, C.J. (2007) *Handbook of Homogeneous Hydrogenation*, Wiley-VCH Verlag GmbH, Weinheim, Germany.
2 Handgraaf, J.-W., Reek, J.N.H. and Meijer, E.J. (2003) *Organometallics* **22**, 3150–7 and references therein.
3 Coffey, R.S. (1967) *Journal of the Chemical Society, Chemical Communications*, 923.
4 Coffey, R.S. (1971) Patent Appl. GB 19670710, to Imperial Chemical Industries, UK.
5 Strohmeier, W. and Steigerwald, H. (1977) *Journal of Organometallic Chemistry*, **129**, C43–6.
6 Kalinkin, M.I., Markosyan, S.M., Kursanov, D.N. and Parnes, Z.N. (1981) *Russian Chemical Bulletin*, **30**, 1759–61.
7 Strohmeier, W., Michel, M., Weigelt, L. (1980) *Zeitschrift für Naturforschung, Teil B: Anorganische Chemie, Organische Chemie*, **35B**, 648–50.
8 Yang, K.J. and Chin, C.S. (1987) *Inorganic Chemistry*, **26**, 2732–3.
9 Chin, C.S. and Park, S.C. (1988) *Bulletin of the Korean Chemical Society*, **9**, 260–1.
10 Chin, C.S., Park, S.C. and Shin, J.H. (1989) *Polyhedron*, **8**, 121–2.
11 Chin, C.S., Lee, B. and Park, S.C. (1990) *Journal of Organometallic Chemistry*, **393**, 131–5.
12 Chin, C.S. and Lee, B. (1991) *Journal of the Chemical Society – Dalton Transactions*, 1323–7.
13 Farnetti, E., Pesce, M., Kaspar, J., Spogliarich, R. and Graziani, M. (1986) *Journal of the Chemical Society, Chemical Communications*, 746–7.
14 Farnetti, E., Kaspar, J., Spogliarich, R. and Graziani, M. (1988) *Journal of the Chemical Society – Dalton Transactions*, 947–52.
15 Spogliarich, R., Farnetti, E. and Graziani, M. (1991) *Tetrahedron*, **47**, 1965–76.
16 Farnetti, E., Pesce, M., Kaspar, J., Spogliarich, R. and Graziani, M. (1987) *Journal of Molecular Catalysis*, **43**, 35–40.
17 Dahlenburg, L., Herbst, K. and Zahl, A. (2000) *Journal of Organometallic Chemistry*, **616**, 19–28.

18 Dahlenburg, L. and Goetz, R. (2004) *European Journal of Inorganic Chemistry*, 888–905.
19 Li, R.-X., Li, X.-J., Wong, N.-B., Tin, K.-C., Zhou, Z.-Y. and Mak, T.C.W. (2002) *Journal of Molecular Catalysis A–Chemical*, **178**, 181–90.
20 Gulyas, H., Benyei, A.C. and Bakos, J. (2004) *Inorganica Chimica Acta*, **357**, 3094–8.
21 (a) Solodar, J.A. (1975) *Chemical Technology*, **5**, 421–3;
(b) US Patent 3883580 (1975) to Monsanto Co. USA.
22 Cesarotti, E., Prati, L., Pallavicini, M., Villa, L., Spogliarich, R., Farnetti, E. and Graziani, M. (1990) *Journal of Molecular Catalysis*, **62**, L29–32.
23 Spogliarich, R., Farnetti, E., Kaspar, J., Graziani, M. and Cesarotti, E. (1989) *Journal of Molecular Catalysis*, **50**, 19–29.
24 Spogliarich, R., Vidotto, S., Farnetti, E., Graziani, M. and Verma Gulati, N. (1992) *Tetrahedron: Asymmetry*, **3**, 1001–2.
25 Zhang, X., Kumobayashi, H. and Takaya, H. (1994) *Tetrahedron: Asymmetry*, **5**, 1179–82.
26 Taketomi, T., Akutagawa, S., Kumobayashi, H., Mashima, K. and Takaya, H. (1992) European Patent Appl. EP479541A1. Takasago International Corp., Japan.
27 Yokosawa, S., Sayou, N., Matsumura, K. and Unrin, H. (1988) Japanese Patent JP10067789 A. Takasago Perfumery Co., Ltd., Japan.
28 Bulliard, M., Laboue, B. and Roussiasse, S. (2002) PCT Int. Appl. WO2002012253 A1, PPG-Sipsy, France.
29 Ferrand, A., Bruno, M., Tommasino, M.L. and Lemaire, M. (2002) *Tetrahedron: Asymmetry*, **13**, 1379–84.
30 Maillet, C., Praveen, T., Janvier, P., Minguet, S., Evain, M., Saluzzo, C., Tommasino, M.L. and Bujoli, B. (2002) *The Journal of Organic Chemistry*, **67**, 8191–6.
31 Dahlenburg, L., Menzel, R. and Heinemann, F.W. (2007) *European Journal of Inorganic Chemistry*, 4364–74.
32 Le Roux, E., Malacea, R., Manoury, E., Poli, R., Gonsalvi, L. and Peruzzini, M. (2007) *Advanced Synthesis Catalysis*, **349**, 309–13.
33 Ohkuma, T., Utsumi, N., Watanabe, M., Tsutsumi, K., Arai, N. and Murata, K. (2007) *Organic Letters*, **9**, 2565–7.
34 (a) Matsumura, K., Hashinguchi, S., Ikariya, T. and Noyori, R. (1997) *Journal of the American Chemical Society*, **119**, 8738–9;
(b) Fujii, A., Hashiguchi, S., Uematsu, N., Ikariya, T. and Noyori, R. (1996) *Journal of the American Chemical Society*, **118**, 2521–2;
(c) Hashiguchi, S., Fujii, A., Takehara, J., Ikariya, T. and Noyori, R. (1995) *Journal of the American Chemical Society*, **117**, 7562–3.
35 Jiang, Y., Jiang, Q. and Zhang, X. (1998) *Journal of the American Chemical Society*, **120**, 3817–18.
36 Zhao, J., Hesslink, H. and Hartwig, J.F. (2001) *Journal of the American Chemical Society*, **123**, 7220–7.
37 Benyei, A. and Joo, F. (1990) *Journal of Molecular Catalysis*, **58**, 151–63.
38 Spogliarich, R., Tencich, A., Kaspar, J. and Graziani, M. (1982) *Journal of Organometallic Chemistry*, **240**, 453–9.
39 Spogliarich, R., Mestroni, G. and Graziani, M. (1984) *Journal of Molecular Catalysis*, **22**, 309–11.
40 Kaspar, J., Graziani, R. and Spogliarich, M. (1982) *Journal of Organometallic Chemistry*, **231**, 71–8.
41 Vinzi, F., Zassinovich, G. and Mestroni, G. (1983) *Journal of Molecular Catalysis*, **18**, 359–66.
42 Visintin, M., Spogliarich, R., Kaspar, J. and Graziani, M. (1984) *Journal of Molecular Catalysis*, **24**, 277–80.
43 Visintin, M., Spogliarich, R., Kaspar, J. and Graziani, M. (1985) *Journal of Molecular Catalysis*, **32**, 349–51.
44 Farnetti, E., Nardin, G. and Graziani, M. (1988) *Journal of the Chemical Society, Chemical Communications*, 1264–5.
45 (a) Bianchini, C., Farnetti, E., Graziani, M., Nardini, G., Vacca, A. and Zanobini, F. (1990) *Journal of the American Chemical Society*, **112**, 9190–7; (b) Bianchini, C., Meli, A., Peruzzini, M., Vizza, F., Frediani, P. and Ramirez, J.A. (1988) *Organometallics*, **7**, 1704.
46 Bianchini, C., Peruzzini, M., Farnetti, E., Kaspar, J. and Graziani, M. (1995) *Journal of Organometallic Chemistry*, **488**, 91–7.

47 Sakaguchi, S., Yamaga, T. and Ishii, Y. (2001) *Journal of Organic Chemistry*, **66**, 4710–12.
48 Hillier, A.C., Lee, H.M., Stevens, E.D. and Nolan, S.P. (2001) *Organometallics*, **20**, 4246–52.
49 Albrecht, M., Miecznikowski, J.R., Samuel, A., Faller, J.W. and Crabtree, R.H. (2002) *Organometallics*, **21**, 3596–604.
50 (a) Miecznikowski, J.R. and Crabtree, R.H. (2004) *Organometallics*, **23**, 629–31; (b) Miecznikowski, J.R., Mata, J.A., Samuel, A. and Crabtree, R.H. (2003) Abstracts of Papers, 226th ACS National Meeting, New York, NY, United States, September 7–11, 2003, INOR-451.
51 Miecznikowski, J.R. and Crabtree, R.H. (2004) *Polyhedron*, **23**, 2857–72.
52 Gnanamgari, D., Moores, A., Rajaseelan, E. and Crabtree, R.H. (2007) *Organometallics*, **26**, 1226–30.
53 Hahn, F.E., Holtgrewe, C., Pape, T., Martin, M., Sola, E. and Oro, L.A. (2005) *Organometallics*, **24**, 2203–9.
54 Mas-Marza, E., Sanau, M. and Peris, E. (2005) *Inorganic Chemistry*, **44**, 9961–7.
55 Wang, C.-Y., Fu, C.-F., Liu, Y.-H., Peng, S.-M. and Liu, S.-T. (2007) *Inorganic Chemistry*, **46**, 5779–86.
56 Clarke, Z.E., Maragh, P.T., Dasgupta, T.P., Gusev, D.G., Lough, A.J. and Abdur-Rashid, K. (2006) *Organometallics*, **25**, 4113–17.
57 Choualeb, A., Lough, A.J. and Gusev, D.G. (2007) *Organometallics*, **26**, 5224–9.
58 Oakley, S.H., Coogan, M.P. and Arthur, R.J. (2007) *Organometallics*, **26**, 2285–90.
59 Ajjou, A.N. and Pinet, J.-L. (2004) *Journal of Molecular Catalysis A–Chemical*, **214**, 203–6.
60 Abura, T., Ogo, S., Watanabe, Y. and Fukuzumi, S. (2003) *Journal of the American Chemical Society*, **125**, 4149–54.
61 Wu, X., Liu, J., Li, X., Zanotti-Gerosa, A., Hancock, F., Vinci, D., Ruan, J. and Xiao, J. (2006) *Angewandte Chemie–International Edition*, **45**, 6718–22.
62 Zassinovich, G., Mestroni, G. and Gladiali, S. (1992) *Chemical Reviews*, **92**, 1051–69.
63 Spogliarich, R., Zassinovich, G., Kaspar, J. and Graziani, M. (1982) *Journal of Molecular Catalysis*, **16**, 359–61.
64 Spogliarich, R., Kaspar, J., Graziani, M. and Morandini, F. (1986) *Journal of Organometallic Chemistry*, **306**, 407–12.
65 Heil, B., Kvintovics, P., Tarszabo, L. and James, B.R. (1985) *Journal of Molecular Catalysis*, **33**, 71–5.
66 Zassinovich, G. and Mestroni, G. (1987) *Journal of Molecular Catalysis*, **42**, 81–90.
67 Zassinovich, G., Bettella, R., Mestroni, G., Bresciani-Pahor, N., Geremia, S. and Randaccio, L. (1989) *Journal of Organometallic Chemistry*, **370**, 187–202.
68 Kaschig, J. (1987) European Patent 0246194 A2, to Ciba-Geigy (Switzerland).
69 Kaschig, J. (1988) European Patent Appl. EP 251994 A1, to Ciba-Geigy A.-G., Switzerland.
70 De Martin, S., Zassinovich, G. and Mestroni, G. (1990) *Inorganica Chimica Acta*, **174**, 9–11.
71 Pavlov, V.A., Vinogradov, M.G., Starodubtseva, E.V., Chel'tsova, G.V., Ferapontov, V.A., Malyshev, O.R. and Heise, G.L. (2001) *Russian Chemical Bulletin*, **50**, 734–5.
72 Li, Y.-Y., Zhang, H., Chen, J.-S., Liao, X.-L., Dong, Z.-R. and Gao, J.-X. (2004) *Journal of Molecular Catalysis A–Chemical*, **218**, 153–6.
73 Chen, J.-S., Li, Y.-Y., Dong, Z.-R., Li, B.-Z. and Gao, J.-X. (2004) *Tetrahedron Letters*, **45**, 8415–18.
74 Dong, Z.-R., Li, Y.-Y., Chen, J.-S., Li, B.-Z., Xing, Y. and Gao, J.-X. (2005) *Organic Letters*, **7**, 1043–5.
75 Xing, Y., Chen, J.-S., Dong, Z.-R., Li, Y.-Y. and Gao, J.-X. (2006) *Tetrahedron Letters*, **47**, 4501–3.
76 Li, B.-Z., Chen, J.-S., Dong, Z.-R., Li, Y.-Y., Li, Q.-B. and Gao, J.-X. (2006) *Journal of Molecular Catalysis A–Chemical*, **258**, 113–17.
77 Chen, G., Xing, Y., Zhang, H. and Gao, J.-X. (2007) *Journal of Molecular Catalysis A–Chemical*, **273**, 284–8.
78 Mueller, D., Umbricht, G., Weber, B. and Pfaltz, A. (1991) *Helvetica Chimica Acta*, **74**, 232–40.
79 Debono, N., Besson, M., Pinel, C. and Djakovitch, L. (2004) *Tetrahedron Letters*, **45**, 2235–8.
80 Inoue, S.-I., Nomura, K., Hashiguchi, S., Noyori, R., Izawa, Y. (1997) *Chemistry Letters*, 957–8.

81 Mashima, K., Abe, T. and Tani, K. (1998) *Chemistry Letters*, 1199–200.

82 Murata, K., Ikariya, T. and Noyori, R. (1999) *The Journal of Organic Chemistry*, **64**, 2186–7.

83 Thorpe, T., Blacker, J., Brown, S.M., Bubert, C., Crosby, J., Fitzjohn, S., Muxworthy, J.P. and Williams, J.M.J. (2001) *Tetrahedron Letters*, **42**, 4041–3.

84 Wu, X., Vinci, D., Ikariya, T., Xiao, J. (2005) *Chemical Communications*, 4447–9.

85 Li, X., Blacker, J., Houson, I., Wu, X. and Xiao, J. (2006) *Synlett*, 1155–60.

86 Hartikka, A., Modin, S.A., Andersson, P.G. and Arvidsson, P.I. (2003) *Organic and Biomolecular Chemistry*, **1**, 2522–6.

87 He, W., Liu, P., Zhang, B.-L., Sun, X.-L. and Zhang, S.-Y. (2006) *Applied Organometallic Chemistry*, **20**, 328–34.

88 He, W., Zhang, B.-L., Jiang, R., Liu, P., Sun, X.-L. and Zhang, S.-Y. (2006) *Tetrahedron Letters*, **47**, 5367–70.

89 (a) Petra, D.G.I., Kamer, P.C.J., Spek, A.L., Schoemaker, H.E. and van Leeuwen, P.W.N.M. (2000) *The Journal of Organic Chemistry*, **65**, 3010–17;
(b) Petra, D.G.I., Kamer, P.C.J., van Leeuwen, P.W.N.M., de Vries, J.-G. and Schoemaker, H.E. (2001) PCT Int. Appl. WO 2001023088 A1. to DSM N.V, The Netherlands.

90 Zhang, X.-Q., Li, Y.-Y., Zhang, H. and Gao, J.-X. (2007) *Tetrahedron: Asymmetry*, **18**, 2049–54.

91 Maire, P., Breher, K., Schoenberg, H. and Gruetzmacher, H. (2005) *Organometallics*, **24**, 3207–18.

92 Carmona, D., Lahoz, F.J., Atencio, R., Oro, L.A., Lamata, M.P., Viguri, F., San Jose, E., Vega, C., Reyes, J., Joo, F. and Katho, A. (1999) *Chemistry – A European Journal*, **5**, 1544–64.

93 Aboulaala, K., Goux-Henry, C., Sinou, D., Safic, M. and Soufiaoui, M. (2005) *Journal of Molecular Catalysis A – Chemical*, **237**, 259–66.

94 Wu, X., Li, X., McConville, M., Saidi, O. and Xiao, J. (2006) *Journal of Molecular Catalysis A – Chemical*, **247**, 153–8.

95 (a) Nishibayashi, Y., Singh, J.D., Arikawa, Y., Uemura, S. and Hidai, M. (1997) *Journal of Organometallic Chemistry*, **531**, 13–18;
(b) Uemura, S. (1998) *Phosphorus, Sulfur, Silicon and the Related Elements*, **136–138**, 219–34.

96 Seo, H., Kim, B.Y., Lee, J.H., Park, H.-J., Son, S.U. and Chung, Y.K. (2003) *Organometallics*, **22**, 4783–91.

97 Khidekel, M.L., Bakhanova, E.N., Astakhova, A.S., Brikenshtein, Kh.A., Savchenko, V.I., Monakhova, I.S., Dorokhov, V.G. (1970) *Russian Chemical Bulletin*, **19**, 461–3.

98 Arimitsu, S., Yanagi, K., Saito, T., Takada, K., Tanaka, K. (1986) Japanese Patent JP 61230741 A. Japan Kokai, Tokyo.

99 Kaspar, J., Graziani, M., Escobar, G.P. and Trovarelli, A. (1992) *Journal of Molecular Catalysis*, **72**, 243–51.

100 Humblot, F., Candy, J.P., Santini, C., Didillon, B., Le Peltier, F. and Boitiaux, J.P. (1994) German Patent DE 4324222 A1. Institut Francais du Petrole, France.

101 De Bruyn, M., Coman, S., Bota, R., Parvulescu, V.I., De Vos, D.E. and Jacobs, P.A. (2003) *Angewandte Chemie - International Edition*, **42**, 5333–6.

102 Fonseca, G.S., Scholten, J.D. and Dupont, J. (2004) *Synlett*, 1525–8.

103 Park, I.S., Kwon, M.S., Kang, K.Y., Lee, J.S. and Park, J. (2007) *Advanced Synthesis Catalysis*, **349**, 2039–47.

104 Azran, J., Buchman, O. and Blum, J. (1981) *Tetrahedron Letters*, **22**, 1925–8.

105 Fache, E., Mercier, C., Pagnier, N., Despeyroux, B. and Panster, P. (1993) *Journal of Molecular Catalysis*, **79**, 117–31.

106 Maillet, C., Janvier, P., Pipelier, M., Praveen, T., Andres, Y. and Bujoli, B. (2001) *Chemistry of Materials*, **13**, 2879–84.

107 Hage, A., Petra, D.G.I., Field, J.A., Schipper, D., Wijnberg, J.B.P.A., Kamer, P.C.J., Reek, J.N.H., van Leeuwen, P.W.N.M., Wever, R. and Schoemaker, H.E. (2001) *Tetrahedron: Asymmetry*, **12**, 1025–34.

5
Catalytic Activity of Cp* Iridium Complexes in Hydrogen Transfer Reactions

Ken-ichi Fujita and Ryohei Yamaguchi

5.1
Introduction

Since the first report by Maitlis and coworkers in 1969 of the synthesis of a trivalent iridium complex bearing η^5-pentamethylcyclopentadienyl (Cp*), [Cp*IrCl$_2$]$_2$ (**1**), [1, 2], a large number of Cp*Ir complexes have been synthesized, starting with **1** [3, 4]. From the time of the first synthesis until recently, the chemistry of Cp*Ir complexes has mainly focused on stoichiometric reactions, and their ability towards C—H bond activation in hydrocarbon molecules was revealed by Bergman *et al* [5–7] and other research groups [8–10]. In contrast, the catalytic chemistry of Cp*Ir complexes remained relatively unexplored until the late 1990s and, indeed, only a few Cp*Ir complex-catalyzed reactions other than simple hydrogenation were reported before 2000 [11–13]. This contrasted with the catalytic chemistry of low-valent iridium complexes, such as [Ir(cod)(PCy$_3$)(py)]PF$_6$ and [Ir(cod)Cl]$_2$, all of which have made significant contributions in the area of catalytic synthetic organic chemistry [14–18]. However, since the recent discovery of the excellent catalytic performance of Cp*Ir complexes in hydrogen transfer reactions, the catalytic chemistry of these compounds has been attracting much interest.

In an earlier report, Maitlis *et al.* showed that **1** could be easily converted into a hydrido complex [Cp*IrHCl]$_2$ (**2**) under ambient conditions by treatment with alcohol and a weak base (Scheme 5.1) [19], probably accompanied by the formation of carbonyl compounds. This fact means that the hydrogen atom in an alcohol can be rapidly transferred to the iridium center in the form of a hydride; but then, if the hydride on the iridium could be re-transferred to another hydrogen acceptor, a new catalytic system using alcohols as substrates might be realized. In fact, a wide variety of Cp*Ir complex-catalyzed hydrogen transfer systems using alcohols as substrates, and based on the above hypothesis, have been reported to date [20].

The aim of this chapter is to review the catalytic hydrogen transfer chemistry wherein Cp*Ir complexes play an important role as catalysts. To date, a number

Iridium Complexes in Organic Synthesis. Edited by Luis A. Oro and Carmen Claver
Copyright © 2009 WILEY-VCH Verlag GmbH & Co. KGaA, Weinheim
ISBN: 978-3-527-31996-1

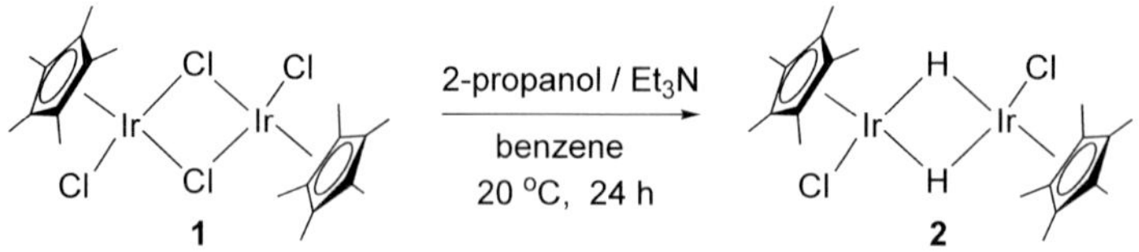

Scheme 5.1

of insightful reviews of transition-metal-catalyzed hydrogen transfer reactions have presented [21–26].

5.2
Hydrogen Transfer Oxidation of Alcohols (Oppenauer-Type Oxidation)

Hydrogen transfer reactions from an alcohol to a ketone (typically acetone) to produce a carbonyl compound (the so-called 'Oppenauer-type oxidation') can be performed under mild and low-toxicity conditions, and with high selectivity when compared to conventional methods for oxidation using chromium and manganese reagents. While the traditional Oppenauer oxidation using aluminum alkoxide is accompanied by various side reactions, several transition-metal-catalyzed Oppenauer-type oxidations have been reported recently [27–29]. However, most of these are limited to the oxidation of secondary alcohols to ketones.

Fujita and Yamaguchi et al. were the first to report that a Cp*Ir complex is an effective catalyst for the Oppenauer-type oxidation of both primary and secondary alcohols, affording aldehydes and ketones [30]. Initially, the group examined the oxidation of benzyl alcohol to benzaldehyde in acetone, using a variety of transition-metal catalysts. Among the catalysts examined, a [Cp*IrCl$_2$]$_2$ (1)/K$_2$CO$_3$ system gave the best results, affording benzaldehyde as a single product without the formation of benzoic acid or other products (Table 5.1; entry 1). Representative results of the oxidation of various primary and secondary alcohols by this catalytic system are summarized in Table 5.1. In addition to benzylic primary alcohols, aliphatic primary alcohols were oxidized to the corresponding aldehydes (entries 1–6), while secondary alcohols were much more easily oxidized to give ketones, using smaller amounts of catalyst 1 (0.50 mol% Ir) and acetone (entries 7–9).

Although the precise mechanism has not yet been clarified, a possible mechanism is shown in Scheme 5.2. First, the iridium alkoxide 3 is produced from 1 and an alcohol, this step being stimulated by a base (K$_2$CO$_3$). A β-hydride elimination of 3 then yields a carbonyl product and the iridium hydride 4. The insertion of acetone into the iridium–hydride bond in 4, giving metal isopropoxide 5, is followed by exchange of the alkoxy moiety to regenerate 3.

Hiroi et al. also reported the Cp*Ir complex-catalyzed Oppenauer-type oxidation of primary alcohols in acetone and butanone [31]. These authors prepared a novel Ir-ligand bifunctional catalyst 6 having an amido-alkoxo ligand, the

Table 5.1 Oxidation of primary and secondary alcohols to aldehydes and ketones catalyzed by [Cp*IrCl$_2$]$_2$ (**1**).[a]

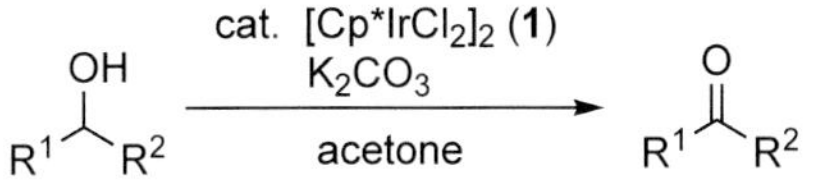

$$\underset{R^1\;\;R^2}{OH} \xrightarrow[\text{acetone}]{\text{cat. [Cp*IrCl}_2]_2\text{ (1)}\atop K_2CO_3} \underset{R^1\;\;R^2}{O}$$

Entry	Alcohol	Conversion (%)[b]	Yield (%)[b,c]
1	PhCH$_2$OH	87	87 (74)
2	4-MeOC$_6$H$_4$CH$_2$OH	100	99 (90)
3	2-MeOC$_6$H$_4$CH$_2$OH	70	67 (63)
4	4-ClC$_6$H$_4$CH$_2$OH	72	70 (61)
5[d]	C$_8$H$_{17}$OH	59	57 (44)
6[d]	Cyclo-C$_6$H$_{11}$CH$_2$OH	51	47
7[e]	PhCH(OH)CH$_3$	100	100 (94)
8[d,e]	CH$_3$CH(OH)C$_6$H$_{13}$	89	88 (77)
9[e]	Cyclopentanol	100	100

a At room temperature for 6 h with alcohol (1.0 mmol), **1** (2.0 mol% Ir), and K$_2$CO$_3$ (10 mol%) in acetone (30 ml).
b Determined by gas chromatography (GC).
c Values in parentheses are isolated yields.
d At reflux temperature.
e At room temperature for 6 h with alcohol (2.0 mmol), **1** (0.5 mol% Ir) and K$_2$CO$_3$ (10 mol%) in acetone (2 ml).

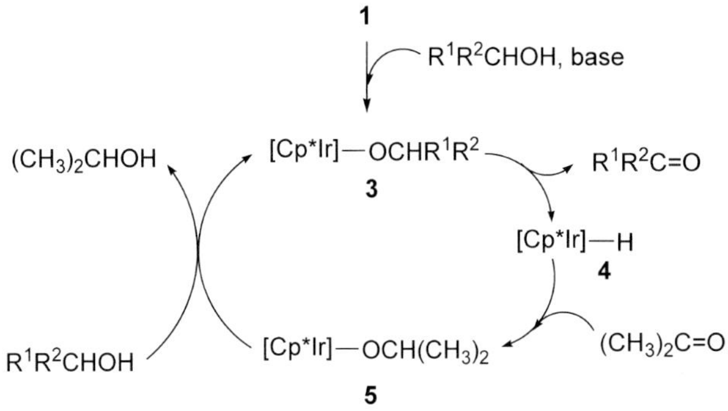

Scheme 5.2

neutral character of which would be effective for the oxidation of primary alcohols without an accompanying aldol reaction. The best result was obtained by performing the reaction in butanone at 80 °C under a high-dilution condition (0.08 M solution). Various benzylic primary alcohols were oxidized to aldehydes in high yields, although the oxidation of aliphatic alcohol resulted in a lower yield (Scheme 5.3).

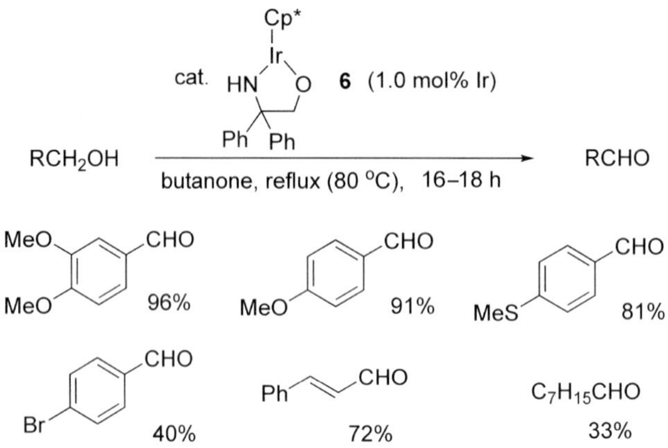

Scheme 5.3

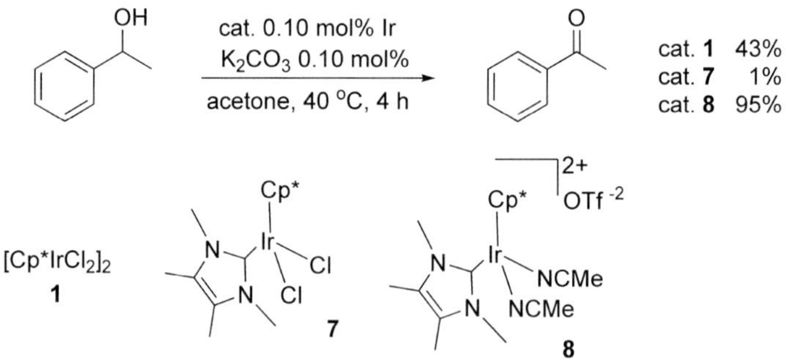

Scheme 5.4

New Cp*Ir complexes bearing N-heterocyclic carbene (NHC) ligands have been synthesized, and have been shown to possess a remarkably high catalytic activity for Oppenauer-type oxidations [32, 33]. The complex **7** was prepared by reaction of [Cp*IrCl$_2$]$_2$ (**1**) with an imidazol-2-ylidene, and further converted to the dicationic complex **8**. Oppenauer-type oxidation reactions of 1-phenylethanol were examined using newly synthesized Cp*Ir NHC catalysts **7** and **8**, and conventional catalyst **1** under the catalyst loading of 0.10 mol% Ir. As shown in Scheme 5.4, complex **1** showed moderate catalytic activity, while the neutral catalyst **7** showed almost no catalytic activity. In contrast, the dicationic catalyst **8** exhibited an extremely high catalytic activity; in fact, when the concentration of **8** was reduced to 0.025 mol% Ir, an 80% yield was obtained and the turnover number (TON) reached up to 3200 after an 8 h reaction time.

Results for the oxidation of several secondary alcohols using catalyst **8** are summarized in Scheme 5.5. The oxidation of benzylic, aliphatic and cyclic secondary

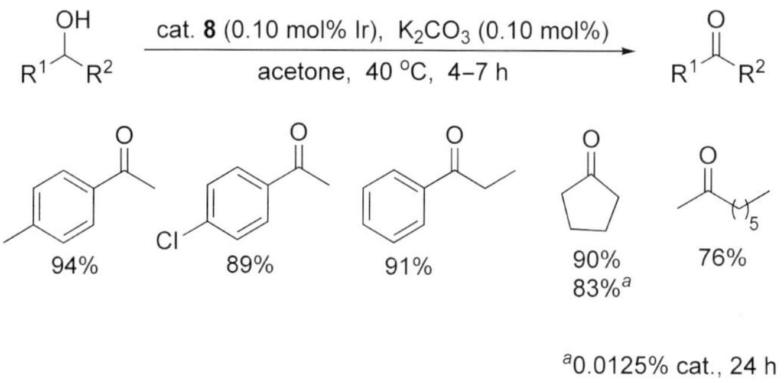

5.2 Hydrogen Transfer Oxidation of Alcohols (Oppenauer-Type Oxidation) | 111

Scheme 5.5

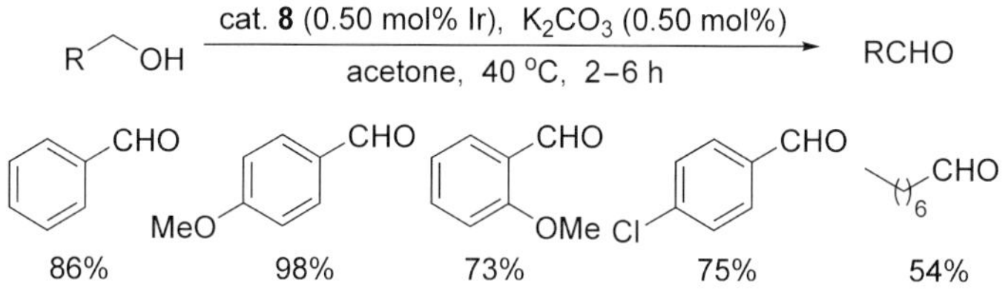

$$R^1R^2CHOH \xrightarrow[\text{acetone, 40 °C, 4–7 h}]{\text{cat. 8 (0.10 mol\% Ir), } K_2CO_3 \text{ (0.10 mol\%)}} R^1R^2C=O$$

- 4′-methylacetophenone: 94%
- 4′-chloroacetophenone: 89%
- propiophenone: 91%
- cyclopentanone: 90%, 83%[a]
- 2-octanone: 76%

[a] 0.0125% cat., 24 h

Scheme 5.6

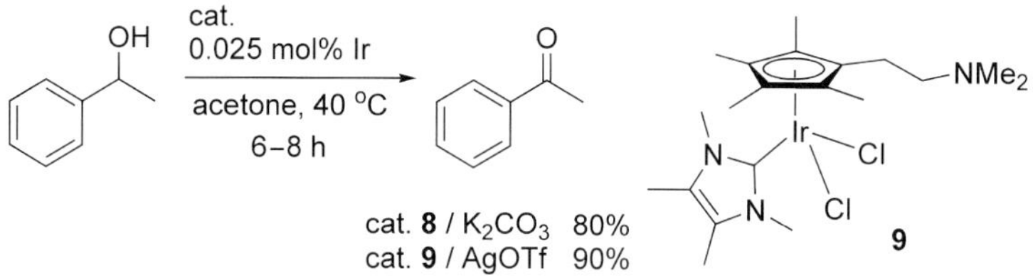

$$R\text{-}CH_2OH \xrightarrow[\text{acetone, 40 °C, 2–6 h}]{\text{cat. 8 (0.50 mol\% Ir), } K_2CO_3 \text{ (0.50 mol\%)}} RCHO$$

- PhCHO: 86%
- 4-MeO-C6H4-CHO: 98%
- 2-MeO-C6H4-CHO: 73%
- 4-Cl-C6H4-CHO: 75%
- C6H13CH(CH3)CHO: 54%

Scheme 5.7

1-phenylethanol $\xrightarrow[\substack{\text{acetone, 40 °C} \\ \text{6–8 h}}]{\text{cat. 0.025 mol\% Ir}}$ acetophenone

cat. **8** / K_2CO_3 80%
cat. **9** / AgOTf 90%

Complex **9**: Cp* with NMe2-tethered Ir-NHC dichloride

alcohols gave the corresponding ketones in good to high yields. In the case of cyclopentanol, the TON reached up to 6640 after 24 h of reaction, thus demonstrating the cationic NHC complex **8** to be the most effective catalyst in homogeneous Oppenauer-type oxidation systems.

Results for the oxidation of primary alcohols using catalyst **8** are summarized in Scheme 5.6. While larger quantities of the catalyst (0.50 mol% Ir) and acetone were required, the oxidation of primary alcohols proceeded selectively in good to high yields.

Yamaguchi *et al.* also reported the use of another new Cp*Ir NHC complex **9** bearing basic 2-(dimethylamino)ethyl group in a Cp* ring as the catalytic precursor in Oppenauer-type oxidation (Scheme 5.7) [34]. Owing to the basic amino

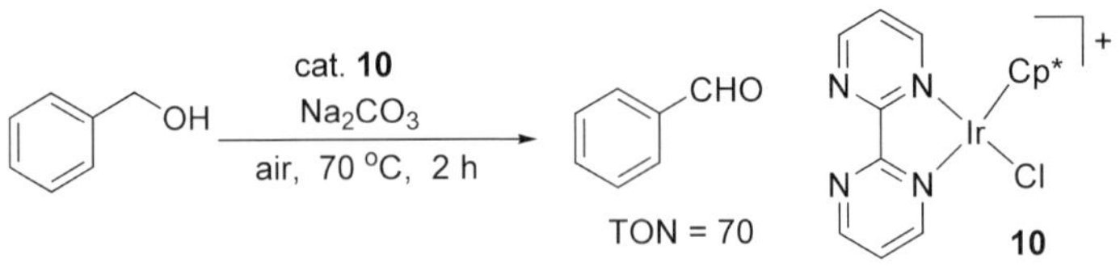

Scheme 5.8

moiety in the ligand, the reaction using **9** could be performed in the absence of an additional base. Compared to the dicationic catalyst **8** and conventional catalyst **1**, the catalytic system composed of **9**/AgOTf without base exhibited a higher activity for the oxidation of 1-phenylethanol. A variety of primary and secondary alcohols were converted to the corresponding carbonyl products, with high TONs.

Gabrielsson *et al.* reported the aerobic oxidation of alcohols catalyzed by a cationic Cp*Ir complexes bearing diamine ligands such as bipyrimidine **10** (Scheme 5.8) [35], the mechanism of which is closely related to the Oppenauer-type oxidation mentioned above. In this reaction, the deprotonation of Ir^{III} hydrido species to afford Ir^{I} species, and the reoxidation of Ir^{I} to Ir^{III} by O_2, are crucial.

5.3
Transfer Hydrogenation of Unsaturated Compounds

5.3.1
Transfer Hydrogenation of Quinolines

The Cp*Ir complex has been found to be a good catalyst for the transfer hydrogenation of nitrogen heteroaromatic compounds using alcohol as hydrogen source [36]. Thus, the Cp*Ir-catalyzed transfer hydrogenation of a variety of quinoline derivatives with 2-propanol proceeds in a regioselective and chemoselective manner to afford various 1,2,3,4-tetrahydroquinoline derivatives, which have themselves attracted considerable attention owing to their importance as synthetic intermediates for pharmaceuticals, agrochemicals and dyes.

The results for the Cp*Ir-catalyzed transfer hydrogenation of a series of quinolines are summarized in Table 5.2. In the presence of catalytic amounts of [Cp*IrCl$_2$]$_2$ (**1**) (1.0 mol% Ir) and HClO$_4$ (10 mol%), the reaction of quinoline in 2-propanol:H$_2$O (95:5) under reflux for 17 h gave 1,2,3,4-tetrahydroquinoline in 93% yield (Table 5.2; entry 1). Other products, such as 1,2-dihydroquinoline, 5,6,7,8-tetrahydroquinoline and decahydroquinoline were not detected, indicating a high regioselectivity of the transfer hydrogenation. The reduction of quinolines bearing electron-withdrawing (NO$_2$, Cl, Br, CO$_2$H) and electron-donating (OMe) substituents proceeded chemoselectively in good to excellent yields (entries 5–10).

Table 5.2 Transfer hydrogenation of quinolines catalyzed by [Cp*IrCl$_2$]$_2$ (**1**).[a]

Entry	R	Catalyst (mol% Ir)	HClO$_4$ (mol%)	Yield (%)[b]
1	H	1.0	10	(93)
2	2-Me	4.0	0	82
3	3-Me	2.0	10	79
4[c]	8-Me	2.0	10	82
5[d]	5-NO$_2$	4.0	10	72
6	6-NO$_2$	2.0	0	94
7	6-Cl	2.0	10	78
8	6-Br	2.0	10	70
9	6-CO$_2$H	4.0	10	64
10	6-OMe	1.9	11	79

a Quinolines (2.0 mmol), **1**, and 60% HClO$_4$(aq.) in solvent (2-propanol 9.5 ml and H$_2$O 0.5 ml) under reflux for 17 h.
b Isolated yield. The values in parentheses are GC yield.
c Reaction for 24 h.
d Reaction for 64 h.

5.3.2
Transfer Hydrogenation of Ketones and Imines

Peris et al. reported the synthesis of new Cp*Ir complexes having N-alkenyl imidazole-2-ylidene ligands, and their application to the catalytic transfer hydrogenation of ketones and imines [37]. The group prepared complex **11** by reaction of [Cp*IrCl$_2$]$_2$ (**1**) with an N-butenylimidazolium salt by the silver-mediated carbene transfer method. The catalytic performance of **11** for transfer hydrogenation was also examined (Table 5.3), whereby aliphatic and aromatic ketones, as well as imines, were easily hydrogenated using 2-propanol as a hydrogen donor in the presence of catalyst **11** (1.0 mol% Ir).

5.4
Asymmetric Synthesis Based on Hydrogen Transfer

5.4.1
Asymmetric Transfer Hydrogenation of Ketones

New Cp*Ir and Cp*Rh complexes having a chiral diamine ligand, which are isoelectric with Noyori's chiral (arene)Ru complex [38], have been synthesized by

Table 5.3 Transfer hydrogenation of ketones and imines catalyzed by **11**.[a]

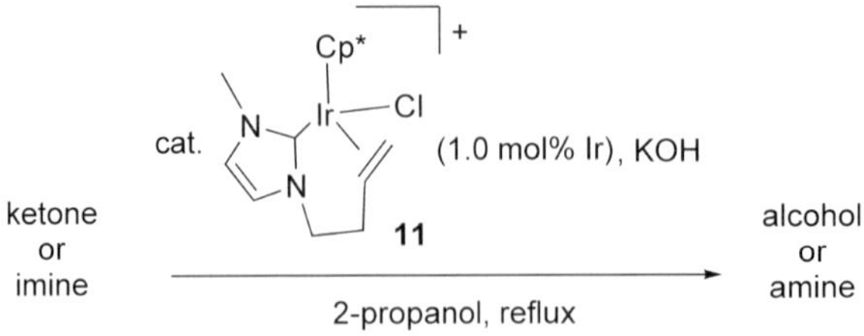

ketone or imine + Cp*Ir(NHC)(Cl) (cat. **11**, 1.0 mol% Ir), KOH, 2-propanol, reflux → alcohol or amine

Entry	Substrate	Product	Time (h)	Yield (%)[b]
1	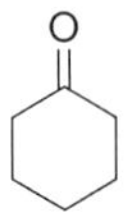 cyclohexanone	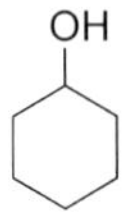 cyclohexanol	0.5	>99
2	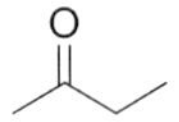 2-butanone	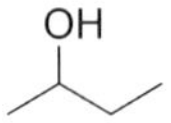 2-butanol	5	>99
3	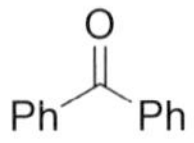 Ph(C=O)Ph	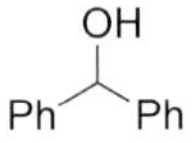 Ph-CH(OH)-Ph	5	>99
4	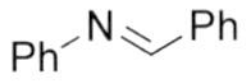 Ph-N=CH-Ph	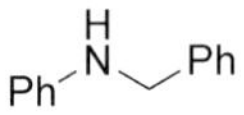 Ph-NH-CH$_2$-Ph	9	>99

a Ketone or imine (1.0 mmol), KOH (0.50 mmol), and **11** in 2-propanol (10 ml) under reflux.
b Determined by ^{1}H NMR.

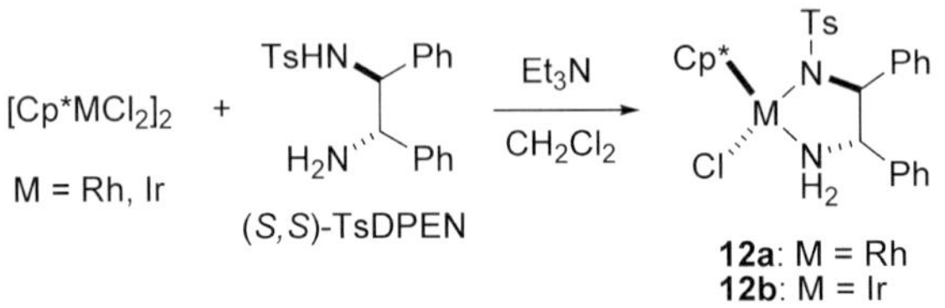

[Cp*MCl$_2$]$_2$ + TsHN-CH(Ph)-CH(Ph)-NH$_2$ (S,S)-TsDPEN — Et$_3$N, CH$_2$Cl$_2$ → Cp*M(Cl)(TsDPEN) complex
M = Rh, Ir
12a: M = Rh
12b: M = Ir

Scheme 5.9

Mashima and Tani *et al.*, and employed in the asymmetric transfer hydrogenation of aromatic ketones [39, 40].

The treatment of [Cp*MCl$_2$]$_2$ (M = Rh and Ir) with (S,S)-TsDPEN gave chiral Cp*Rh and Cp*Ir complexes (**12a** and **12b**; Scheme 5.9). An asymmetric transfer hydrogenation of aromatic ketones using complex **12** was carried out in 2-propanol in the presence of aqueous KOH (1 equiv.); the results obtained are summarized in Table 5.4. In all of the reactions, the (S)-alcohols were obtained with more than 80% enantiomeric excess (ee) and in moderate to excellent yields. The rhodium catalyst **12a** was shown to be considerably more active than the iridium catalyst

Table 5.4 Asymmetric transfer hydrogenation of prochiral ketones catalyzed by **12** giving (S)-alcohols.[a]

Entry	Catalyst	Ketone	KOH[b]	Yield (%)	ee (%)
1	12a	Acetophenone	2	80	90
2	12a		1	95	84
3	12b		2	58	90
4	12b		1	89	88
5	12a	2-Acetonaphthone	1	82	85
6	12b		1	67	81
7	12a	1-Tetralone	1	79	97
8	12a		2	28	95
9	12b		1	68	96
10	12b		2	27	90
11	12a	1-Indanone	1	47	99
12	12b		1	41	91

a 0.1 M solution of ketone, **12** (1.0 mol% metal) in 2-propanol at room temperature for 48 h.
b Molar ratio of KOH : [**12**].

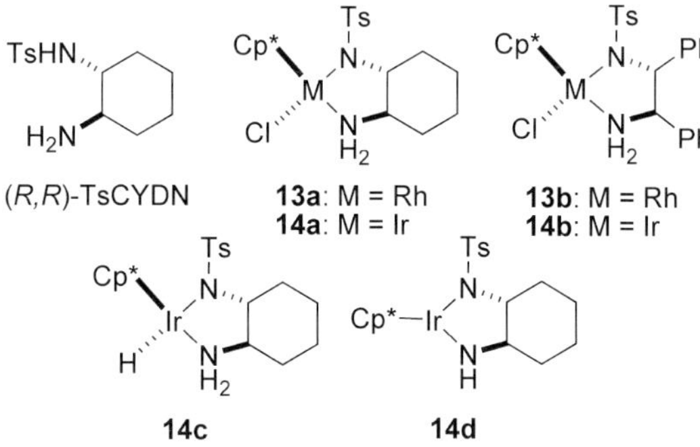

Scheme 5.10

12b under the same reaction conditions, whereas the catalytic activity and enantioselectivity of **12** were less than those of the chiral ruthenium catalyst [41].

Ikariya and Noyori *et al.* also reported the synthesis of new chiral Cp*Rh and Cp*Ir complexes (**13** and **14**) bearing chiral diamine ligands [(R,R)-TsCYDN and (R,R)-TsDPEN] (Scheme 5.10); these are isoelectronic with the chiral Ru complex mentioned above, and may be used as effective catalysts in the asymmetric transfer hydrogenation of aromatic ketones [42]. The Cp*Ir hydride complex [Cp*IrH(R,R)-Tscydn] (**14c**) and 5-coordinated amide complex (**14d**), both of which would have an important role as catalytic intermediates, were also successfully prepared.

Table 5.5 Asymmetric transfer hydrogenation of aromatic ketones catalyzed by preformed chiral catalysts and KO^{t}Bu system in 2-propanol.[a]

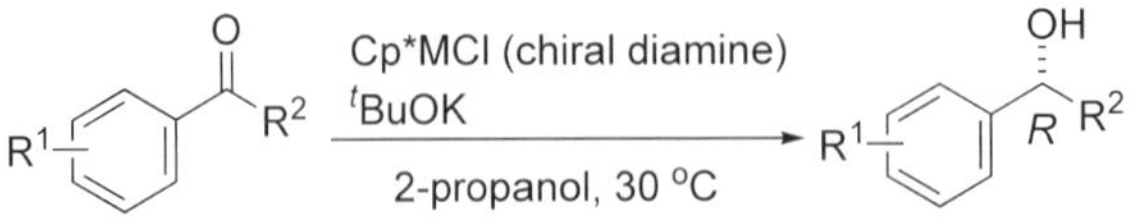

R^1–C$_6$H$_4$–CO–R^2 + Cp*MCl (chiral diamine) / ^{t}BuOK, 2-propanol, 30 °C → R^1–C$_6$H$_4$–CH(OH)–R^2 (R)

Entry	Catalyst	Ketone		Time (h)	Conversion (%)[b]	ee (%)[c]
		R^1	R^2			
1	13a	H	Me	12	85	97
2	13b	H	Me	12	14	90
3	14a	H	Me	12	36	96
4	13a	m-CF$_3$	Me	12	>99	97
5	14a	m-CF$_3$	Me	24	99	94
6	13a	o-CF$_3$	Me	12	>99	96
7	13a	o-F	Me	12	97	91
8	13a	o-Me	Me	24	20	94
9	13a	p-Et	Me	24	58	>99
10	13a	p-OMe	Me	24	22	>99
11	13a	1-Tetralone		24	53	95[d]
12	13a	1-Indanone		24	43	97[d]

a At 30 °C using a 0.1 M solution of the ketone in 2-propanol. Ketone:cat.:^{t}BuOK = 200:1:1.2.
b Determined by GC analysis.
c Determined by chiral GC analysis.
d Determined by HPLC analysis.

Results for the asymmetric transfer hydrogenation of various aromatic ketones using these complexes as catalysts are summarized in Table 5.5. The TsCYDN complexes provide higher reactivity than the TsDPEN complexes (entries 1 and 2), while the Rh complex with TsCYDN is a better catalyst in terms of reaction rate and enantioselectivity than the similar Ir complex (entries 1 and 3). The analogous preformed chiral Ru complex with the TsDPEN ligand has a somewhat higher reactivity, with a comparable to lower enantioselectivity (92% yield, 94% ee) for the reduction of acetophenone. The hydride complex **14c** and amide complex **14d** catalyzed the asymmetric transfer hydrogenation with the same ee-value as shown above, which suggests that this asymmetric reduction takes place by the action of **14c** or **14d** as the catalyst or intermediate.

Analogous water-soluble Cp*Rh and Cp*Ir complexes were prepared by Williams *et al.*, and used in the asymmetric transfer hydrogenation of aromatic ketones under aqueous conditions [43]. These catalyst complexes contain water-soluble chiral diamine ligands (Scheme 5.11), and were prepared *in situ* by reacting [Cp*MCl$_2$]$_2$ (M = Rh, Ir) with ligands **15a** or **15b** in the presence of a base, and used immediately. The results of the asymmetric transfer hydrogenation of

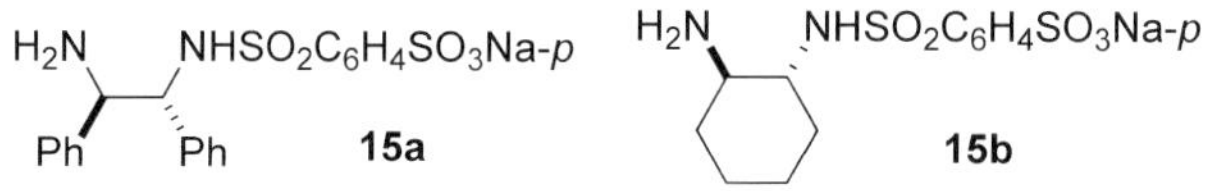

5.4 Asymmetric Synthesis Based on Hydrogen Transfer

Scheme 5.11

Table 5.6 Asymmetric transfer hydrogenation of aromatic ketones catalyzed by polar Cp*Rh and Cp*Ir complexes with polar ligands **15**.[a]

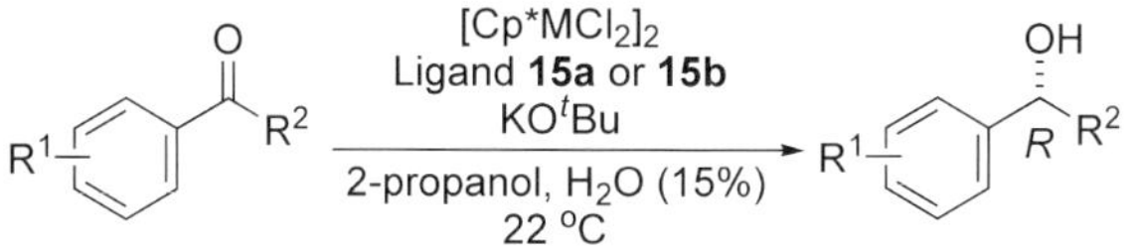

Entry	M	Ligand	Ketone		Time (h)	Conversion (%)[b]	ee (%)[b]
			R^1	R^2			
1	Rh	15a	H	Me	24	92	97
2	Rh	15b	H	Me	18	94	95
3	Rh	15b	m-CF$_3$	Me	4	99	94
4	Rh	15b	p-OMe	Me	42	65	95
5	Ir	15a	H	Me	140	90	82
6	Ir	15b	H	Me	26	88	96
7	Ir	15b	m-CF$_3$	Me	4	98	93
8	Ir	15b	m-F	Me	26	99	94
9	Ir	15b	p-OMe	Me	141	80	95
10	Ir	15b	p-Br	Me	20	99	95
11	Ir	15b	1-Indanone		45	55	97

a 0.2 M solution of 2-propanol containing water (15%) using ketone:metal:ligand:KO^{t}Bu = 200:2:8:20.
b Determined by GC using Supelco beta-dex.

aromatic ketones are summarized in Table 5.6. The Cp*Rh complexes were superior catalysts in terms of rate and enantioselectivity, while the ligand **15b** systems provided higher reactivities (entries 1, 2 and 5, 6).

The asymmetric transfer hydrogenation of acetophenone using the Cp*Ir complex with flexible chiral ligands (**16**) has been also reported by Furegati et al. (Scheme 5.12) [44]. Among the metal complexes examined, [CpIrHCl]$_2$ (**2**) showed the best activity, and the reaction was carried out using various ligands with different concentrations of a base (^{i}PrOK) in a 0.1 M solution in 2-propanol. While the reaction rate was generally high compared to the catalytic system using the rigid ligand TsCYDN, the enantioselectivity was moderate. It should be noted that this catalytic system showed an unusual base-dependent enantioselectivity; using a lesser amount of base than the chiral catalyst resulted in the formation of (R)-1-phenylethanol, while using an excess amount gave the (S)-enantiomer.

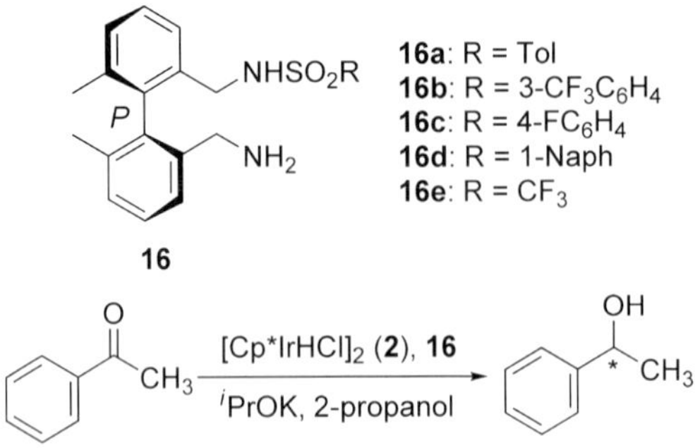

Scheme 5.12

16
- **16a**: R = Tol
- **16b**: R = 3-CF$_3$C$_6$H$_4$
- **16c**: R = 4-FC$_6$H$_4$
- **16d**: R = 1-Naph
- **16e**: R = CF$_3$

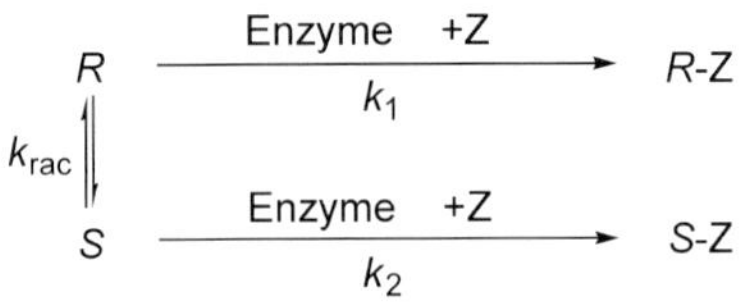

Acetophenone + CH$_3$ → [Cp*IrHCl]$_2$ (**2**), **16**, ^{i}PrOK, 2-propanol → 1-phenylethanol

acetophenone : **16** : **2** : ^{i}PrOK
= 100 : 1.1 : 0.5 : 0.6, 1, or 10

Scheme 5.13

R ⇌ S (k$_{rac}$)
R + Enzyme + Z → R-Z (k$_1$)
S + Enzyme + Z → S-Z (k$_2$)

5.4.2
Dynamic Kinetic Resolution

Dynamic kinetic resolution (DKR) is an attractive protocol for the production of enantiopure compounds from racemic mixtures [45]. The concept of DKR is illustrated in Scheme 5.13. In many cases, DKRs are accomplished by the combination of enzymatic resolution and transition-metal-catalyzed racemization based on hydrogen transfer. Thus, the use of Cp*Ir complexes as catalysts for racemization in DKR can be anticipated.

Page et al. reported the DKR system for secondary amines using a Cp*Ir complex coupled with an enzyme [46]. Since, initially, it was found that the racemization reaction of (S)-1-methyl-1,2,3,4-tetrahydroisoquinoline (**17**) proceeded successfully when using [Cp*IrI$_2$]$_2$ (**18**) as a catalyst (Scheme 5.14), Page's group decided to investigate whether the racemization and the enzymatic resolution could be combined. Consequently, when the racemic amine **19** was reacted with carbonate **20** in the presence of **18** (0.4 mol% Ir) and *Candida rugosa* lipase at 40 °C for 23 h, a 90% conversion to carbamate **21** resulted, with 96% ee.

Saunders et al. reported the DKR system for secondary alcohols using Cp*Ir complexes bearing a NHC ligand as racemization catalysts [47]. As shown in Scheme 5.15, the reaction of racemic 1-phenylethanol with isopropenyl acetate in the presence of catalyst **22** (0.1 mol% Ir) and Novozyme 435 at 70 °C for 8 h gave

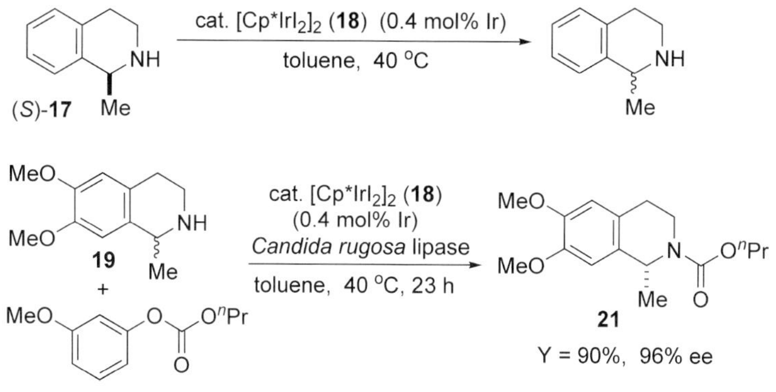

Scheme 5.14

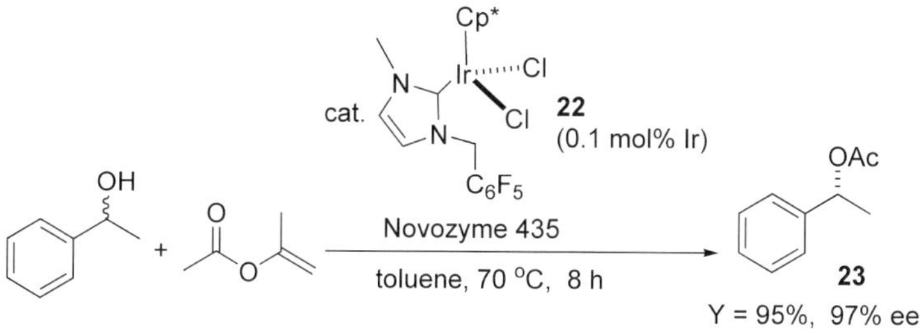

Scheme 5.15

ester **23** in 95% yield with 97% ee. By using this system, DKRs of racemic 3,3-dimethyl-2-butanol and 1-cyclohexylethanol were also achieved.

5.5
Hydrogen Transfer Reactions in Aqueous Media

Ogo, Makihara and Watanabe et al. reported the synthesis of water-soluble Cp*Ir complexes and their pH-dependent catalytic activities for the transfer hydrogenation of carbonyl compounds [48]. Initially, the water-soluble complex $[Cp^*Ir(H_2O)_3]^{2+}$ (**24**) was quantitatively synthesized by the reaction of $[Cp^*IrCl_2]_2$ (**1**) with Ag_2SO_4 at pH 2.3 in water. Because the active catalyst, a dinuclear μ-hydride complex $[(Cp^*Ir)_2(\mu\text{-H})(\mu\text{-OH})(\mu\text{-HCOO})]^+$ (**25**) was generated from the reaction of **24** with HCOONa at pH 3.2 in the highest yield, the rate of transfer hydrogenation showed a sharp maximum at pH 3.2. The results of the transfer hydrogenation of various carbonyl compounds with HCOONa using **24** as a precatalyst at pH 3.2 are summarized in Table 5.7. Aldehydes were reduced faster than ketones; although the aldehyde group of glyoxylic acid was reduced very easily, the ketone group of pyruvic acid was not reduced at all.

Table 5.7 Transfer hydrogenation of carbonyl compounds with [Cp*Ir(H$_2$O)$_3$]$^{2+}$ (**24**) and HCOONa in water at pH 3.2.^{a}

Entry	Substrate	HCOONa (equiv.)	Product	TOF^b
1	propanal	1	1-propanol	0.6
2	propanal	5	1-propanol	1.5
3	acetone (propan-2-one)	5	2-propanol	0.5
4	glyoxylic acid (OHC-COOH)	5	glycolic acid (HOCH$_2$-COOH)	4.3
5	pyruvic acid (CH$_3$COCOOH)	5	no reaction	–

a **24**·SO$_4$ (1 μmol), substrate (10 μmol), HCOONa (10 or 50 μmol), H$_2$O (1 ml), 25 °C. The pH was adjusted to 3.2 by the addition of 0.1 M HOTf (aq.) or 0.1 M NaOH (aq.). The pH did not change during the reaction.
b Turnover frequency.

A mechanism proposed for this pH-dependent hydrogen transfer reaction in water is shown in Scheme 5.16. At pH 3.2, the catalyst precursor **24** is in equilibrium with the dinuclear complex **26**, while the addition of HCOONa to the equilibrium mixture generates the active catalyst **25** through β-hydrogen elimination with the evolution of CO$_2$. The active catalyst **25** then reacts with the substrates to give the products, during which hydride transfer to the substrate and Ir–H regeneration occur by utilizing two Ir centers of the dinuclear unit.

Three water-soluble Cp*Ir aqua complexes with different Lewis acidity have been prepared and employed as catalysts for the pH-dependent transfer hydrogenation, reductive amination and dehalogenation of water-soluble substrates [49]. The Lewis acidity of the complexes, [Cp*Ir(H$_2$O)$_3$]$^{2+}$ (**24**), [(Cp^Py)Ir(H$_2$O)$_2$]$^{2+}$ (**27**) and [Cp*Ir(bpy)(H$_2$O)]$^{2+}$ (**28**), was in the following order: **24** > **27** > **28**. These complexes were reversibly deprotonated to form the catalytically inactive hydroxo complexes [(Cp*Ir)$_2$(μ-OH)$_3$]$^+$ (**26**), [{(Cp^Py)Ir}$_2$(μ-OH)$_2$]$^{2+}$ (**29**) and [Cp*Ir(bpy)(OH)]$^+$ (**30**) at pH-values of about 2.8, 4.5 and 6.6, respectively. The results of pH-dependent transfer hydrogenation, reductive amination and dehalogenation using HCOONa and HCOONH$_4$ as hydrogen donors catalyzed by **24**, **27** and **28** are summarized in Table 5.8. The transfer hydrogenation of the carbonyl compounds using **24** and **27** proceeded, whereas that using **28** did not (entries 1–3). The reductive amination using **28** proceeded much more efficiently than that using **24** or **27** (entry 4), because the pK_a value of NH$_4^+$ is 4.7 and, therefore, free

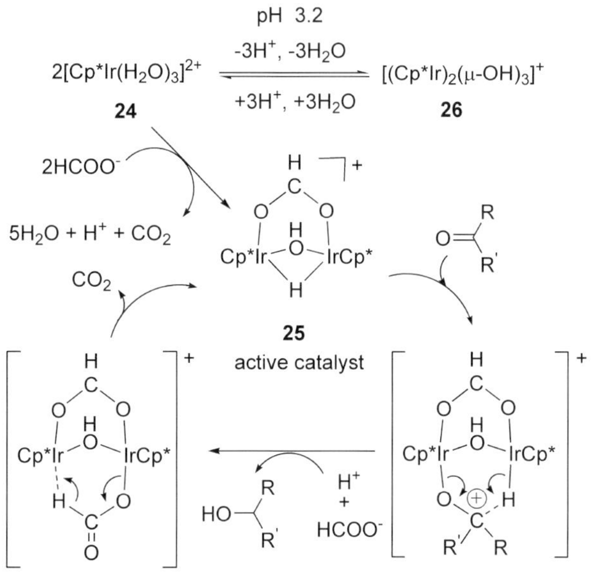

5.5 Hydrogen Transfer Reactions in Aqueous Media

$$2[Cp^*Ir(H_2O)_3]^{2+} \underset{+3H^+, +3H_2O}{\overset{pH\ 3.2\ \ -3H^+, -3H_2O}{\rightleftharpoons}} [(Cp^*Ir)_2(\mu\text{-}OH)_3]^+$$

24   **26**

Scheme 5.16

Table 5.8 Transfer hydrogenation, reductive amination and dehalogenation of water-soluble compounds with **24, 27 28**, and hydrogen donors.[a]

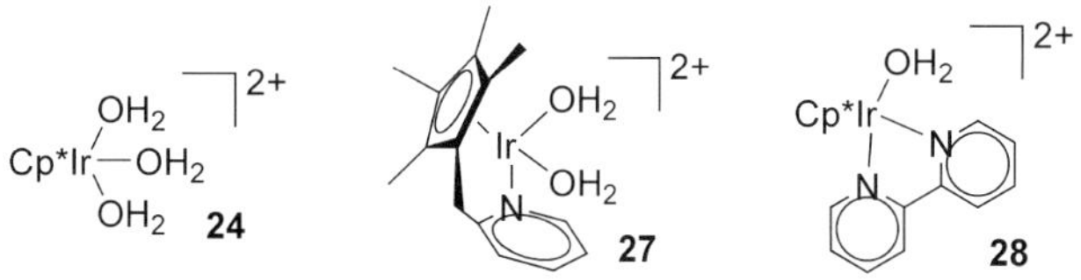

Entry	Substrate	Hydrogen donor	Product	Initial TOF[b]		
				24 (pH 2.8)	27 (pH 3.5)	28 (pH 5.0)
1	HCOPr	HCO$_2$Na	1-BuOH	1.5	0.5	0.0
2	MeCOEt	HCO$_2$Na	2-BuOH	0.5	0.1	0.0
3	HCOCO$_2$H	HCO$_2$Na	HOCH$_2$CO$_2$H	4.3	1.8	0.0
4	HCOPr	HCO$_2$NH$_4$	1-BuNH$_2$	0.1	0.2	2.1
			1-BuOH	1.0	0.2	0.0
5	MeCHBrCO$_2$H	HCO$_2$Na	EtCO$_2$H	0.0	0.8	6.3

a Conditions: **24, 27** or **28**, 1 µmol; substrates, 10 µmol; HCOONa and HCOONH$_4$, 50 µmol; H$_2$O, 1 ml; 25 °C.
b The initial TOF is equal to (mol of products)/(mol of **24, 27** or **28**) after 1 h of the reaction.

NH$_3$ is generated above pH 4.7. In these reactions, the hydride complexes [(Cp*Ir)$_2$(μ-H)(μ-OH)(μ-HCOO)]$^+$ (**25**) and [Cp*Ir(bpy)(H)]$^+$ (**31**), which would be generated from the reactions of **24** and **28** with HCOO$^-$, would be key catalytic intermediates. The dehalogenation of the alkyl halides using **24** did not occur, most likely due to the bulkiness of the active catalyst **25** compared to **31** in S$_N$2-type reactions (entry 5).

The catalytically active species **31** was isolated and its structure definitively determined using X-ray analysis. The complex **31** was seen to be quite stable below 70 °C under an argon atmosphere at pH 2.0–6.0 and in the absence of any reducible carbonyl compounds [50]. When the reducing ability of isolated **31** in acidic media was examined, the catalytic reactions using cyclohexanone and acetophenone (**31** : ketones : HCOOH = 1 : 200 : 1000) yielded the corresponding alcohols quantitatively in 2 h at pH 2.0 at 70 °C (Scheme 5.17).

A pH-dependent chemoselective catalytic reductive amination of α-keto acids, affording α-amino acids with HCOONH$_4$ in water, was achieved using the complex **31** or its precursor **28** as the catalyst [51]. The formation rates of alanine and lactic acid from pyruvic acid exhibited a maximum value around pH 5 and pH 3, respectively, and therefore, alanine was obtained quite selectively (96%) with a small amount of lactic acid (4%) at pH 5 (Scheme 5.18). A variety of nonpolar, uncharged polar and charged polar amino acids were also synthesized in high yields.

The highly efficient catalytic system for the chemoselective transfer hydrogenation of aldehydes was reported by Xiao et al. [52]. This system consisted of [Cp*IrCl$_2$]$_2$ (**1**), a diamine and HCOONa, and worked on water and in air. A wide range of aromatic aldehydes were reduced to the corresponding primary alcohols in a highly chemoselective manner; some representative examples are summarized in Table 5.9.

Various α,β-unsaturated aldehydes were also selectively reduced to give allylic alcohols by this catalytic system (Scheme 5.19). The transfer hydrogenation of ali-

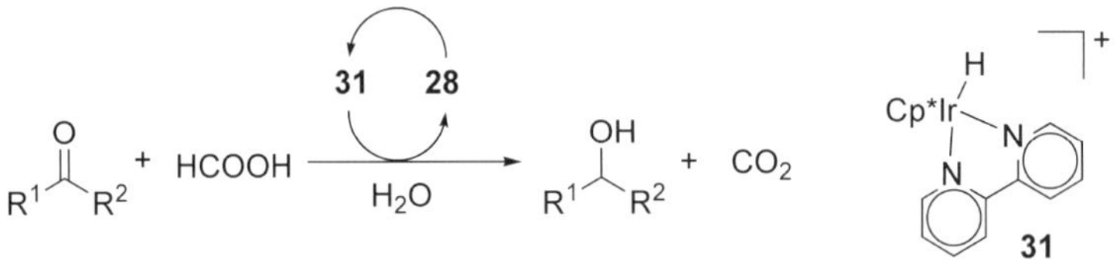

Scheme 5.17

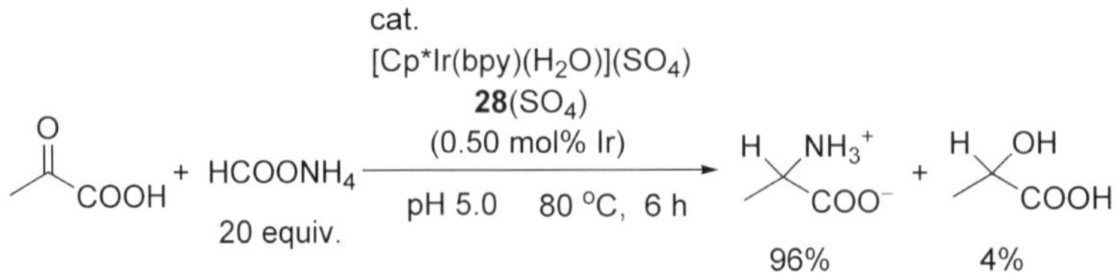

Scheme 5.18

Table 5.9 Transfer hydrogenation of aromatic aldehydes with HCOONa in water.[a]

ArCHO $\xrightarrow[\text{HCOONa (5 equiv.), H}_2\text{O, 80 °C}]{\text{[Cp*IrCl}_2]_2 \text{ (1)} \text{ (0.020 mol% Ir)}, \text{ H}_2\text{N-NHSO}_2\text{Tol-}p}$ ArCH$_2$OH

Entry	Substrate	Time (h)	Conversion (%)
1	C$_6$H$_5$CHO	0.6	>99
2[b]	C$_6$H$_5$CHO	0.6	>99
3	p-BrC$_6$H$_4$CHO	0.67	99
4	p-CF$_3$C$_6$H$_4$CHO	3	98
5	p-MeOC$_6$H$_4$CHO	0.5	>99
6[c]	p-MeSC$_6$H$_4$CHO	0.5	>99
7	m-NO$_2$C$_6$H$_4$CHO	0.5	91
8	o-BrC$_6$H$_4$CHO	7	99
9	2,6-(MeO)$_2$C$_6$H$_3$CHO	3	96
10	p-AcC$_6$H$_4$CHO	0.5	99
11[c]	2-thienyl-CHO	0.5	>99

a 80°C, 1/diamime catalyst, HCOONa (5 equiv.) at substrate:catalyst (S/C) ratio 5000:1 in water.
b Reaction in air.
c S/C ratio 1000:1.

phatic aldehydes under normal conditions gave alcohols in very low yields, but this difficulty could be avoided by lowering the concentration of substrates in order to disfavor the aldol reaction of the aldehydes. Thus, various aliphatic aldehydes were reduced when the aldehyde was slowly added portionwise.

5.6
Carbon–Nitrogen Bond Formation Based on Hydrogen Transfer

5.6.1
N-Alkylation of Amines with Alcohols

The development of versatile and efficient methods for the synthesis of amines has long been an active area of research, mainly because a wide variety of amines play important roles in many fields of organic chemistry. The N-alkylation of amines with alcohols represents an attractive method for synthesizing various amines because it does not generate any wasteful byproducts (H$_2$O is the only stoichiometric coproduct). Although several catalytic systems for this process have been studied using transition-metal catalysts [53–57], most of these require a high reaction temperature (>150°C).

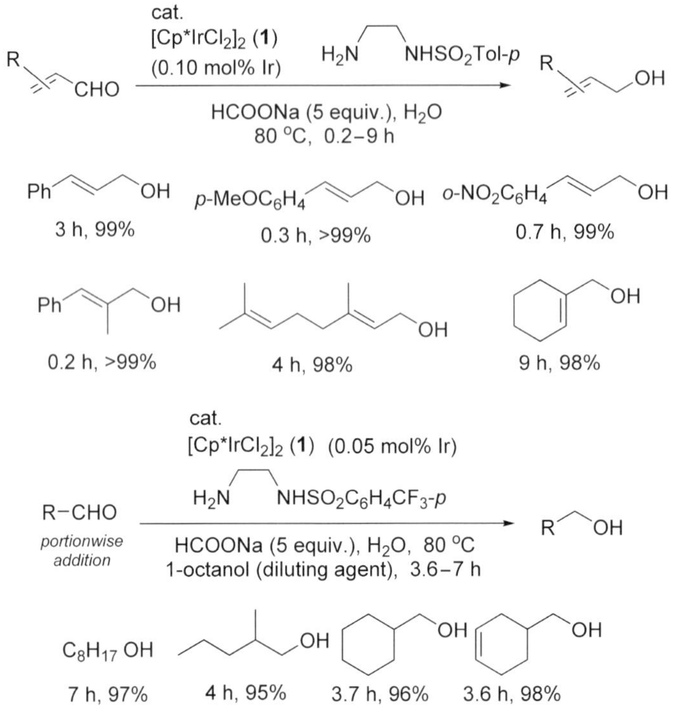

Scheme 5.19

[Scheme 5.19 shows reduction of α,β-unsaturated aldehydes (R-CH=CH-CHO) to allylic alcohols (R-CH=CH-CH₂OH) using:
cat. [Cp*IrCl₂]₂ (1) (0.10 mol% Ir), H₂N-CH₂CH₂-NHSO₂Tol-p, HCOONa (5 equiv.), H₂O, 80 °C, 0.2–9 h]

Ph-CH=CH-CH₂OH 3 h, 99%
p-MeOC₆H₄-CH=CH-CH₂OH 0.3 h, >99%
o-NO₂C₆H₄-CH=CH-CH₂OH 0.7 h, 99%

Ph-C(CH₃)=CH-CH₂OH 0.2 h, >99%
(geraniol-type) 4 h, 98%
(cyclohexenylmethanol) 9 h, 98%

Second reaction:
R-CHO (portionwise addition) →
cat. [Cp*IrCl₂]₂ (1) (0.05 mol% Ir), H₂N-CH₂CH₂-NHSO₂C₆H₄CF₃-p,
HCOONa (5 equiv.), H₂O, 80 °C, 1-octanol (diluting agent), 3.6–7 h
→ R-CH₂OH

C₈H₁₇OH 7 h, 97%
n-pentanol 4 h, 95%
cyclohexylmethanol 3.7 h, 96%
cyclohexenylmethanol 3.6 h, 98%

Scheme 5.19

Fujita and Yamaguchi et al. developed a new and efficient system for the N-alkylation of amines with alcohols under mild conditions, based on the high catalytic performance of the Cp*Ir complex in hydrogen transfer reactions [58]. Results of the N-alkylation of primary amines with a variety of alcohols, catalyzed by the [Cp*IrCl₂]₂ (1)/K₂CO₃ system, are summarized in Table 5.10. As shown in entry 1, the reaction of equimolar amounts of aniline and benzyl alcohol in the presence of 5.0 mol% Ir catalyst 1 and base (K₂CO₃) at 110 °C gave N-benzylaniline as a single product, in good yield. Benzyl alcohols with either electron-donating or electron-withdrawing groups, as well as other primary and secondary alcohols, could be also used as alkylating reagents (entries 2–9). In addition to aniline, benzylamine and other aliphatic amines were also alkylated by this N-alkylation system (entries 10 and 11).

The N-alkylation of secondary amines was also examined under the same catalytic conditions. Reactions of N-methylaniline and N-methylbenzylamine with benzyl alcohol gave the corresponding tertiary amines in good yields (Scheme 5.20).

A possible mechanism for the N-alkylation of primary amines is shown in Scheme 5.21. The first step of the reaction involves the oxidation of an alcohol to a carbonyl intermediate, accompanied by the generation of an iridium hydride.

Table 5.10 N-Alkylation of primary amines with various primary and secondary alcohols catalyzed by [Cp*IrCl$_2$]$_2$ (**1**).[a]

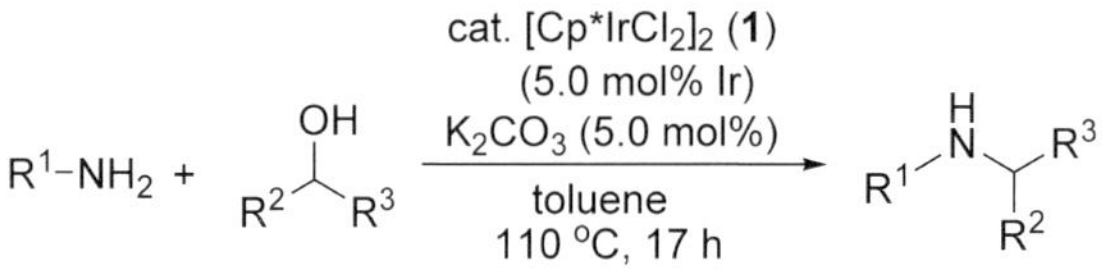

$$R^1-NH_2 + \underset{R^2 \; R^3}{\overset{OH}{\diagup}} \xrightarrow[\substack{\text{toluene} \\ 110\,°C,\,17\,h}]{\substack{\text{cat. [Cp*IrCl}_2\text{]}_2\text{ (1)} \\ (5.0\,\text{mol\% Ir}) \\ K_2CO_3\,(5.0\,\text{mol\%})}} R^1-\underset{R^2}{\overset{H}{N}}-R^3$$

Entry	Amine	Alcohol	Yield (%)[b]
	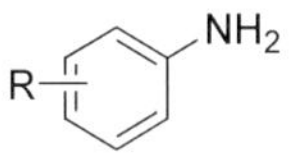 R—C$_6$H$_4$—NH$_2$	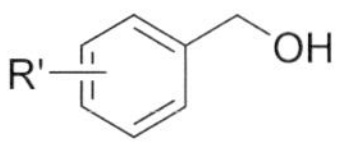 R'—C$_6$H$_4$—CH$_2$OH	
1	R=H	R'=H	88
2	R=H	R'=4-OMe	95
3	R=H	R'=4-Cl	83
4[c]	R=H	R'=4-NO$_2$	86
5	R=4-OMe	R'=H	93
6	R=4-Cl	R'=H	95
7	R=H	1-Octanol	79
8[c]	R=H	2-Octanol	69
9	R=H	Cyclohexanol	92
10	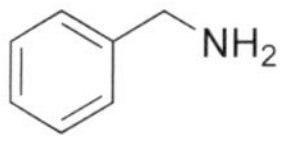 PhCH$_2$NH$_2$	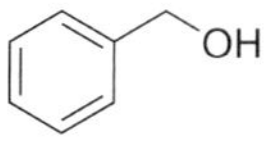 PhCH$_2$OH	67
11[d]	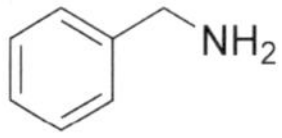 PhCH$_2$NH$_2$	1-Octanol	88

a Amines (1.0 mmol), alcohols (1.0 mmol), **1** (5.0 mol% Ir), and K$_2$CO$_3$ (5.0 mol%) in toluene (0.5 ml) at 110 °C for 17 h.
b Isolated yield.
c Reaction for 40 h.
d Reaction at 90 °C.

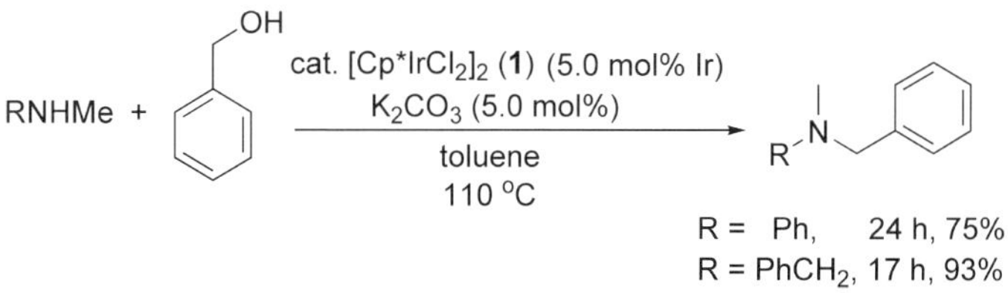

$$\text{RNHMe} + \text{PhCH}_2\text{OH} \xrightarrow[\substack{\text{toluene} \\ 110\,°C}]{\substack{\text{cat. [Cp*IrCl}_2\text{]}_2\text{ (1) (5.0 mol\% Ir)} \\ K_2CO_3\,(5.0\,\text{mol\%})}} R-N(Me)-CH_2Ph$$

R = Ph,   24 h, 75%
R = PhCH$_2$, 17 h, 93%

Scheme 5.20

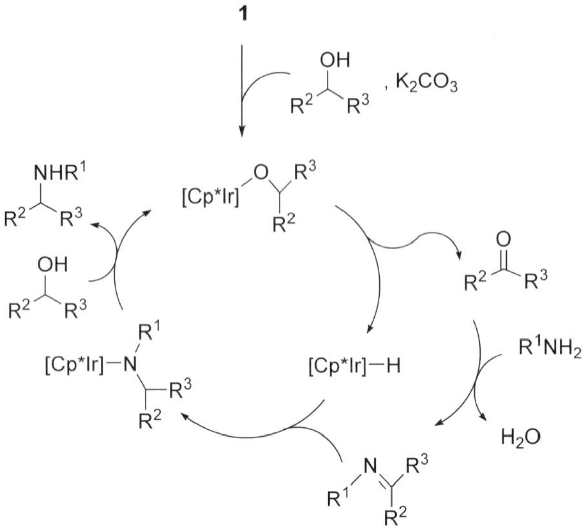

Scheme 5.21

The carbonyl intermediate then reacts readily with a primary amine to afford an imine and water. A subsequent addition of the iridium hydride to the C=N double bond of the imine, followed by amide-alkoxide exchange, would then occur to release the product.

5.6.2
Cyclization of Amino Alcohols

The cyclization of amino alcohols should be an attractive method for the synthesis of N-heterocyclic compounds, mainly because they can be obtained in a single step and without the generation of wasteful byproducts. Carbon–nitrogen bond formation catalyzed by Cp*Ir complexes has been extended to the synthesis of N-heterocyclic compounds by the cyclization of amino alcohols.

Fujita, Yamamoto and Yamaguchi et al. demonstrated the cyclization of amino alcohols to benzo-fused N-heterocycles catalyzed by the [Cp*IrCl$_2$]$_2$ (**1**)/K$_2$CO$_3$ system [59]. The reaction of 2-aminophenethyl alcohol under toluene reflux in the presence of 5.0 mol% Ir catalyst gave indole in 80% yield. 2-Aminophenethyl alcohol derivatives having a substituent on the aromatic ring or methylene chain were also converted into the corresponding indoles in moderate to high yields (Scheme 5.22).

As shown in Scheme 5.23, 1,2,3,4-tetrahydroquinoline derivatives were synthesized using 3-(2-aminophenyl)propanols as starting materials and catalyzed by the **1**/K$_2$CO$_3$ system. This catalytic system was also applicable to the synthesis of 2,3,4,5-tetrahydro-1-benzazepine from 4-(2-aminophenyl)butanol.

Eary and Clausen reported the cyclization of anilino alcohols to give tetrahydroquinoxalines catalyzed by the **1**/K$_2$CO$_3$ system [60]; the results are summarized in

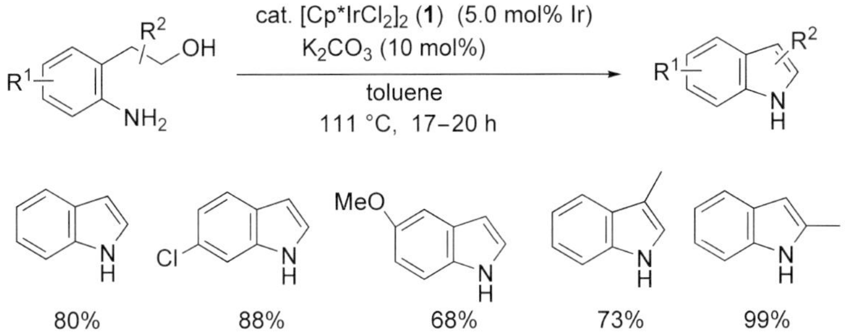

5.6 Carbon–Nitrogen Bond Formation Based on Hydrogen Transfer | 127

Scheme 5.22

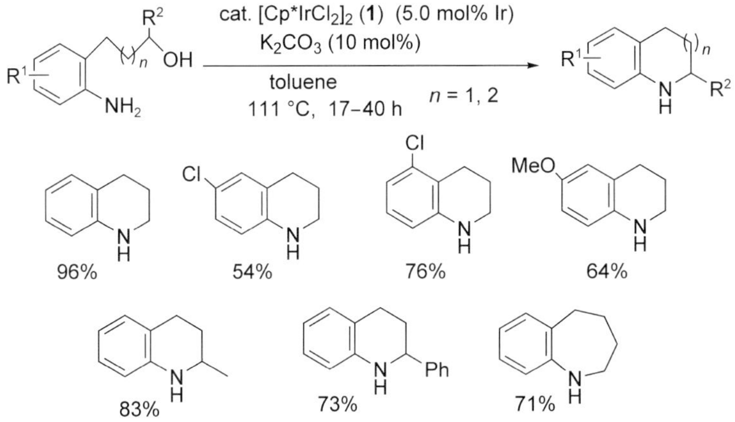

Scheme 5.23

Scheme 5.24. The reaction of 2-(2-aminophenylamino)ethanol at 110 °C for 17 h in the presence of 25% catalyst gave N-methyl-1,2,3,4-tetrahydroquinoxaline in 80% yield. Similar reactions of 2-(2-aminophenylamino)ethanol derivatives gave corresponding tetrahydroquinoxaline products in moderate to high yields, although a longer reaction time and a higher catalyst loading were required. 1-Methyl-2,3,4,5-tetrahydro-1H-benzo[b][1,4]diazepine was also obtained in 68% yield after the reaction for seven days.

5.6.3
Cyclization of Primary Amines with Diols

Fujita and Yamaguchi *et al.* reported a new method for the N-heterocyclization of primary amines with diols catalyzed by the **1**/NaHCO$_3$ system, and its application to the asymmetric synthesis of (S)-2-phenylpiperidine [61]. The representative results of the reaction of primary amines with diols are summarized in Table 5.11. As shown in entry 1, the reaction of benzylamine with 1,4-butanediol at 110 °C for

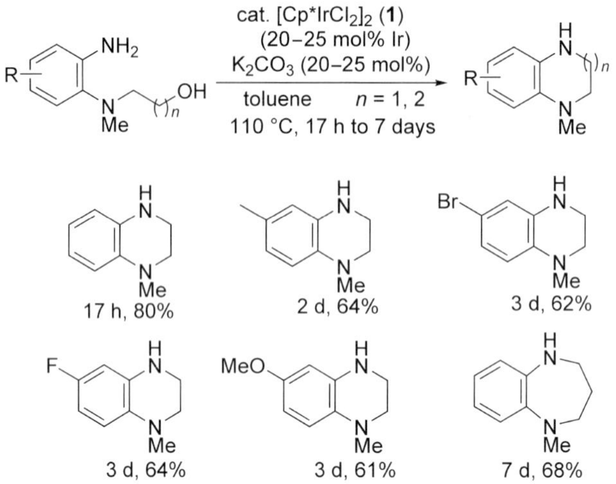

Scheme 5.24

17 h in the presence of 1.0 mol% Ir catalyst gave N-benzylpyrrolidine in 72% yield. A variety of five- to seven-membered N-heterocyclic compounds were synthesized in good to excellent yields, using easily available benzylamines and diols as starting materials. In addition to benzylamine, other aromatic and aliphatic primary amines such as aniline and octylamine were successfully used. The synthesis of N-benzylpiperidine on a 100 mmol scale and the double N-heterocyclization producing 1,4-bis(N-pyrrolidinylmethyl)benzene were also reported [62, 63].

The same group also demonstrated an efficient, two-step asymmetric synthesis of (S)-2-phenylpiperidine as an extension of the N-heterocyclization of primary amines with diols; the results are illustrated in Scheme 5.25. First, the reaction of enantiomerically pure (R)-1-phenylethylamine and 1-phenyl-1,5-pentanediol was conducted to produce a diastereomeric mixture of the corresponding N-(1-phenylethyl)-2-phenylpiperidines **32** and **33** with 92% diastereomeric excess (de). Hydrogenation of this diastereomeric mixture of **32** and **33** with Pd/C catalyst then gave (S)-2-phenylpiperidine in 96% yield (78% ee).

5.6.4
Amidation of Alcohols with Hydroxylamine

Williams *et al.* reported a unique system for converting alcohols into amides in a one-pot process catalyzed by **1** [64]. First, they identified a rearrangement of oximes into amides catalyzed by **1**, a proposed mechanism for which is shown in Scheme 5.26. Investigations were then conducted with the aim of combining this rearrangement with the hydrogen transfer oxidation of alcohols catalyzed by the same catalyst **1**. The reaction of benzyl alcohols with styrene (hydrogen acceptor) in the presence of **1** (5.0 mol% Ir) and Cs_2CO (5.0 mol%) under reflux for 24–36 h, followed by the addition of hydroxylamine hydrochloride and further reflux for 16 h, gave benzamides in moderate to high yields (Scheme 5.26).

Table 5.11 N-Heterocyclization of primary amines with a variety of diols by [Cp*IrCl$_2$]$_2$ (1).[a]

$$R^1NH_2 + HO\underset{n}{\overset{R^2}{\diagup\!\!\!\diagdown}}OH \xrightarrow[\text{toluene, 110 °C, 17 h} \atop n=1-3]{\text{cat. [Cp*IrCl}_2\text{]}_2 \text{ (1)} \atop \text{NaHCO}_3} R^1\text{-N}\underset{n}{\diagup\!\!\!\diagdown}R^2$$

Entry	Amine	Diol	Catalyst (mol% Ir)	Yield[b] (%)
1	PhCH$_2$NH$_2$	HO(CH$_2$)$_4$OH	1.0	72
2[c]	PhCH$_2$NH$_2$	HO(CH$_2$)$_5$OH	1.0	91
3[d]	PhCH$_2$NH$_2$	HO(CH$_2$)$_6$OH	2.0	73
4	PhCH$_2$NH$_2$	HOCH(Ph)(CH$_2$)$_3$OH	4.0	78[e]
5[f,g]	PhCH$_2$NH$_2$	1,2-C$_6$H$_4$(CH$_2$OH)$_2$	2.0	63
6	PhCH$_2$NH$_2$	1,2-C$_6$H$_4$(CH$_2$CH$_2$OH)(CH$_2$OH)	2.0	76
7	4-MeO-C$_6$H$_4$-CH$_2$NH$_2$	HOCH$_2$CH$_2$OCH$_2$CH$_2$OH	2.0	76
8[h,i]	PhNH$_2$	HO(CH$_2$)$_4$OH	5.0	70
9	C$_8$H$_{17}$NH$_2$	HO(CH$_2$)$_4$OH	4.0	81[e]

a Amine (3.0 mmol), diol (2.0 mmol), 1, and NaHCO$_3$ in toluene (1 ml) at 110 °C for 17 h.
b Isolated yield.
c At 90 °C.
d Toluene (3 ml).
e GC yield.
f Amine (2.0 mmol).
g Without base.
h At 130 °C.
i 40 h.

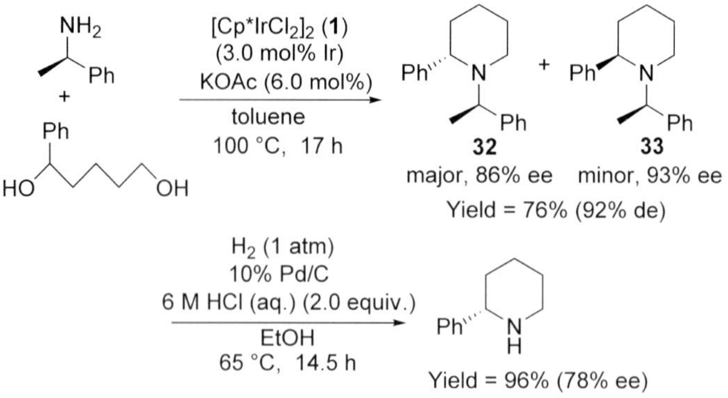

5 Catalytic Activity of Cp* Iridium Complexes in Hydrogen Transfer Reactions

Scheme 5.25

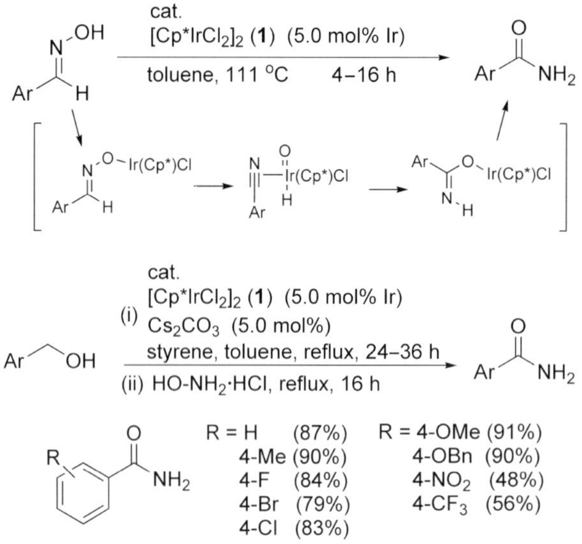

Scheme 5.26

5.7
Carbon–Carbon Bond Formation Based on Hydrogen Transfer

5.7.1
β-Alkylation of Secondary Alcohols

Although a large number of methods for the synthesis of alcohols have been devised, the generation of a variety of alcohols having intricate structures through the alkylation of simple alcohols usually requires tedious processes in which many reagents must be employed. For example, the β-alkylation of a secondary alcohol is usually accomplished via three-step transformations; oxidation, alkylation and reduction (Scheme 5.27). In this context, if the surplus hydrogen generated during

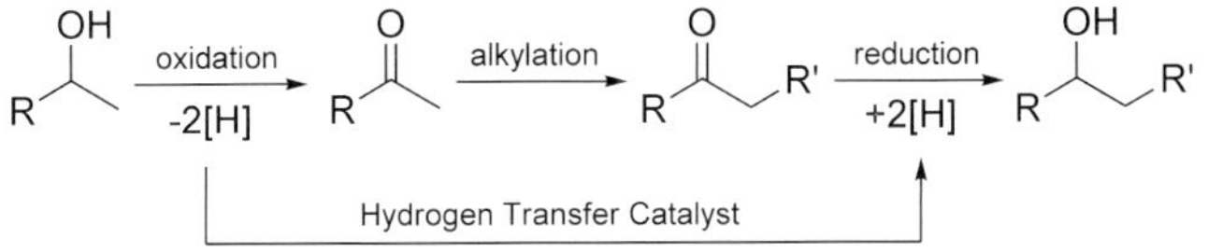

$$R\overset{OH}{\underset{}{\diagup}} \xrightarrow[-2[H]]{\text{oxidation}} R\overset{O}{\underset{}{\diagup}} \xrightarrow{\text{alkylation}} R\overset{O}{\underset{}{\diagup}}R' \xrightarrow[+2[H]]{\text{reduction}} R\overset{OH}{\underset{}{\diagup}}R'$$

Hydrogen Transfer Catalyst

Scheme 5.27

the oxidation stage could be consumed during the reduction stage with the aid of a hydrogen transfer catalyst, then a highly atom-economical system could be anticipated.

Fujita and Yamaguchi et al. reported a direct β-alkylation reaction of alcohols based on the high catalytic performance of [Cp*IrCl$_2$]$_2$ (**1**) for hydrogen transfer [65]. The results of β-alkylation of secondary alcohols with primary alcohols catalyzed by the **1**/NaO^{t}Bu system are summarized in Table 5.12. When the reaction of 1-phenylethanol with 1-butanol was performed at 110 °C for 17 h in the presence of 1.0 mol% Ir catalyst and NaO^{t}Bu, 1-phenylhexane-1-ol was obtained in 88% yield. A variety of aliphatic and benzylic primary alcohols were successfully used as an alkylating reagent. In addition to 1-phenylethanol, substituted 1-arylethanol, aliphatic secondary alcohols and cyclic secondary alcohols can be utilized as substrates.

A possible mechanism for the β-alkylation of secondary alcohols with primary alcohols catalyzed by a **1**/base system is illustrated in Scheme 5.28. The first step of the reaction involves oxidation of the primary and secondary alcohols to aldehydes and ketones, accompanied by the transitory generation of a hydrido iridium species. A base-mediated cross-aldol condensation then occurs to give an α,β-unsaturated ketone. Finally, successive transfer hydrogenation of the C=C and C=O double bonds of the α,β-unsaturated ketone by the hydrido iridium species occurs to give the product.

Ishii and colleagues reported the β-alkylation of two molecules of primary alcohols (Guerbet reaction) catalyzed by **1** [66]. First, they examined the selfcondensation of 1-butanol into 2-ethyl-1-hexanol by using a variety of iridium catalysts. Thus, they found that **1**/KO^{t}Bu was a viable choice as a catalyst, and that the addition of 1,7-octadiene as a hydrogen acceptor greatly improved the efficiency of the reaction. The results of the reactions of various primary alcohols under the optimized conditions are summarized in Scheme 5.29. When the reaction of 1-butanol was performed at 120 °C for 4 h in the presence of **1** (2.0 mol% Ir), KO^{t}Bu (40 mol%) and 1,7-octadiene (10 mol%), 2-ethyl-1-hexanol was formed in 93% yield. Other aliphatic primary alcohols were also applicable to this system to produce higher alcohols.

5.7.2
Alkylation of Active Methylene Compounds with Alcohols

The carbon–carbon bond-forming reactions based on hydrogen transfer catalyzed by Cp*Ir complex have been extended to the alkylation of active methylene compounds. Grigg et al. reported the alkylation of arylacetonitriles catalyzed by the

Table 5.12 β-Alkylation of secondary alcohols with primary alcohols catalyzed by [Cp*IrCl$_2$]$_2$ (1).[a]

$$R^1\text{-CH(OH)-CH}_3 + R^2\text{-CH}_2\text{OH} \xrightarrow[\text{toluene, 110 °C, 17 h}]{\substack{\text{cat. [Cp*IrCl}_2\text{]}_2 \text{ (1)} \\ \text{(1.0–4.0 mol% Ir)} \\ \text{base (1 equiv.)}}} R^1\text{-CH(OH)-CH}_2\text{-CH}_2\text{-}R^2$$

Entry	Secondary alcohol	Primary alcohol (R–CH$_2$OH)	Catalyst (mol% Ir)	Base	Yield (%)[b]
1	Ph-CH(OH)-CH$_3$	R=Pr	1.0	NaO^{t}Bu	88
2	Ph-CH(OH)-CH$_3$	R=heptyl	2.0	NaOH	77
3	Ph-CH(OH)-CH$_3$	R=^{i}Pr	2.0	NaOH	75
4[c]	Ph-CH(OH)-CH$_3$	R=Ph	2.0	NaO^{t}Bu	75
5[c]	Ph-CH(OH)-CH$_3$	R=4-MeOC$_6$H$_4$	4.0	NaO^{t}Bu	81
6[c]	Ph-CH(OH)-CH$_3$	R=4-ClC$_6$H$_4$	4.0	NaO^{t}Bu	80
7	Ph-CH(OH)-CH$_3$	R=benzyl	2.0	NaO^{t}Bu	74
8	Ph-CH(OH)-CH$_3$	R=phenethyl	2.0	NaOH	83
9	Ph-CH$_2$-CH(OH)-CH$_3$	R=Pr	4.0	NaO^{t}Bu	58
10	^{i}Pr-CH(OH)-CH$_3$	R=Ph	2.0	NaO^{t}Bu	78
11	1-tetralinol	R=Pr	4.0	NaO^{t}Bu	65[d]

a Secondary alcohol (3.0 mmol), primary alcohol (3.6 mmol), 1, and base in toluene (0.3 ml) at 110 °C for 17 h.
b Isolated yield.
c Toluene (3.0 ml).
d Mixture of diastereomers (cis:trans = 54:46).

1/KOH system [67]; representative results of the reactions of arylacetonitriles with primary alcohols are summarized in Table 5.13. As shown in entry 1, the reaction of phenylacetonitrile with an excess amount of benzyl alcohol at 100 °C for 13 h using the catalyst 1 (5.0 mol% Ir) and KOH (15 mol%) under solvent-free conditions gave 2,3-diphenylpropanenitrile in high yield (88%). A variety of arylacetonitriles having electron-withdrawing and electron-donating groups, and heteroarylnitriles were efficiently alkylated with benzyl alcohol. Some of these reactions were performed under microwave irradiation, which caused the reaction to be vastly accelerated.

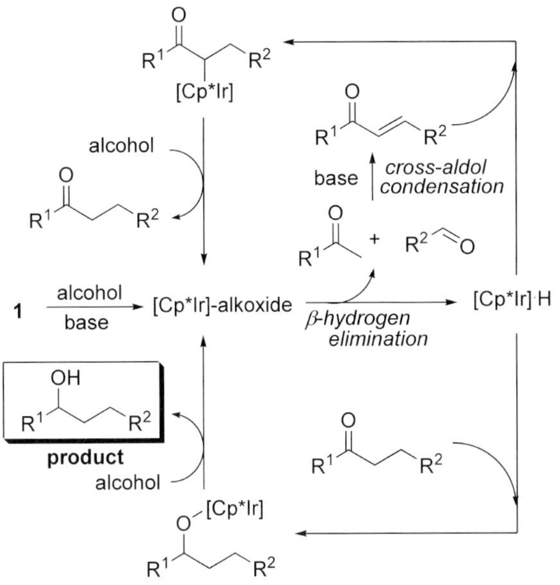

5.7 Carbon–Carbon Bond Formation Based on Hydrogen Transfer | 133

Scheme 5.28

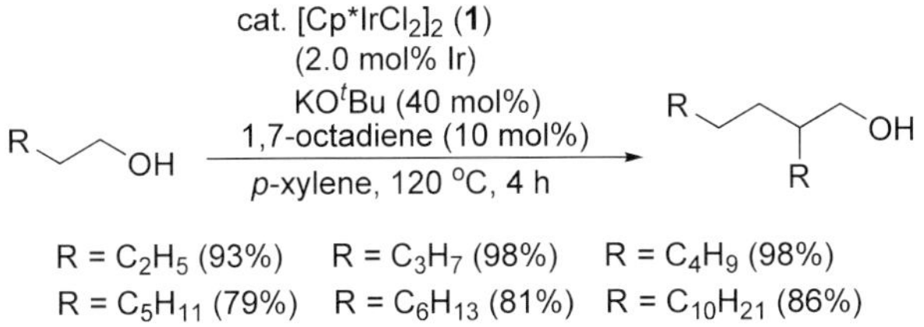

Scheme 5.29

R = C$_2$H$_5$ (93%) R = C$_3$H$_7$ (98%) R = C$_4$H$_9$ (98%)
R = C$_5$H$_{11}$ (79%) R = C$_6$H$_{13}$ (81%) R = C$_{10}$H$_{21}$ (86%)

A possible mechanism for the alkylation of arylacetonitriles is shown in Scheme 5.30. The reaction would proceed via successive hydrogen transfer and Knoevenagel condensation as follows:

1. Hydrogen transfer oxidation of an alcohol to give an aldehyde and an iridium hydride.
2. Base-promoted Knoevenagel condensation to give an alkylated arylacrylonitrile.
3. Transfer hydrogenation of the arylacrylonitrile by the iridium hydride to give the product.

By using the same catalytic system, alkylations of 1,3-dimethylbarbituric acid with alcohols were also accomplished (Scheme 5.31) [68]. The Cp*Ir-catalyzed alkylation using 2-iodobenzyl alcohol, followed by palladium-catalyzed carbon–carbon bond formation with allene, gave spirocyclic barbituric acid derivatives in a one-pot process.

Table 5.13 Alkylation of arylacetonitriles with benzyl alcohols catalyzed by [Cp*IrCl$_2$]$_2$ (**1**).[a]

Entry	Nitrile	Alcohol	Time	Yield (%)[b]
1	PhCH$_2$CN	PhCH$_2$OH	13 h	88
2	4-MeO-C$_6$H$_4$-CH$_2$CN	PhCH$_2$OH	16 h	93
3	3-pyridyl-CH$_2$CN	PhCH$_2$OH	12 h	88
4[c]	3-pyridyl-CH$_2$CN	3,4-(MeO)$_2$-C$_6$H$_3$-CH$_2$OH	10 min	82
5[c]	3-pyridyl-CH$_2$CN	3,4-(OCH$_2$O)-C$_6$H$_3$-CH$_2$OH	10 min	86
6[c]	3-pyridyl-CH$_2$CN	4-Cl-C$_6$H$_4$-CH$_2$OH	10 min	85
7[c]	3-pyridyl-CH$_2$CN	3-pyridyl-CH$_2$OH	10 min	81

a Arylacetonitrile (1.0 mmol), alcohol (3.0 mmol), **1** (5.0 mol% Ir), and KOH (15 mol%) at 100 °C.
b Isolated yield.
c Under microwave condition at 110 °C.

Grigg et al. also reported the alkylation of indole at the C3-position with alcohols, catalyzed by the **1**/KOH system (Scheme 5.32) [69]. A variety of indole derivatives having an alkyl substituent at the C3-position were synthesized by this methodology. The same group also developed a Cp*Ir-catalyzed process for successive carbon–nitrogen and carbon–carbon bond formation, starting with 2-aminophenethyl alcohol (also see Scheme 5.22).

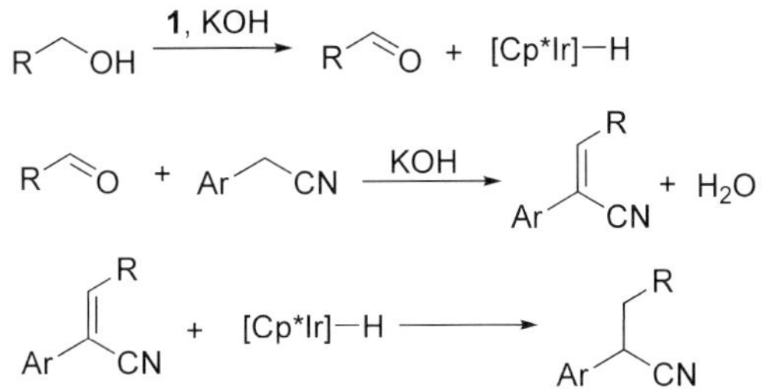

5.8 Carbon–Oxygen Bond Formation Based on Hydrogen Transfer

Scheme 5.30

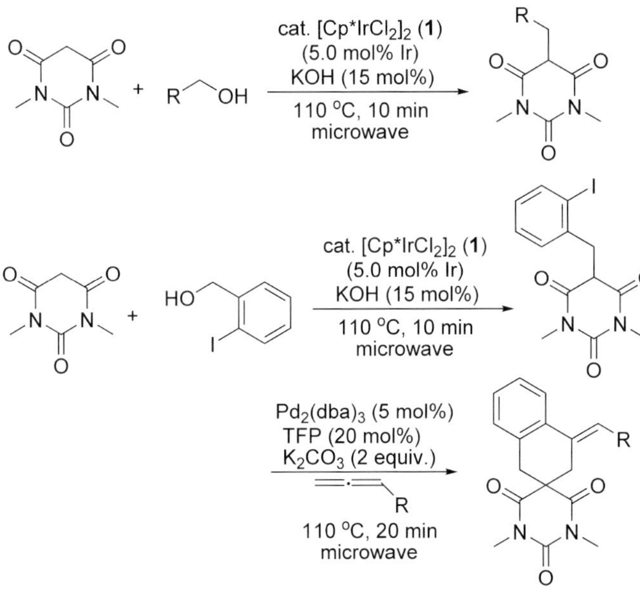

Scheme 5.31

5.8
Carbon–Oxygen Bond Formation Based on Hydrogen Transfer

5.8.1
Oxidative Lactonization of Diols

The environmentally benign synthesis of lactones has attracted attention because of their importance in natural product chemistry. The oxidative cyclization of diols via carbon–oxygen bond formation is the most well-known approach for the synthesis of lactones [70].

Hiroi et al. developed a new system for the oxidative lactonization of diols using acetone as cooxidant and catalyzed by the Cp*Ir complex bearing an

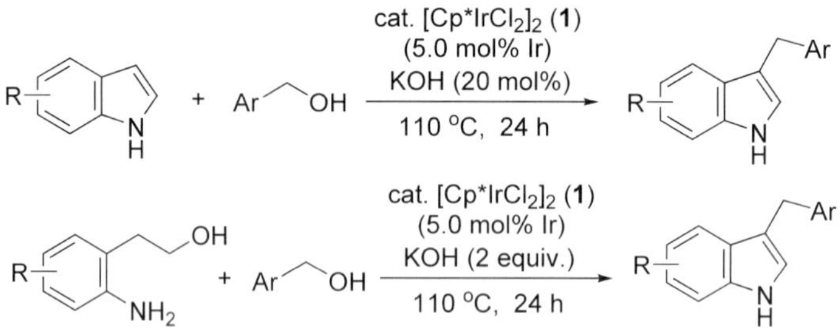

Scheme 5.32

Table 5.14 Oxidative lactonization of diols catalyzed by **6**.[a]

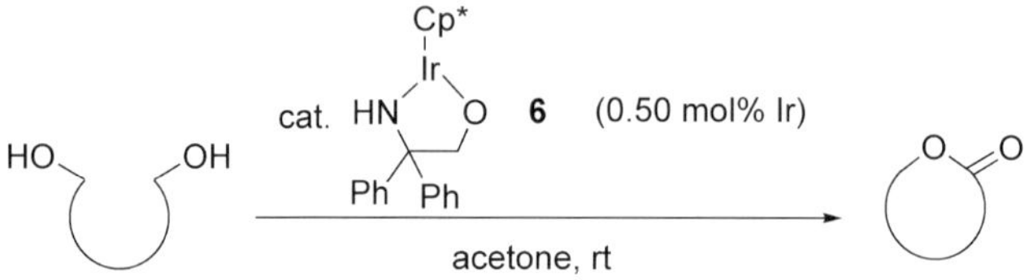

Entry	Diol	Time (h)	Product	Yield (%)[b]
1	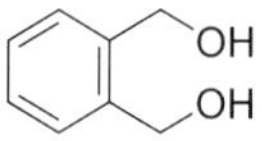 benzene-1,2-diyldimethanol	4	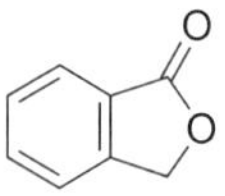 phthalide	>99
2	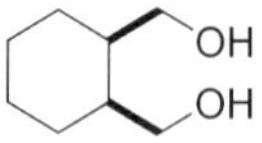 cis-cyclohexane-1,2-diyldimethanol	36	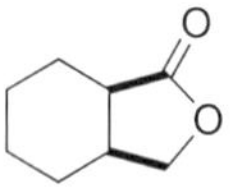 cis-hexahydroisobenzofuran-1(3H)-one	97
3	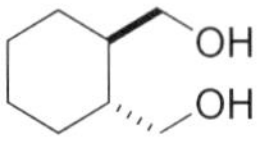 trans-cyclohexane-1,2-diyldimethanol	36	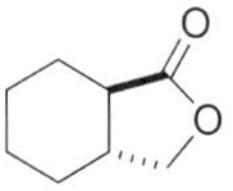 trans-hexahydroisobenzofuran-1(3H)-one	98
4	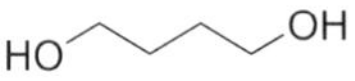 1,4-butanediol	20	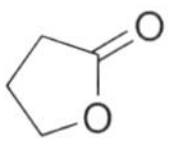 γ-butyrolactone	96
5	 1,5-pentanediol	24	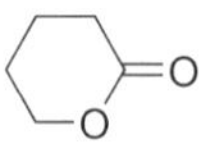 δ-valerolactone	95
6[c]	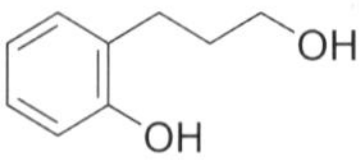 2-(2-hydroxyphenyl)ethanol derivative	5	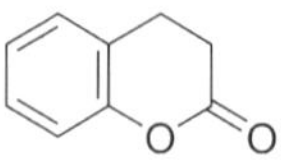 chroman-2-one	95

a Diol (1.0 mmol, 1.0 M solution in acetone), **6** (0.50 mol% Ir) at room temperature.
b Isolated yield.
c 2.0 M solution of diol in acetone.

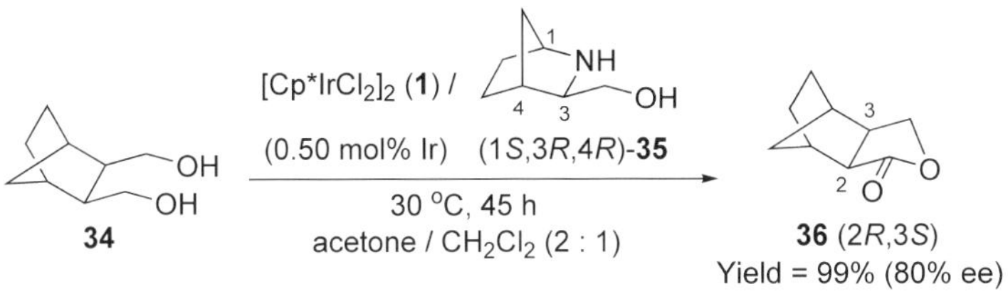

Scheme 5.33

amido-alkoxo ligand [71]; the representative results are summarized in Table 5.14. As shown in entry 1, the reaction of 1,2-bis(hydroxymethyl)benzene in acetone in the presence of the Cp*Ir catalyst **6** (0.50 mol% Ir) at room temperature for 4 h resulted in the formation of phthalide in quantitative yield. By using this catalytic system, a variety of 1,4- and 1,5-diols were transformed to the corresponding lactones in excellent yields.

Hiroi *et al.* also reported the asymmetric lactonization of meso-diols catalyzed by a Cp*Ir complex bearing a chiral amido-alkoxo ligand (Scheme 5.33) [72]. The reaction of meso-diol **34** in acetone/dichloromethane solvent at 30 °C for 45 h in the presence of a Cp*Ir catalyst bearing a chiral amido-alkoxo ligand **35** gave a corresponding lactone **36** in quantitative yield, with 80% ee.

5.8.2
Inter- and Intra-Molecular Tishchenko Reactions

The coupling of two molecules of aldehydes into esters (Tishchenko reaction) has been used as an efficient method for the industrial preparation of dimeric esters. Although a number of systems for such reactions using transition-metal catalysts have been reported [73], there is still great room for improvement of the synthetic efficiency.

Suzuki and Katoh *et al.* reported an atom-economical system for the Tishchenko reaction of aliphatic and aromatic aldehydes catalyzed by a Cp*Ir complex **37** having an amino-alkoxo ligand prepared *in situ* from **6** and 2-propanol (Scheme 5.34) [74]. The reaction of 3-phenylpropanal in acetonitrile at room temperature for 13 h in the presence of preformed **37** (1.0 mol% Ir) and K_2CO_3 (20 mol%) gave 3-phenylpropyl 3-phenylpropanoate in 91% yield. Both, aliphatic and aromatic aldehydes, could be converted to the corresponding dimeric esters in high yields. The same group has also reported the intramolecular Tishchenko reaction of δ-ketoaldehydes leading to 3,4-dihydroisocoumarin, catalyzed by preformed **37** (Scheme 5.34) [75].

A proposed mechanism for the Cp*Ir-catalyzed Tishchenko reaction is illustrated in Scheme 5.35. In this reaction, hydrogen transfer from the hemiacetal to aldehyde catalyzed by the Cp*Ir complex would be crucial.

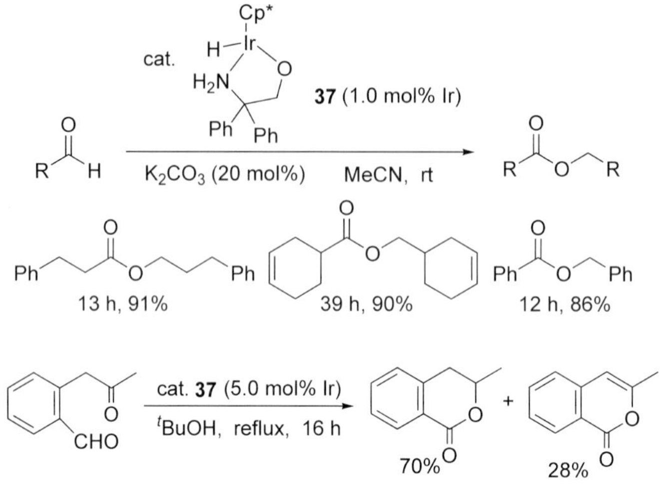

Scheme 5.34

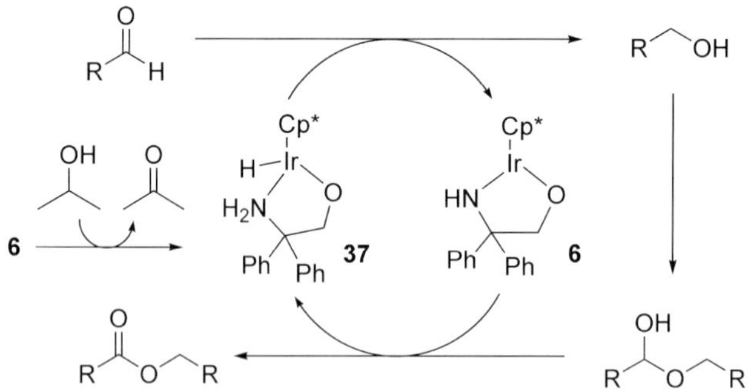

Scheme 5.35

5.9
Dehydrogenative Oxidation of Alcohols

From the viewpoint of atom efficiency and safety of the reaction, the oxidation of alcohols to produce carbonyl compounds without any oxidant must represent an ideal method [76–79]. Fujita and Yamaguchi et al. developed a new system for the oxidant-free oxidation of secondary alcohols to ketones by using a Cp*Ir catalyst bearing hydroxypyridine as a functional ligand [80]. First, a new catalyst **38** was prepared by the reaction of [Cp*IrCl$_2$]$_2$ (**1**) with 2-hydroxypyridine. The catalyst **38** exhibited high catalytic activity for the oxidant-free oxidation of secondary alcohols to ketones with hydrogen evolution (H$_2$); some representative results are summarized in Table 5.15. A variety of aromatic, aliphatic and cyclic secondary alcohols were oxidized to ketones in good to excellent yields, using 0.2–1.0 mol% Ir of **38**.

A possible mechanism for the oxidation of secondary alcohols catalyzed by **38** is shown in Scheme 5.36. The first step of the reaction involves formation of the

Table 5.15 Oxidant-free oxidation of various secondary alcohols to ketones catalyzed by **38**.[a]

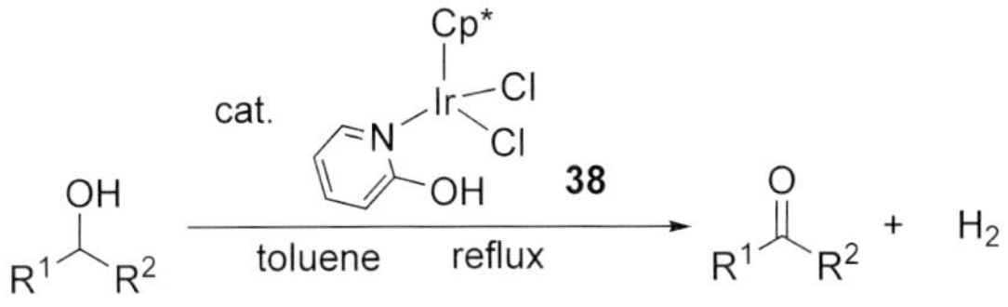

Entry	Alcohol	Catalyst (mol% Ir)	Time (h)	Yield (%)[b]
1	R=H	0.20	20	95
2	R=4-Me	0.20	20	82
3	R=4-OMe	0.20	20	94
4	R=4-Br	0.20	50	82
5	R=4-NO$_2$	0.33	50	86
6	2-Octanol	0.33	50	93[c]
7	1-Phenylpropan-1-ol	0.20	20	92
8	Cyclohexanol	1.0	50	85[c]
9	Indanol	0.20	20	97

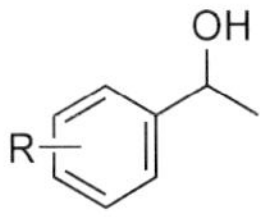

a   Secondary alcohol (1.0–5.0 mmol) and **38** (0.2–1.0 mol% Ir) in toluene under reflux.
b   Isolated yield.
c   Determined by GC.

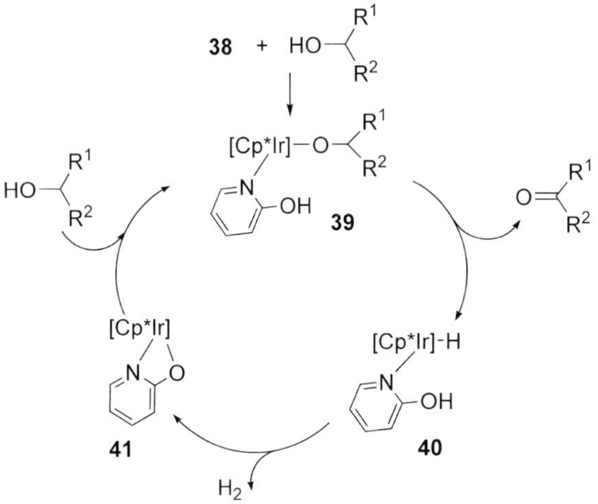

Scheme 5.36

alkoxo iridium **39**, and subsequent β-hydrogen elimination could afford the product, accompanied by formation of the iridium hydride species **40**. Reaction of the hydride on iridium with the protic hydroxyl proton on the ligand would release a hydrogen molecule, with concomitant generation of the chelated intermediate **41**; this would then be subjected to the addition of alcohol to regenerate **39**. Indeed, **41** was not only prepared separately but also exhibited high catalytic activity comparable to **38**, thereby supporting the proposed mechanism.

5.10
Conclusions

In this chapter we have reviewed recent developments in the catalytic chemistry of Cp* iridium complexes for organic transformations brought about by their extremely high performance for hydrogen transfer reactions. It is highly likely that these catalytic reactions proceed through trivalent iridium hydride species as key intermediates, with such intermediates being stabilized with aid of favorable steric and electron-donating effects of the five methyl groups of the Cp* ligand. Thus, Cp*Ir complexes exhibit an excellent catalytic performance in a variety of hydrogen transfer reactions. Because the hydrogen transfer reactions catalyzed by Cp*Ir complexes can be carried out under mild conditions (as described above), utilization of their catalytic ability in environmentally benign organic synthesis is highly promising. Hence, in the future, greater attention will be paid to the catalytic chemistry of Cp*Ir complexes.[1]

References

1 Kang, J.W., Moseley, K. and Maitlis, P.M. (1969) *Journal of The American Chemical Society*, **91**, 5970.
2 White, C., Yates, A. and Maitlis, P.M. (1992) *Inorganic Synthesis*, **29**, 228.
3 Leigh, G.J. and Richards, R.L. (1982) *Comprehensive Organometallic Chemistry, Vol. 5, Iridium* (eds G. Wilkinson, F.G.A. Stone and E.W. Abel), Pergamon, Oxford, p. 541.
4 Atwood, J.D. (1995) *Comprehensive Organometallic Chemistry II, Vol. 8, Iridium* (eds E.W. Abel, F.G.A. Stone and G. Wilkinson), Pergamon, Oxford, p. 303.
5 Janowicz, A.H. and Bergman, R.G. (1982) *Journal of the American Chemical Society*, **104**, 352.
6 Buchanan, J.M., Stryker, J.M. and Bergman, R.G. (1986) *Journal of the American Chemical Society*, **108**, 1537.
7 Arndtsen, B.A. and Bergman, R.G. (1995) *Science*, **270**, 1970.
8 Hoyano, J.K. and Graham, W.A.G. (1982) *Journal of the American Chemical Society*, **104**, 3723.
9 Chetcuti, P.A., Knobler, C.B. and Hawthorne, M.F. (1988) *Organometallics*, **7**, 650.

1) Other than hydrogen transfer reactions, catalytic applications of Cp*Ir complexes for the deuteration of organic molecules [81–84], asymmetric Diels–Alder reactions [85, 86], carbon–carbon bond cleavage and formation [87–89], hydrogenation of CO_2 [90], hydrosilylation [91] and hydroboration [92, 93] have been reported recently.

10 Fujita, K., Nakaguma, H., Hamada, T. and Yamaguchi, R. (2003) *Journal of the American Chemical Society*, **125**, 12368.

11 Maitlis, P.M. (1978) *Accounts of Chemical Research*, **11**, 301.

12 Carmona, D., Lahoz, F.J., Elipe, S., Oro, L.A., Lamata, M.P., Viguri, F., Mir, C., Cativiela, C. and López-Ram de Víu, M.P. (1998) *Organometallics*, **17**, 2986.

13 Iverson, C.N. and Smith, M.R. III (1999) *Journal of the American Chemical Society*, **121**, 7696.

14 Crabtree, R.H., Felkin, H. and Morris, G.E. (1977) *Journal of Organometallic Chemistry*, **141**, 205.

15 Crabtree, R.H. and Davis, M.W. (1986) *The Journal of Organic Chemistry*, **51**, 2655.

16 Takeuchi, R. and Kezuka, S. (2006) *Synthesis*, 3349 and references cited therein.

17 Ishii, Y. and Sakaguchi, S. (2004) *Bulletin of the Chemical Society of Japan*, **77**, 909 and references cited therein.

18 Ishiyama, T. and Miyaura, N. (2003) *Journal of Organometallic Chemistry*, **680**, 3 and references cited therein.

19 Gill, D.S. and Maitlis, P.M. (1975) *Journal of Organometallic Chemistry*, **87**, 359.

20 Fujita, K. and Yamaguchi, R. (2005) *Synlett*, 560 and references cited therein.

21 Brieger, G. and Nestrick, T.J. (1974) *Chemical Reviews*, **74**, 567.

22 Johnstone, R.A.W., Wilby, A.H. and Entwistle, I.D. (1985) *Chemical Reviews*, **85**, 129.

23 Zassinovich, G., Mestroni, G. and Gladiall, S. (1992) *Chemical Reviews*, **92**, 1051.

24 Carmona, D., Lamata, M.P. and Oro, L.A. (2002) *European Journal of Inorganic Chemistry*, 2239.

25 Hamid, M.H.S.A., Slatford, P.A. and Williams, J.M.J. (2007) *Advanced Synthesis Catalysis*, **349**, 1555.

26 Guillena, G., Ramón, D.J. and Yus, M. (2007) *Angewandte Chemie – International Edition*, **46**, 2358.

27 Wang, G.-Z. and Bäckvall, J.-E. (1992) *Journal of the Chemical Society, Chemical Communications*, 337.

28 Almeida, M.L.S., Beller, M., Wang, G.-Z. and Bäckvall, J.-E. (1996) *Chemistry - A European Journal*, **2**, 1533.

29 Gauthier, S., Scopelliti, R. and Severin, K. (2004) *Organometallics*, **23**, 3769.

30 Fujita, K., Furukawa, S. and Yamaguchi, R. (2002) *Journal of Organometallic Chemistry*, **649**, 289.

31 Suzuki, T., Morita, K., Tsuchida, M. and Hiroi, K. (2003) *The Journal of Organic Chemistry*, **68**, 1601.

32 Hanasaka, F., Fujita, K. and Yamaguchi, R. (2004) *Organometallics*, **23**, 1490.

33 Hanasaka, F., Fujita, K. and Yamaguchi, R. (2005) *Organometallics*, **24**, 3422.

34 Hanasaka, F., Fujita, K. and Yamaguchi, R. (2006) *Organometallics*, **25**, 4643.

35 Gabrielsson, A., van Leeuwen, P. and Kaim, W. (2006) *Chemical Communications*, 4926.

36 Fujita, K., Kitatsuji, C., Furukawa, S. and Yamaguchi, R. (2004) *Tetrahedron Letters*, **45**, 3215.

37 Corberán, R., Sanaú, M. and Peris, E. (2007) *Organometallics*, **26**, 3492.

38 Haack, K.-J., Hashiguchi, S., Fujii, A., Ikariya, T. and Noyori, R. (1997) *Angewandte Chemie – International Edition in English*, **36**, 285.

39 Mashima, K., Abe, T. and Tani, K. (1998) *Chemistry Letters*, 1199.

40 Mashima, K., Abe, T. and Tani, K. (1998) *Chemistry Letters*, 1201.

41 Hashiguchi, S., Fujii, A., Takehara, J., Ikariya, T. and Noyori, R. (1995) *Journal of the American Chemical Society*, **117**, 7562.

42 Murata, K., Ikariya, T. and Noyori, R. (1999) *The Journal of Organic Chemistry*, **64**, 2186.

43 Thorpe, T., Blacker, J., Brown, S.M., Bubert, C., Crosby, J., Fitzjohn, S., Muxworthy, J.P. and Williams, J.M.J. (2001) *Tetrahedron Letters*, **42**, 4041.

44 Furegati, M. and Rippert, A.J. (2005) *Tetrahedron: Asymmetry*, **16**, 3947.

45 Huerta, F.F., Minidis, A.B.E. and Bäckvall, J.-E. (2001) *Chemical Society Reviews*, **30**, 321.

46 Stirling, M., Blacker, J. and Page, M.I. (2007) *Tetrahedron Letters*, **48**, 1247.

47 Marr, A.C., Pollock, C.L. and Saunders, G.C. (2007) *Organometallics*, **26**, 3283.

48 Ogo, S., Makihara, N. and Watanabe, Y. (1999) *Organometallics*, **18**, 5470.
49 Ogo, S., Makihara, N., Kaneko, Y. and Watanabe, Y. (2001) *Organometallics*, **20**, 4903.
50 Abura, T., Ogo, S., Watanabe, Y. and Fukuzumi, S. (2003) *Journal of the American Chemical Society*, **125**, 4149.
51 Ogo, S., Uehara, K., Abura, T. and Fukuzumi, S. (2004) *Journal of the American Chemical Society*, **126**, 3020.
52 Wu, X., Liu, J., Li X., Zanotti-Gerosa, A., Hancock, F., Vinci, D., Ruan, J. and Xiao, J. (2006) *Angewandte Chemie – International Edition*, **45**, 6718.
53 Watanabe, Y., Tsuji, Y. and Ohsugi, Y. (1981) *Tetrahedron Letters*, **22**, 2667.
54 Murahashi, S.-I., Kondo, K. and Hakata, T. (1982) *Tetrahedron Letters*, **23**, 229.
55 Grigg, R., Mitchell, T.R.B., Sutthivaiyakit, S. and Tongpenyai, N. (1981) *Journal of the Chemical Society, Chemical Communications*, 611.
56 Naota, T., Takaya, H. and Murahashi, S.-I. (1998) *Chemical Reviews*, **98**, 2599 and references cited therein.
57 Hollmann, D., Tillack, A., Michalik, D., Jackstell, R. and Beller, M. (2007) *Chemistry – An Asian Journal*, **2**, 403.
58 Fujita, K., Li, Z., Ozeki, N. and Yamaguchi, R. (2003) *Tetrahedron Letters*, **44**, 2687.
59 Fujita, K., Yamamoto, K. and Yamaguchi, R. (2002) *Organic Letters*, **4**, 2691.
60 Eary, C.T. and Clausen, D. (2006) *Tetrahedron Letters*, **47**, 6899.
61 Fujita, K., Fujii, T. and Yamguchi, R. (2004) *Organic Letters*, **6**, 3525.
62 Fujita, K., Enoki, Y. and Yamaguchi, R. (2006) *Organic Syntheses*, **83**, 217.
63 Fujita, K., Fujii, T., Komatsubara, A., Enoki, Y. and Yamaguchi, R. (2007) *Heterocycles*, **74**, 673.
64 Owston, N.A., Parker, A.J. and Williams, J.M.J. (2007) *Organic Letters*, **9**, 73.
65 Fujita, K., Asai, C., Yamaguchi, T., Hanasaka, F. and Yamaguchi, R. (2005) *Organic Letters*, **7**, 4017.
66 Matsu-ura, T., Sakaguchi, S., Obora, Y. and Ishii, Y. (2006) *The Journal of Organic Chemistry*, **71**, 8306.
67 Löfberg, C., Grigg, R., Whittaker, M.A., Keep, A. and Derrick, A. (2006) *The Journal of Organic Chemistry*, **71**, 8023.
68 Löfberg, C., Grigg, R., Keep, A., Derrick, A., Sridharan, V. and Kilner, C. (2006) *Chemical Communications*, 5000.
69 Whitney, S., Grigg, R., Derrick, A. and Keep, A. (2007) *Organic Letters*, **9**, 3299.
70 For example: Fetizon, M., Golfier, M., and Louis, J.-M. (1975) *Tetrahedron*, **31**, 171.
71 Suzuki, T., Morita, K., Tsuchida, M. and Hiroi, K. (2002) *Organic Letters*, **4**, 2361.
72 Suzuki, T., Morita, K., Matsuo, Y. and Hiroi, K. (2003) *Tetrahedron Letters*, **44**, 2003.
73 Seki, T., Nakajo, T. and Onaka, M. (2006) *Chemistry Letters*, **35**, 824.
74 Suzuki, T., Yamada, T., Matsuo, T., Watanabe, K. and Katoh, T. (2005) *Synlett*, 1450.
75 Suzuki, T., Yamada, T., Watanabe, K. and Katoh, T. (2005) *Bioorganic and Medicinal Chemistry Letters*, **15**, 2583.
76 Dobson, A. and Robinson, S.D. (1975) *Journal of Organometallic Chemistry*, **87**, C52.
77 Ligthart, G.B.W.L., Meijer, R.H., Donners, M.P.J., Meuldijk, J., Vekemans, J.A.J.M. and Hulshof, L.A. (2003) *Tetrahedron Letters*, **44**, 1507.
78 Zhang, J., Gandelman, M., Shimon, L.J.W., Rozenberg, H. and Milstein, D. (2004) *Organometallics*, **23**, 4026.
79 Adair, G.R.A. and Williams, J.M.J. (2005) *Tetrahedron Letters*, **46**, 8233.
80 Fujita, K., Tanino, N. and Yamaguchi, R. (2007) *Organic Letters*, **9**, 109.
81 Golden, J.T., Andersen, R.A. and Bergman, R.G. (2001) *Journal of the American Chemical Society*, **123**, 5837.
82 Klei, S.R., Golden, J.T., Tilley, T.D. and Bergman, R.G. (2002) *Journal of the American Chemical Society*, **124**, 2092.
83 Yung, C.M., Skaddan, M.B. and Bergman, R.G. (2004) *Journal of the American Chemical Society*, **126**, 13033.
84 Corberán, R., Sanaú, M. and Peris, E. (2006) *Journal of the American Chemical Society*, **128**, 3974.
85 Carmona, D., Lahoz, F.J., Elipe, S., Oro, L.A., Lamata, M.P., Viguri, F., Sánchez, F., Martínez, S., Cativiela, C. and López-Ram

de Víu, M.P. (2002) *Organometallics*, **21**, 5100.

86 Carmona, D., Lamata, M.P., Viguri, F., Rodríguez, R., Lahoz, F.J., Dobrinovitch, I.T. and Oro, L.A. (2007) *Dalton Transactions*, 1911.

87 Hou, Z., Koizumi, T., Fujita, A., Yamazaki, H. and Wakatsuki, Y. (2001) *Journal of the American Chemical Society*, **123**, 5812.

88 Fujita, K., Nonogawa, M. and Yamaguchi, R. (2004) *Chemical Communications*, 1926.

89 Ueura, K., Satoh, T. and Miura, M. (2007) *The Journal of Organic Chemistry*, **72**, 5362.

90 Himeda, Y., Onozawa-Komatsuzaki, N., Sugihara, H., Arakawa, H. and Kasuga, K. (2004) *Organometallics*, **23**, 1480.

91 Sridevi, V.S., Fan, W.Y. and Leong, W.K. (2007) *Organometallics*, **26**, 1157.

92 Wynberg, N.A., Leger, L.J., Conrad, M.L., Vogels, C.M., Decken, A., Duffy, S.J. and Westcott, S.A. (2005) *Canadian Journal of Chemistry*, **83**, 661.

93 Vogels, C.M., Decken, A. and Westcott, S.A. (2006) *Tetrahedron Letters*, **47**, 2419.

6
Iridium-Catalyzed Hydroamination
Romano Dorta

6.1
Introduction

The hydroamination of alkenes and alkynes is a 100% atom-economical route to value-added amines (Equation 6.1) and enamines/imines (Equation 6.2), respectively. Although hydroamination as C—N bond-forming method has great economic potential, it is still considered a difficult reaction. This is particularly true for the olefin hydroamination (OHA) reaction. Hydroamination is therefore of great interest to both, industry and academia [1]. The ΔG and ΔH values for these reactions are generally favorable, whereas ΔS is not, and activation barriers are high [2]. This makes catalytic approaches necessary in order to overcome the activation barrier. OHAs are catalyzed by Group 1, 2, 4, 13 and f-elements on the one hand, and by late transition metals on the other hand. Homogeneous, metal-catalyzed OHA furthermore opens the possibility of enantiocontrol by the judicial design of chiral ligands when prochiral olefins are used [3]. Moreover, the particular interest in late transition metals is due to the perceived advantage of improved functional group tolerance they offer.

$$\underset{R^4 \quad R^3}{\overset{R^1 \quad R^2}{\diagup\!\!\!\!\diagdown}} + H{-}NR^5R^6 \longrightarrow \underset{R^4 \quad R^3}{\overset{H \quad NR^5R^6}{R^1{\diagup\!\!\!\!\diagdown}R^2}} \qquad (6.1)$$

$$R^1{\equiv}R^2 + H{-}N{\overset{R^3}{\underset{R^4}{\diagdown}}} \longrightarrow R^1{\overset{H \quad R^3}{\underset{R^2}{\diagup\!\!\!\!\diagdown}}}N{\diagdown}R^4 \;\overset{R^3 = H}{\underset{\longleftarrow}{\longrightarrow}}\; R^1{\overset{N}{\underset{R^2}{\diagup\!\!\!\!\diagdown}}}R^4 \qquad (6.2)$$

From a historic point of view, metal-catalyzed or metal-promoted hydroaminations were first achieved with alkali metals [4]. The use of soluble transition-metal complexes as catalysts for the OHA reaction was pioneered by DuPont workers during the 1970s, the best results being obtained with Rh and Ir salts [5]. Later, the finding that electron-rich Ir(I) species cleanly activated N—H bonds to form Ir–amido-hydrido species [6] opened the way to study the reactivity of these amides

Iridium Complexes in Organic Synthesis. Edited by Luis A. Oro and Carmen Claver
Copyright © 2009 WILEY-VCH Verlag GmbH & Co. KGaA, Weinheim
ISBN: 978-3-527-31996-1

towards olefins. This led to the development of the first soluble organometallic Ir-based OHA catalysis by Casalnuovo, Calabrese and Milstein (CMM) at DuPont, who demonstrated the feasibility of the catalytic addition of aniline to norbornene [7]. After an enantioselective intramolecular OHA reaction had been discovered by using chiral organolanthanide catalysts [8], an intermolecular version was achieved with chiral Ir complexes based on the CMM model reaction [9]. In general, catalytic OHA systems tend to be highly substrate-specific, during the past few years the Ir-catalyzed hydroamination of alkynes (alkyne hydroamination; AHA) has experienced a dynamic development. As will be discussed in the following sections, the mechanism of AHA is fundamentally different from that of OHA. AHA catalyst precursors are mostly trivalent Ir species that are believed to maintain their oxidation state during catalysis, whereas the Ir-catalyzed OHA reaction is only possible with electron-rich Ir(I) precursors. AHA catalysis relies on electrophilic alkyne activation, while the key step in OHA is amine activation via oxidative addition of the N—H bond. This chapter is divided into four sections: the first two describe the catalytic OHA and AHA reactions, followed by a section outlining proposed mechanisms and catalytic cycles. A final section discusses the stoichiometric reactions and Ir complexes that are relevant to hydroamination catalysis and the proposed mechanisms.

6.2
Iridium-Catalyzed Olefin Hydroamination (OHA)

6.2.1
The Ir(III)/Secondary Amines/Ethylene System

In 1971, Coulson at DuPont reported the first example of an OHA reaction catalyzed by soluble Rh and Ir complexes [5]. Secondary amines such as dimethylamine, pyrrolidine and piperidine were effectively added to ethylene, while primary amines, ammonia and heavier olefins were essentially unreactive (see Equation 6.3). $IrCl_3 \cdot 3H_2O$ proved to be an equally effective catalyst precursor in these reactions. It is probable that, under the conditions employed in this study, the Rh(III) and Ir(III) salts are reduced to monovalent, electron-rich species such as **3** (see Equation 6.6).

$$\text{pyrrolidine-NH} + \text{CH}_2=\text{CH}_2 \xrightarrow[\text{THF (25 ml)}]{\substack{1\% \text{ mol} \\ \text{RhCl}_3 \\ 453 \text{ K, 3 h}}} \text{N-ethylpyrrolidine} \quad (6.3)$$

17 g, 200 mmol + 9 g, 300 mmol → 14.6 g, 65% yield

6.2.2
The Ir(I)/ZnCl$_2$/Aniline/Norbornene System

The first example of a catalytic OHA reaction that was shown to proceed via N—H activation, was published by Casalnuovo, Calabrese and Milstein (CCM) in

1988 [7], and represents a milestone in iridium-catalyzed OHA. The best catalytic system for the model reaction of norbornene with aniline to form *exo*-2-phenylaminonorbornane (**1**) [10] was found to be a combination of the electron-rich complex IrCl(PEt$_3$)$_2$(C$_2$H$_4$)$_2$ and the Lewis acid ZnCl$_2$ (see Equation 6.4). *Exo* selectivity in the C—N bond-formation step was almost complete. A continuous flow of inert gas was apparently necessary to remove ethylene from the catalyst precursor. TlPF$_6$ as cocatalyst instead of ZnCl$_2$ was also effective. The mechanistic and structural details of this reaction are described in Sections 6.4.1 and 6.5.

$$\text{norbornene} + \text{PhNH}_2 \xrightarrow[\text{THF (10 ml)}]{\substack{0.192 \text{ mmol IrCl(PEt}_3)_2(\text{C}_2\text{H}_4)_2 \\ + \\ 0.037 \text{ mmol ZnCl}_2 \\ \text{3d reflux}}} \mathbf{1} \quad (6.4)$$

10.6 mmol   1.92 mmol   2 to 6 turnovers

6.2.3
The Chiral Ir(I)/'Naked Fluoride'/Norbornene/Aniline System

Based on the CCM system described above, an enantioselective version thereof was presented by Togni and coworkers a decade later [9]. The use of a new class of catalyst precursors led to substantially improved reactivity, and the system now was truly catalytic, with turnover frequency (TOF) values reaching up to 3.37 h^{-1} at 348 K (Equation 6.5; Table 6.1). The absolute configuration of the resultant amine was determined by internal comparison of the X-ray crystal structure of the

Table 6.1 Effect of fluoride on activity and stereoselectivity in the Ir-catalyzed addition of aniline to norbornene (Equation 6.5).[a]

Catalyst precursor	Temperature (°C)	(F$^-$)/[Ir] ratio	Yield (%)[b] (N$_t$, h^{-1})	ee (%) (abs. conf.)
5[c]	50	0	12 (0.17)	51 (2S)
5[c]	50	0.25	76 (1.05)	31 (2R)
5[c]	50	1	81 (1.11)	50 (2R)
5[c]	50	4	51 (0.71)	16 (2R)
5[c]	75	1	81 (3.37)[d]	38 (2R)
4[e]	50	0	12 (0.08)	57 (2R)
4[e]	75	2	45 (0.31)	78 (2R)
4[e]	75	4	22 (0.15)	95 (2R)
7[e]	75	4	24 (0.17)	92 (2S)

a Reaction conditions: [IrCl(PP)]$_2$ + commercially available P2-fluoride (0.5 M in benzene), no solvent, work-up after 72 h.
b Determined after flash chromatography.
c 1 mol% Ir.
d Reaction time = 24 h.
e 2 mol% Ir.

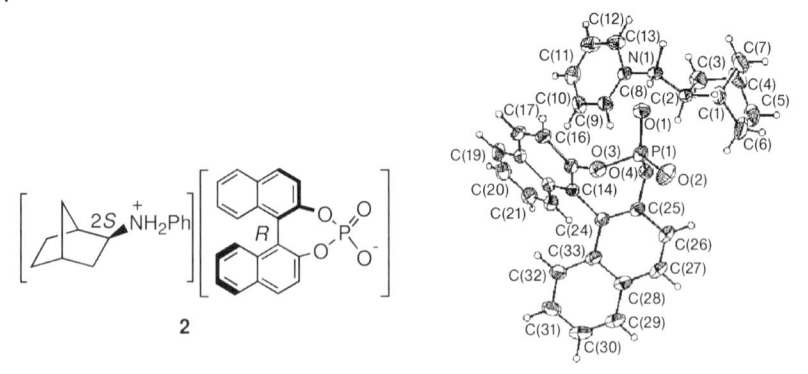

Figure 6.1 *exo*-2-(S)-Phenylammoniumnorbornan-(R)-1,1'-binaphthalene-2,2'-diyl hydrogen phosphate (**2**) and its solid-state structure (reproduced with permission from ACS) for determination of the absolute stereochemistry of the chiral product **1**.

salt **2** that was obtained by reacting an enriched sample (92% ee) of (−)-**7** with commercially available, optically pure (R)-(−)-1,1'-binaphthyl-2,2'-diyl hydrogen phosphate (Figure 6.1).

$$\text{norbornene} + \text{PhNH}_2 \xrightarrow[\text{"F}^-\text{"}]{1\,\text{mol\%}\; 4-9} \text{(2S)-1 NHPh} + \text{(2R)-1 NHPh} \qquad (6.5)$$

The design of a new catalyst precursor was based on the original mechanistic proposal by Casalnuovo, Calabrese and Milstein, as outlined in Scheme 6.1 that suggested a 14-electron species IrCl(PEt$_3$)$_2$ (**19**) to be the true catalyst. In the CCM system, **19** is thought to form by successive ethylene dissociation from the precursor IrCl(PEt$_3$)$_2$(C$_2$H$_4$)$_2$. However, analogues of **19** ought as well form by the dissociation of chloro-bridged, olefin-free Ir dimers such as **4–9** (Chart 6.1) [11]. Indeed, such dimers are possible resting states of the catalyst. The use of this class of complexes eliminates the problem associated with the liberation of unwanted equivalents of olefin and/or phosphines that may act as poisons in the catalytic cycle. There are surprisingly few structurally characterized complexes of this type (probably due to their sometimes pronounced moisture and air sensitivities), and the X-ray single crystal structure of **4** is a rare example (Figure 6.2) [12]. The Ir centers have square-planar coordination and the Ir$_2$Cl$_2$ core is butterfly-shaped. The corresponding (R)-BIPHEMP complex (**7**) also catalyzed the addition of aniline to norbornene with up to 92% enantiomeric excess (ee). When C_1 symmetric ligands such as ferrocenyldiphosphines were used, mixtures of *cis* and *trans* isomers formed in varying ratios, and it was necessary to use the Ir precursor **3** [13] for their synthesis (Equation 6.6). Chart 6.1 shows only the *trans* isomers of the Josiphos-type complexes **5, 6, 8** and **9**. Complex **5** was the most reactive catalyst,

6.2 Iridium-Catalyzed Olefin Hydroamination (OHA)

4 ([IrCl((S)-BINAP)]$_2$)

trans isomers of **5**
(R=Cy, [IrCl((R)-(S)-Josiphos)]$_2$)
and **6** (R = *tert*-Bu)

7 ([IrCl((R)-Biphemp)]$_2$), R = Me, Ar = Ph
8 ([IrCl((R)-triMeO-MeOBiphep)]$_2$),
R = MeO, Ar = 3,4,5-trimethoxyphenyl

trans isomers of **9** (R=Cy) and **10** (R=*t*-Bu)

11 ([IrCl((R)-DM-Segphos)]$_2$),
Ar = 3,5-dimethylphenyl
12 ([IrCl((R)-DTBM-Segphos)]$_2$),
Ar = 3,5-di-*tert*-butyl-4-methoxyylphenyl

Chart 6.1 Catalyst precursors for the enantioselective addition of anilines to norbornenes [9,16,17].

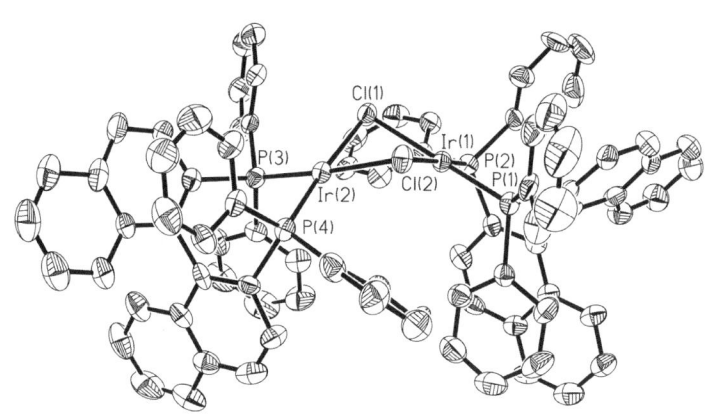

Figure 6.2 ORTEP-view of **4** (50% probability ellipsoids, reproduced with permission of the American Chemical Society). Selected interatomic distances (Å) and angles (°) are as follows: Ir(1)–P(1) 2.196(2), Ir(1)–P(2) 2.200(2), Ir(1)–Cl(1) 2.398(2), Ir(1)–Cl(2) 2.425(2), Ir(2)–P(3) 2.202(2), Ir(2)–P(4) 2.205(2), Ir(2)–Cl(1) 2.425(2), Ir(2)–Cl(2) 2.417(2), Ir(1)–Ir(2) 3.322(2); P(1)–Ir(1)–P(2) 91.16(8), Cl(1)–Ir(1)–Cl(2) 79.27(7), Ir(1)–Cl(1)–Ir(2) 87.00(6), Ir(1)–Cl(2)–Ir(2) 86.58(7), Cl(1)–Ir(2)–Cl(2) 78.89(7), P(3)–Ir(2)–P(4) 90.93(8).

producing TOFs of up to 3.37 h^{-1} at 348 K. Higher temperatures led to the formation of large amounts of dimerized norbornene.

$$\text{3} \quad + \quad *\binom{P^1}{P^2} \quad \xrightarrow[195\,K]{-4\,C_2H_4} \quad 0.5 \quad \begin{bmatrix} *(\underset{P^2}{\overset{P^1}{\text{Ir}}}\underset{Cl}{\overset{Cl}{\text{Ir}}}\underset{P^2}{\overset{P^1}{}})^* \\ cis \\ + \\ *(\underset{P^2}{\overset{P^1}{\text{Ir}}}\underset{Cl}{\overset{Cl}{\text{Ir}}}\underset{P^1}{\overset{P^2}{}})^* \\ trans \end{bmatrix} \quad (6.6)$$

The selectivities and activities of this reaction were greatly improved by running the catalysis solvent-free and by adding cocatalytic amounts of Schwesinger-type fluorides ('naked fluoride', e.g. phosphacenium-fluoride-P$_2$ **13**) [14]. The fluoride effect is pronounced both in terms of activity and enantioselectivity (even inverting the sense of enantioselection in one example; see Table 6.1).

$$\begin{bmatrix} (H_3C)_2N & \oplus & N(CH_3)_2 \\ (H_3C)_2N-P=N=P-N(CH_3)_2 \\ (H_3C)_2N & & N(CH_3)_2 \end{bmatrix} \begin{bmatrix} \ominus \\ F \end{bmatrix}$$

13

In a further development of the norbornene/aniline OHA reaction, Salzer and coworkers used planar chiral arene–chromium-tricarbonyl-based diphosphines for the *in situ* formation of *cis-trans* mixtures of complexes **9** and **10** that gave enantioselectivities of 51% and 70%, respectively, at 333 K and with a 40-fold excess of 'naked fluoride', but activities were very low. In the same paper complex **6** was shown to be superior in both activity and enantioselectivity (64% ee) to the corresponding Josiphos compound **5** [15]. The activated N-H bond of benzamide was also stereoselectively added across the double bond of norbornene to afford *N*-benzoyl-*exo*-aminonorbornane in up to 50% yield and 73% *ee* in the presence of 0.5 mol% [IrCl((*R*)-MeO-bipheb)]$_2$ at 373 K [16].

6.2.4
The Chiral Ir(I)/Organic Base/Anilines/Olefins System

Zhou and Hartwig recently discovered the beneficial effect of added potassium hexamethyldisilazanide (KHMDS) base for the asymmetric addition of anilines to norbornenes, thereby widening the synthetic scope of the original CMM system (see Table 6.2) [17]. [IrCl(COE)$_2$]$_2$ and two equivalents of variants of the Segphos and Biphep ligands first presumably form complexes **8**, **11**, and **12** *in situ* (see Chart 1) and then in combination with co-catalytic KHMDS generate the catalytically active species (see Table 6.2 and Section 6.4 for a discussion of the mechanism).

It should be recognized that already Togni and co-workers had observed a similar base effect in the context of the intramolecular hydroamination of 2-

6.2 Iridium-Catalyzed Olefin Hydroamination (OHA)

Table 6.2 Widening the synthetic scope of the CMM system.

Ar-NH$_2$ + 2 equiv olefin $\xrightarrow[\text{no solvent, 343 K, 12 h}]{\substack{\text{1 mol\% [IrCl(COE)}_2\text{]}_2 \\ \text{2 mol\% ligand} \\ \text{2 mol\% KHMDS}}}$ R-norbornyl-NHAr

Ar	Olefin	Ligand (complex formed *in situ*)	Yield (%)	ee (%)
p-t-BuC$_6$H$_4$[a]	norbornene	(R)-DM-Segphos (11)	85	92
p-BrC$_6$H$_4$		(R)-DTBM-Segphos (12)	91	96, *exo*-(2S)
p-MeOC$_6$H$_4$[a,b]		(R)-triMeO-MeOBiphep (8)	75	98
p-CF$_3$C$_6$H$_4$		(R)-DTBM-Segphos (12)	77	91
m-Xylyl[c]	norbornadiene	(R)-DTBM-Segphos (12)	94	99
p-MeOC$_6$H$_4$[b]		(R)-DTBM-Segphos (12)	88	99
m-Xylyl[a,c,d]	benzonorbornadiene	(R)-DTBM-Segphos (12)	90	99
m-Xylyl[a,c,e]	N-methylmaleimide-cyclopentadiene adduct	(R)-DTBM-Segphos (12)	84	98

a 100 °C.
b 0.5 mol % [Ir]; 40 h.
c 1 mol % [Ir]; 12 h.
d 1.2 equiv of olefin.
e 1.2 equiv of *m*-xylylamine.

(propen-3-yl)sulfonylaniline (see Equation 6.7) [18]. Conversion to the cyclic product occured only when the amine was activated by a sulfonyl group and when a base (typically triethylamine) was added as a co-catalyst. However, competitive olefin isomerization to the corresponding 2-(propen-1-yl) derivative constituted the major reaction path in this reaction. Thus, the hydroamination product is obtained in up to 40% yield and 67% ee, the highest selectivity being obtained with the Josiphos ligand.

$$\text{2-allyl-NHSO}_2\text{Tol} \xrightarrow[\text{Benzene, 353 K}]{\substack{\text{0.5 mol\% [IrCl((R)-(S)-Josiphos)]}_2 \\ \text{2 mol\% NEt}_3}} \text{2-methyl-1-(SO}_2\text{Tol)-indoline} \quad (6.7)$$

6.2.5
The Ir(I)/Piperidine/Methacrylonitrile System

The addition of piperidine to methacrylonitrile is catalyzed by the phosphine-free cationic complex [Ir(COD)$_2$]BF$_4$, according to Equation 6.8 [19]. An acceleration of the reactions was observed when phosphines such as PPh$_3$ (2 equiv.), BINAP (1 equiv.) and/or DPPE (1 equiv.) were added *in situ*. In these cases the reactions were complete after 2 h, compared to 50% conversion with [Ir(COD)$_2$]BF$_4$ in the same timespan.

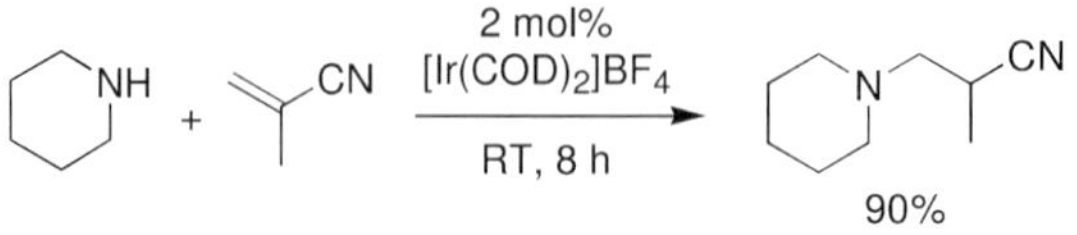

$$\text{piperidine-NH} + \text{CH}_2=\text{C(CH}_3\text{)CN} \xrightarrow[\text{RT, 8 h}]{\text{2 mol\% [Ir(COD)}_2\text{]BF}_4} \text{piperidine-N-CH(CH}_3\text{)CN} \quad 90\% \qquad (6.8)$$

6.3
Iridium-Catalyzed Alkyne Hydroamination (AHA)

6.3.1
Intramolecualar Aliphatic Systems

Cationic Rh(I) and Ir(I) complexes such as **14** and **15** bearing bidentate nitrogen-donor ligands (see Chart 6.2) were used by Field, Messerle and coworkers for the cyclization of 4-pentyneamine to 2-methylpyrroline (Equation 6.9), and the activities of the Ir catalyst precursors are summarized in Table 6.3 [20]. While catalyst precursors **16** and **17** bearing phosphine–N-heterocyclic carbene (NHC) ligands showed comparably low activities (see Table 6.3) [21], the use of phosphine-pyrazolyl ligated cationic complexes led to a breakthrough in terms of activity [22]. In particular, complex **18** gave the best results with a TOF of 3100 h^{-1} at 50% conversion, thus being by far the most active transition-metal catalyst for this specific reaction. The penta-coordinated carbonyl adduct of **18** showed a much lower catalytic activity. Analogues of **18** that had the isopropyl groups substituted for H, Me

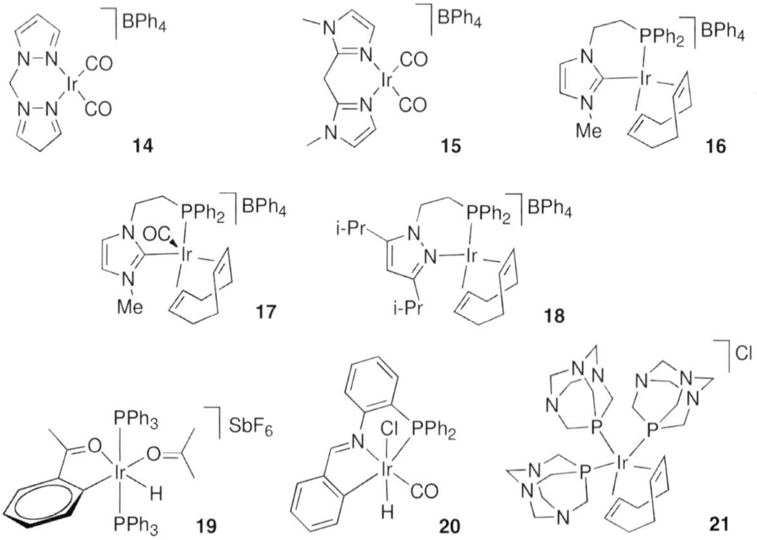

Chart 6.2 Complexes used for the cyclization of alkyneamines.

6.3 Iridium-Catalyzed Alkyne Hydroamination (AHA)

Table 6.3 Turnover frequencies (TOFs) for the reaction of Equation 6.9 at 50% conversion.

Complex	14	15	16	17	18
TOF (h^{-1}) (Loading, solvent)	5 (1.5 mol%, THF-D_8)	50 (1.5 mol%, THF-D_8)	12 (1.4 mol%, THF-D_8); 15 (1.3 mol%, $CDCl_3$)	19 (0.9 mol%, THF-D_8); 28 (1.4 mol%, $CDCl_3$)	3100 (1.4 mol%, $CDCl_3$)

or Ph were less reactive. The methyl-substituted analogue of complex **18**, as well as the carbonyl complex **20**, were structurally characterized (see Figures 6.3 and 6.4, respectively). Higher homologues of pentyneamine and phenylalkyneamines were also used as substrates [23].

$$\text{HC≡C-CH}_2\text{CH}_2\text{-NH}_2 \xrightarrow[333\,K]{\{\text{complexes 14-18}\}} \text{2-methyl-1-pyrroline} \quad (6.9)$$

Water-soluble PTA-ligated Ir(I) complexes (PTA = 1,3,5-triaza-7-phosphaadamantane) such as **21** were used by Krogstad to perform the cyclization of 4-pentyneamine (Equation 6.8) in water [24]. The TOF at 50% conversion was 0.33 h^{-1} for complex **21**, while the carbonyl adducts [IrCl(CO)(PTA)$_3$] and [Ir(CO)(PTA)$_4$]Cl produced TOFs of 0.36 and 0.32 h^{-1}, respectively. The X-ray crystal structure of complex **21** is shown in Figure 6.5.

6.3.2
Indoles via Intramolecular AHA

Crabtree and coworkers used the cyclometallated Ir(III)-hydride complex **19** for the intramolecular hydroalkoxylation and hydroamination of aromatic alkynes (Equation 6.10) [25], while Liu and coworkers used the neutral Ir(III)-hydride complex **20** in combination with cocatalytic amounts of NaB[3,5-$C_6H_3(CF_3)_2$]$_4$ for the synthesis of a range of indoles (Table 6.4) [26].

$$\text{(2-alkynyl-N-phenylaniline)} \xrightarrow[\text{RT, 2 h, 96\%}]{0.5\%\ \mathbf{19}} \text{(1-phenyl-2-}^n\text{Pr-indole)} \quad (6.10)$$

6.3.3
Intermolecular Alkyne Hydroamination

In the above-mentioned report [26], it was also shown that complex **20** catalyzed the intermolecular tandem alkyne hydroamination/hydrosilylation

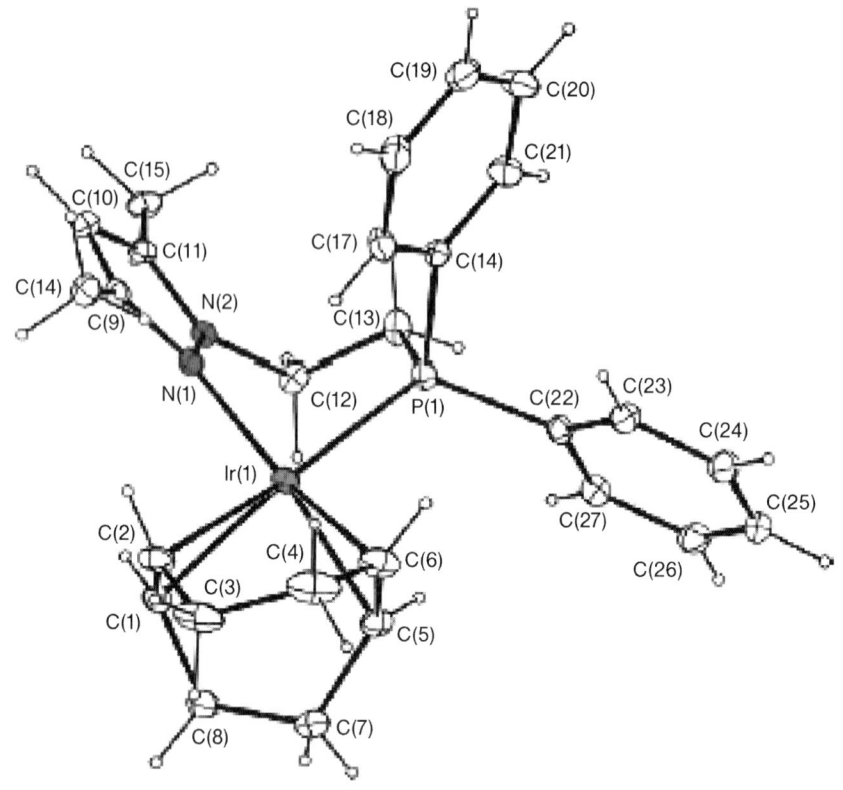

Figure 6.3 ORTEP view of the cationic fragment of [Ir(Me$_2$PyP)(COD)]BPh$_4$ (20% probability ellipsoids, reproduced with permission from the American Chemical Society). Selected bond lengths (Å) and angles (°): Ir(1)–P(1) 2.2882(6); Ir(1)–N(1) 2.1045(16); Ir(1)–C(1) 2.2072(2); Ir(1)–C(2) 2.33(2); Ir(1)–C(5) 2.132(7); Ir(1)–C(6) 2.140(2); C(1)–C(2) 1.379(3); C(5)–C(6) 1.408(3); P(1)–Ir(1)–N(1) 84.31(5); P(1)–Ir(1)–C(5) 93.73(7); P(1)–Ir(1)–C(6) 95.99(6); N(1)–Ir(1)–C(1) 91.43(7); N(1)–Ir(1)–C(2) 96.69(8); C(1)–Ir(1)–C(6) 81.05(8); C(2)–Ir(1)–C(5) 80.62(10); C(1)–Ir(1)–C(5) 96.92(9); C(2)–Ir(1)–C(6) 87.92(9).

(Equation 6.11), which yielded 95% of purified free amine after hydrolysis of the silylated amine.

(6.11)

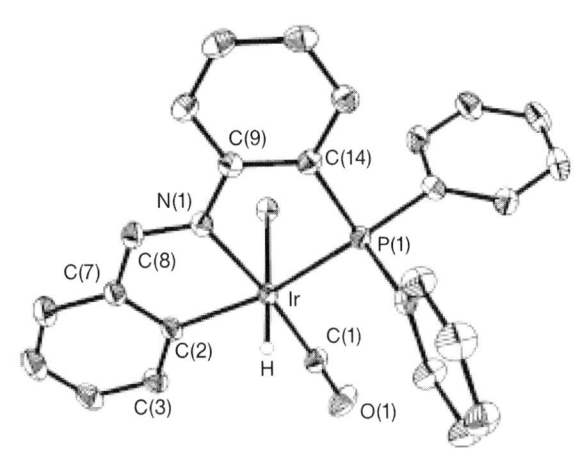

Figure 6.4 ORTEP view of complex **20** (30% probability ellipsoids, reproduced with permission from the American Chemical Society). Selected bond lengths (Å) and angles (°): Ir–N(1) 2.088(2), Ir–P(1) 2.3469(7), Ir–C(2) 2.069(3), N(1)–C(8) 1.30(2), N(1)–C(9) 1.437(4), Ir–C(1) 1.855(3), Ir–Cl(1) 2.4875(7), C(1)–O(1) 1.146(4), N(1)–Ir–C(2) 79.3(1), N(1)–Ir–P(1) 82.54(7), P(1)–Ir–C(2) 160.70(9), Ir–N(1)–C(9) 121.8(2), Ir–N(1)–C(8) 115.1(2).

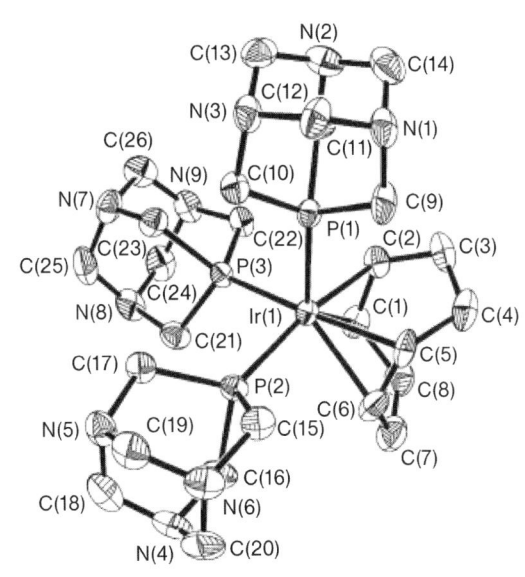

Figure 6.5 ORTEP view of the cation of complex **21** (50% probability ellipsoids, reproduced with permission from Elsevier). Selected bond lengths (Å) and angles (°): Ir–P1 2.349(2), Ir–C1 2.208(7), Ir–P2 2.3007(18), Ir–C2 2.194(7), Ir–P3 2.3022(18), Ir–C5 2.226(7), C1–C2 1.431(11), Ir–C6 2.236(7), C5–C6 1.404(12), P1–Ir–P2 99.65(7), P1–Ir–C5 88.1(2), P1–Ir–P3 95.93(7), P2–Ir–C5 98.2(2), P2–Ir–P3 89.95(6), P3–Ir–C2 92.4(2), P1–Ir–C2 90.3(2), C2–Ir–C5 78.7(3).

Table 6.4 Intramolecular hydroamination of aromatic alkynes.

R¹	R²	Z	Time (h)	Yield (%)
H	Ph	H	12	95
H	H	H	24	10
H	n-Bu	H	6	96
H	n-Bu	Me	6	96
CH_3CO	n-Bu	H	24	21
H	Ph	NO_2	24	16
H	n-Bu	Cl	24	84
H	Ph	Cl	24	27

6.4
Proposed Mechanisms

6.4.1
Olefin Hydroamination

Despite the fact that the CMM system (see Section 6.2.2, Equation 6.4) showed modest activities, it nevertheless demonstrated the feasibility of catalytic OHA governed by classic organometallic reactivity, namely oxidative addition, migratory insertion and reductive elimination. The authors were able to propose a catalytic cycle as outlined in Scheme 6.1 based on the following experimental observations:

- The catalyst precursor $IrCl(PEt_3)_2(C_2H_4)_2$ readily lost ethylene to form the intermediate **22** (or its dimer).

- The initial step of the cycle forming **23** was supported by separate stoichiometric reactions involving similar Ir(I) complexes that oxidatively added the N—H bond of aniline (see Section 6.5.2). It should be noted that norbornene does not form stable adducts with electron-rich Ir(I) complexes such as those used in this catalysis (R. Dorta and A. Togni, unpublished results).

- The formal migratory insertion of the C=C double bond into the Ir-amide function to form the new C—N bond in the key intermediate **24** was conclusively proven by an X-ray crystal structure analysis (see Figure 6.11).

- The last step, reductive C—H elimination from the Ir(III) intermediate, was shown by kinetic studies of the decomposition of isolated **24** to proceed rapidly in the presence of chloride abstractors such as $ZnCl_2$ or $TlPF_6$. Based on this

Scheme 6.1 The CCM olefin hydroamination cycle.

observation, the authors postulated a pentacoordinated cationic intermediate **25** that precedes reductive elimination.

The origin of the fluoride effect in the enantioselective addition of aniline to norbornene mentioned in Section 6.2.3 remains a matter of speculation. Togni and coworkers proposed the formation of Ir(I)–fluoride intermediates that would augment the reactivity of the metal center [27] and improve stereoselectivity thanks to a directing effect of the hydrogen bridge that would form between the incoming amine and the fluoride ligand (adduct **26**, see Scheme 6.2) [27]. A σ-bond metathesis-type process could lead to the Ir(I)–amido–HF adduct **28** which would give way to a formal oxidative addition of HF leading to Ir(III)–amido-hydrido-fluoride complex **29**. The importance of hydrogen bridging between a coordinated amine and halogen ligands in Ir complexes is known [28, 29], and might explain the improved selectivity of the Ir/fluoride system, while enhanced reactivity [40] towards oxidative addition [27] due to fluoride ligation has also precedents (*vide infra*).

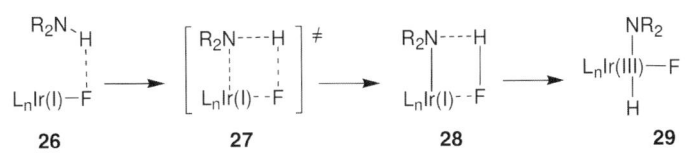

Scheme 6.2 The proposed role of 'naked fluoride' in Ir-catalyzed olefin hydroamination.

In the recent development of the CMM reaction by Hartwig a catalytic cycle based on a monomeric Ir(I) amide as active species was postulated (**31**, Scheme 6.3)[16]. In the reaction mixture **31** is thought to form from *in situ* generated chloro bridged dimeric complexes of type **30** (*cf.* Chart 1 above) upon metathetical exchange of the chloride with anilide ligands. Potassium anilide is produced from substrate aniline and the KHMDS additive. The ensuing reactivity follows the CCM cycle of Scheme 6.1, *i.e.* oxidative addition of the N—H bond of the substrate aniline to form **32**, followed by the migratory insertion of norbornene (**33**) and final reductive C—H elimination to close the cycle. Catalyst deactivation is postulated to be due to the formation of the dimer **34**. A catalytically equally active

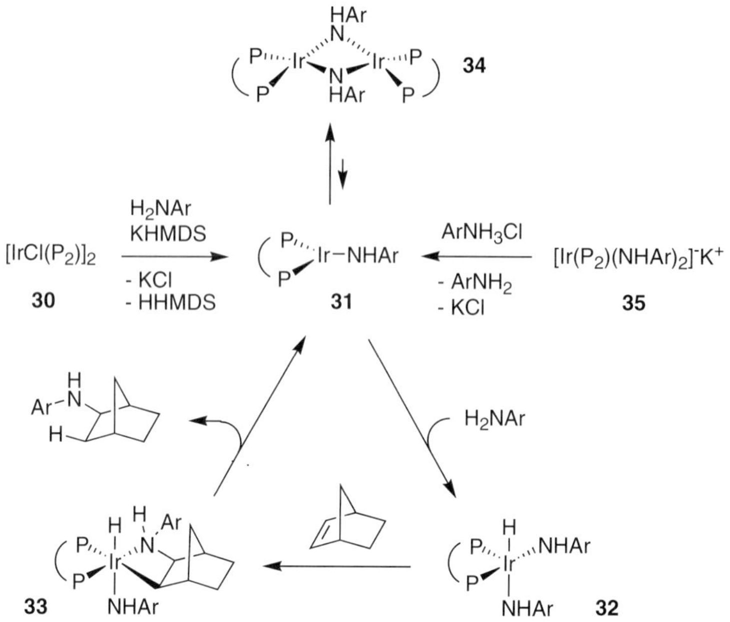

Scheme 6.3 Proposed role of KHMDS in Ir-catalyzed OHA.

system was obtained by protonating the iridato complex **35** with an equimolar amount of an anilinium salt. However, in the absence of substrate alkene the neutral dimer **34** formed. We note that the only difference between this cycle and the CMM cycle in Scheme 6.1 is the presence of an anilide ancillary ligand instead of chloride. The higher basicity of the anilide ligand may lead to faster N—H activation since the Ir centre is richer in electrons. On the other hand, the use of substituted anilines in combination with the Segphos and Biphep ligands bearing bulky aryl subsituents seem both to displace the equilibrium from dimer **34** towards the active monomer **31**. When comparing the "naked fluoride" with the KHDMS effect it is tempting to explain the former just with its comparatively high Brœnstedt basicity. However, the observation that the Ir/BINAP system works well with F⁻ but not with KHMDS as co-catalysts points to the formation of different catalytically active species. Complexes analogous to **34** and **35** and their conversion were discribed in a separate study (see Section 6.5 below).

6.4.2
Alkyne Hydroamination

Crabtree and coworkers proposed a catalytic cycle for the reaction outlined in Equation 6.10. The mechanism is based on labeling and kinetic studies, and is outlined in Scheme 6.4 [25]. Adduct **36** was observed in nuclear magnetic resonance (NMR) spectra and appears to be a catalyst resting state. It should be noted that there is no change in the oxidation state of Ir, and that the key step is thought

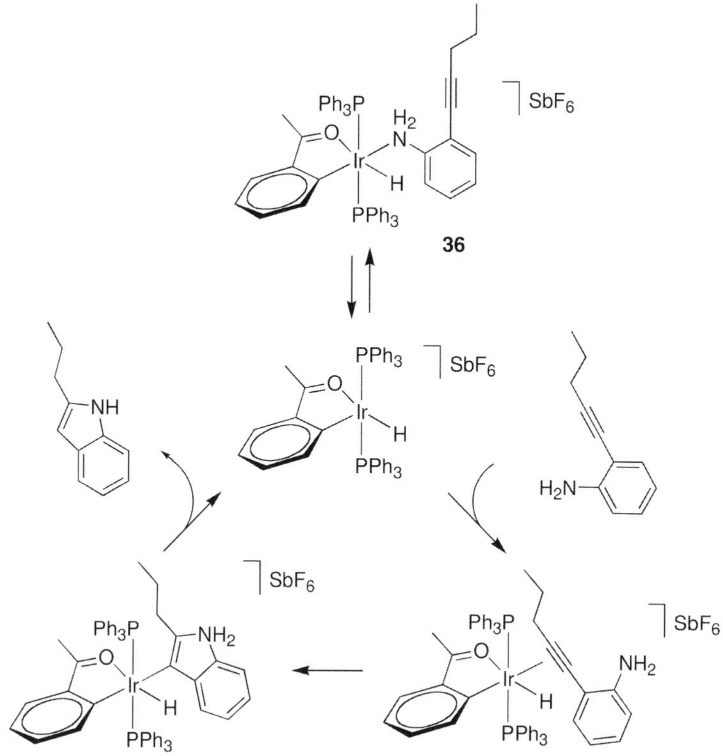

Scheme 6.4 The proposed mechanism for the intramolecular alkane hydroamination/cyclization reaction (Equation 6.10) [24, 25].

to be the electrophilic activation of the alkyne function by the Lewis acidic Ir(III) center towards nucleophilic attack by the amine function. An analogous mechanism for intermolecular AHA was proposed by Liu and coworkers, and is depicted in Scheme 6.5.

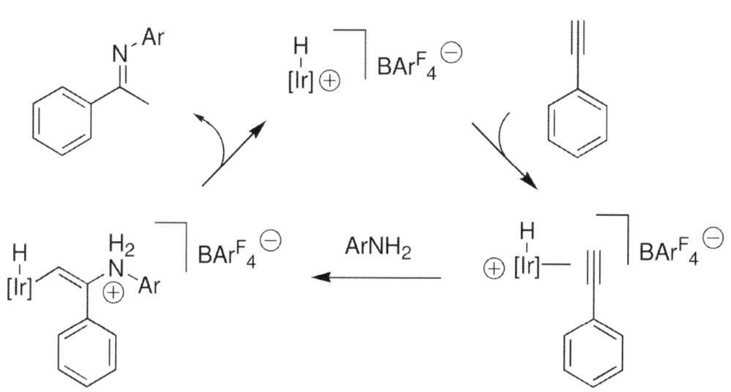

Scheme 6.5 The proposed mechanism for the intermolecular alkane hydroamination reaction [25].

6.5
Complexes and Reactions of Ir Relevant to Hydroamination

6.5.1
Ir(I)–Amine Complexes

The reaction preceding the actual activation of an incoming amine would be its coordination to the electron rich Ir(I) center. However, the isolation of such complexes is not simple. First, the acidity of the metal center is low compared to an Ir(III) center to coordinate the incoming amine base, and second there is a soft/hard mismatch between the metal acid and the amine base. It is therefore not surprising that the number of well-characterized Ir(I)–amine complexes is not large, and only few of them have been structurally characterized [30]. A compound that serves as model for the first step in the CMM cycle (preceding intermediate **23** of Scheme 6.1) was synthesized by reacting the chloro-bridged Ir dimer **37** with NEt_2H to form the adduct **38**, according to Equation 6.12, and its the crystal structure is shown in Figure 6.6 [31].

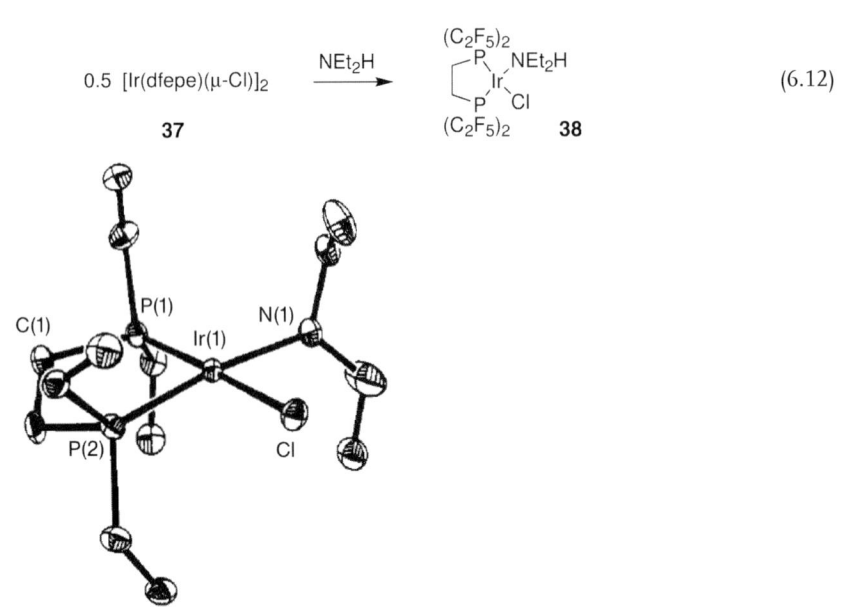

$$0.5 \ [Ir(dfepe)(\mu\text{-}Cl)]_2 \xrightarrow{NEt_2H} \begin{array}{c}(C_2F_5)_2\\ \text{P}\diagdown\ \diagup NEt_2H \\ \quad Ir \\ \text{P}\diagup\ \diagdown Cl \\ (C_2F_5)_2\end{array} \qquad (6.12)$$

37   **38**

Figure 6.6 Molecular structure of (dfpe)Ir(NEt$_2$H)Cl (**38**, thermal ellipsoids drawn at the 30% level, reproduced with permission from the American Chemical Society). Selected bond lengths (Å) and angles (°): Ir(1)–P(1) 2.151(2), Ir(1)–P(2) 2.165(2), Ir(1)–Cl 2.356(2), Ir(1)–N(1) 2.185(8), P(1)–Ir(1)–P(2) 84.3(1), P(1)–Ir(1)–Cl 174.7(1), P(1)–Ir(1)–N(1) 98.0(2), P(2)–Ir(1)–Cl 91.4(1), P(2)–Ir(1)-N(1) 172.9(2), Ir(1)–P(1)–C(1) 114.3(3), Ir(1)–P(1)–C(3) 119.5(3), Ir(1)–P(1)–C(5) 122.4(3), C(1)–P(1)–C(3) 103.4(4), C(1)–P(1)–C(5) 99.4(4), C(3)–P(1)–C(5) 93.9(4), Ir(1)–P(2)–C(2) 113.1(3), Ir(1)–P(2)–C(7) 119.7(3), Ir(1)–P(2)–C(9) 121.4(3), C(2)–P(2)–C(7) 101.0(4), C(2)–P(2)–C(9) 99.0(4), C(7)–P(2)–C(9) 98.9(4), Ir(1)–N(1)–C(11) 121.0(8), Ir(1)–N(1)–C(21) 116.7(6), C(11)–N(1)–C(21) 117.1(10).

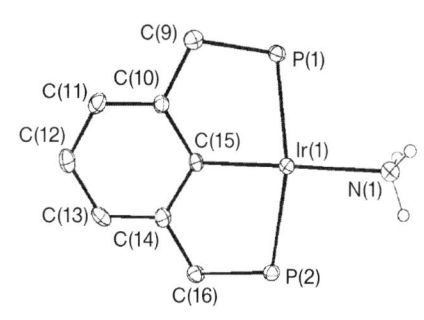

Figure 6.7 ORTEP view of complex **39** (50% probability ellipsoids, reproduced with permission from the American Chemical Society). Selected bond lengths (Å) and angles (°): Ir(1)–N(1) 2.215(5), Ir(1)–C(15) 2.013(4), Ir(1)–P(2) 2.2610(13), Ir(1)–P(1) 2.2737(14), C(15)–Ir(1)–N(1) 175.87(19), N(1)–Ir(1)–P(2) 97.05(13), N(1)–Ir(1)–P(1) 97.93(13).

The fact that complex **38** does not react further – that is, it does not oxidatively add the N—H bond – is due to the comparatively low electron density present on the Ir center. However, in the presence of more electron-rich phosphines an adduct similar to **38** may be observed *in situ* by NMR (see Section 6.5.3; see also below), but then readily activates N—H or C—H bonds. Amine coordination to an electron-rich Ir(I) center further augments its electron density and thus its propensity to oxidative addition reactions. Not only accessible N—H bonds are therefore readily activated but also C—H bonds [32] (cf. cyclo-metallations in Equation 6.14 and Scheme 6.10 below). This latter activation is a possible side reaction and mode of catalyst deactivation in OHA reactions that follow the CMM mechanism. Phosphine-free cationic Ir(I)–amine complexes were also shown to be quite reactive towards C—H bonds [30a]. The stable Ir–ammonia complex **39**, which was isolated and structurally characterized by Hartwig and coworkers (Figure 6.7) [33], is accessible either by thermally induced reductive elimination of the corresponding Ir(III)–amido-hydrido precursor or by an acid–base reaction between the 14-electron Ir(I) intermediate **53** and ammonia (see Scheme 6.9).

6.5.2
Ir(I)-Anilido Complexes

Ir(I) anilides are believed to be the catalytically active species in the OHA of norbornenes with anilines when dimeric [IrCl(P_2)]$_2$ (P_2 = Segphos, Biphep) complexes are used as catalyst precursors in combination with KHMDS (Section 6.2.4, *vide supra*). The reactive species are postulated to be monomeric three-coordinate Ir(I) anilido complexes with formally 14 valence electrons (**31**, Scheme 6.3). Indeed, **31** was also generated by *in situ* protonation of the iridate catalyst precursor K[Ir(L$_{-2}$)(NH*m*Xylyl)$_2$] (**35**, Scheme 6.3) following the general protocol for the synthesis of Ir(I) anilido complexes as outlined in Scheme 6.6 [45]. Lithium iridates such as complex **40** were isolated in excellent yields and controlled alcoholysis led to the corresponding neutral dimer **41**. The solid state structure of this complex is depicted in Figure 6.8 revealing the square planar coordination environment of the Ir centers and the butterfly shaped Ir$_2$N$_2$ core.

Scheme 6.6 Preparation of anionic and neutral anilido bridged Ir(I) complexes.

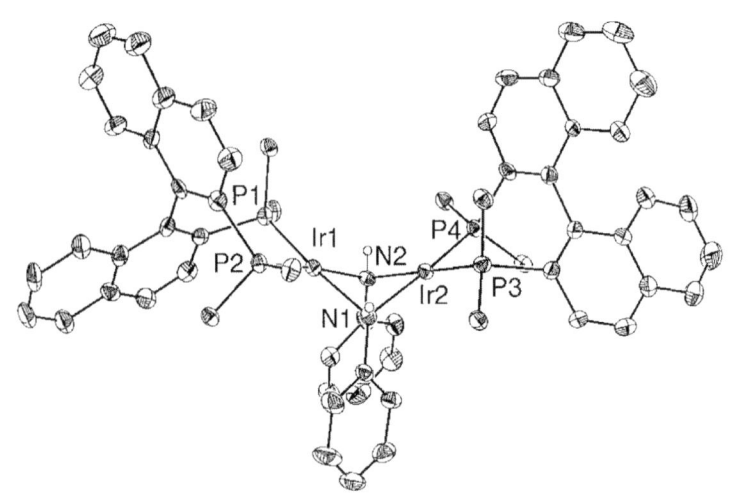

Figure 6.8 ORTEP view of complex **41** (30% probability ellipsoids, with permission from VCH). Selected bond lengths (Å) and angles (deg): Ir1-N1 2.166(7), Ir1-N2 2.128(8), Ir1-P1 2.229(3), Ir1-P2 2.213(3), Ir1-Ir2 3.0215(5), Ir2-N1 2.168(7), Ir2-N2 2.156(7), Ir2-P3 2.214(2), Ir2-P4 2.198(2), N1-Ir1-N2 78.0(3), P1-Ir1-P2 89.08(10), N1-Ir2-N2 77.4(3), P3-Ir2-P4 89.77(9), Ir1-N1-Ir2 88.4(2), Ir1-N2-Ir2 89.7(3).

6.5.3
N—H Bond Activation Leading to Ir(III)–Amido-Hydrido Complexes

An understanding of N—H activation via oxidative addition to Ir(I) fragments – and of its microscopic reverse, reductive elimination – is of fundamental importance

6.5 Complexes and Reactions of Ir Relevant to Hydroamination

to Ir-catalyzed OHA. Indeed, it should be considered the key step and *conditio sine qua non* of Ir-catalyzed OHA. The N—H oxidative addition results in Ir(III)–hydrido-amido species that are anticipated to be highly reactive due to the mismatch of orbital energies caused by the combination of a soft late transition-metal center with hard amide ligands.

The Ir(III)–amido-hydrido complex **43** was isolated by reacting the electron-rich neutral phosphine–cyclooctene complex **42** according to Equation 6.13 [7] (note the *cis* arrangement of the hydride and amide function). Likewise, the cationic complex **44** reacted in neat aniline to afford a 50/50 mixture of the N—H (**45**) and C—H (**46**) activation products that was isolated as a light orange powder (Equation 6.14). Compound **45** was separated from **46** and purified in 49% overall yield.

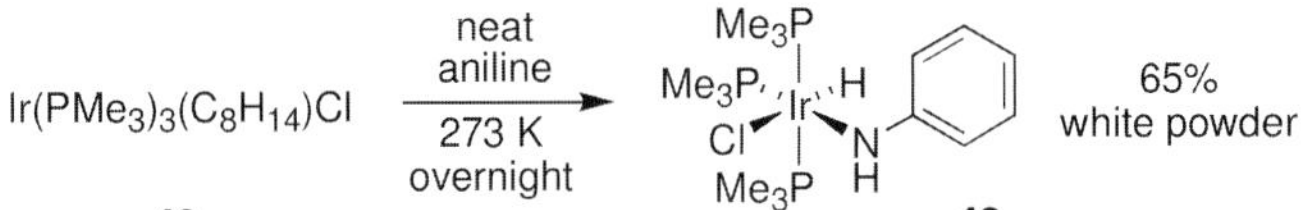

$$\text{Ir(PMe}_3)_3(\text{C}_8\text{H}_{14})\text{Cl} \xrightarrow[\text{273 K overnight}]{\text{neat aniline}} \text{43} \quad 65\% \text{ white powder} \quad (6.13)$$

42

$$[\text{Ir(PMe}_3)_4]\text{PF}_6 \xrightarrow[\substack{348 \text{ K} \\ 36 \text{ h}}]{\text{neat aniline}} \text{45} + \text{46} \quad (6.14)$$

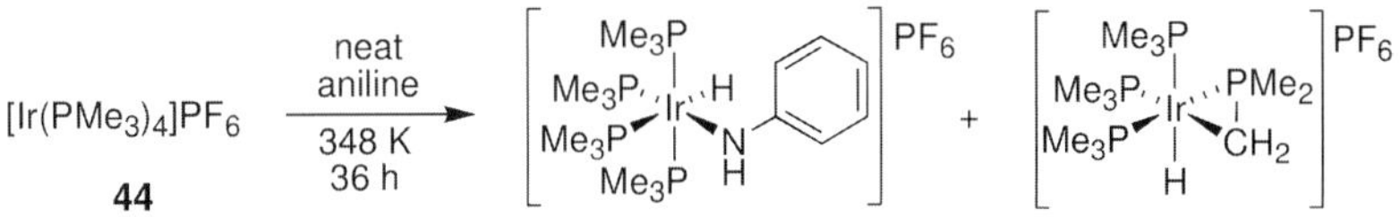

44   50% **45**   50% **46**

Ammonia represents the biggest challenge for functionalization by oxidative addition [31] since its bond dissociation energy of 107 kcal mol^{-1} is between 10 and 20% higher when compared to primary and secondary amines [34]. Early work by DuPont researchers showed that the electron-rich Ir(I) center of complex **1** is indeed capable of oxidatively adding the N—H bond of ammonia to form dinuclear amido-bridged Ir(III) complexes, as outlined in Scheme 6.7 [36], and the dinuclear structures of **48** and **49** were confirmed by single-crystal X-ray diffraction studies (Figure 6.9) [6].

The phosphine-free complex [IrCl(COE)$_2$]$_2$ (COE = cyclooctene) was also shown to activate ammonia to form mixtures of Ir–hydride and Ir–amide complexes [37].

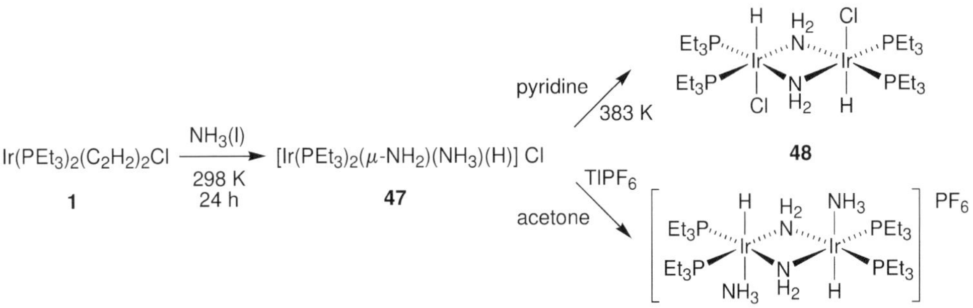

$$\text{Ir(PEt}_3)_2(\text{C}_2\text{H}_2)_2\text{Cl} \xrightarrow[\substack{298 \text{ K} \\ 24 \text{ h}}]{\text{NH}_3(\text{l})} [\text{Ir(PEt}_3)_2(\mu\text{-NH}_2)(\text{NH}_3)(\text{H})]\text{Cl}$$

1   **47**

pyridine / 383 K → **48**

TlPF$_6$ / acetone → **49**

Scheme 6.7 Oxidative addition of ammonia to Ir(I).

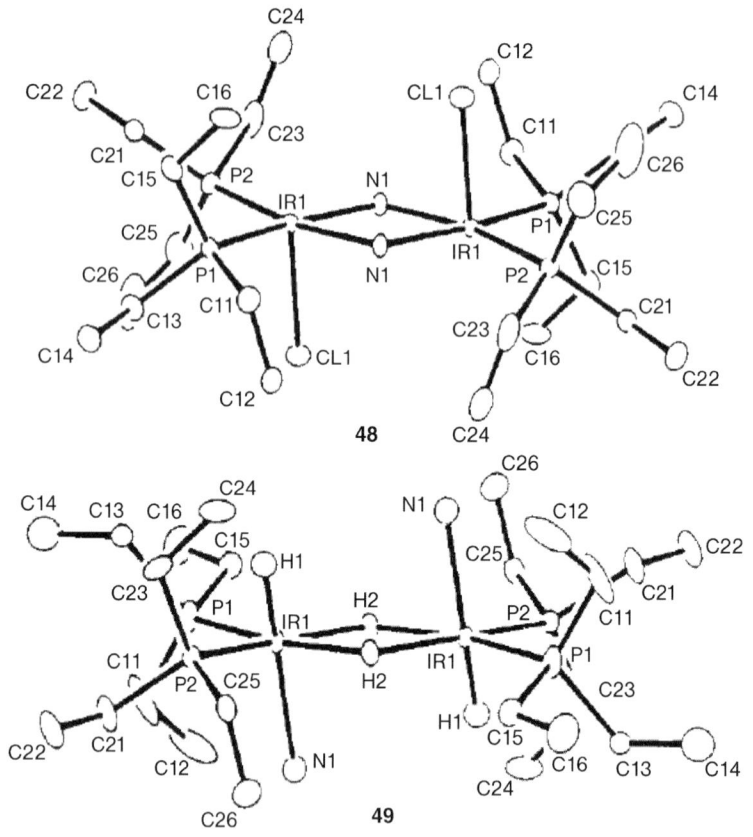

Figure 6.9 ORTEP view of **48** and the cation of **49** (thermal ellipsoids at the 20% probability level, reproduced with permission from the American Chemical Society). Selected bond lengths (Å) and angles (°) for **48**: Ir1–N1 2.10(1), Ir1–N1 2.13(1), Ir1–P1, 2.264(4), Ir1–P2 2.264(4), Ir1–Cl1 2.504(4), Cl1–Ir1–P1 96.4(1), Cl1–Ir1–P2 94.3(1), Cl1–Ir1–N 87.8 (3), Cl1–Ir1–N1 84.4(3), P1–Ir1–N1 168.6(3), P2–Ir1–N1 168.7(3), N1–Ir1–N1 75.3(5). For **49**: Ir–N1 2.244(4), Ir1–N2 2.128(3), Ir1–N2 2.127(3), Ir1–H1 1.61(5), Ir1–P1 2.274(1), Ir1–P2 2.280(1), N2–Ir1–N2 77.0(1), N1–Ir1–N2 84.3(1), N1–Ir1–N2 84.5(1), P1–Ir1–N2 170.1(1), P2–Ir1–N1 169.9(1), H1–Ir1–N1 157(2).

In a further development, Hartwig and coworkers studied a reversible Ir(I) pincer system **50** that led to the Ir–amide **52**, via a 14-valence electron complex **51** (Scheme 6.6) [38]. Note the electronic similarity of intermediate **51** with the postulated active catalyst of the CMM cycle (*vide supra*). The 14-valence electron

Scheme 6.8

intermediate **53** generated *in situ* and bearing an aromatic PCP pincer ligand also oxidatively added the N—H bond of aniline to form the anilide **54**, but not the N—H bond of ammonia, which in this case led to the Lewis base adduct **39**. Amide **55** was isolated in a separate synthesis at low temperature and shown to reductively eliminate ammonia and to form its adduct **39** [33]. Moreover, it was shown that the N—H bond of aniline is vastly favored for oxidative addition to the aromatic (PCP)Ir(I) fragment over the C—H bond of benzene. This is likely due to stronger ionic and pπ–dπ interactions between the Ir fragment and the amido ligand [33]. Monomeric amides such as **52** and **54** are potentially more reactive than corresponding dimeric structures in catalytic applications [39]. Their crystal structures are shown in Figures 6.10 and 6.11, respectively, and the structure of **39** is shown in Figure 6.7 (see also Section 6.5.1).

6.5.4
Alkyl–Amino-Hydrido Complexes of Ir(III)

Once the N—H bond has been oxidatively added to the Ir(I) complex (in the context of the CCM cycle, *vide supra*), the resultant Ir(III) intermediate is a Lewis acid that is thought to coordinate the olefin. A synergistic effect between the coordinated electrophilically activated olefin and the highly nucleophilic nature of the amido function is believed to facilitate the C—N bond formation within the coordination sphere of the Ir center (see **56**). Alkyl–amino–Ir(III) complexes, such as the key intermediate **24** of the CCM system (as described in Section 6.2.1) are of paramount importance to better understand Ir-catalyzed hydroaminations. Complex

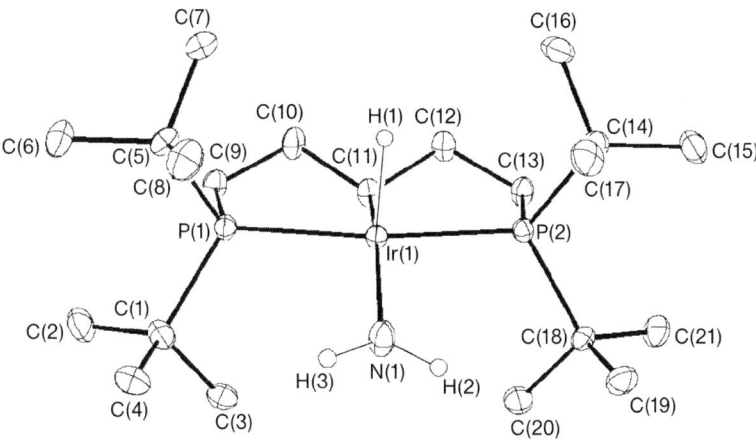

Figure 6.10 ORTEP view of complex **52** (50% probability ellipsoids, reproduced with permission from Sciencemag). Selected bond lengths (Å) and angles (°): Ir(1) H(1) 1.51(3), Ir(1) N(1) 1.999(4), Ir(1) C(11) 2.128(4), Ir(1) P(2) 2.2978(11), Ir(1) P(1) 2.2995(11), N(1) H(3) 0.90(5), N(1) H(2) 0.91(6), H(1) Ir(1) N(1) 139.4(13), H(1) Ir(1) C(11) 66.8(13), N(1) Ir(1) C(11) 153.69(17), H(1) Ir(1) P(2) 82.7(15), N(1) Ir(1) P(2) 96.66(13), C(11) Ir(1) P(2) 82.82(11), H(1) Ir(1) P(1) 90.4(15), N(1) Ir(1) P(1) 96.50(13), C(11) Ir(1) P(1) 83.22(11), P(2) Ir(1) P(1) 165.97(4).

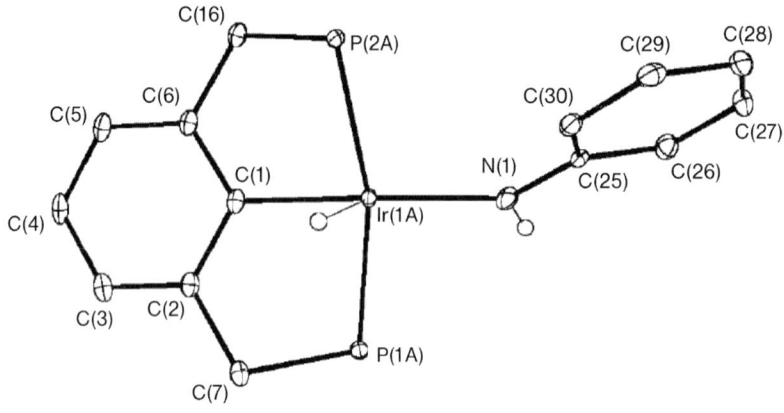

Figure 6.11 ORTEP view of complex **54** (50% probability ellipsoids, reproduced with permission from the American Chemical Society). Selected bond lengths (Å) and angles (°): Ir(1A)–N(1) 2.082(2), C1–Ir(1A) 2.049(2), Ir(1A)–P(1A) 2.2917(14), Ir(1A)–P(2A) 2.3429(11), C(25)–N(1)–Ir(1A) 133.27(16), C(1)–Ir(1A)–N(1) 164.49(9), N(1)–Ir(1A)–P(1A) 92.78(8), N(1)–Ir(1A)–P(2A) 102.69(7).

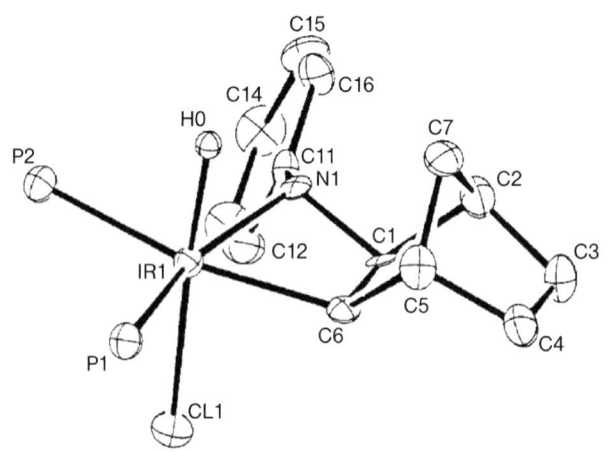

Figure 6.12 ORTEP drawing of complex **24** (reproduced with permission from the American Chemical Society). Selected distances (Å) and angles (°) are: Ir(1)–Cl(1) 2.515 (5), Ir(1)–P(1) 2.238(2), Ir(1)–P(2) 2.339(2), Ir(1)–N(1) 2.202(6), Ir(1)–C(6) 2.120(7), Ir(1)–H 1.705 (75), N(1)–C(11) 1.465(9), N(1)–C(1) 1.520(9); Cl(1)–Ir(1)–H 175(2), P(1)–Ir(1)–P(2) 100.02(7), P(1)–Ir(1)–N(1) 162.1(2), P(2)–Ir(1)–N(1) 96.3(2), P(1)–Ir(1)–C(6) 96.4(2), P(2)–Ir(1)–C(6) 163.5 (2), N(1)–Ir(1)–C(6) 67.2(2), Ir(1)–N(1)–C(1) 91.4(4), Ir(1)–N(1)–C(11) 130.3(5).

24 was isolated as a white powder in 50% yield and 90% purity, starting from norbornene and aniline according to Equation 6.15, and an X-ray crystal structure analysis thereof was undertaken [7]. The structure and a selection of bond distances and angles is shown in Figure 6.12 (note the *exo* orientation of the formed C(1)–N(1) bond).

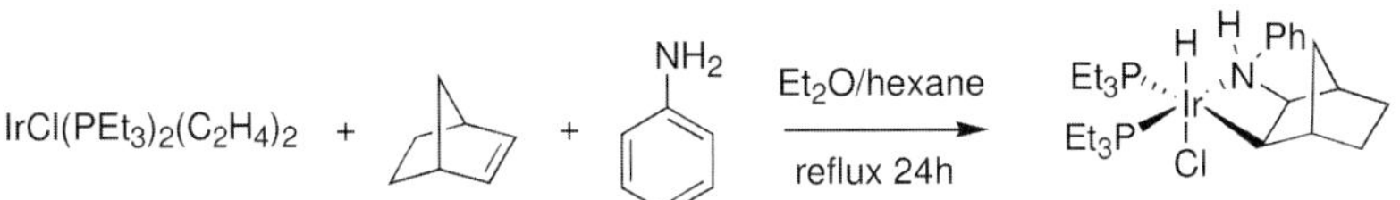

6.5 Complexes and Reactions of Ir Relevant to Hydroamination | 167

$$IrCl(PEt_3)_2(C_2H_4)_2 \; + \; \text{[norbornene]} \; + \; \text{[PhNH}_2\text{]} \xrightarrow[\text{reflux 24h}]{Et_2O/\text{hexane}} \; \textbf{24}$$

(6.15)

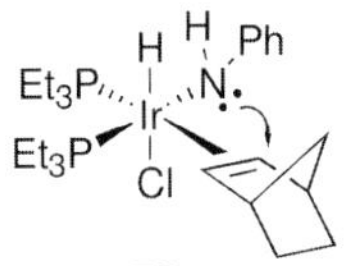

56

Complex **24** was used in thermal and Lewis acid-catalyzed decomposition experiments. Intramolecular C—H reductive elimination from **24** to form *exo*-2-phenylaminonorbornane was demonstrated with labeling experiments [7].

In closely related experiments it was shown that sp^3 C—H activation takes place reversibly within the coordination sphere of the electron-rich Ir(I)–diphosphine complex **58** (Scheme 6.9) to form an alkyl-amino-hydrido derivative **57** reminiscent of the CCM intermediate **24**; the solid-state structure of **57** is shown in Figure 6.13 [40]. It appears that C—H activation only takes place after coordination of the amine function to the Ir(I) center (complex **58**, NMR characterized). Amine coordination allows to break the chloro bridge of **59** and to augment the electron density of the metal center, thus favoring oxidative addition of the C—H bond. Most importantly, the microscopic reverse of this C—H activation process (i.e. C—H reductive elimination) models the final step of the CCM cycle (see Scheme 6.1); indeed, the reaction of Scheme 6.10 is cleanly reversible at 373 K.

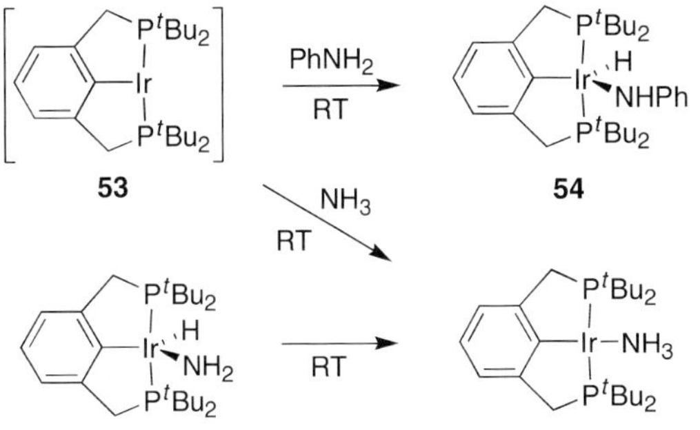

Scheme 6.9

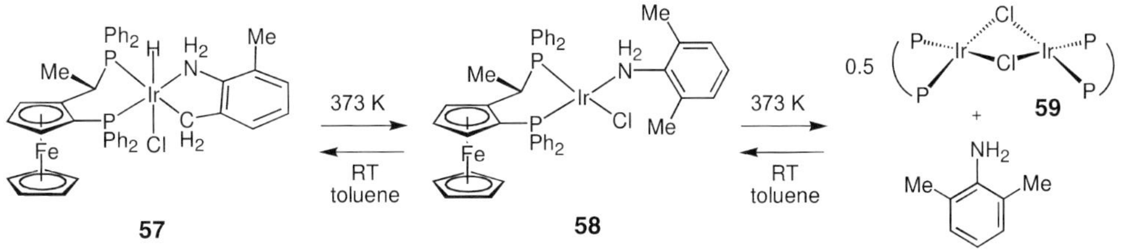

Scheme 6.10 Reversible intramolecular C—H activation in an Ir(I)–amine complex.

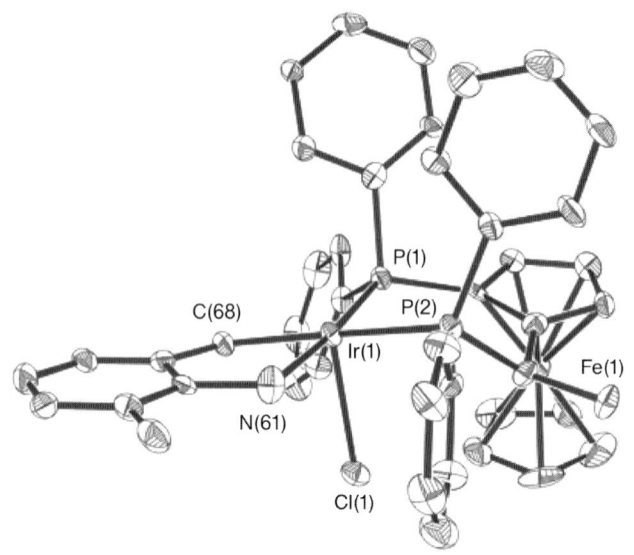

Figure 6.13 ORTEP drawing of complex **57** (reproduced with permission from the American Chemical Society). Selected distances (Å) and angles (°) are: Ir(1)–P(1) 2.224(3), Ir(1)–P(2) 2.325(3), Ir(1)–Cl(1) 2.543(3), Ir(1)–N(61) 2.110(10), Ir(1)–C(68) 2.108(12); P(1)–Ir(1)–P(2) 90.18(12), P(2)–Ir(1)–N(61) 93.1(3), P(1)–Ir(1)–Cl(1) 98.74(11), P(2)–Ir(1)–Cl(1) 91.52(11), P(1)–Ir(1)–C(68) 96.1(3), N(61)–Ir(1)–C(68) 80.6(4), C(68)–Ir(1)–Cl(1) 88.9(3), N(61)–Ir(1)–Cl(1) 85.2(3).

Fryzuk demonstrated the high reactivity of the amide function present in the P–amide–P pincer Ir complex **60** towards dihydrogen [28]. In fact, heterolytic dihydrogen activation by the Ir(III)–amide bond present in the pincer complex **60** led to the stable Ir(III)–alkyl-amino-hydrido complex **61**, as outlined in Equation 6.16. The X-ray crystal structure of the amine product is depicted in Figure 6.14.

6.5.5
Iridium–Fluoride Complexes

Although fluorides play a crucial Role in some Ir-catalyzed reactions [9, 41], experimental information regarding relevant Ir–fluoride complexes is scarce. Vaska-type fluoride complexes have been reported but are not well characterized [42], while structurally characterized Ir–fluorides are even rarer [43]. Crabtree and coworkers estimated the hydrogen bridge strength of the F … H interaction in complexes **62** and **63** (see below) to be approximately 5–7 kcal mol^{-1} [29]. On the other hand,

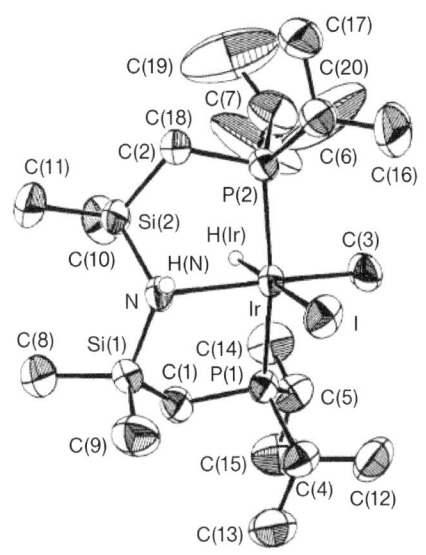

Figure 6.14 Molecular structure and numbering scheme for **61** (reproduced with permission from the American Chemical Society). Selected distances (Å) and angles (°) are: Ir–I 2.8176(4), Ir–P(1) 2.3070(13), Si(2)–N 1.763(5), Ir–P(2) 2.3152(14), Ir–N 2.368(4), Ir–C(3) 2.127(6), Ir–H(Ir) 1.65, N–H(N) 0.70(8), I–Ir–P(1) 97.29(3), I–Ir–P(2) 94.88(4), I–Ir–N 86.59(11), I–Ir–C(3) 91.9(2), I–Ir–H 172, P(1)–Ir–P(2) 167.30(5), P(1)–Ir–N 88.72(11), P(1)–Ir–C(3) 90.5(2).

Caulton and coworkers found that fluoride ligands in certain Ir complexes promote oxidative addition reactions [44]. This group's results showed that the fluoride complex Ir(H)$_2$F(P^{t}Bu$_2$Ph)$_2$ rapidly activated C–H bonds under dehydrogenation conditions. The reactive intermediate in these reactions may be a fluoro-bridged analogue of compounds **4–12**, namely [Ir(μ-F)(P^{t}Bu$_2$Ph)$_2$]$_2$. This would explain the improved reactivity in the Ir-catalyzed OHA reaction in the presence of cocatalytic 'naked fluoride'.

62 L = PPh$_3$ **63**

6.6 Conclusions

Late transition-metal hydroamination is the method of choice for the atom economical and functional group-tolerant construction of C–N bonds, and in this context Ir plays a central role (indeed, homogenous transition-metal-catalyzed OHA was discovered with Rh and Ir). However, there is a strong need for the development of better OHA catalyst systems that are applicable to a wider range of substrates and conditions. The characteristics of current Ir based catalyst systems to function **via** N–H bond activation, though, is a potential handicap to achieve this goal, since it implies highly reactive Ir intermediates that are prone to secondary reactions (e.g. C—H activation) and functional group intolerance. In this context, olefin rather than amine activation seems to represent the more promising strategy, as was demonstrated by Hartwig with, *inter alia*, Pd(II)-based systems. In contrast, Ir-catalyzed alkyne hydroamination that relies on alkyne activation has experienced some interesting developments during recent years.

References

1 For leading reviews on hydroamination, see: (a) Brunet, J.J. and Neibecker, D. (2001) *Catalytic Heterofunctionalization* (eds A. Togni and H. Grützmacher), Wiley-VCH Verlag GmbH, Weinheim, pp. 91–141;
(b) Müller, T.E. and Beller, M. (1998) *Chemical Reviews*, **98**, 675;
(c) Alonso, F., Beletskaya, I.P. and Yus, M. (2004) *Chemical Reviews*, **104**, 3079, and references cited therein.

2 Steinborn, D. and Taube, R. (1986) *Zeitschrift für Chemie*, **26**, 349.

3 For recent reviews, see: (a) Hultzsch, K.C. (2005) *Advanced Synthesis Catalysis*, **347**, 367;
(b) Roesky, P.W. and Müller, T.E. (2003) *Angewandte Chemie – International Edition*, **42**, 2708.

4 (a) Gresham, W.F., Brooks, R.E. and Bruner, W.M. (1950) US Patent 2,501,509; assigned to E.I. du Pont de Nemours and Co.;
(b) A recent variant thereof may be: Beller, M., Breindl, C., Riermeier, T.H. and Tillack, A. (2001) *The Journal of Organic Chemistry*, **66**, 1403.

5 (a) Coulson, D.R. (1971) *Tetrahedron Letters*, **5**, 429;
(b) For the first example of an addition of a primary amine catalyzed by Rh, see: Diamond, S.E., Szalkiewicz, A. and Mares, F. (1978) *Journal of the American Chemical Society*, **101**, 490.

6 Casalnuovo, A.L., Calabrese, J.C. and Milstein, D. (1987) *Inorganic Chemistry*, **26**, 971.

7 Casalnuovo, A.L., Calabrese, J. and Milstein, D. (1988) *Journal of the American Chemical Society*, **110**, 6738.

8 (a) Giardello, M.A., Conticello, V.P., Brard, L., Gagné, M.R. and Marks, T.J. (1994) *Journal of the American Chemical Society*, **116**, 10241;
(b) Li, Y. and Marks, T.J. (1996) *Organometallics*, **15**, 3770;
(c) Li, Y. and Marks, T.J. (1996) *Journal of the American Chemical Society*, **118**, 9295, and references cited therein.

9 Dorta, R., Egli, P., Zürcher, F. and Togni, A. (1997) *Journal of the American Chemical Society*, **119**, 10857.

10 The aniline/norbornene model system was recently employed to show the feasibility of $TiCl_4$-catalyzed OHA: Ackermann, L., Kaspar, L.T. and Gschrei, C.J. (2004) *Organic Letters*, **6**, 2515.

11 (a) Werner, H., Wolf, J. and Höhn, A. (1985) *Journal of Organometallic Chemistry*, **287**, 395;
(b) Hitchcock, P.B., Morton, S. and Nixon, J.F. (1985) *Journal of the Chemical Society - Dalton Transactions*, 1295;
(c) Fryzuk, M.D., Mc Conville, D.H. and Rettig, S. (1993) *Journal of Organometallic Chemistry*, **445**, 245.

12 Yamagata, T., Iseki, A. and Tani, K. (1997) *Chemistry Letters*, 1215.

13 Onderdelinden, A.L. and van der Ent, A. (1972) *Inorganica Chimica Acta*, **6**, 420.

14 1,1,1,3,3,3-Hexakis-(dimethylamino)-diphosphazenium fluoride. See (a) Schwesinger, R. (1991) *Angewandte Chemie*, **103**, 1376;
(b) Schwesinger, R. (1991) *Angewandte Chemie – International Edition in English*, **29**, 1372;
(c) Wollenweber, M., Pinkos, R., Leonhardt, J. and Prinzbach, H. (1994) *Angewandte Chemie*, **106**, 84;
(d) Wollenweber, M., Pinkos, R., Leonhardt, J. and Prinzbach, H. (1994) *Angewandte Chemie - International Edition in English*, **33**, 117.

15 Vasen, D., Salzer, A., Gerhards, F., Gais, H.-J., Stürmer, R., Bieler, N.H. and Togni, A. (2000) *Organometallics*, **19**, 539.

16 Auf den Blatten, R., Diezi, S. and Togni, A. (2000) *Monatsheft Für Chemiker*, **131**, 1345.
17 Zhou, J., Hartwig, J.F. (2008) *Journal of the American Chemical Society*, ASAP Article, DOI: 10.1021/ja803523z.
18 Togni, A., Bieler, N., Burckhardt, U., Köllner, G., Pioda, G., Schneider, R. and Schnyder, A. (1999) *Pure and Applied Chemistry*, **71**, 1531.
19 Kawatsura, M. and Hartwig, J.F. (2001) *Organometallics*, **20**, 1960.
20 Burling, S., Field, L.D., Messerle, B.A. and Turner, P. (2003) *Organometallics*, **23**, 1714.
21 Field, L.D., Messerle, B.A., Vuong, K.Q. and Turner, P. (2005) *Organometallics*, **24**, 4241.
22 (a) Field, L.D., Messerle, B.A., Vuong, K.Q., Turner, P. and Failes, T. (2007) *Organometallics*, **26**, 2058;
(b) For a one-pot tandem hydroamination/hydrosialation version of the reaction outlined in Equation 6.5, see Field, L.S., Messerle, B.A. and Wren, S.L. (2003) *Organometallics*, **22**, 4393.
23 Burling, S., Field, L.D., Messerle, B.A. and Rumble, S.L. (2007) *Organometallics*, **26**, 4335.
24 Krogstad, D.A., DeBoer, A.J., Ortmeier, W.J., Rudolf, J.W. and Halfen, J.A. (2005) *Inorganic Chemistry Communications*, **8**, 1141.
25 Li, X., Chianese, A.R., Vogel, T. and Crabtree, R.H. (2005) *Organic Letters*, **7**, 5437.
26 Lai, R.-Y., Surekha, K., Hayashi, A., Ozawa, F., Liu, Y.-H., Peng, S.-M. and Liu, S.-T. (2007) *Organometallics*, **26**, 1062.
27 Togni, A., Dorta, R., Köllner, C. and Pioda, G. (1998) *Pure and Applied Chemistry*, **70**, 1477.
28 Fryzuk, M.D., MacNeil, P.A. and Rettig, S.J. (1987) *Journal of the American Chemical Society*, **109**, 2803.
29 (a) Peris, E., Lee, J.C., Rambo, J.R., Eisenstein, O. and Crabtree, R.H. (1995) *Journal of the American Chemical Society*, **117**, 3485;
(b) Patel, B.P. and Crabtree, R.H. (1996) *Journal of the American Chemical Society*, **118**, 13105.
30 (a) Dorta, R., Broggini, D., Kissner, R. and Togni, A. (2004) *Chemistry – A European Journal*, **10**, 4546;
(b) Roundhill, D.M., Bechtold, R.A. and Roundhill, S.G.N. (1980) *Inorganic Chemistry*, **19**, 284;
(c) Bianchini, C., Farnetti, E., Graziani, M., Nardin, G., Vacca, A. and Zanobini, F. (1990) *Journal of the American Chemical Society*, **112**, 9190;
(d) Van der Zeijden, A.A.H., van Koten, G., Luijk, R., Nordemann, R.A. and Spek, A.L. (1988) *Organometallics*, **7**, 1549.
31 Schnabel, R.C. and Roddick, D.M. (1993) *Inorganic Chemistry*, **32**, 1513.
32 Schulz, M. and Milstein, D. (1993) *Journal of the Chemical Society, Chemical Communications*, 318.
33 Kanzelberger, M., Zhang, X., Emge, T.J., Goldman, A.S., Zhao, J., Incarvito, C. and Hartwig, J.F. (2003) *Journal of the American Chemical Society*, **125**, 13644.
34 For a recent Highlight article, see: Braun, T. (2005) *Angewandte Chemie - International Edition*, **44**, 5012.
35 Roundhill, D.M. (1992) *Chemical Reviews*, **92**, 1.
36 For a hydrido-amido-Ir(III) complex that is not the result of oxidative addition, see: Matsuzaka, H., Ariga, K, Kase, H., Kamura, T., Kondo, M., Kitagawa, S. and Yamasaki, M. (1997) *Organometallics*, **16**, 4514.
37 Koelliker, R. and Milstein, D. (1991) *Angewandte Chemie – International Edition*, **30**, 707.
38 Zhao, J., Goldman, A.S. and Hartwig, J.F. (2005) *Science*, **307**, 1080.
39 (a) Ir(I) anilide bridged complexes do not catalyze the addition of aniline to norbornene. Dorta, R. and Togni, A. (2000) *Helvetica Chimica Acta*, **83**, 119;
(b) Tejel, C., Ciriano, M.A., Bordonaba, M., López, J.A., Lahoz, F.J. and Oro, L.A. (2002) *Chemistry – A European Journal*, **8**, 3128.
40 Dorta, R. and Togni, A. (1998) *Organometallics*, **17**, 3423.
41 Cooper, A.C. and Caulton, K.G. (1996) *Inorganica Chimica Acta*, **251**, 41.
42 (a) Vaska, L. (1971) *Transactions of the New York Academy of Sciences*, **33**, 70;
(b) Vaska, L. and Peone, J. (1971) *Journal of the Chemical Society – Dalton Transactions*, 418.

43 (a) Veltheer, J.E., Burger, P., Bergman, R.G. (1995) *Journal of the American Chemical Society*, **117**, 12478;
(b) Bourgeois, C.J., Garratt, S.A., Hughes, R.P., Larichev, R.B., Smith, J.M., Ward, A.J., Willemsen, S., Zhang, D., DiPasquale, A.G., Zakharov, L.N. and Rheingold, A.L. (2006) *Organometallics*, **25**, 3474.

44 Cooper, A.C., Folting, K., Huffman, J.C. and Caulton, K.G. (1997) *Organometallics*, **16**, 505.

45 Dorta, R. and Togni, A. (2000) *Helvetica Chimica Acta*, **83**, 119; For dynamic and structural studies of similar, phosphine free complexes, see: Tejel, G., Giriano, M.A., Bordonaba, M., López, J.A. and Oro, L.A. (2002) *Chemistry – A European Journal*, **8**, 3128.

7
Iridium-Catalyzed Boron-Addition
Elena Fernández and Anna M. Segarra

7.1
Introduction

The main advantages of using organoboranes as intermediates for synthetic organic purposes [1] are their stability, relatively low toxicity and easy accessibility. In particular, the catalytic activation of unsaturated carbon–carbon bonds through boron addition may be considered as a platform for introducing functionality with particular emphasis on the selective control of C—B formation and the retention of configuration in the functionalization process from the organoborane intermediates towards the targeted products [2]. Transition-metal complexes face the catalytic new C—B bond formation, and in particular iridium complexes have been shown essentially to be efficient, with a distinctive stability and catalytic control.

7.2
Iridium–Boryl Complexes

The remarkable stability of iridium–boryl complexes, as a function of the substituents on boron, is most likely responsible for the unique behavior of iridium in metal-promoted B-addition to unsaturated molecules.

Although boranes can be easily added to low valence–late transition metals by means of oxidative additions [3], the first example of this conceptual advantage was reported in the iridium complex $[Ir(\mu\text{-}Cl)(C_8H_{14})]_2$ intramolecular B—H activation of 1-dimethylphosphino-1,2-carborane [4] (Scheme 7.1), and extended to intermolecular B—H [5] and B—X [6] activation (Schemes 7.2 and 7.3). The great stability of these iridium complexes made it possible to determine the first crystal structure of a transition metal–boranyl complex in $[IrBr_2(B_5H_8)(CO)(PMe_3)_2]$ (**1**) [7]. At that time, it was claimed that a σ-bonded boron had a stronger *trans*-lengthening influence than a σ-bonded carbon.

It is no coincidence that the first X-ray structure analysis of a transition metal–boryl complex [8] was made on the iridium complex *fac*-$[IrH_2(BC_8H_{14})(PMe_3)_3]$ (**2**) [9] derived from the B—H activation of the 9-borabicyclo[3.3.1]-nonane (9-BBN)

Iridium Complexes in Organic Synthesis. Edited by Luis A. Oro and Carmen Claver
Copyright © 2009 WILEY-VCH Verlag GmbH & Co. KGaA, Weinheim
ISBN: 978-3-527-31996-1

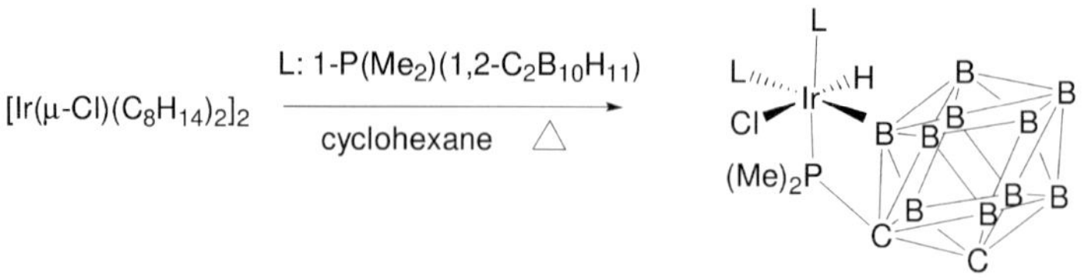

Scheme 7.1

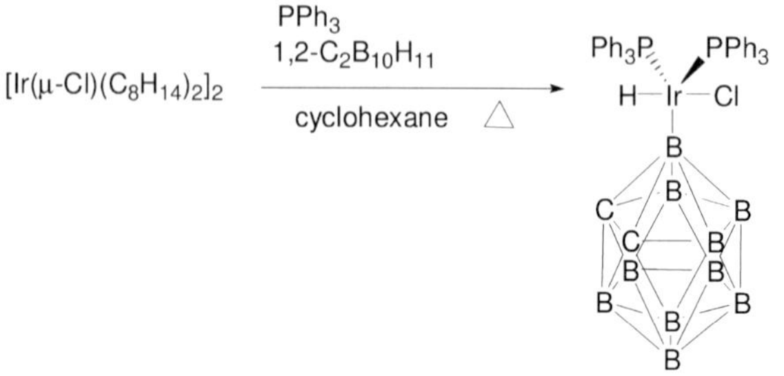

$[Ir(\mu\text{-Cl})(C_8H_{14})_2]_2 \xrightarrow[\text{cyclohexane } \triangle]{\text{L: 1-P(Me}_2)(1,2\text{-C}_2B_{10}H_{11})}$ [Ir complex with L, H, Cl, (Me)₂P, and carborane cage]

Scheme 7.2

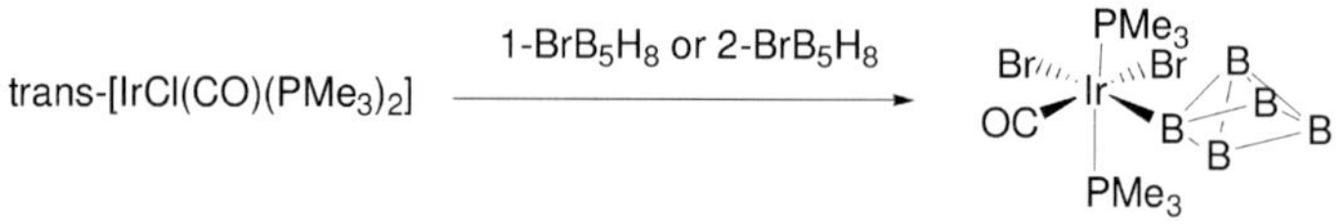

$[Ir(\mu\text{-Cl})(C_8H_{14})_2]_2 \xrightarrow[\text{cyclohexane } \triangle]{\substack{\text{PPh}_3 \\ 1,2\text{-C}_2B_{10}H_{11}}}$ [Ir complex with Ph₃P, PPh₃, H, Cl and carborane cage]

Scheme 7.3

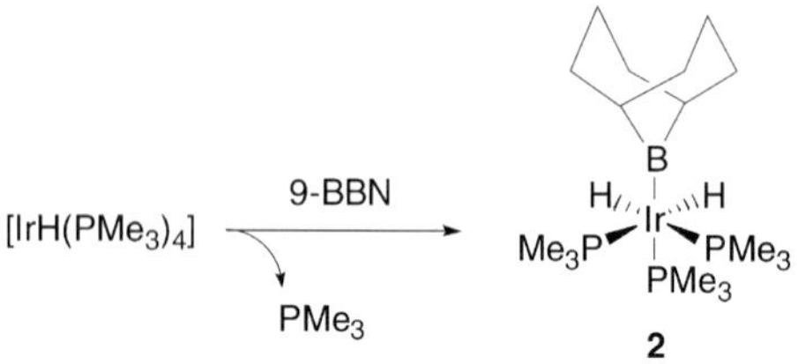

$\text{trans-}[IrCl(CO)(PMe_3)_2] \xrightarrow{1\text{-BrB}_5H_8 \text{ or } 2\text{-BrB}_5H_8}$ complex **1** [Ir with Br, Br, OC, PMe₃, PMe₃, and B₅ cage]

Scheme 7.4

$[IrH(PMe_3)_4] \xrightarrow[\text{PMe}_3]{9\text{-BBN}}$ complex **2** [Ir with H, H, Me₃P, PMe₃, PMe₃, and 9-BBN boryl group]

with [IrH(PMe₃)₄] [10] (Scheme 7.4). The use of catechol group (cat = 1,2-O₂C₆H₄) as a substituent on boron facilitated synthesis of the contemporary iridium complex [IrClH(Bcat)(PMe₃)₃] (**3**) [11] from the B—H activation of catecholborane and [IrCl(COE)(PMe₃)₃], (COE = cyclooctene) (Scheme 7.5). Both complexes **2** and **3** have a distorted octahedral geometry about the Ir(III) center, but with a facial and meridional arrangement of PMe₃ in **2** and **3**, respectively. The Ir—B bond distance for **2** (2.093(7) Å) is similar to the Ir—B bond distances in **1**, (2.071(14) Å), thus suggesting the weakness of the dπ-pπ backbonding from Ir to B. This is conceptually important because the catalyzed B-addition and related reactions are based on the facile cleavage of the metal–boron bonds as an indication of the kinetic lability of boryl complexes [12]. However, a tendency towards the formation of π back-

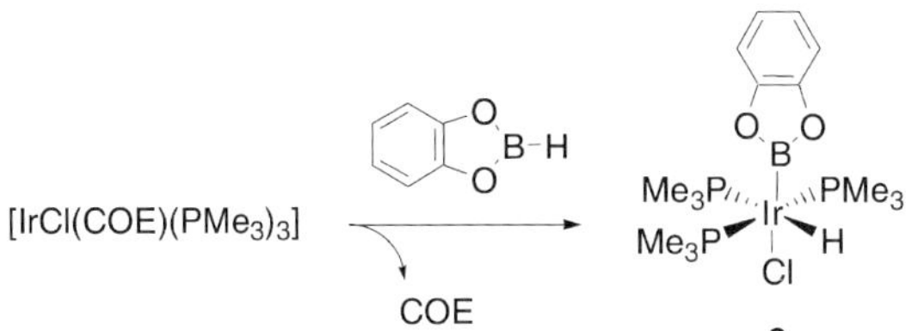

7.2 Iridium–Boryl Complexes

Scheme 7.5

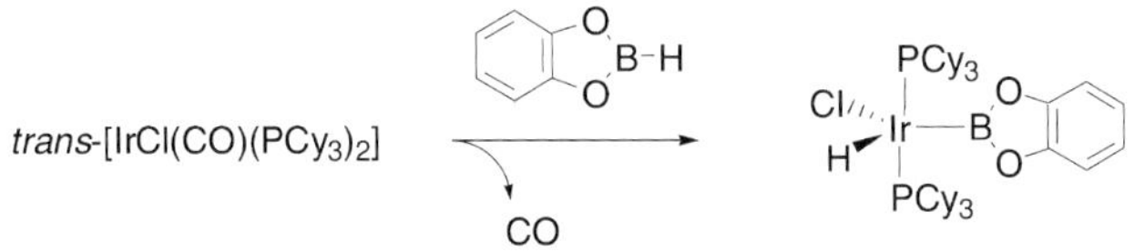

[IrCl(COE)(PMe$_3$)$_3$] + HBcat → [Ir(Bcat)(H)(Cl)(PMe$_3$)$_3$] (3) + COE

Scheme 7.6

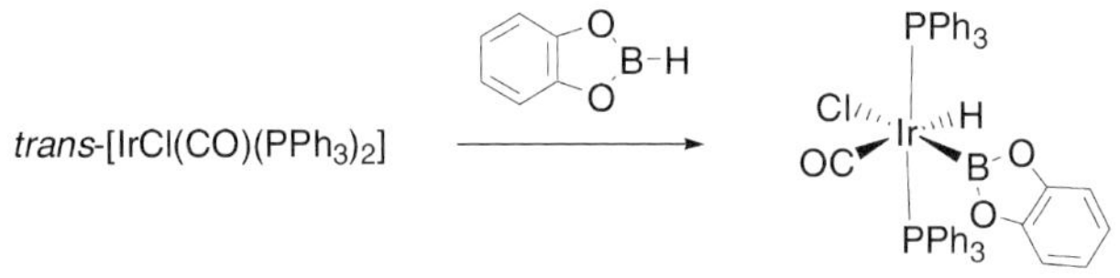

trans-[IrCl(CO)(PCy$_3$)$_2$] + HBcat → [IrClH(Bcat)(PCy$_3$)$_2$] (4) + CO

Scheme 7.7

trans-[IrCl(CO)(PPh$_3$)$_2$] + HBcat → trans-[IrClH(Bcat)(CO)(PPh$_3$)$_2$] (5)

bonding could depend on the substituents bonded to boron and the relative position of the boryl ligand in the metal complex. This additional stabilization is observed in complex **3**, where the catechol group enables backbonding from iridium to boron (Ir—B bond distance = 2.023(7) Å), and justifies the broad applicability of catecholborane derivatives in the synthesis of boryl–metal complexes. Also, the equatorial position of the boryl ligand in the related complex [IrClH(Bcat)(PCy$_3$)$_2$] (**4**) [13] (Scheme 7.6), could explain why the iridium–boron bond length is shortened with respect to the sum of the covalent radii, indicating the presence of π backbonding.

The presence of an ancillary CO ligand has been reported to facilitate the oxidative addition of HBcat to a metal center. Thus, the saturated iridium boryl complex trans-[IrClH(Bcat)(CO)(PPh$_3$)$_2$] (**5**) was prepared in high yield by reacting HBcat with Vaska's complex trans-[IrCl(CO)(PPh$_3$)$_2$] (Scheme 7.7). This was the first example of a metal–carbonyl boryl complex characterized by crystal X-ray diffraction [14]. The CO ligand played two significant roles: (i) it inhibited the PPh$_3$ orthometallation, hence guaranteeing a coordination site required for the oxidative addition of HBcat; and (ii) it destabilized the resulting Ir—B bond as a consequence of its electron-withdrawing nature as a π acid ligand. As the knowledge of transition metal–ligand covalent bond energies became more important in understanding catalysis [15], the first information on metal–boron bond energy was provided for the iridium–catecholboryl linkage of [IrClH(Bcat)(CO)(PPh$_3$)$_2$] (**5**) [16]. The

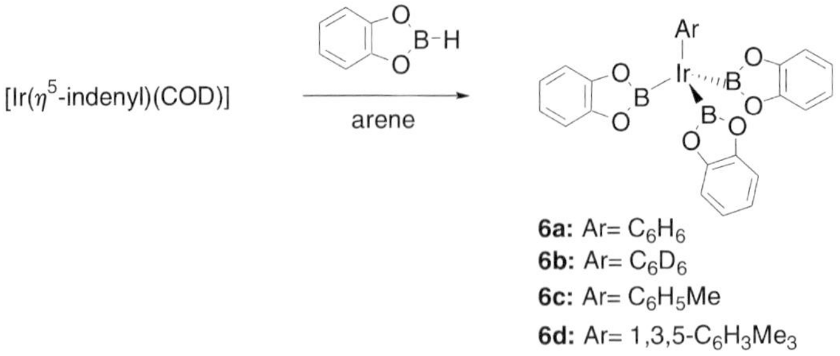

Scheme 7.8

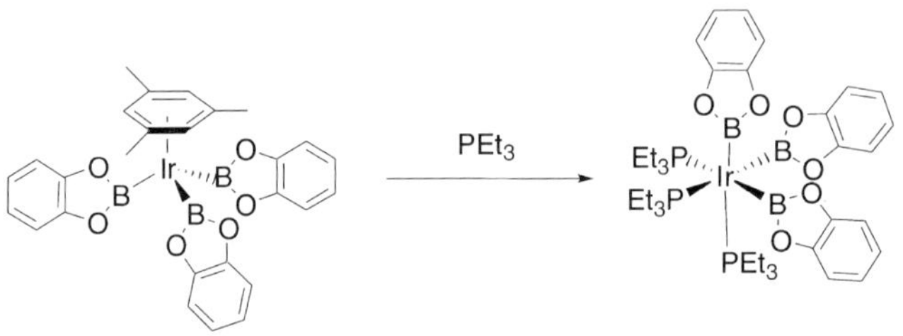

6a: Ar= C_6H_6
6b: Ar= C_6D_6
6c: Ar= C_6H_5Me
6d: Ar= 1,3,5-$C_6H_3Me_3$

Scheme 7.9

iridium–boron bond strength value was found to exceed those of both corresponding alkyls and hydrides.

The pronounced stability of iridium–boryl complexes can be observed in the synthesis of trisboryl complexes such as [Ir(η^6-arene)(Bcat)$_3$] (**6**), (arene: **6a** = C_6H_6, **6b** = C_6D_6, **6c** = C_6H_5Me, **6d** = 1,3,5-$C_6H_3Me_3$), from [Ir(η^5-indenyl)(COD)], (COD = 1,5-cyclo-octadiene), with an excess of catecholborane in arene solvents (Scheme 7.8). The electron density at Ir can be finely tuned by arene ring displacement using various phosphines. Therefore, compound **6d** undergoes complete displacement of the mesitylene ligand when treated with 3 equiv. of PEt$_3$, to form fac-[Ir(Bcat)$_3$(PEt$_3$)$_3$] (**7**) [17] (Scheme 7.9). Analogously to the catecholboryl complex, the iridium complex fac-[Ir(Bpin)$_3$(PMe$_3$)$_3$] (**8**) [18], with the pinacolboryl ligand (pin = OCMe$_2$CMe$_2$O) was isolated, even though compounds with substituents on boron other than the catechol group are rather rare [19].

7.3
Hydroboration

The catalytic H—B addition of organoboron compounds to unsaturated molecules is synthetically interesting in main-group chemistry, as a wide range of functional groups can be transformed from the selective new C—B bond [1]. Although considerable energy has been expended to use rhodium(I) complexes to catalyze the

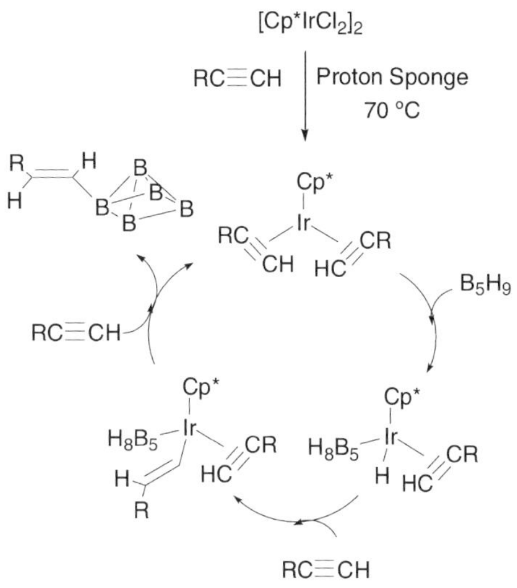

Scheme 7.10

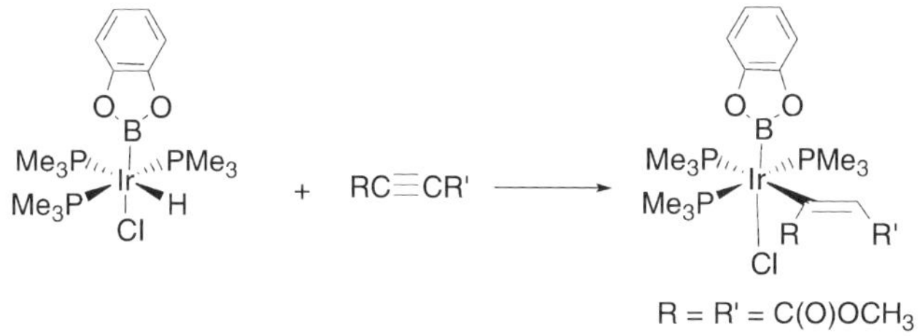

Scheme 7.11

hydroboration of unsaturated compounds [20], analogous iridium complexes have also been shown to be active through alternative mechanistic insights [21].

Initially, Sneddon and coworkers [22] reported that iridium complex [IrCl$_2$(η^5-C$_5$Me$_5$)]$_2$ catalyzed the H—B addition of polyhedral boranes to alkynes to give a variety of alkenylboranes. The same group also suggested a plausible catalytic cycle involving the oxidative addition of borane [23], followed by the insertion of an alkyne into a metal–hydride bond to form a metallovinyl complex which, after reductive elimination [24], would form alkenylpentaborane and regenerate the catalytic species (Scheme 7.10). Merola *et al.* [11] suggested that iridium complexes could be a useful system for modeling the catalytic hydroboration mechanism, as vinyliridium intermediates were observed and characterized when the vinyl group contained strong electron-withdrawing substituents (Scheme 7.11). Evans and Fu [25] subsequently used [Ir(COD)(PCy$_3$)(py)]PF$_6$ (py = pyridine) as the catalyst precursor in the first example of a catalytically directed hydroboration reaction in the amide-directed H—B addition to alkenes, with catecholborane (HBcat) as the boron reagent (Scheme 7.12). It is noteworthy that cationic iridium catalysts were essentially more effective in the delivered reaction than the analogous cationic rhodium complexes. The amide-directing group was capable of turning over the

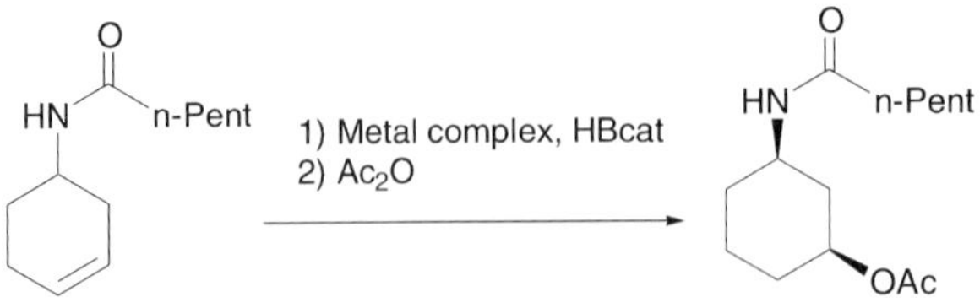

Scheme 7.12

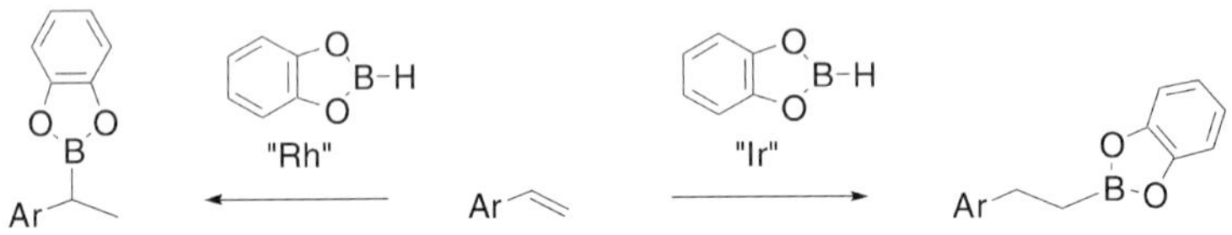

Scheme 7.13

expected regioselectivity of the iridium-catalyzed reaction of terminal alkenes, because although 1-hexene can be hydroborated with 98:2 selectivity in favor of the formation of the primary alcohol, the reaction of the 3-butenamide affords the secondary alcohol as the major product. Under optimized conditions, the highest selectivity for the 1,3-*cis* isomer could be achieved by modifying the iridium complex with electron-deficient indenyl ligands [26]. The remarkable diastereoselectivity achieved has been correlated to the ability of indenyl ligands to 'slip' to a lower coordination form [27] and to remain attached to the catalyst during the hydroboration reaction; this is in contrast to other related η^3-ligands, which are destroyed during the catalytic H—B addition [17].

The catalytic hydroboration of vinylarenes has also been well studied and, depending on the rhodium or iridium catalytic system used, the product distribution can be tuned. $[IrCl_2(\eta^5\text{-}C_5Me_5)]_2$ catalyzed the hydroboration of 4-vinylanisole in the presence of HBcat with the exclusive formation of the terminal hydroboration product, in contrast to the analogue rhodium complexes which mainly afford the branched alkylboronate ester (Scheme 7.13) [14].

The reaction parameters responsible for the high levels of regioselectivity can only be understood by studying the mechanism of the catalytic cycle [28]. Hartwig and Nolan quantified individual borane bond dissociative energies and measured the enthalpy of the main catalytic steps [16]. Interestingly, the strength of the iridium–boron bond could provide the necessary driving force for the oxidative addition of H—B in catecholborane (Scheme 7.14). Alkene insertion into the iridium–hydride seems to be slightly less favorable thermodynamically than alkene insertion into the iridium–boron bond. Eventually, B—C reductive elimination of the hydroborated product from the alkyliridium complex is more exothermic than the H—C reductive elimination. Another example supporting the favored B—C reductive elimination was observed by Baker *et al.* [9], in the synthesis of *fac*-$[IrH(PMe_3)_3(\eta^2\text{-}CH_2BHC_8H_{14})]$ (Scheme 7.15).

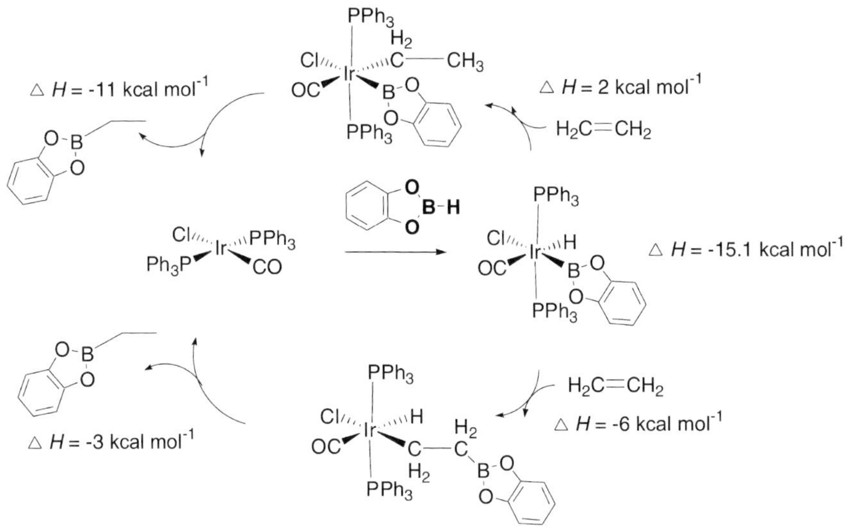

Scheme 7.14

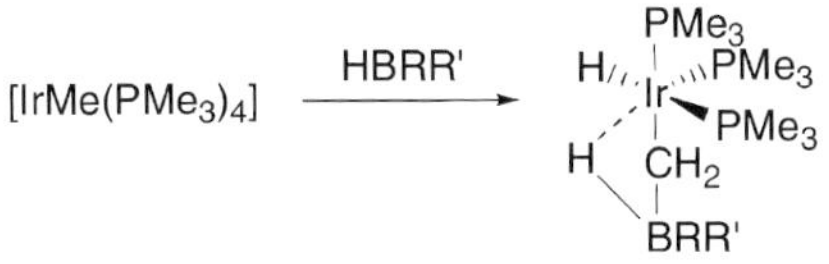

Scheme 7.15

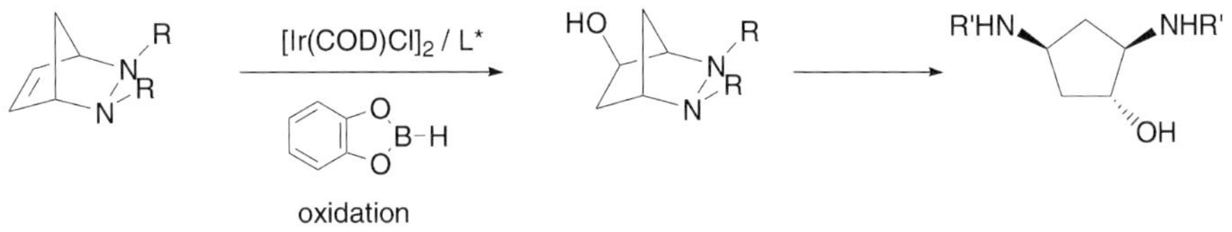

Scheme 7.16

The most striking catalytic pathway seems to be the insertion of the unsaturated molecule in the metal–H [29] or the metal–B bonds [30, 31c]. The similarity of the two enthalpic terms seems to be consistent with the apparent competition between the two insertion processes [31], implying that kinetic factors may dominate the selectivity. In this context, the first example of the asymmetric iridium-catalyzed hydroboration reaction brought new insights into the catalytic cycle of this reaction. Bonin and Micouin [32] reported the desymmetrization of *meso* hydrazines through rhodium- and iridium-catalyzed asymmetric hydroboration. This was considered to be the key step in the straightforward access to polysubstituted diamino cyclopentanes (Scheme 7.16), with enantiomeric excess (ee) values up to 64%, when Josiphos was used as ligand. Bonin and Micouin found that the enantioselectivity was completely reversed in the Rh and Ir systems, and suggested that the stabilization of the transition state of the migratory insertion by two σ-nitrogen

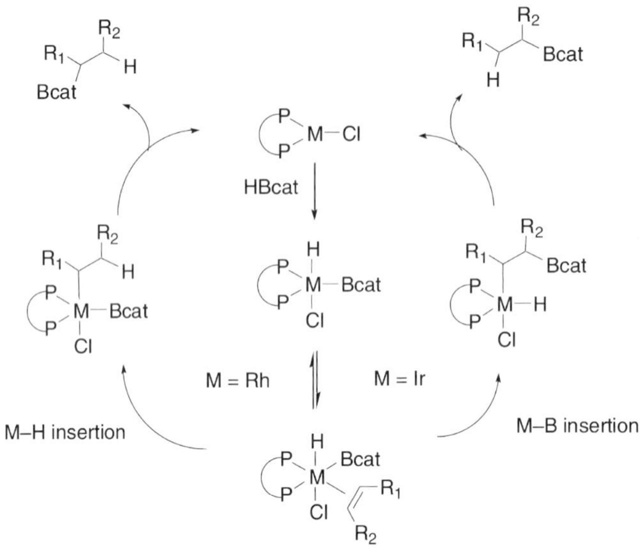

Scheme 7.17

lone pair interactions was the first irreversible step in the cycle. This led them to postulate that – at least with *meso* substrates – rhodium-catalyzed hydroboration involved a metal–H insertion, and iridium-catalyzed hydroboration a boryl migration (Scheme 7.17). Further studies have shown that chiral nonracemic monodentate phosphoramidite or phosphite ligands were efficient in the Ir-catalyzed asymmetric hydrogenation of *meso*-hydrazines [33]. However, the benefits of the Josiphos ligand in iridium-mediated hydroboration reaction were initially highlighted by Togni *et al*. [34], based on the highest ee-values (77%) afforded for 1-phenylethanol with this metal, although previous data on regiocontrol were not provided.

In an attempt to rationalize the factors that control selectivity in the Rh- and Ir-catalyzed hydroboration reactions, Fernández and Bo [35] carried out experimental and theoretical studies on the H—B addition of catecholborane to vinylarenes with [M(COD)(R-QUINAP)]BF$_4$, (QUINAP = 1-(2-diphenylphosphino-1-naphthyl)isoquinoline). A considerable difference was found in the stability of the isomers when the substrate was coordinated to the iridium(I) or rhodium(I) complexes. In particular, the difference between pro-R B1 and pro-S B2 isomers was not so great when the metal center was iridium and not rhodium (Figure 7.1), which explains the low ee-values observed experimentally when asymmetric iridium-catalyzed hydroboration was performed. Structurally, the energy analysis of the π2 and π3 interactions [36] seems to be responsible for the extra stabilization of the B2 isomer in the iridium intermediates (Figure 7.1). The coordination and insertion of alkenes, then, could be considered key steps in the enantiodifferentiation pathway.

Unlike the selectivity of the hydroboration of terminal and internal alkenes, which depends on the metal catalysts and hydroboration reagents [20], the boron

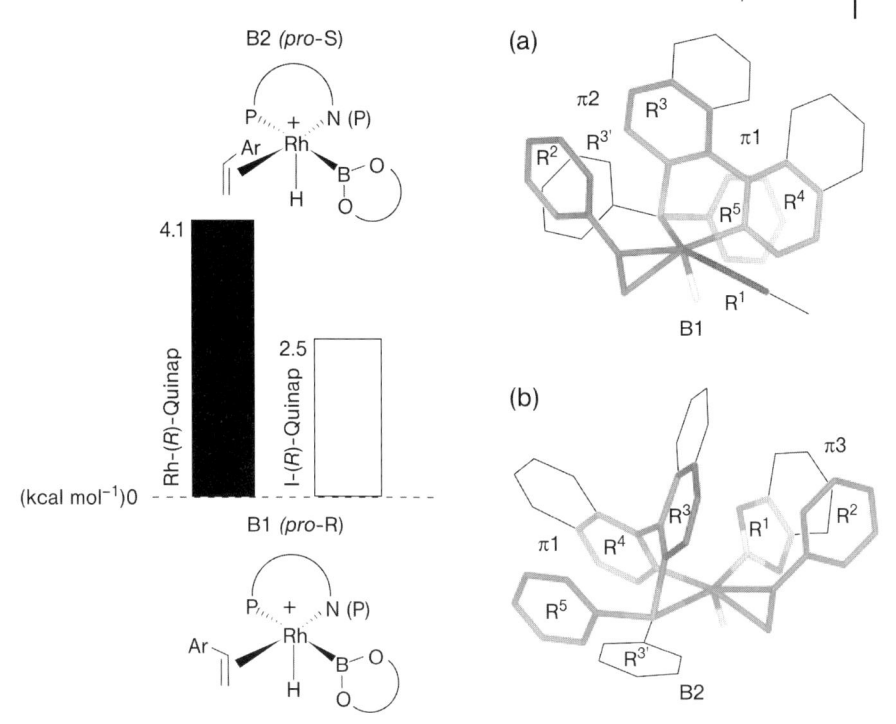

Figure 7.1 Relative stability of the most stable isomers in the hydroboration of vinylarenes with cationic metal complexes modified with QUINAP.

atom can be selectively added in the presence of iridium complexes and bulky hydroboration reagents. Pinacolborane (HBpin), has recently been found to be an excellent alternative to catecholborane because it is more stable and easy to handle than borane [37]. Crudden et al. [38] found that HBpin was quantitatively added towards the linear isomer in the hydroboration of vinylarenes with [Ir(μ-Cl)(COD)]$_2$/P-P (P-P = 1,2-(bis(diphenylphosphino)benzene (DPPB)), independently of the electron-donating or electron-withdrawing substituted substrates. Miyaura et al. [39] reported that neutral iridium(I)–phosphine complexes such as [Ir(μ-Cl)(COD)]$_2$/P-P (P-P = bis(diphenylphosphino)methane; DPPM and bis(diphenylphosphino)ethane; DPPE) are excellent catalysts for H—B addition to terminal and internal alkenes with an aliphatic or aromatic substituent on the vinylic carbon, when pinacolborane is used. Only the anti-Markovnikov addition product was formed in the hydroboration of perfluorovinylalkenes, whereas a mixture of products was formed in the rhodium-catalyzed H—B addition from catecholborane and pinacolborane (Scheme 7.18) [40]. Iridium(I)-catalyzed hydroboration of internal alkenes with pinacolborane afforded total formation of pinacol 1-alkylboronates, as a result of the complete isomerization to the terminal carbon [39]. Of particular interest was the selectivity observed by Westcott et al. [41] in the hydroboration of 1-phenylpropene with HBpin. Here, while the branched product was favored when

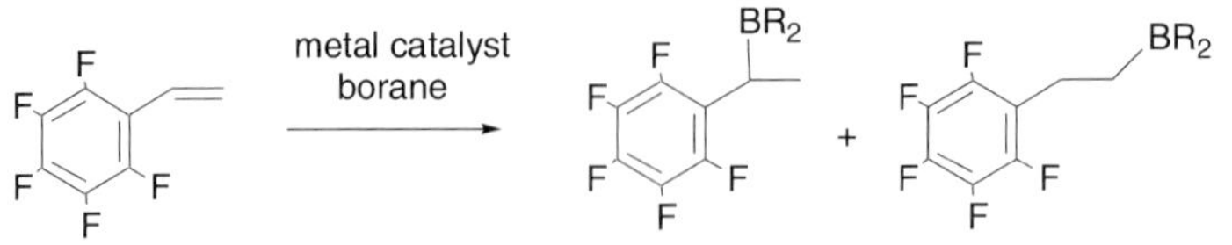

Scheme 7.18

catalyst	borane	branched	linear
[Ir(COD)Cl]$_2$ / 2 dppe	HBpin	0	100
RhCl(PPh$_3$)$_3$	HBpin	29	71
RhCl(PPh$_3$)$_3$	HBcat	79	21

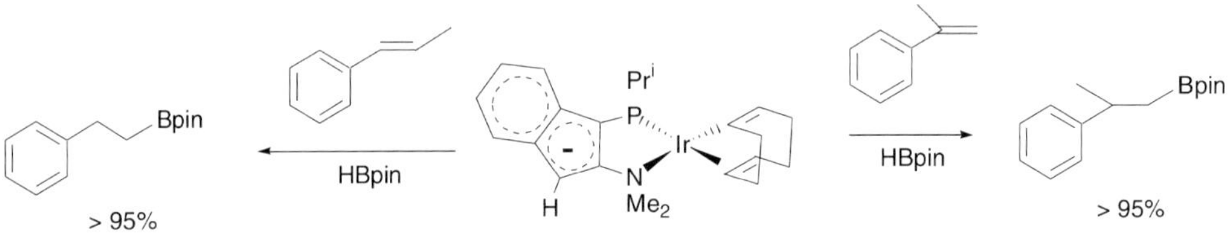

Scheme 7.19

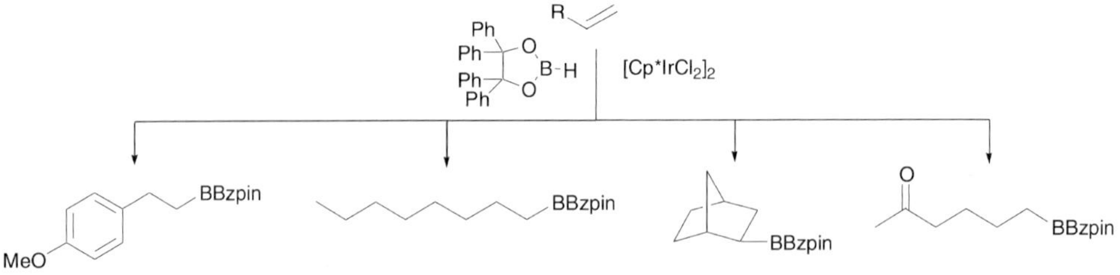

Scheme 7.20

using cationic or zwitterionic rhodium complexes, the linear product was exclusively formed in reactions mediated by either the cationic or zwitterionic iridium catalysts. This product distribution presumably arises from an initial metal-mediated isomerization of the double bond from 1-phenylpropene to give the corresponding α-olefin (3-phenylpropene), followed by a regioselective hydroboration step (Scheme 7.19). Formally, zwitterionic iridium catalytic systems provided improved catalytic performance compared to their cationic analogue complexes in the hydroboration of 2-phenylpropene with HBpin, exhibiting only the linear addition product (Scheme 7.19).

The combination of iridium(I)-catalyzed hydroboration with the bulkiest borane 4,4,5,5-tetraphenyl-1,3,2-dioxaborolane (HBBzpin) allowed the selective synthesis of a series of novel stable organoboronates, in contrast to the rhodium-catalyzed version (Scheme 7.20) [42].

The iridium-catalyzed hydroboration of heteroatom-containing substrates (vinyl sulfides, sulfoxides, sulfones and sulfonates) was examined by Westcott and Baker [43]. These authors observed the selective formation of the linear boronate ester

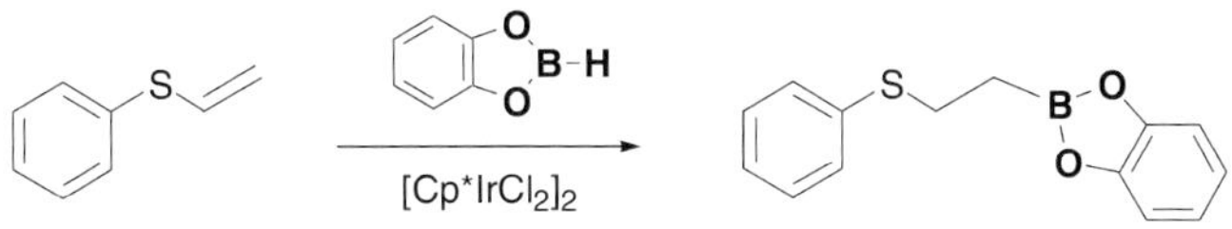

Scheme 7.21

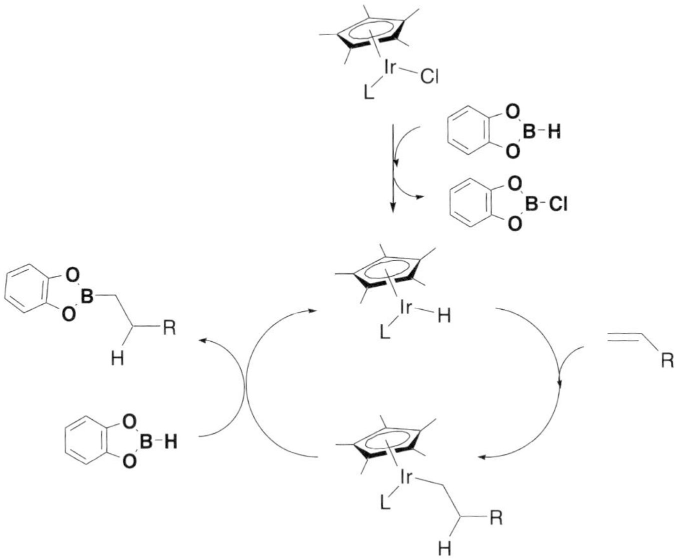

Scheme 7.22

product in the H—B addition of catecholborane to phenyl vinyl sulfide (Scheme 7.21), despite the directing effect of the sulfur heteroatom to form the branched isomer. Hence, they suggested an alternative mechanism for the exclusive formation of the linear products, in agreement with related lanthanum-catalyzed hydroboration of alkenes (Scheme 7.22) [44].

Westcott et al. have also observed the exclusive linear product 3-$O_2NC_6H_4CH_2CH_2Bcat$ in the hydroboration of 3-nitrostyrene by $[IrCl_2(\eta^5\text{-}C_5Me_5)]_2$ and HBcat [45]. Similarly, the linear isomer 4-$(Bcat)_2NC_6H_4CH_2CH_2Bcat$ could be preferentially obtained in the iridium-mediated hydroboration of 4-vinylaniline as an example of a mild route to aniline derivatives containing boronate esters. Westcott et al. claimed a mechanism that may not necessarily proceed via conventional pathways that invoke initial oxidative addition of HBcat to the metal center.

A formal trans-hydroboration of terminal alkynes with catecholborane and pinacolborane to yield cis-1-alkenylboronates has also been carried out in the presence of Rh(I) and Ir(I). The dominant factors for reversing the conventional cis-hydroboration to the trans-hydroboration were the use of NEt_3 and the use of bulk phosphines such as P^iPr_3 and PCy_3, and an excess of alkyne in front of the boron reagent [46].

7.4
Diboration

The most convenient mode of difunctionalization of multiple carbon–carbon bonds through boron-containing intermediates, involves the metal-mediated addition of a diboron reagent to an unsaturated carbon–carbon bonds [47]. While considerable efforts has been made to use metal–phosphine complexes to catalyze the B—B additions, a series of N-heterocyclic carbene ligand (NHC)-based metal complexes have recently been proved to be a convenient alternative [48]. Therefore, the unique reported attempt to diborate unsaturated substrates by cationic and zwitterionic iridium complexes modified with P,N ligands, afforded a mixture of products [41], while [Ir(η^5-C$_5$Me$_5$)(NHC)] afforded chemoselective organodiboronate compounds in the B—B addition to vinylarenes [49] (Figure 7.2). An extra interest to evaluate the asymmetric induction [50], on the reaction products of the catalytic diboration reaction, was considered by using the chiral-at-metal center and at-NHC ligand, complex [Ir(η^5-C$_5$Me$_5$)(NHC)].

Although the [Ir(η^5-C$_5$Me$_5$)(NHC)] complex was initially tested in the diboration of styrene, this did not lead to the formation of any diboronate esters when bis(catecholato)diboron (B$_2$cat$_2$) was added to a solution of the catalyst and styrene in tetrahydrofuran (THF) under argon (Scheme 7.23). However, the addition of

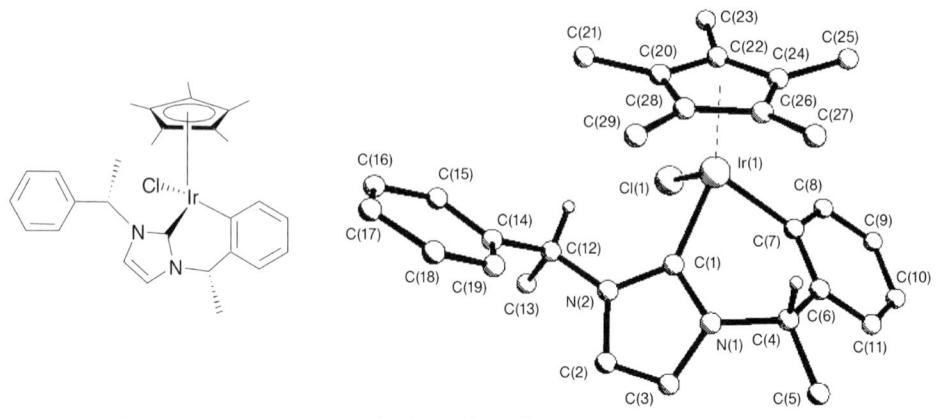

Figure 7.2 [Cp*Ir(NHC)] complex for catalytic diboration reaction.

Scheme 7.23

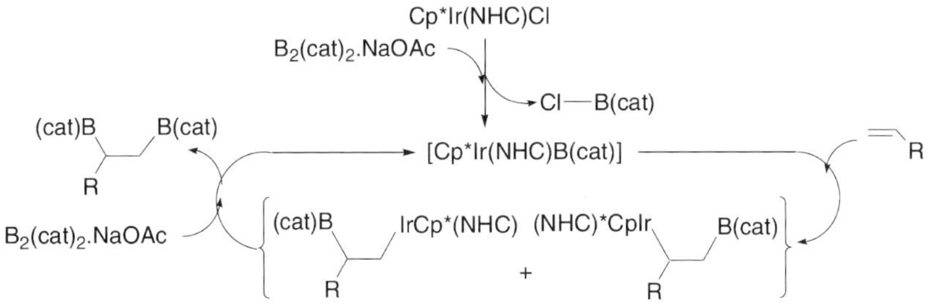

Scheme 7.24

NaOAc and an excess of diboron reagent afforded an almost quantitative conversion of the styrene, and an extraordinarily high chemoselectivity on the diborated product, with values up to 99.2%. Other neutral and cationic Ir(I) complexes modified with (S)-Quinap were less active under such conditions, despite the effectiveness of their Rh analogue complexes [50]. The specific effectiveness of the NHC ligands towards the high chemoselectivities on the diboronate product has been shown in the comparative diboration reaction catalyzed by [Ir(μ-Cl)Cl(η^5-C$_5$Me$_5$)]$_2$ and [Ir(I^{Me})(Cl)$_2$(η^5-C$_5$Me$_5$)] (I^{Me} = 1,3-dimethylimidazolylidene). The scope of the reaction included vinyl and aliphatic alkenes. Although, under these reaction conditions asymmetric induction was not achieved, the addition of AgBF$_4$ to the reaction mixture in order to promote removal of the halide from the Ir metal sphere favored efficient catalytic diboration with total conversion, high chemoselectivity and ee-values of about 10% for the corresponding acetal derivative in the catalytic diboration/oxidation protocol. A tentative mechanistic pathway has been suggested for the specific role of NaOAc which could favor the heterolytic cleavage of B$_2$cat$_2$, thus justifying formation of the Ir—B with no need for oxidative addition of the diboron to the Ir(III) complex. This means that the metal could remain in the same oxidation state throughout the reaction process. Subsequent alkene coordination and migratory insertion into the Ir—B bond could provide the Ir-alkyboronate formation, which could finally transmetallate with B$_2$cat$_2$ to generate the diborated product (Scheme 7.24). A similar mechanism has been postulated for the first catalytic diboration reaction with palladium complexes, in agreement with the findings of experimental and theoretical studies [51]. The interaction of the base with the diboron reagent, together with the strong exothermicity of the transmetallation step in the presence of the base, could very well serve as the driving force to complete the reaction.

7.5
Borylation

Since the pioneering studies of Hartwig and colleagues [52] and Smith and coworkers [53], the transition-metal-catalyzed activation and functionalization of unreactive C—H bonds of hydrocarbons [54] have led to increasing attention being paid

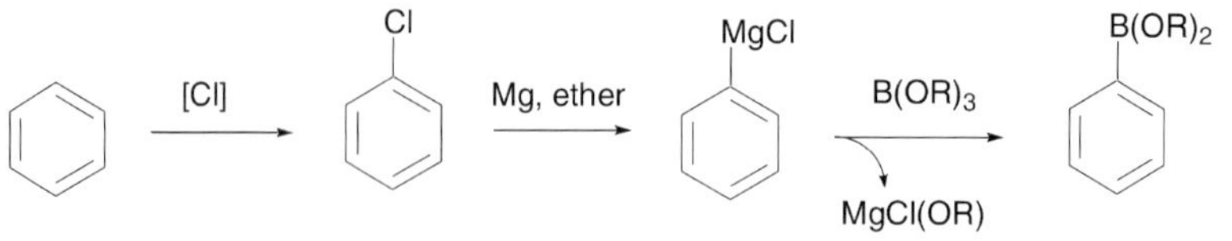

7 Iridium-Catalyzed Boron-Addition

Scheme 7.25

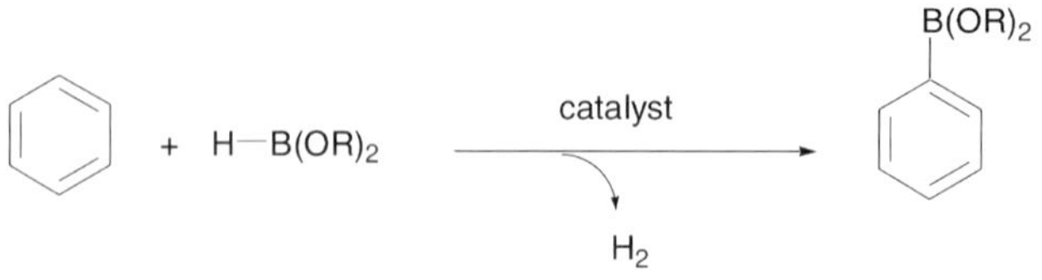

Scheme 7.26

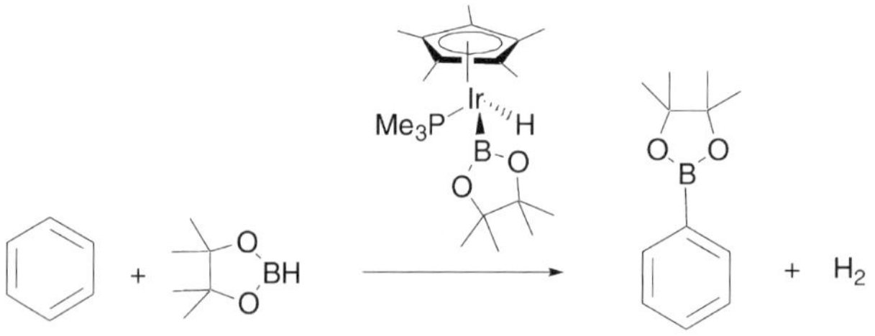

Scheme 7.27

to the C—H borylation reaction for the synthesis of organoboron compounds [55]. In this context, arylboron compounds possess intriguing properties and are important building blocks for chemical synthesis, such as C—C bond-forming strategies through crosscoupling reactions [56]. Iridium catalysts are considered to be highly selective for C—H activation [57] and do not interfere with subsequent *in situ* transformation. Because of their favorable activities and exceptional selectivities, these iridium catalysts have become a genuine alternative to the classical multistep process (Scheme 7.25) for the direct synthesis of arylboron compounds from aromatic hydrocarbons and boranes (Scheme 7.26).

Iverson and Smith were the first to explore the catalytic viability of forming the B—C bond from inactivated hydrocarbons in the presence of iridium complexes [Ir(H)(R)(η^5-C$_5$Me$_5$)(PMe$_3$)] (Scheme 7.27), and showed that borane substituents are critical for effecting catalytic conversion [53a]. In particular, thermal B—C bond formation involved a catalytic cycle for pinacolborane, HBpin, in contrast to the photochemical methods developed by Hartwig and coworkers in the end-functionalization of alkanes through borylation reaction with bis(pinacolato)diboron, B$_2$pin$_2$ [52c]. In subsequent studies, the iridium complex [Ir(H)(BR$_2$)(η^5-C$_5$Me$_5$)(PMe$_3$)] proved to be less active than the rhodium complex [Rh(η^5-C$_5$Me$_5$)(η^4-C$_6$Me$_6$)] [52d], but much more selective towards arene C—H activation (e.g. the arene versus benzylic activation in *m*-xylene borylations [53b]). The major metal-containing product in the arene borylation reaction with iridium complexes was [Ir(H)(η^5-C$_5$Me$_5$)(Bpin)(PMe$_3$)], which turned out to be stable in benzene solutions after prolonged thermolysis. Mechanistically, a simple phosphine dissociative pathway was excluded [58]. In fact, phosphines or related donor ligands seemed to be required for the catalytic borylation, because of the lack of catalytic activity provided by [Ir[(η^6-mesitylene)(Bpin)$_3$] [53d] and the beneficial effects of the excess of phosphine to the metal. The analogous [Ir[(η^6-arene)(Bcat)$_3$] complex [59] also showed discrete toluene borylation products included in the supplementary material. An alternative means for generating active precatalysts showed that [Ir[(η^5-C$_9$H$_7$)(COD)] was highly active for benzene borylation in the presence of PMe$_3$ or chelating diphosphines such as dppe and dmpe (dmpe = 1,2-bis(dimethylphosphino)ethane), where the effective turnover number (TON) of 4500 was improved by more than 1000-fold over the precatalyst [Ir(H)(η^5-C$_5$Me$_5$)(Bpin)(PMe$_3$)]. The *in situ* iridium complex prepared from [Ir(COD)]$_2$ and indenyl lithium [60] also had similar activity in the presence of phosphines.

Iridium catalysts are compatible with the entire range of aryl halides, preferentially activating the C—H bonds, unlike the dehalogenations observed in rhodium-mediated borylations [53b, 61]. This remarkable functional group tolerance made it possible to carry out consecutive metal-catalyzed cascade reactions, such as a borylation/Miyaura–Suzuki crosscoupling reaction for one-pot biaryl synthesis (Scheme 7.28). Smith and coworkers [53d] suggested a mechanistic catalytic cycle in which iridium species Ir(III) and Ir(V) could be potentially consistent with the experimental observations (Scheme 7.29). Some of these data involved: (i) the borylation products of iodobenzene were not obtained from Ir(I) sources, whereas Ir(III) species affected both stoichiometric and catalytic borylations; (ii) the catalytic activity improved when chelating biphosphines were involved, which supports the presence of 18-electron biphosphine Ir(V) resting state; and (iii) the complex [IrH$_5$(PMe$_3$)] became an effective precatalyst for borylation.

The beneficial effects of chelating ligands were also demonstrated by Hartwig, Ishiyama and Miyaura [62]. This group isolated the iridium(I) complex [Ir(Bpin)$_3$(COE)(DTBPY)] modified with simple 2,2′-bipyridine ligands (such as 4,4′-di-*tert*-butyl-2,2′-bipyridine; DTBPY), which seemed to be responsible for the first catalytic C—H borylation at room temperature (Scheme 7.30). An extension

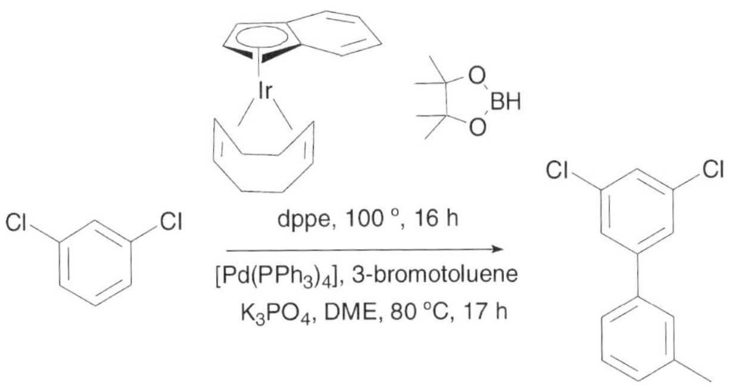

Scheme 7.28

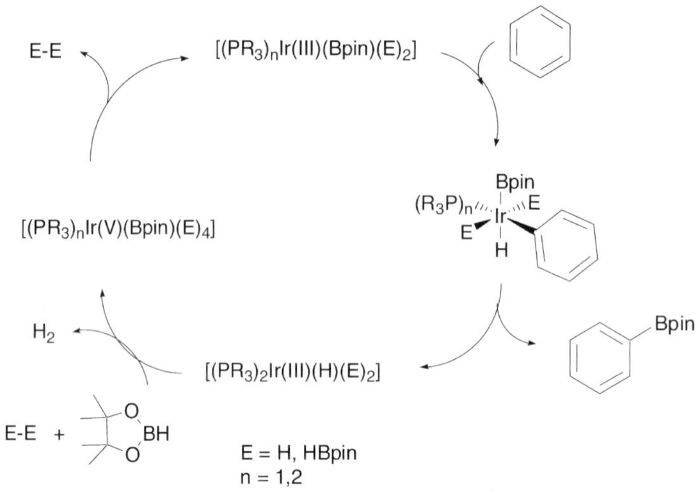

Scheme 7.29

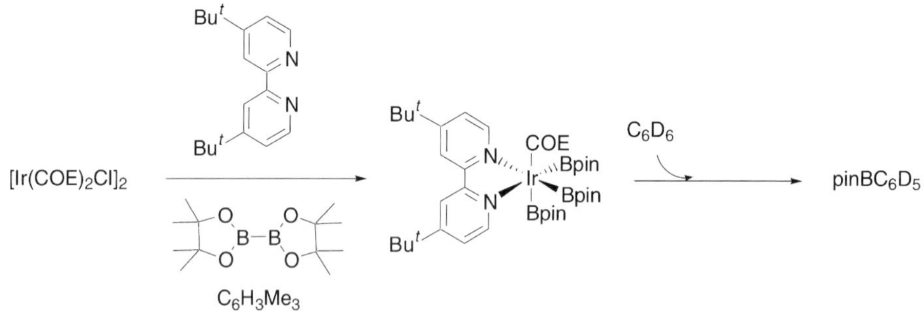

Scheme 7.30

of these studies involved the C—H coupling of aromatic heterocycles with bis(pinacolato)diboron, where the reactions with five-membered substrates exclusively produced 2-borylated products, while reactions with six-membered heterocycles selectively occurred at the 3-position (Scheme 7.31) [63]. The borylation of these heteroatomic substrates most likely proceeded through tris(boryl)iridium (III) intermediates [53b, 59, 62], which gave arylboronates upon thermolysis with arenes. The C—H activation of the 2-position in five-membered substrates by the tris(boryl)iridium (III) intermediate, followed by C—B bond formation, seems to be a reasonable mechanism for their borylation. A small steric hindrance of a planar bipyramidal ligand, as well as its electron donation to the metal center, is also critical for the formation of such sterically hindered Ir(V) intermediates. Although these processes have been confirmed by the results of recent theoretical studies conducted by Sakaki [64], the perplexing regioselectivity of the borylation of pyridine and quinoline to give 3-borylated products suggested that the iridium complex or a boron compound bound reversibly to the basic nitrogen; indeed, both

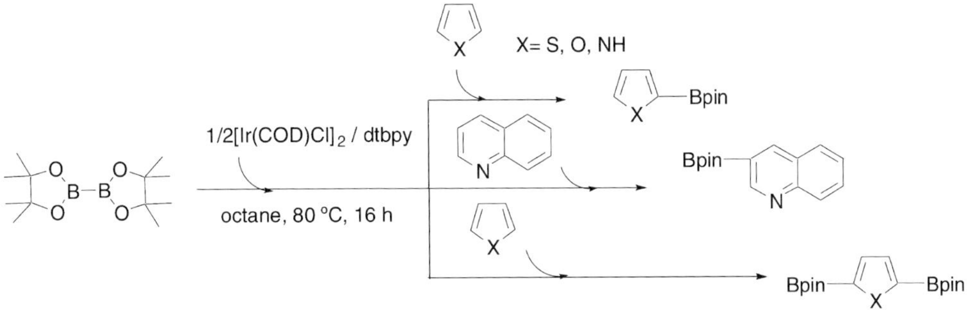

Scheme 7.31

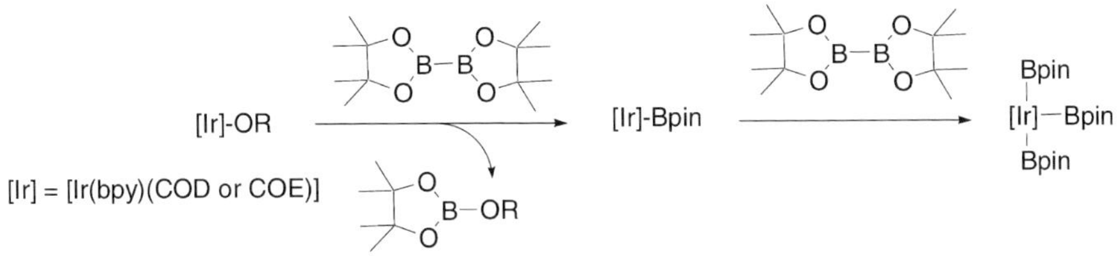

Scheme 7.32

were seen to activate the substrate for reaction blocking borylation at the 2-position. A regioselective synthesis of bis(boryl)heteroatomics was also achieved by using an almost equimolar amount of substrates and the diboron (Scheme 7.31).

Subsequent studies on iridium-mediated borylation demonstrated the effect of varying the anionic ligands on iridium(I) precursors [65]. Whereas, halide and cationic iridium complexes do not catalyze the reaction, the borylation was completed within 4 h at room temperature if the iridium(I) precursor possessed an OH, OPh or OMe ligand. Therefore, the high catalyst efficiency of the (hydroxo)- or (alkoxo)iridium complexes could be attributed to the more facile formation of (boryl)iridium complexes (Scheme 7.32). The mono-(boryl)iridium(I) complex could be generated by the oxidative addition of bis(pinacolato)diboron to (alkoxo)iridium(I) complexes, followed by the reductive elimination of pinBOR; alternatively, it may be generated by transmetallation between (alkoxo)iridium(I) complexes and B_2pin_2. Under these optimized conditions, the synthesis of unsymmetrical biaryls was explored by means of an efficient one-pot, two-step procedure [66] in an inert solvent. The aromatic borylation of arenes and heteroarenes with HBpin, but without an excess of substrate or reagent, provided aryl- and heteroarylboronates which could be converted *in situ* towards the corresponding biaryl (Scheme 7.33).

Iridium-catalyzed borylation has also proved to develop the first general approach to functionalized unprotected indoles at the 7-position [67]. This selectivity can be explained by the nitrogen-directed aromatic borylation pathway in the mechanistic steps (Scheme 7.34).

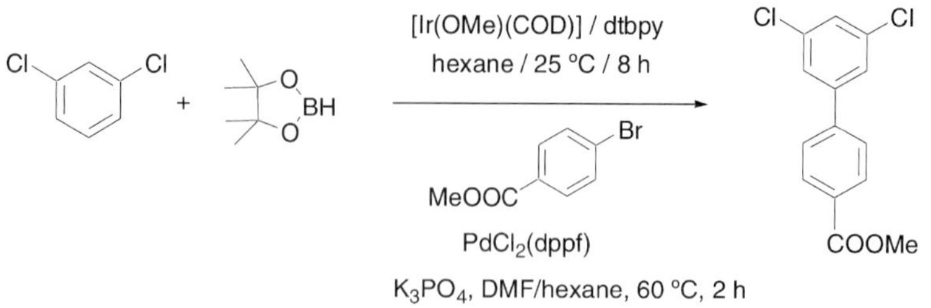

Scheme 7.33

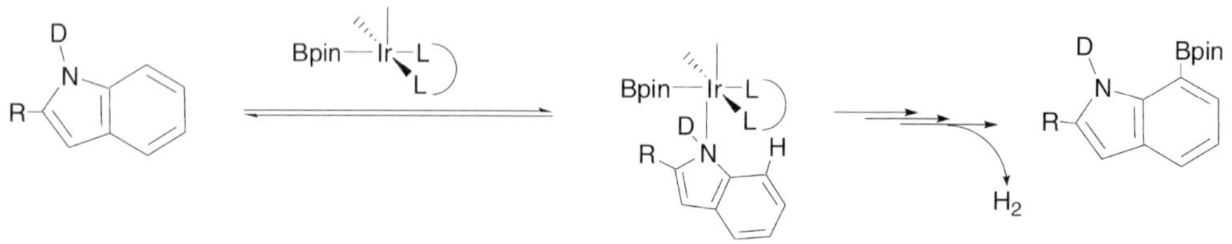

Scheme 7.34

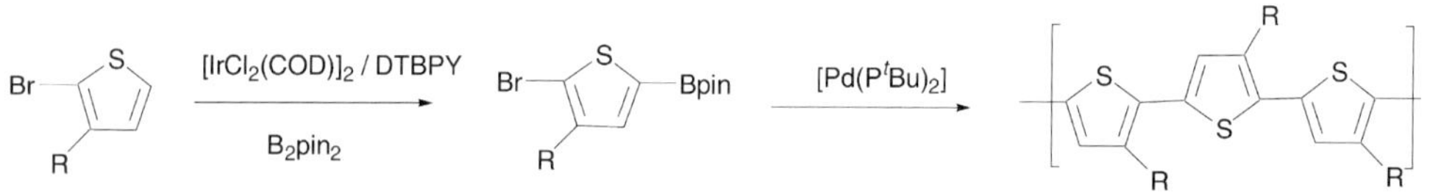

Scheme 7.35

A convenient direct route has recently been described for obtaining regioregular polyalkylthiophenes using a tandem iridium-catalyzed borylation to produce the monomer, and a palladium-mediated coupling to produce the polymer [68]. The treatment of substituted thiophenes with B_2pin_2 in the presence of $[IrCl_2(COD)]_2$/ 4,4′-di-*tert*-butyl-2,2′-bipyridine (DTBPY) provided the expected monomer in 97% yield (Scheme 7.35).

Selective C—C bond formation by the activation of vinylic C—H bonds has become a challenging task in organic synthesis, mainly because it raises selectivity issues as both $C(sp^2)$—H and $C(sp^3)$—H bonds are available for functionalization. Szabó has shown that selective C—C bond formation can be achieved in cycloalkenes by using an iridium-catalyzed C—H activation borylation reaction [69]. Transient organoboranes can react further with either aldehydes or aryliodides in a one-pot sequence (Scheme 7.36). Marder and coworkers suggested that alkenes can undergo dehydrogenative borylation, which involves inserting the substrate into the metal–boron bond of the active catalyst, followed by β-hydride elimination to give vinylboronates [70]. A similar conceptual mechanism could be involved in the catalytic borylation of unactivated cycloalkenes, starting from the fact that the double bond of cyclohexene could be inserted into the Ir—B bond to provide intermediate C. Assuming that the insertion was proceeded by a *syn* mechanism, the C—B and C—Ir bonds would be in *cis* position in the metal intermediate complex C. Because of the hindered rotation of the C—C bond, only the allylic C—H bond

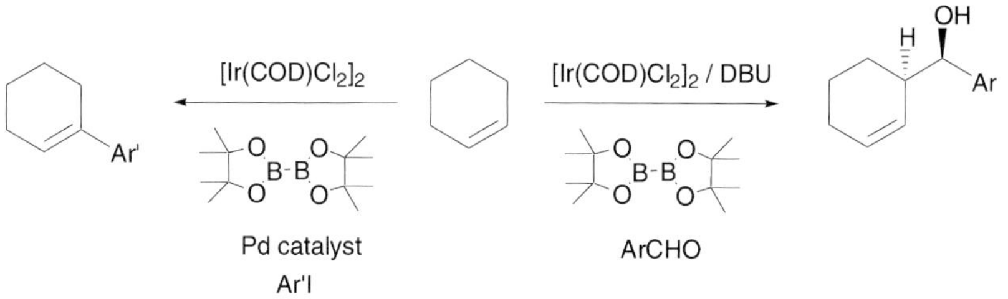

Scheme 7.36

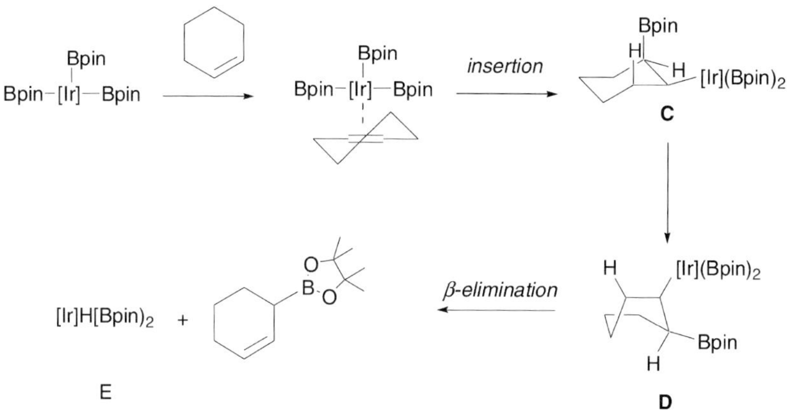

Scheme 7.37

could adapt to the *syn* conformation D required for β-hydride elimination, and this would explain the initial formation of the allyl product. Subsequent allyl rearrangement leads to the formation of vinylboronate, which can be retarded by applying of 1,8-diazabicyclo[5.4.0]undecane (DBU). The molecular geometry required for the β-hydride elimination could be most easily realized in the cyclohexane derivative D to provide the E intermediate, which undergoes reductive elimination to give pinacolborane and regenerate the catalyst (Scheme 7.37).

References

1 (a) Brown, H.C. (1962) *Hydroboration*, W.A. Benjamin, New York;
(b) Brown, H.C. (1972) *Boranes in Organic Chemistry*, Cornell University Press, Ithaca;
(c) Brown, H.C. (1975) *Organic Synthesis via Boranes*, John Wiley & Sons, Inc., New York.;
(d) Pelter, A., Smith, K. and Brown, H.C. (1988) *Borane Reagents*, Academic Press, New York;
(e) Cragg, G.M.L. (1973) *Organoboranes in Organic Synthesis*, Dekker, New York;
(f) Onak, T. (1975) *Organoborane Chemistry*, Academic Press, New York.

2 For catalytic hydroboration reaction: (a) Männig, D. and Nöth, H. (1985) *Angewandte Chemie – International Edition in English*, **24**, 878;
(b) Beletskaya, I. and Pelter, A. (1997) *Tetrahedron*, **53**, 4957;
(c) Burgess, M.J. and Ohlmeyer, M.J.

(1988) *The Journal of Organic Chemistry*, **53**, 5178;
(d) Brown, J.M., Doucet, H., Fernández, E., Heeres, H.E., Hooper, M.W., Hulmes, D.I., Knight, F.I., Layzell, T.P. and Lloyd-Jones, G.C. (1999) *Transition Metal-Catalysed Reactions* (eds S.I. Murahashi and S.G. Davies), Blackwell Science, Oxford, UK, p. 465;
(e) Brown, J.M. (2004) *Modern Rhodium-Catalyzed Organic Reactions* (ed. P.A. Evans), Wiley-VCH Verlag GmbH, Weinheim, p. 33;
(f) For catalytic diboration reaction: Ishiyama, T., Nishijima, K., Miyaura, N. and Suzuki, A. (1993) *Journal of the American Chemical Society*, **115**, 7219;
(g) Marder, T.B. and Norman, N.C. (1998) *Topics in Catalysis*, **5**, 63;
(h) Dembitsky, V.M., Abu Ali, H. and Srebnik, M. (2003) *Applied Organometallic Chemistry*, **17**, 327;
(i) Ishiyama, T. and Miyaura, N. (2004) *The Chemical Record*, **3**, 271;
(j) Beletskaya, I. and Moberg, C. (2006) *Chemical Reviews*, **106**, 2320–54.

3 Schmid, G., Petz, W., Arloth, W. and Nöth, H. (1967) *Angewandte Chemie – International Edition in English*, **6**, 696.

4 Hoel, E.L. and Hawthorne, M.F. (1973) *Journal of the American Chemical Society*, **95**, 2712–3.

5 Hoel, E.L. and Hawthorne, M.F. (1974) *Journal of the American Chemical Society*, **96**, 6770–1.

6 Churchill, M.R., Hackbarth, J.J., Davison, A., Traficante, D.D. and Wreford, S.S. (1974) *Journal of the American Chemical Society*, **96**, 4041–2.

7 Churchill, M.R. and Hackbarth, J.J. (1975) *Inorganic Chemistry*, **14**, 2047–51.

8 (a) Schmid, G. and Nöth, H. (1963) *Angewandte Chemie – International Edition in English*, **2**, 623; (b) Schmid, G. (1970) *Angewandte Chemie – International Edition in English*, **9**, 819.

9 Baker, R.T., Ovenall, D.W. and Calabrese, J.C. (1990) *Journal of the American Chemical Society*, **112**, 9399–400.

10 Thorn, D.J. and Tulip, T.H. (1982) *Organometallics*, **1**, 1580.

11 Knorr, J.R. and Merola, J.S. (1990) *Organometallics*, **9**, 3008–10.

12 He, X. and Hartwig, J.F. (1996) *Organometallics*, **15**, 400–7.

13 Einertshofer, C. (1992) Dissertation, Ludwig-Maximilians-Universität, München.

14 Westcott, S.A., Marder, T.B., Baker, R.T. and Calabrese, J.C. (1993) *Canadian Journal of Chemistry*, **71**, 930.

15 Simöes, J.A.M. and Beauchamp, J.L. (1990) *Chemical Reviews*, **90**, 629–88.

16 Rablen, P.R., Hartwig, J.F. and Nolan, S.P. (1994) *Journal of the American Chemical Society*, **116**, 2712–13.

17 Nguyen, P., Blom, H.P., Westcott, S.A., Taylor, N.J. and Marder, T.B. (1993) *Journal of the American Chemical Society*, **115**, 9329–30.

18 Cho, J.-Y., Tse, M.K., Holmes, D., Maleczka, R.E., Jr and Smith, M.R., III (2002) *Science*, **295**, 305.

19 (a) Braunschweig, H. (1998) *Angewandte Chemie – International Edition*, **37**, 1786–801;
(b) Braunschweig, H., Kollann, C. and Rais, D. (2006) *Angewandte Chemie – International Edition*, **45**, 5254.

20 (a) Männig, D. and Nöth, H. (1985) *Angewandte Chemie – International Edition*, **24**, 878–9;
(b) Burgess, M.J. and Ohlmeyer, M.J. (1988) *The Journal of Organic Chemistry*, **53**, 5178.

21 Crudden, C.M. and Edwards, D. (2003) *European Journal of Organic Chemistry*, **24**, 4695.

22 Mirabelli, M.G.L. and Sneddon, L.G. (1988) *Journal of the American Chemical Society*, **110**, 449.

23 Kono, H., Ito, K. and Nagai, Y. (1975) *Chemistry Letters*, **4**, 1095.

24 Huang, X. and Lin, Z. (2002) *Computational Modeling of Homogeneous Catalysis*, Kluwer Academic, Amsterdam, pp. 189–212.

25 (a) Evans, D.A. and Fu, G.C. (1991) *Journal of the American Chemical Society*, **113**, 4042;
(b) Evans, D.A., Fu, G.C. and Hoveyda, A.H. (1992) *Journal of the American Chemical Society*, **114**, 6671.

26 Brinkman, J.A., Nguyen, T.T. and Sowa, J.R., Jr (2000) *Organic Letters*, **2**, 981.

27 Garret, L.E. and Fu, G.C. (1998) *The Journal of Organic Chemistry*, **63**, 1370.

28 Wadepohl, H. (1997) *Angewandte Chemie – International Edition*, **36**, 2441.
29 (a) Evans, D.A., Fu, G.C. and Anderson, B.A. (1992) *Journal of the American Chemical Society*, **114**, 6679;
(b) Dorigo, A.E. and Von Ragué Schleyer, P. (1995) *Angewandte Chemie – International Edition*, **34**, 115.
30 Musaev, D.G., Mebel, A.M. and Morokuma, K. (1994) *Journal of the American Chemical Society*, **116**, 10693.
31 (a) Lynch, A.T. and Sneddon, L.G. (1989) *Journal of the American Chemical Society*, **111**, 6201;
(b) Brown, J.M. and Lloyd-Jones, G.C. (1992) *Journal of the Chemical Society, Chemical Communications*, 710;
(c) Burgess, K., van der Donk, W.A., Westcott, S.A., Marder, T.B., Baker, R.T. and Calabrese, G.C. (1992) *Journal of the American Chemical Society*, **114**, 9350;
(d) Westcott, S.A., Marder, T.B. and Baker, R.T. (1993) *Organometallics*, **12**, 975.
32 Pérez Luna, A., Bonin, M., Micouin, L. and Husson, H.-P. (2002) *Journal of the American Chemical Society*, **124**, 12098–9.
33 Alexakis, A., Polet, G., Bournaud, Ch., Bonin, M. and Micouin, L. (2005) *Tetrahedron: Asymmetry*, **16**, 3672.
34 Togni, A., Breutel, C., Schnyder, A., Spindler, F., Landert, H. and Tijani, A. (1994) *Journal of the American Chemical Society*, **116**, 4062.
35 Segarra, A.M., Daurá-Oller, E., Claver, C., Poblet, J.M., Bo, C. and Fernández, E. (2004) *Chemistry – A European Journal*, **10**, 6456–67.
36 Daurá-Oller, E., Segarra, A.M., Poblet, J.M., Claver, C., Fernández, E. and Bo, C. (2004) *The Journal of Organic Chemistry*, **69**, 2669–80.
37 Tucker, C.E., Davidson, J. and Knochel, P. (1992) *The Journal of Organic Chemistry*, **47**, 3482.
38 Crudden, C.M., Hleba, Y.B. and Chen, A.C. (2004) *Journal of the American Chemical Society*, **126**, 9200–1.
39 Yamamoto, Y., Fujikawa, R., Umemoto, T. and Miyaura, N. (2004) *Tetrahedron*, **60**, 10695–700.
40 (a) Ramachandran, P., Jennigs, M.P. and Brown, H.C. (1999) *Organic Letters*, **1**, 1399;
(b) Segarra, A.M., Claver, C. and Fernandez, E. (2004) *Chemical Communications*, 464–5.
41 Cipot, J., Vogels, Ch.M., McDonald, R., Westcott, S.A. and Stradiotto, M. (2006) *Organometallics*, **25**, 5965–8.
42 Fritschi, C.B., Wernitz, S.M., Vogels, C.M., Shaver, M.P., Decken, A., Bell, A. and Westcott, S.A. (2008) *European Journal of Inorganic Chemistry*, **5**, 799.
43 (a) Carter, C.A.G., Vogels, C.M., Harrison, D.J., Gagnon, M.K.J., Norman, D.W., Langler, R.F., Baker, R.T. and Westcott, S.A. (2001) *Organometallics*, **20**, 2130;
(b) Webb, J.D., Harrison, D.J., Norman, D.W., Blacquiere, J.M., Vogels, Ch.M., Decaen, A., Bates, C.G., Venkataraman, D., Baker, R.T. and Westcott, S.A. (2007) *Journal of Molecular Catalysis A – Chemical*, **275**, 91–100.
44 Harrison, K.N. and Marks, T.J. (1992) *Journal of the American Chemical Society*, **114**, 9220.
45 Vogels, C.M., Decken, A. and Westcott, S.A. (2006) *Tetrahedron Letters*, **47**, 2419.
46 Ohmura, T., Yamamoto, Y. and Miyaura, N. (2000) *Journal of the American Chemical Society*, **122**, 4990.
47 (a) Ishiyama, T., Nishijima, K., Miyaura, N. and Suzuki, A. (1993) *Journal of the American Chemical Society*, **115**, 7219;
(b) Marder, T.B. and Norman, N.C. (1998) *Topics in Catalysis*, **5**, 63;
(c) Dembitsky, V.M., Abu Ali, H. and Srebnik, M. (2003) *Applied Organometallic Chemistry*, **17**, 327;
(d) Ishiyama, T. and Miyaura, N. (2004) *The Chemical Record*, **3**, 271;
(e) Beletskaya, I. and Moberg, C. (2006) *Chemical Reviews*, **106**, 2320–54.
48 (a) Ramirez, J., Corberan, R., Sanau, M., Peris, E. and Fernandez, E. (2005) *Chemical Communications*, 3056;
(b) Corberan, R., Ramirez, J., Poyatos, M., Peris, E. and Fernandez, E. (2006) *Tetrahedron: Asymmetry*, **17**, 1759;
(c) Lillo, V., Mata, J., Ramirez, J., Peris, E. and Fernandez, E. (2006) *Organometallics*, **25**, 5829;
(d) Lillo, V., Fructos, M.R., Ramírez, J., Braga, A.A.A., Maseras, F., Díaz Requejo, M.M., Pérez, P.J. and Fernandez, E. (2007) *Chemistry – A European Journal*, **13**, 2614–21.

49 Corberan, R., Lillo, V., Mata, J.A., Fernandez, E. and Peris, E. (2007) *Organometallics*, **26**, 4350.

50 (a) Ramírez, J., Lillo, V., Segarra, A.M. and Fernandez, E. (2007) *Comptes Rendus Chimie*, **10**, 138–51;
(b) Trudeau, S., Morgan, J.B., Shrestha, M. and Morken, J.P. (2005) *The Journal of Organic Chemistry*, **70**, 9538;
(c) Ramirez, J., Segarra, A.M. and Fernandez, E. (2005) *Tetrahedron: Asymmetry*, **16**, 1289;
(d) Morgan, J.B., Miller, S.P. and Morken, J.P. (2003) *Journal of the American Chemical Society*, **125**, 8702.

51 Lillo, V., Más-Marzá, E., Segarra, A.M., Carbó, J.J., Bo, C., Peris, E. and Fernández, E. (2007) *Chemical Communications*, 3380–2.

52 (a) Waltz, K.M., He, X., Muhoro, C. and Hartwig, J.F. (1995) *Journal of the American Chemical Society*, **117**, 11357–8;
(b) Waltz, K.M. and Hartwig, J.F. (1997) *Science*, **277**, 211–13;
(c) Chen, H. and Hartwig, J.F. (1999) *Angewandte Chemie – International Edition in English*, **38**, 3391;
(d) Chen, H., Schlecht, S., Semple, T.C. and Hartwig, J.F. (2000) *Science*, **287**, 1995.

53 (a) Iverson, C.N. and Smith, M.R., III (1999) *Journal of the American Chemical Society*, **121**, 7696;
(b) Cho, J.-Y., Iverson, C.N. and Smith, M.R., III (2000) *Journal of the American Chemical Society*, **122**, 12868;
(c) Tse, M.K., Cho, J.-Y. and Smith, M.R., III (2001) *Organic Letters*, **3**, 2831;
(d) Cho, J.-Y., Tse, M.K., Holmes, D., Msleczka, R.E., Jr and Smith, M.R., III (2002) *Science*, **295**, 305.

54 Arndtsen, B.A., Bergman, R.G., Mobley, T.A. and Peterson, T.H. (1995) *Accounts of Chemical Research*, **28**, 154–62.

55 Ishiyama, T. and Miyaura, N. (2003) *Journal of Organometallic Chemistry*, **680**, 3.

56 (a) Miyaura, N. and Suzuki, A. (1995) *Chemical Reviews*, **95**, 2457;
(b) Suzuki, A. (1998) *Metal-Catalyzed Cross-Coupling Reactions* (eds F. Diederich and P.J. Stang), Wiley-VCH Verlag GmbH, Weinheim, p. 49.

57 (a) Bergman, R.G. (1984) *Science*, **223**, 902;
(b) Jones, W.D. and Feher, F.J. (1989) *Accounts of Chemical Research*, **22**, 91.

58 Kawamura, K. and Hartwig, J.F. (2001) *Journal of the American Chemical Society*, **123**, 8422.

59 Nguyen, P., Blom, H.P., Westcott, S.A., Taylor, N.J. and Marder, T.B. (1993) *Journal of the American Chemical Society*, **115**, 9329–30.

60 Merola, J.S. and Kacmarcik, R.T. (1989) *Organometallics*, **8**, 778.

61 Ezbiansky, K. Djurovich, P.I., LaForest, M., Sinning, D.J., Zayes, R. and Berry, D.H. (1998) *Organometallics*, **17**, 1455.

62 Ishiyama, T., Takagi, J., Ishida, K., Miyaura, N., Anastasi, N.R. and Hartwig, J.F. (2002) *Journal of the American Chemical Society*, **124**, 390–1.

63 Takagi, J., Sato, K., Hartwig, J.F., Ishiyama, T. and Miyaura, N. (2002) *Tetrahedron Letters*, **43**, 5649–51.

64 (a) Yamazaki, H., Tamura, H., Sugimoto, M., Sato, H. and Sakaki, S. (2002) 49th Symposium on Organometallic Chemistry, Japan; Kobe, Japan, PA119;
(b) Wan, X., Wang, X., Luo, Y., Takami, S., Kubo, M. and Miyamoto, A. (2002) *Organometallics*, **21**, 3703.

65 (a) Ishiyama, T., Takagi, J., Hartwig, J.F. and Miyaura, N. (2002) *Angewandte Chemie – International Edition in English*, **41**, 3056–8;
(b) Ishiyama, T., Takagi, J., Yonekawa, Y., Hartwig, J.F. and Miyaura, N. (2003) *Advanced Synthesis Catalysis*, **345**, 1103–6.

66 Ishiyama, T., Nobuta, Y., Hartwig, J.F. and Miyaura, N. (2003) *Chemical Communications*, 2924–5.

67 Paul, S., Chotana, G.A., Holmes, D., Reichle, R.C., Maleczka, R.E., Jr and Smith, M.R., III (2006) *Journal of the American Chemical Society*, **128**, 15552–3.

68 Liversedge, I.A., Higgins, S.J., Giles, M., Heeney, M. and McCulloch, I. (2006) *Tetrahedron Letters*, **47**, 5143–6.

69 Olsson, V.J. and Szabó, K.J. (2007) *Angewandte Chemie – International Edition in English*, **46**, 6891.

70 Westcott, S.A., Marder, T.B. and Baker, R.Th. (1993) *Organometallics*, **12**, 975.

8
Iridium-Catalyzed Methanol Carbonylation
Philippe Kalck and Philippe Serp

8.1
Introduction

Today, acetic acid is the most important commodity chemical produced by homogeneous catalysis, with a regular annual growth in production of 2.4%. Some 40% of this acetic acid is used in the manufacture of vinyl acetate to produce textile fibers, adhesives and paints; it is also used directly as a solvent in the synthesis of terephthalic acid (ca. 10%) and in the form of acetate esters for solvents. Acetic acid is also used directly in the pharmaceutical and food industries. One further major use involves the manufacture of acetic anhydride (by high-temperature dehydration into ketene, which is further condensed with acetic acid), the main use of which is in cellulose acetate production. Together, the oxidation of fermentation-derived ethanol, the distillation of hardwood, the oxidation of acetaldehyde (obtained by the oxidative hydration of ethylene; the Wacker process), and the oxidation of light petroleum fractions (1 ton of acetic acid is obtained from 3 tons of oxygenated products) accounts for an annual acetic acid production in excess of 8 million metric tons [1]. In fact, two-thirds of this production arises from the carbonylation of methanol, catalyzed by a late transition-metal complex. Thus, acetic acid is issued from methanol and, as a consequence, from carbon monoxide and hydrogen, and represents the larger second-generation intermediate in chemistry. As the CO/H_2 couple (syngas) can be generated either by the steam reforming of natural gas or light petroleum fractions or from coal (and perhaps in the near future by hardwood gasification), methanol will in time constitute the most abundant raw material for acetic acid production. At present, the annual worldwide production of methanol exceeds 45 million tons; hence, a highly selective catalytic process in which methanol is carbonylated to acetic acid (Equation 8.1) represents the most elegant synthetic pathway for low-cost acetic acid production.

In 1913, methanol was identified by A. Mittasch at the BASF Company [1] as a byproduct in the synthesis of ammonia. Subsequently, in 1923, the first industrial unit went on stream to convert syngas, in the presence of significant amounts

Iridium Complexes in Organic Synthesis. Edited by Luis A. Oro and Carmen Claver
Copyright © 2009 WILEY-VCH Verlag GmbH & Co. KGaA, Weinheim
ISBN: 978-3-527-31996-1

of CO_2, into methanol on a heterogeneous copper-based catalyst. During the 1920s methanol carbonylation became the subject of intense investigation, not only at BASF but also at British Celanese. Reppe and colleagues at BASF found that cobalt diiodide, operating at 680 bar (68×10^4 hPa) and 250 °C, catalyzed the reaction [2, 3]. Unfortunately, the process encountered major problems with corrosion until, in 1950, a highly resistant molybdenum/nickel alloy (Hastelloy™; this is used as a bulk material to build industrial reactors) was discovered and commercialized [1]. When, in 1960, BASF built the first unit to produce acetic acid, at Ludwigshafen, Germany, the process proved to be very nonselective since, for every ton of acetic acid produced, about 40 kg of byproducts (ethanol, propanal, propanoic acid, butanal, butanol, etc.) was created. The result was that five distillation columns were required to obtain CH_3COOH with adequate (99.8%) purity. Moreover, selectivity towards methanol was only 90%, and to CO 70% [4], due to the abundant quantities of CO_2 produced by the water gas shift (WGS) reaction (Equation 8.2).

$$CH_3OH + CO \rightarrow CH_3COOH \tag{8.1}$$

Thus, the carbonylation, which can be simply written as in Equation 8.1

$$CO + H_2O \leftrightarrows CO_2 + H_2 \tag{8.2}$$

requires a late transition-metal as catalyst and an iodide as promoter in order to activate methanol and transform it into the more reactive CH_3I species. During the mid-1960s, Paulik and Roth at the Monsanto Company [5] found that rhodium and an iodide promoter were more catalytically active than cobalt, with selectivities of 99% and 85% with regards to methanol and CO, respectively, and operated under significantly milder conditions [40–50 bar ($4-5 \times 10^4$ hPa) pressure; approx. 190 °C]. Despite rhodium being 1000 times more expensive than cobalt at the time, Monsanto chose to develop the rhodium-based catalyst system mainly for concerns of selectivity. During the same period, Paulik and Roth [5] demonstrated the efficiency of iridium in the reaction of Equation 8.1, and noted that it was somewhat less rapid – but more stable – in the low CO partial pressure zones of the industrial unit [6].

In 1970, the first rhodium-based acetic acid production unit went on stream in Texas City, with an annual capacity of 150 000 tons. Since that time, the Monsanto process has formed the basis for most new capacities such that, in 1991, it was responsible for about 55% of the total acetic acid capacity worldwide. In 1986, B.P. Chemicals acquired the exclusive licensing rights to the Monsanto process, and 10 years later announced its own carbonylation iridium/ruthenium/iodide system [7, 8] (Cativa™). Details of this process, from the viewpoint of its reactivity and mechanism, are provided later in this chapter. A comparison will also be made between the iridium- and rhodium-based processes. Notably, as the iridium system is more stable than its rhodium counterpart, a lower water content can be adopted which, in turn, leads to higher reaction rates, a reduced formation of byproducts, and a better yield on CO.

8.2
Rhodium-Based Processes

In terms of coordination chemistry, the reactivity of rhodium and iridium are comparable, with the coordination sphere of the two metals being similar for much of the time, notably when phosphorus- or nitrogen-containing ligands or CO are coordinated. However, the iridium metal center is more nucleophilic as its valence orbitals are more diffuse, and it has a greater tendency to adopt penta-coordinated geometries. Although this chapter is focused on the iridium-catalyzed carbonylation of methanol, it is commonplace for research groups to compare the reactivity of iridium with that of rhodium. Hence, some brief details will be provided here on the mechanism of the rhodium system.

8.2.1
The Monsanto Process

The Monsanto process [9] refers to a rhodium system operating with a high level of water, as it is necessary to maintain at least a 14 wt/wt% concentration to ensure stability of the rhodium complex in the low CO pressure zones of the process (the 'flashing zone'). The active species is the $[H_3O]^+[RhI_2(CO)_2]^-$ entity, which reacts with CH_3I in the rate-determining step to provide the methyl rhodium $[RhI_3(COCH_3)(CO)_2]^-$ species. This then evolves rapidly by a migratory CO insertion mechanism towards the acetyl species $[RhI_3(COCH_3)(CO)]^-$, which coordinates a CO ligand to give $[RhI_3(CH_3)(CO)_2]^-$ and then restores the active species $[RhI_2(CO)_2]^-$ by reductive elimination of $CH_3(CO)I$. Both, experimental data and density functional theory (DFT) calculations agree with this reaction pathway [10–12]. The net result of catalysis is:

$$CH_3I + CO \rightarrow CH_3(CO)I \tag{8.3}$$

The catalytic cycle [13, 14] is summarized in Figure 8.1.

The role of the iodide promoter is to activate methanol and to produce iodomethane, generally by direct reaction of HI. The organic reactions which take place in the medium are:

$$CH_3OH + HI \rightarrow CH_3I + H_2O \tag{8.4}$$

$$CH_3(CO)I + H_2O \rightarrow CH_3COOH + HI \tag{8.5}$$

$$CH_3COOH + CH_3OH \rightleftharpoons CH_3COOCH_3 + H_2O \tag{8.6}$$

$$CH_3COOCH_3 + HI \rightarrow CH_3I + CH_3COOH \tag{8.7}$$

The WGS reaction (Equation 8.2) plays a significant role in the process, with the species $[H_3O]^+[RhI_2(CO)_2]^-$ starting another catalytic cycle by reacting with hydriodic acid to give $[RhI_4(CO)_2]^-$ and H_2 (see Figure 8.2); this stable rhodium(III) complex then reacts with CO and water to regenerate $[RhI_2(CO)_2]^-$ with the

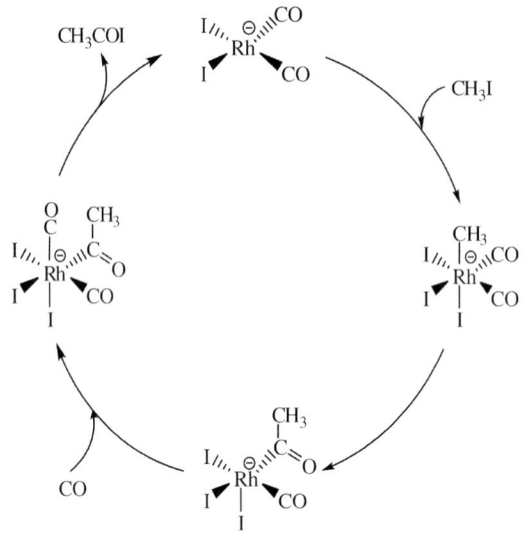

Figure 8.1 Catalytic cycle for the rhodium-catalyzed methanol carbonylation.

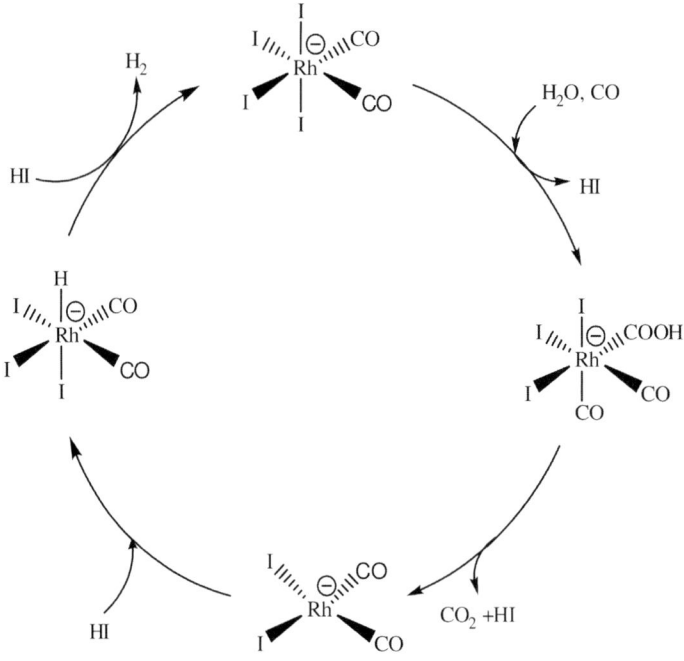

Figure 8.2 Catalytic cycle for the rhodium-catalyzed water gas shift (WGS) reaction.

formation of CO_2. Although Eisenberg and coworkers studied the mechanism and kinetics of this reaction in depth during the 1980s [15, 16], the catalytic cycle is for most of the time represented in an elliptical fashion, and the intimate succession of some of the steps remains to be fully clarified. The selectivity of the carbonylation reaction is governed by the capacity of the active $[RhI_2(CO)_2]^-$ species to follow exclusively the main catalytic cycle.

It is of interest to note that the carbonylation rate is independent of both the CO partial pressure and the methanol concentration. However, the rate is first-order in the rhodium and methyl iodide concentrations, consistent with the CH_3I oxidative addition to $[RhI_2(CO)_2]^-$ being the rate-determining step.

8.2.2
The Celanese Process

A large part of the process energy consumption arises from the dehydrating distillation column in which water and acetic acid (as well as small amounts of propanoic acid that require the use of a special distillation column) are tediously separated. An improvement to the rhodium process was made by the Hoechst Celanese Company, which was the first Monsanto licensee [17]. In 1986, this company patented [18] a low-water content technology based on the addition of significant amounts of LiI to promote and stabilize the rhodium active complex and to operate at a water content of about 5%. First, the formation of $Li[RhI_2(CO)_2]$ permits stabilization of the $[RhI_2(CO)_2]^-$ active species and prevents the precipitation of insoluble RhI_3. In addition, at low water and high LiI contents, the methyl acetate concentration plays a determinant role, mainly reducing the concentration of HI, which would promote the WGS cycle (Figure 8.2). Second, LiI reacts with methyl acetate to produce CH_3I and lithium acetate, which can further react with hydriodic acid to produce acetic acid and regenerate LiI (Equations 8.8 and 8.9):

$$LiI + CH_3OAc \rightleftharpoons CH_3I + LiOAc \qquad (8.8)$$

$$HI + LiOAc \rightleftharpoons CH_3COOH + LiI \qquad (8.9)$$

The high concentration of LiOAc in the medium was proposed to induce the formation of a $[RhI_2(CO)_2(OAc)]^{2-}$ pentacoordinated dianionic species, which would activate the methyl iodide more rapidly than would the $[RhI_2(CO)_2]^-$ species [19]. The net result of the Hoechst Celanese process is a better selectivity with regards to CO, since the WGS reaction is reduced by one order of magnitude [17].

Beyond these two industrial catalytic processes, many studies have been devoted to the introduction of more electron-donating ligands in the rhodium coordination sphere in order to increase the nucleophilic character of the metal center. Most of these ligands include mono-, di-phosphines or mixed phosphine ligands containing P = O or P = S terminations [20]. Although for most of the time the reaction rates appear to be largely higher than that of $[RhI_2(CO)_2]^-$, recycling experiments have demonstrated a progressive decrease in activity to reach that of $[RhI_2(CO)_2]^-$, presumably due to the formation of phosphonium salts [21].

8.3
Iridium Reactivity in the Methanol Carbonylation Reaction

Shortly after Paulik and Roth's short communication [5] that $IrCl_3 \cdot 3H_2O$ and [IrCl(CO)(PPh$_3$)$_2$] were giving rise to high reaction rates at low pressure for the carbonylation of methanol in acetic acid, Brodzki et al. [22] reported in a detailed study the activity of various iridium salts as precursors in batch experiments. The maximum activity was obtained at 190 °C and 60 bar (6×10^4 hPa), starting from the [Ir(η^4-C$_8$H$_{12}$)(η^5-C$_5$H$_5$)] precursor with a half-reaction time of 50 min, which was largely superior to [Ir$_2$(μ-Cl)$_2$((η^4-C$_8$H$_{12}$)$_2$] ($t_{1/2}$ = 114 min). Matsumoto et al. [23] reported that, when starting with Ir(IV) chloride, and operating in acetophenone as solvent at 30 bar (3×10^4 hPa) and 173 °C, methyl acetate and acetic acid were obtained, with reaction rates that were independent of the CO pressure (>15 bar; >1.5×10^4 hPa) and the amount of methyl iodide present (above a CH$_3$I : Ir ratio of 20). By adding triphenylphosphine to the frozen solution extracted from the autoclave, the authors intercepted the acetyl complex [IrI$_2$(COCH$_3$)(CO)(PPh$_3$)$_2$] and suggested that methanolysis of the neutral acetyl iridium(III) species was the rate-determining step.

In parallel to all of his studies on rhodium reactivity, Forster recognized at an early stage that [AsPh$_4$][IrI$_2$(CO)$_2$] underwent a very rapid reaction with methyl iodide at room temperature, to yield the methyl complex more rapidly than did the corresponding rhodium system [24]. Similarly, [AsPh$_4$][IrI$_2$(CO)$_2$] was seen to react with an excess of HI to produce [AsPh$_4$][IrI$_4$(CO)$_2$], and CH$_3$COI to give [AsPh$_4$][Ir(COCH$_3$)I$_3$(CO)$_2$]. Further kinetic and spectroscopic studies have highlighted the differences between rhodium and iridium chemistry. For example, the identification of both anionic and neutral iridium species along the methanol carbonylation catalytic cycle [25] allowed Forster to propose a general scheme involving two interconnected neutral and anionic catalytic cycles, either for the methanol carbonylation or for the WGS reaction (Figures 8.3a and b).

The main factor controlling the anionic or neutral catalytic pathway is the iodide concentration. Indeed, when the latter is maintained at low level by using methylacetate and limited water amounts (low methanol and methyl iodide concentrations) the predominant species is the neutral [IrI(CO)$_3$] complex. In contrast, as the [I$^-$] concentration is increased the major species becomes [Ir(CH$_3$)I$_3$(CO)$_2$]$^-$. Although the neutral complex [IrI(CO)$_3$] is largely less reactive towards the oxidative addition of CH$_3$I than the anionic species [IrI$_2$(CO)$_2$]$^-$, the formation of the acetyl complex is faster from [Ir(CH$_3$)I$_2$(CO)$_2$]$_{1\,or\,2}$ or under CO from [Ir(CH$_3$)I$_2$(CO)$_3$] than from [Ir(CH$_3$)I$_3$(CO)$_2$]$^-$. Forster has proposed the following seminal concept to rationalize the relative I$^-$ and CO effects on the rate of formation of the acyl species:

$$[Ir(CH_3)I_3(CO)_2]^- + CO \longrightarrow [Ir(CH_3)I_2(CO)_3] + I^-$$
$$\hookrightarrow \quad Ir(CH_3CO) \text{ species}$$

More recently, several studies have confirmed the initial observations of Forster, and some of the species involved in the catalytic cycle of Figure 8.3 have been

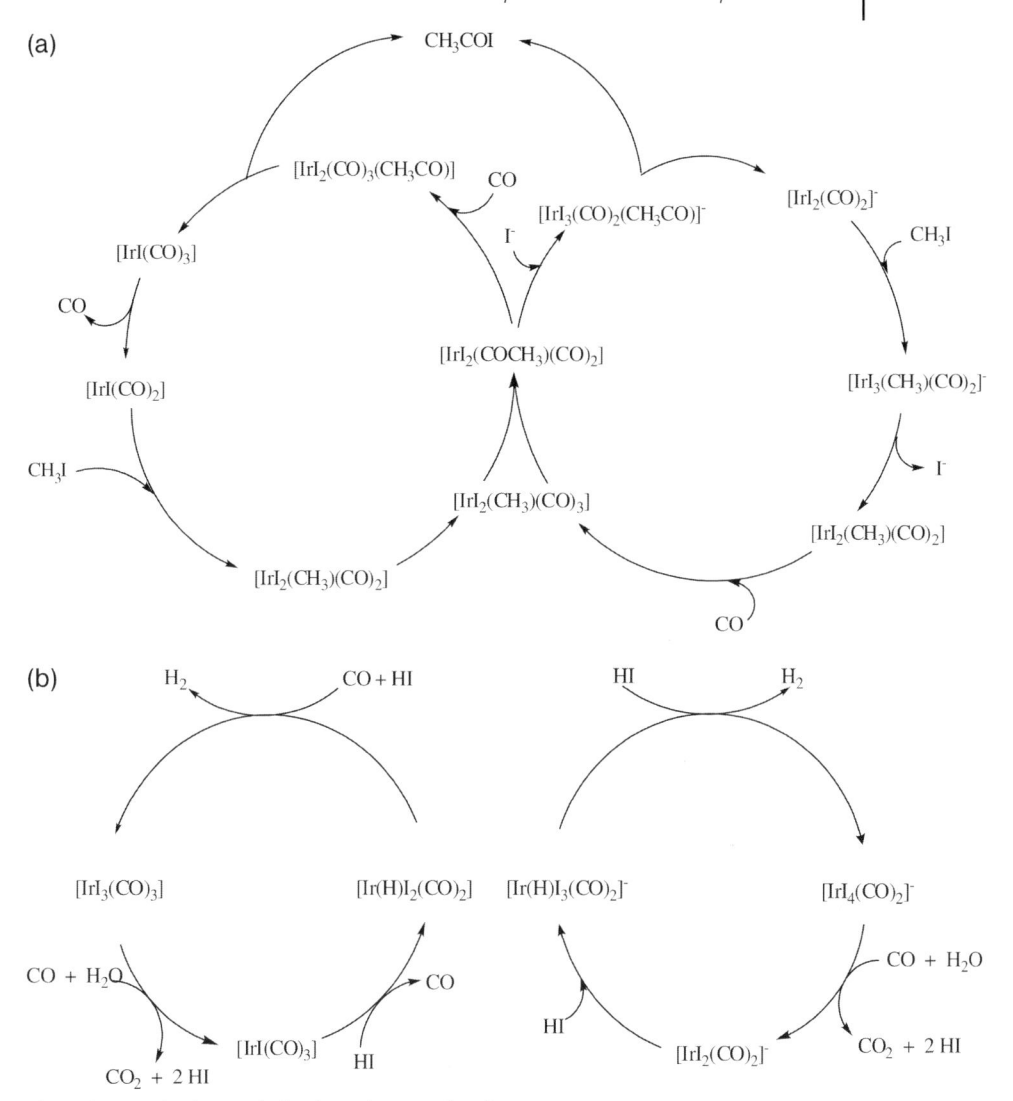

Figure 8.3 (a) Catalytic cycle for the iridium-catalyzed methanol carbonylation; (b) catalytic cycle for the iridium-catalyzed water gas shift (WGS) reaction. Both as originally proposed by D. Forster (adapted from Ref. [25]).

isolated and fully characterized. The X-ray crystal structure of the [PPN][IrI$_2$(CO)$_2$] complex has been solved, and shows the iridium(I) metal center as being in a square-planar environment, with the two CO ligands in a *cis*-geometry [26]. Theoretical calculations have revealed that the free energy of the *cis*-species is 10.4 kcal mol^{-1} lower than that of the *trans*-isomer [27]. High-pressure infrared and nuclear magnetic resonance (NMR) studies have been carried out on the [IrI$_2$(CO)$_2$]$^-$ species, and the reactivity of the rhodium analogue compared [28]. It has been shown that, under 50 bar (5×10^4 hPa) of CO the major proportion of $[IrI_2(CO)_2]^-$ is converted to $[IrI(CO)_3]$ by the substitution of one iodide ligand by CO. This reaction is two orders of magnitude slower on $[RhI_2(CO)_2]^-$.

Hydriodic acid reacts readily with $[PPN][IrI_2(CO)_2]$, even in the presence of methyl iodide, to give the complex *fac,cis*-$[PPN][Ir(H)I_3(CO)_2]$, in which the two CO ligands are still in a *cis*-arrangement and the three iodo ligands are on one face of the pseudo-octahedron [26]. Even at high temperature, the oxidative addition of CH_3I occurs relatively slowly and results in the complex *fac,cis*-$[PPN][Ir(CH_3)I_3(CO)_2]$; however, as this reaction was conducted in a nonpolar and nonreactive solvent it can be expected that, under the methanol carbonylation conditions, HI would react rapidly with methanol to produce iodomethane. Nevertheless, high concentrations of HI in the medium should be avoided in order to prevent the WGS cycle. The calculated free energies of activation for the oxidative addition of methyl iodide to *cis*-$[RhI_2(CO)_2]^-$ and *cis*-$[IrI_2(CO)_2]^-$ are 20.8 and 16.9 kcal mol^{-1}, respectively [12], thus highlighting the higher reactivity of the more nucleophilic iridium species. Experimental data have correlated these theoretical values, since activation energies of 26 and 21 kcal mol^{-1} have been reported for rhodium and iridium, respectively [29, 30]. Additionally, a neutral methyl—iridium complex has been prepared by I^- abstraction from $[Ir(CH_3)I_3(CO)_2]^-$ by InI_3 to produce $[Ir(CH_3)I_2(CO)_2]$ (and InI_4^-), which dimerizes into the stable complex $[Ir_2(\mu\text{-}I)_2I_2(CH_3)_2(CO)_4]$, the X-ray crystal structure of which has been resolved [31].

The next step in the iridium catalytic cycle is the migratory CO insertion which leads to the corresponding acyl species. In this case, two isomers have been observed, with the kinetic product having been isolated as the *fac,cis*-$[AsPh_4]$ $[Ir(COCH_3)I_3(CO)_2]$ compound [32]. The thermodynamic product, the *mer,trans*-$[Ir(COCH_3)I_3(CO)_2]^-$ anionic complex, can be obtained either by photolysis of the complex *fac,cis*-$[AsPh_4][Ir(CH_3)I_3(CO)_2]$ at 366 nm under 1 atm (10^3 hPa) of CO [33], or by reacting *fac,cis*-$[PPN][Ir(CH_3)I_3(CO)_2]$ with 5 bar (0.5×10^4 hPa) of CO in a CH_2Cl_2/methanol mixture containing significant amounts of methanol [34]. Under long-term photolysis of $[Ir(CH_3)I_3(CO)_2]^-$ the dimeric complex $[Ir_2(\mu\text{-}I)_2I_4(COCH_3)_2(CO)_2]^{2-}$ has also been isolated and its X-ray crystal structure resolved [33].

The results of DFT studies [12] on the migratory CO insertion [12, 35] have shown the methyl—iridium bond to be stronger than the methyl—rhodium bond, and the free energy of activation for the methyl migration on a vicinal CO ligand to be 29.2 kcal mol^{-1} for Ir and 18.4 kcal mol^{-1} for Rh; these data are in good agreement with the experimental measurements of 30.6 kcal mol^{-1} [36] and 19.3 kcal mol^{-1} [10], respectively. Interestingly, *ab initio* molecular dynamics simulations have shown that, in the transition state, the iridium—iodine bond *trans* to the CO ligand which will be attacked by the methyl group begins to break, and is completely dissociated before the acyl bond is finally formed [35]. Such a dissociation of the M—I bond during the migratory CO insertion does not occur with rhodium. The dissociation of an iodide ligand in $[Ir(CH_3)I_3(CO)_2]^-$ in methanol-containing solvents is endothermic, and the corresponding enthalpy is about 8 kcal mol^{-1}. The I^- ligand can be easily replaced by a soft ligand such as CO, but not by methanol or acetic acid. Similarly, instead of CO a phosphite ligand can be introduced into the

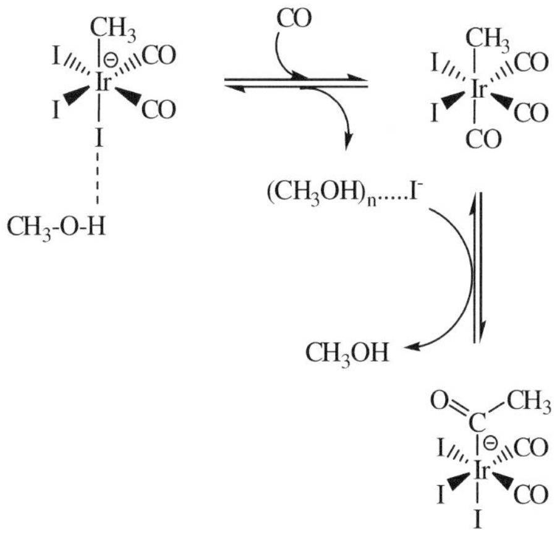

Figure 8.4 Methanol-assisted migratory CO insertion in the iridium-catalyzed methanol carbonylation (adapted from Refs [37, 38]).

coordination sphere of iridium, which causes the migratory CO insertion rate to be significantly increased [35, 36].

Interestingly, Maitlis et al. have reported that the rate of the migratory CO insertion is dramatically accelerated on the addition of small amounts of methanol [37]. Whereas in chlorobenzene, the oxidative addition of CH_3I on $[IrI_2(CO)_2]^-$ is 120-fold faster than that on $[RhI_2(CO)_2]^-$, the methyl migration is 10^5-fold faster for rhodium than for iridium. However, the addition of methanol as solvent provides a more polar and protic medium, and this assists in the iodide abstraction through a hydrogen interaction between the hydroxy group and one iodide ligand [37, 38]. The general mechanism proposed by Maitlis et al. for this methanol-assisted migratory CO insertion is shown in Figure 8.4. In neat chlorobenzene, the values at 25 °C for the activation parameters are $\Delta G^{\neq} = 30.6\,kcal\,mol^{-1}$, and $22\,kcal\,mol^{-1}$ in methanol/chlorobenzene mixtures. In this situation, the authors have noted a strong entropic effect.

Thus, the migratory CO insertion reaction on the iridium center, which is the rate-determining step of the iridium catalytic cycle, is accelerated by I^- ligand dissociation. These results tend to rationalize the seminal and inspired observations of Forster at Monsanto Co. [25].

The reductive elimination step has undergone much less examination, with the majority of authors considering that the acyl species produces CH_3COI and regenerates the active anionic $[IrI_2(CO)_2]^-$ species. When a DFT study was carried out by Kinnunen and Laasonen [39], the *fac,cis*-$[Ir(COCH_3)I_3(CO)_2]^-$ isomer was seen to be the dominant intermediate for the anionic route, whereas for the neutral pathway the *mer,cis*-$[Ir(COCH_3)I_2(CO)_3]$ isomer allowed a faster reductive elimination reaction.

Here, it is worthy of note that dihydrogen, which is present in the medium via the WGS reaction, can also react in its own right with the iridium species

(Equations 8.10 and 8.11). Thus, Merbach and coworkers have shown from high-pressure (HP) NMR experiments that the oxidative addition of H_2 to $[IrI_2(CO)_2]^-$ [50 bar (5×10^4 hPa) H_2, $-20\,°C$] provides the anionic dihydride species $[Ir(H)_2I_2(CO)_2]^-$ [28]. The reactivity of $[Ir(CH_3)I_3(CO)_2]^-$ towards dihydrogen has also been investigated by Maitlis *et al.* [40], who have shown that methane is produced at $110\,°C$ and 1 bar (0.1×10^4 hPa) of dihydrogen, giving rise to the hydride $[Ir(H)I_3(CO)_2]^-$ (Equation 8.11).

$$[IrI_2(CO)_2]^- + H_2 \rightarrow [Ir(H)_2I_2(CO)_2]^- \tag{8.10}$$

$$[Ir(CH_3)I_3(CO)_2]^- + H_2 \rightarrow [Ir(H)I_3(CO)_2]^- + CH_4 \tag{8.11}$$

Recent mechanistic studies using HP infrared equipment, as well as HP-NMR measurements involving the use of ^{13}CO and $^{13}CH_3I$, have allowed the iridium intermediates which are present in solution as methyl acetate and water, and are consumed to produce acetic acid [32, 34, 41–43], to be followed. All of these observations can be rationalized by a single catalytic cycle (see Figure 8.5), in which equilibria exist between the neutral and anionic complexes for all species. The main species involved in the carbonylation, which are detected in batch mode under carbonylation conditions [34], and correspond to the slower steps of catalysis, are the methyl—iridium and acetyl-iridium complexes $[Ir(CH_3)I_3(CO)_2]^-$ and $[Ir(COCH_3)I_3(CO)_2]^-$ respectively.

8.4
The Iridium-Based Cativa Process

In 1993, B.P. Chemicals patented the use of iodo carbonyl ruthenium complexes or their osmium analogues, to promote the iridium-catalyzed carbonylation of methanol to acetic acid [44]. Under the same experimental conditions, iridium alone gives a reaction rate of $12.2\,mol\,l^{-1}\,h^{-1}$, whereas in the presence of ruthenium (Ru:Ir = 10) the rate is $23.9\,mol\,l^{-1}\,h^{-1}$. Following the announcement of this discovery in the literature [7, 8], the performances of the novel system [44–47] were detailed later [48, 49]. For example, for a Ru:Ir molar ratio of 5, the carbonylation rate is increased by a factor of 2.6 when $[RuI_2(CO)_4]$ is introduced. The addition of SnI_2 also accelerates also the catalysis, though less efficiently [37]. Although InI_3, GaI_3 or ZnI_2 can each promote catalysis, ionic iodides such as LiI or $[NBu_4]I$ present an inhibitory effect. The role of the promoters is mainly to remove an iodide ligand in $[Ir(CH_3)I_3(CO)_2]^-$ and to form $[RuI_3(CO)_3]^-$, or $[SnI_3]^-$, $[InI_4]^-$, $[GaI_4]^-$, $[ZnI_3]^-$ and consequently the neutral methyl complex $[Ir(CH_3)I_2(CO)_2]$. These promoters also accelerate the reaction of HI with methyl acetate to produce methyl iodide [50]. In fact, they exert a key role in moderating the concentration of free iodide, which inhibits the migratory CO insertion, and in reducing the concentration of HI which can react with $[IrI_2(CO)_2]^-$ to enter the WGS catalytic

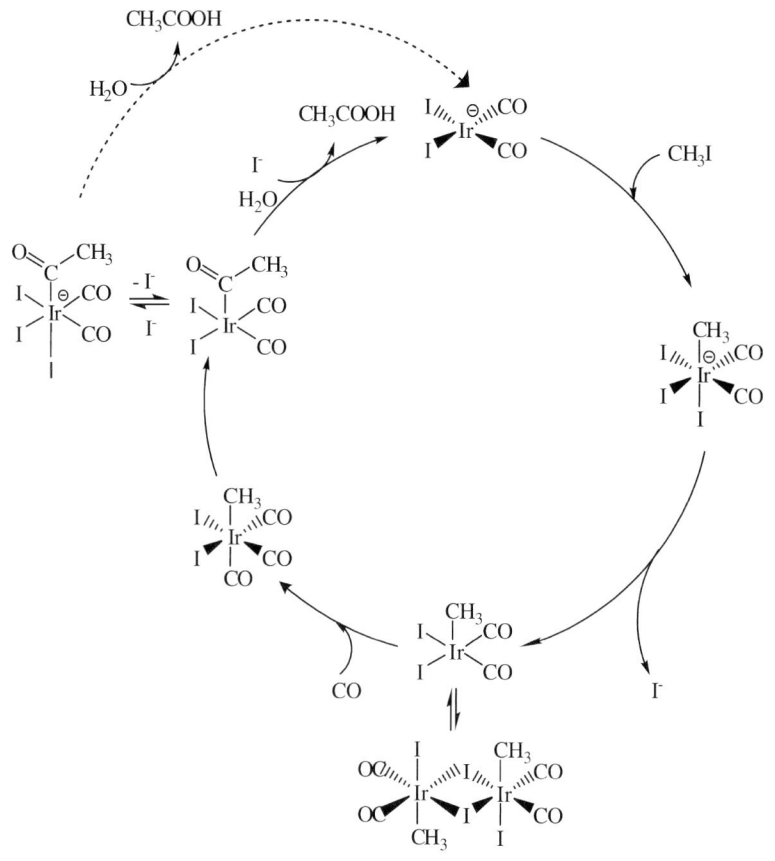

Figure 8.5 General catalytic cycle for the iridium-catalyzed methanol carbonylation, as proposed by Haynes et al. (adapted from Ref. [32]).

cycle. As the process operates at a water content of about 5% wt/wt, the [I$^-$] concentration remains at a low level so that the neutral iridium species are favored [32, 41–43, 50]. The stability of the iridium catalyst is maintained over a broad range of experimental conditions, especially of the CO pressure, since – in contrast to rhodium – no metal precipitation occurs in the flashing zone.

Kinetic analyses of the effect of ruthenium on the rate of the carbonylation reaction [32] have shown that the addition of [Ru$_2$(μ-I)$_2$I$_2$(CO)$_6$], [RuI$_2$(CO)$_4$], or [RuI$_2$(CO)$_2$]$_n$ in small amounts (Ru:Ir molar ratio < 0.2) provides significant accelerations, whereas introduction of the anionic complexes [NBu$_4$][RuI$_3$(CO)$_3$] or [NBu$_4$]$_2$[RuI$_4$(CO)$_2$] results in an absence of promotion of catalysis, though not in its inhibition. Similarly, the inhibiting effect of LiI (I$^-$:Ir molar ratio = 1) is compensated by the addition of [RuI$_2$(CO)$_4$]. At least at low concentrations, the effect of ruthenium on the carbonylation rate is proportional to its concentration. All of these observations are consistent with the ability of the promoter to remove an iodide ligand from the anionic $[Ir(CH_3)I_3(CO)_2]^-$ iridium intermediate and, indeed, that it is its key role [32, 41, 42, 51]. Thus, under a CO atmosphere, $[Ir(CH_3)I_3(CO)_2]^-$ evolves rapidly towards $[Ir(CH_3)I_2(CO)_3]$, before the methyl cis-migration occurs to provide the neutral acetyl—iridium species. The mechanism by which ruthenium draws an iodide ligand has been elucidated by Haynes et al. [50], and Heaton et al. [51]: the two $[Ir(CH_3)I_3(CO)_2]^-$ and $[RuI_2(CO)_4]$ complexes react together to give the intermediate heterobimetallic species $[IrI_2(CH_3)(CO)_2(\mu\text{-}I)RuI_2(CO)_3]^-$, which has been characterized by NMR. Under 6 bar (0.6×10^4 hPa) CO, this species gives $[Ir(CH_3)I_2(CO)_3]$ and $[RuI_3(CO)_3]^-$. As the undercoordinated $[Ir(CH_3)I_2(CO)_2]$ intermediate is formed, its dimerization provides $[Ir_2(\mu\text{-}I)_2I_2(CH_3)_2(CO)_4]$, or its reaction with the starting iridium material produces $[Ir_2(\mu\text{-}I)I_3(CH_3)_2(CO)_4]^-$. The neutral dimer has been characterized by an X-ray crystal structure, and shows the two methyl ligands being in the equatorial plane and in mutual trans-positions [26, 31]. It is cleaved by ^{13}CO to give $[Ir(CH_3)I_2(CO)_2(^{13}CO)]$, the labeled ligand being specifically in trans position to the methyl ligand. The next step of the intrasphere nucleophilic attack of the methyl group concerns exclusively one of the two nonlabeled CO ligands [32].

8.5
The Iridium–Platinum-Based Process

Acetex Chimie, in collaboration with our laboratory, has patented the use of platinum to promote the iridium-catalyzed carbonylation of methanol [52]. The addition of a small amount of the preformed carbonyl iodoplatinum(II) complex $[Pt_2(\mu\text{-}I)_2I_2(CO)_2]$ gives rise to high reactions rates with performances comparable to those of the [Ir—Ru] system. High-pressure infrared and NMR analyses have been carried out to provide a deeper insight into the mechanism by which the cocatalyst assists the reaction [34]. Starting from the stable complex $[PPN][IrI_2(CO)_2]$ led us to isolate the two products resulting from the oxidative addition of HI and CH_3I (see Section 8.3). We were also able to identify, by using NMR and fast-atom bombardment (FAB) mass spectrometry, the two products which are formed when the pentacoordinated species $[Ir(CH_3)I_2(CO)_2]$ is produced by iodide abstraction from $[Ir(CH_3)I_3(CO)_2]^-$ by adding $[Pt_2(\mu\text{-}I)_2I_2(CO)_2]$. The neutral dimer $[Ir_2(\mu\text{-}I)_2I_2(CH_3)_2(CO)_4]$ has been identified, as well as the homodinuclear anionic complex $[Ir_2(\mu\text{-}I)I_3(CH_3)_2(CO)_4]^-$. In addition, the [Ir—Pt] key intermediate $[IrI_2(CH_3)(CO)_2(\mu\text{-}I)PtI_2(CO)]^-$ has been observed, which is the signature of the removal of I^- from $[Ir(CH_3)I_3(CO)_2]^-$ by $[PtI_2(CO)_2]$ under a CO atmosphere. Thus, here also, the key role of the cocatalyst is to reduce the electron density on the iridium center, by abstracting an iodo ligand, and therefore promoting the migratory CO insertion. Under optimized conditions, and in particular with a Ir:Pt molar ratio close to 0.7:0.3, catalysis can occur at rates which are some 20% faster than those of the rhodium Monsanto process, and for water concentrations close to 5%.

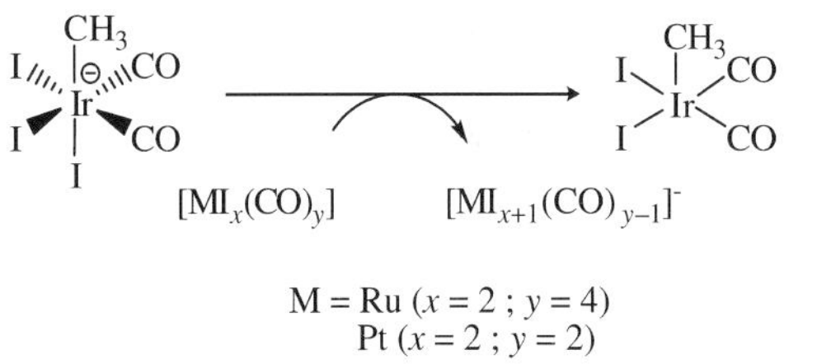

$$\text{[structure: } I_{\cdots}\overset{CH_3}{\underset{I}{\overset{|}{\underset{|}{Ir}}}}\overset{\ominus}{\underset{CO}{\cdots}}CO \longrightarrow I\overset{CH_3}{\underset{I}{\overset{|}{\underset{}{Ir}}}}\overset{CO}{\underset{CO}{\cdots}}\text{]}$$

$$[MI_x(CO)_y] \qquad [MI_{x+1}(CO)_{y-1}]^-$$

$$M = Ru \ (x = 2 \ ; \ y = 4)$$
$$Pt \ (x = 2 \ ; \ y = 2)$$

Figure 8.6 Main step showing the role of the cocatalyst in the iridium-catalyzed methanol carbonylation reaction.

8.6
The Iridium–Cocatalyst Mechanism, and Conclusions

The B.P. Chemicals process, which currently is operating in five industrial units, and the Acetex Chimie process, both use iridium in the presence of a promoting agent. As shown in Figure 8.3a, from the $[Ir(CH_3)I_3(CO)_2]^-$ species which is formed very easily by oxidative addition of CH_3I to the active species $[IrI_2(CO)_2]^-$, it is necessary to generate efficiently the neutral $[Ir(CH_3)I_2(CO)_3]$ intermediate. Either $[RuI_2(CO)_4]$ or $[PtI_2(CO)_2]$ are able to abstract an I^- ligand from the anionic methyl—iridium complex, under the conditions of catalysis. The corollary is the reformation of these two promoters from $[RuI_3(CO)_3]^-$ or $[PtI_3(CO)]^-$, respectively. Under the correct operating conditions this occurs during the catalytic process by the substitution of an I^- ligand by CO, to regenerate the neutral promoters and ensure an efficient catalytic rate. Figure 8.6 shows the key step where the cocatalyst operates.

It is interesting to note that, when operating at approximately 5% water content, and with good reaction rates, it is possible to save about 30% of the energy costs of the process when compared to the Monsanto process. In addition, the current price of rhodium is 20-fold that of iridium. Even with annual unit capacity productions as large as 500 000 tons, it is still possible to improve catalysis and to achieve significantly better performances.

Acknowledgments

We gratefully acknowledge the strong collaboration between our laboratory and Acetex-chimie since 1996, and in particular we thank Daniel Thiébaut. This [Ir—Pt] research project was also supported by the Ministère de la Recherche et de la Technologie (especially Prof. Roger Guilard) during the period 1999–2002 as a joint project *Equipe de Recherche et de Technologie* between the Institut National Polytechnique and Acetex-chimie. We are also indebted to our coworkers, PhD students and postdoctoral fellows during this 12-year period, including Brigitte Zahner, Carole Le Berre, Laurent Azam, Roberto Giordano, Samuel Gautron and Nicolas Lassauque.

References

1 Weissermel, K. and Harpe, H.J. (1997) *Industrial organic Chemistry*, 3rd edn, Wiley-VCH Verlag GmbH, Weinheim, Germany.
2 Reppe, W. (1956) US Patent 2 729 651, assigned to BASF.
3 Reppe, W. (1957) US Patent 2 789 137., assigned to BASF.
4 Mullen, A. (1980) *New Syntheses with Carbon Monoxide* (ed. J. Falbe), Springer-Verlag, Berlin, pp. 243–308.
5 Paulik, F.E. and Roth, J.F. (1968) *Journal of the Chemical Society, Chemical Communications*, 1578.
6 Roth, J.F., Craddock, J.H., Hershman, A. and Paulik, F.E. (1971) *CHEMTECH*, 600–5.
7 Vercauteren, C.J.E., Clode, K.E. and Watson, D.J. (1994) European Patent 616 997.
8 (a) Anonymous (1996) *Chemistry in Britain*, **32**, 7; (b) Anonymous (1996) *Chemistry and Industry (London)*, 483.
9 Paulik, F.E., Hershman, A., Knox, W.R. and Roth, J.F. (1968) GB Patent 1 234 641, assigned to Monsanto.
10 Haynes, A., Mann, B.E., Morris, G.E. and Maitlis, P.M. (1993) *Journal of the American Chemical Society*, **115**, 4093–100.
11 Cheong, M., Schmid, R. and Ziegler, T. (2000) *Organometallics*, **19**, 1973–82.
12 Kinnunen, T. and Laasonen, K. (2001) *Journal of Molecular Structure*, **542**, 273–88.
13 Forster, D. (1975) *Journal of the American Chemical Society*, **97**, 951–2.
14 Forster, D. (1976) *Journal of the American Chemical Society*, **98**, 846–8.
15 Cheng, C.-H., Hendricksen, D.E. and Eisenberg, R. (1977) *Journal of the American Chemical Society*, **99**, 2791–2.
16 Baker, E.C., Hendricksen, D.E. and Eisenberg, R. (1980) *Journal of the American Chemical Society*, **102**, 1020–7.
17 Gauss, M., Seidel, A., Torrence, P. and Heymanns, P. (1996) *Applied Homogeneous Catalysis with Organometallic Compounds* (eds B. Cornils and W.A. Herrmann), Wiley-VCH Verlag GmbH, Weinheim, pp. 104–38.
18 Smith, B.L., Torrence, G.P., Aguilo, A. and Alder, J.S. (1991) US Patent 5 001 259 (priority June 03, 1986), assigned to Hoechst Celanese Corporation.
19 Smith, B.L., Torrence, G.P., Murphy, M.A. and Aguilo, A. (1987) *Journal of Molecular Catalysis*, **39**, 115–36.
20 Haynes, A. (2006) *Topics in Organometallic Chemistry*, **18**, 179–205 (and references quoted therein).
21 Rankin, J., Benyei, A.C., Poole, A.D. and Cole-Hamilton, D.J. (1999) *Journal of the Chemical Society–Dalton Transactions*, 3771–82.
22 Brodzki, D., Denise, B. and Pannetier, G. (1977) *Journal of Molecular Catalysis*, **2**, 149–61.
23 Matsumoto, T., Mizoroki, T. and Ozaki, A. (1978) *Journal of Catalysis*, **51**, 96–100.
24 Forster, D. (1972) *Inorganic Chemistry*, **11**, 473–5.
25 Forster, D. (1979) *Journal of the Chemical Society–Dalton Transactions*, 1639–45.
26 Gautron, S., Giordano, R., Le Berre, C., Jaud, J., Daran, J.-C., Serp, P. and Kalck, P. (2003) *Inorganic Chemistry*, **42**, 5523–30.
27 Kinnunen, T. and Laasonen, K. (2001) *Journal of Molecular Structure*, **540**, 91–100.
28 Churlaud, R., Frey, U., Metz, F. and Merbach, A.E. (2000) *Inorganic Chemistry*, **39**, 304–7.
29 Bassetti, M., Monti, D., Haynes, A., Pearson, J.M., Stanbridge, I.A. and Maitlis, P.M. (1992) *Gazzetta Chimica Italiana*, **122**, 391.
30 Ellis, P.R., Pearson, J.M., Haynes, A., Adams, H., Bailey, N.A. and Maitlis, P.M. (1994) *Organometallics*, **13**, 3215–26.
31 Ghaffar, T., Adams, H., Maitlis, P.M., Sunley, G.J., Backer, M.J. and Haynes, A. (1998) *Chemical Communications*, 1023–1024.
32 Haynes, A., Maitlis, P.M., Morris, G.E., Sunley, G.J., Adams, H., Badger, P.W., Bowers, C.M., Cook, D.B., Elliott, P.I.P., Ghaffar, T., Green, H., Griffin, T.R., Payne, M., Pearson, J.M., Taylor, M.J., Vickers, P.W. and Watt, R.J. (2004) *Journal*

of the American Chemical Society, **126**, 2847–61.

33 Volpe, M., Wu, G., Iretskii, A. and Ford, P.C. (2006) *Inorganic Chemistry*, **45**, 1861–70.

34 Gautron, S., Lassauque, N., Le Berre, C., Azam, L., Giordano, R., Serp, P., Laurenczy, G., Daran, J.-C., Duhayon, C., Thiébaut, D. and Kalck, P. (2006) *Organometallics*, **25**, 5894–905.

35 Cheong, M., Schmid, R. and Ziegler, T. (2000) *Organometallics*, **19**, 1973–82.

36 Haynes, A., Pearson, J.M., Vickers, P.W., Charmant, J.P.H. and Maitlis, P.M. (1998) *Inorganica Chimica Acta*, **270**, 382–91.

37 Pearson, J.M., Haynes, A., Morris, G.E., Sunley, G.J. and Maitlis, P.M. (1995) *Journal of the Chemical Society, Chemical Communications*, 1045–6.

38 Maitlis, P.M., Haynes, A., Sunley, G.J. and Howard, M.J. (1996) *Journal of the Chemical Society – Dalton Transactions*, 2187–96.

39 Kinnunen, T. and Laasonen, K. (2001) *Journal of Organometallic Chemistry*, **628**, 222–32.

40 Ghaffar, T., Charmant, J.P.H., Sunley, G.J., Morris, G.E., Haynes, A. and Maitlis, P.M. (2000) *Inorganic Chemistry Communications*, **3**, 11–12.

41 Haynes, A. (2006) Acetic acid synthesis by catalytic carbonylation of methanol, in *Topics in Organometallic Chemistry*, Vol. 18 (ed. M. Beller), Springer-Verlag, Berlin, pp. 179–205.

42 Maitlis, P.M., Haynes, A., James, B.R., Catellani, M. and Chiusoli, G.P. (2004) *Dalton Transactions*, 3409–19.

43 Morris, G. (2005) Carbonylation of methanol to acetic acid and methyl acetate to acetic anhydride, in *Mechanisms in Homogeneous Catalysis, A Spectroscopic Approach* (ed. B. Heaton), Wiley-VCH Verlag GmbH, Weinheim, pp. 195–230.

44 Garland, C.S., Giles, M.F. and Sunley, J.G. (1995) European Patent 0643034 (priority GB 93-18809; 09 October 1993), assigned to BP Chemicals.

45 Baker, M.J., Giles, M.F., Garland, C.S. and Rafeletos, G. (1996), European Patent 0749948 (priority GB 95-12606; 21 June 1995), assigned to BP Chemicals.

46 Ditzel, E.J. and Sunley, G.J. (1998), European Patent 0849248 (priority GB 96-26429; 19 December 1996), assigned to BP Chemicals.

47 Ditzel, E.J. and Sunley, G.J. (1998), European Patent 849249 (priority GB 96-26317; 19 December 1996), assigned to BP Chemicals.

48 Sunley, G.J. and Watson, D.J. (2000) *Catalysis Today*, **58**, 293–307.

49 Jones, J.H. (2000) *Platinum Metals Review*, **44**, 94–105.

50 Haynes, A. (2005) The use of high pressure spectroscopy to study catalytic mechanism, in *Mechanisms in Homogeneous Catalysis, A Spectroscopic Approach* (ed. B. Heaton), Wiley-VCH Verlag GmbH, Weinheim, pp. 107–50.

51 Whyman, R., Wright, A.P., Iggo, J.A. and Heaton, B.T. (2002) *Dalton Transactions*, 771–7.

52 Le Berre, C., Serp, P., Kalck, P., Layeillon, L. and Thiebaut, D. (1998) French Patent 98.13954 (priority 05.11.1998), assigned to Acetex Chimie.

9
Iridium-Catalyzed Asymmetric Allylic Substitutions
Günter Helmchen

9.1
Introduction

Transition-metal-catalyzed asymmetric allylic substitution is an important tool in organic synthesis [1]. Both, the enantioselectivity and regioselectivity of this reaction are determined by many parameters, including the metal ion, the leaving group, the nucleophile and the substituents at the allylic moiety. Most often, Pd-catalysts are used in combination with symmetrically substituted allylic derivatives as substrates. Synthetically more easily accessible monosubstituted allylic substrates (Scheme 9.1) are increasingly the domain of Ir-catalysis, which reliably allows branched chiral products to be obtained with a high degree of regioselectivity. With Pd catalysts, linear products are generally produced. Catalysts and conditions that allow the branched product to be preferentially obtained have been found only recently [2, 3].

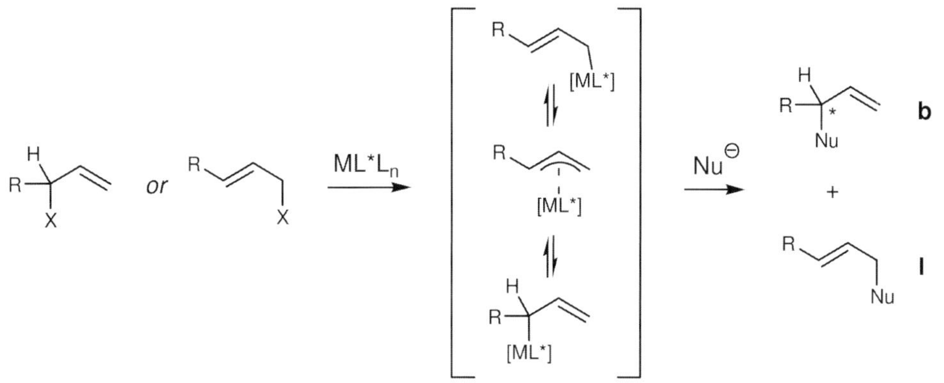

Scheme 9.1 Allylic substitution of monosubstituted allylic substrates.

Iridium Complexes in Organic Synthesis. Edited by Luis A. Oro and Carmen Claver
Copyright © 2009 WILEY-VCH Verlag GmbH & Co. KGaA, Weinheim
ISBN: 978-3-527-31996-1

Allylic substitutions catalyzed by Ir-complexes were first carried out by Takeuchi in 1997 [4]. In the same year, the first asymmetric variant was published by the present author's group [5], and two years later phosphoramidites were first used as ligands in the Ir-catalyzed allylic alkylation [6a]. Since then the field has rapidly developed, with contributions of crucial importance being due to Hartwig and coworkers, in particular concerning allylic amination [7] and mechanistic aspects [8], and to Alexakis, Polet and colleagues who identified the so-far best chiral ligand in this area [9]. As a result, it is today possible to prepare a wide range of C-, N- and O-substitution products with a very high degree of regioisomeric and enantiomeric purity. A considerable number of applications in natural products synthesis and medicinal chemistry have already been reported. Due to the rapid progress in this area, several reviews describing the complete field have been published [4, 10]. The aim of this chapter was to cover *all* publications up to the end of 2007.

9.2
Ir-Catalyzed Allylic Substitutions: Fundamentals

9.2.1
Reactivity and Regioselectivity

The following general features were found for Ir-catalyzed allylic substitutions with achiral catalysts [6, 11]. Points (b)–(e) are illustrated by Scheme 9.2.

(a) A large number of Ir-complexes have been screened for catalytic activity [11]. The complex [Ir(COD)Cl]$_2$ was found to be best suited; it still is the preferred precatalyst.

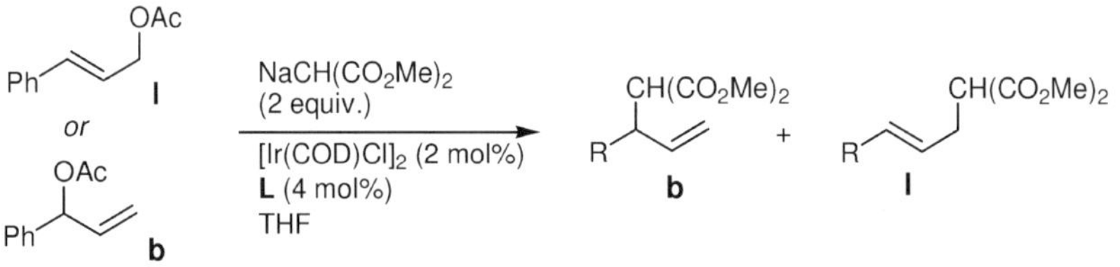

Acetate	L	T (°C)	Time (h)	Yield (%)	b : l
l	P(OPh)$_3$	rt	3	98	98 : 2
l	–	65	24	89	32 : 68
l	PPh$_3$	65	24	58	64 : 36
l	dppe^a	65	16	18	39 : 61
b	P(OPh)$_3$	rt	3	99	98 : 2
b	–	rt	3	98	98 : 2
b	PPh$_3$	rt	3	15	98 : 2

^{a} NaCH(CO$_2$Et)$_2$ was the pronucleophile.

Scheme 9.2 Ir-catalyzed allylic alkylations of an *(E)*-linear (**l**) and a branched (**b**) acetate.

(b) The parent complex [Ir(COD)Cl]$_2$ is catalytically active, but the regioselectivity of the substitution at the linear *(E)*-acetate is low. Reaction rate and regioselectivity are enhanced by an electron-poor ligand.

(c) Optimal results were obtained with a 1:1 ratio of a monodentate ligand and Ir. Additional ligand does not affect regioselectivity but leads to a decrease in rate.

(d) The reaction rate is distinctly higher for the branched substrate (**b**) than for the linear substrate (**l**). This indicates the formation of an (allyl)Ir intermediate via S_N2' substitution. A similar reaction course has been invoked for Rh-catalyzed substitutions [12, 13].

(e) Regioselectivity in favor of the branched substitution product is generally higher with the branched substrate than with the linear allylic substrate.

(f) *(Z)*-Allylic substrates react with a high degree of stereoconservation to give *(Z)*-products (Scheme 9.3).

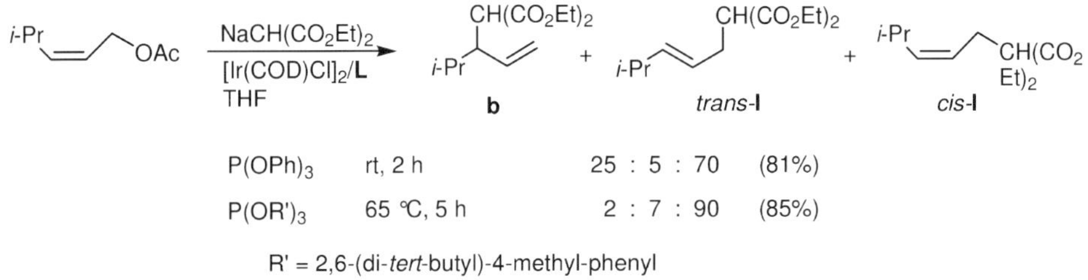

Ligand	Conditions	b : trans-l : cis-l	(yield)
P(OPh)$_3$	rt, 2 h	25 : 5 : 70	(81%)
P(OR')$_3$	65 °C, 5 h	2 : 7 : 90	(85%)

R' = 2,6-(di-*tert*-butyl)-4-methyl-phenyl

Scheme 9.3 Allylic substitution at a *(Z)*-allylic substrate.

The catalyst [Ir(COD)Cl]$_2$/P(OPh)$_3$ was highly effective also for allylic aminations. Branched monoallylation products were mainly obtained with primary amines as nucleophiles and linear *(E)*-allylic substrates. In contrast, mixtures of linear mono- and disubstitution products are usually produced with Pd-catalysts. Many types of amine could be used, for example benzylamine, piperidine and aniline [14]. In terms of allylic substrates, carbonates were more suitable than acetates. With regards to the solvent, the best results were obtained with ethanol, with complete conversion typically being achieved after a reaction time of 3 h at 50 °C. The reactions of *(Z)*-allylic carbonates to give linear *(Z)*-propenylamines proceeded with perfect stereospecificity.

A significant observation concerning the mechanism of the Ir-catalyzed allylic substitution was made when the preparation of an (allyl)(P(OPh)$_3$)Ir-intermediate was attempted [6b]. Mixing [Ir(COD)Cl]$_2$ and 2 equiv. of P(OPh)$_3$ yields the complex **K1**, which is a coordinatively unsaturated d^8–Ir^I complex [16 valence electrons (VEs)]; surprisingly, this did not react with typical allylic substrates (Scheme 9.4); rather, a reaction was started upon the addition of NaCH(CO$_2$Me)$_2$. The nucleophile also acts as base, with C—H activation affording an Ir^{III} complex, which eliminates HCl to produce a 16-VE Ir^I complex. The subsequent addition of

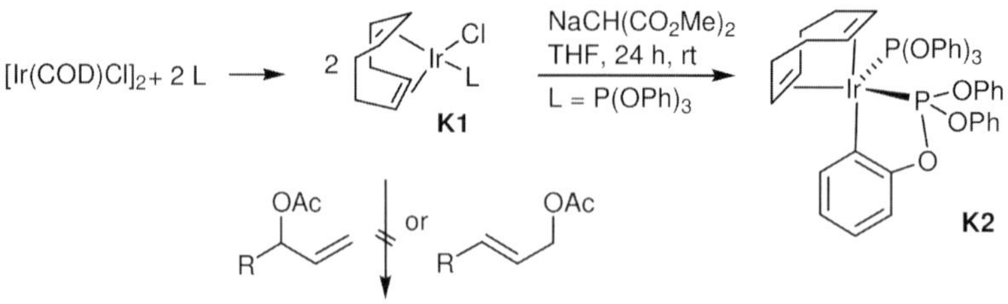

Scheme 9.4 Base-promoted C–H activation of P(OPh)$_3$.

P(OPh)$_3$ leads to the coordinatively saturated complex **K2** [15]. As complex **K2** is coordinatively saturated, P(OPh)$_3$ must dissociate in order to obtain a catalytically active 16-VE d^8–Ir complex. Similar C—H activation was later found for (phosphoramidite)Ir complexes.

9.2.2
Steric Course

The Pd-catalyzed allylic substitution with soft nucleophiles proceeds via two substitutions with inversion – that is, with a net retention of configuration. By using standard tests, the same steric course was found for the Ir-catalyzed alkylation [6b].

Pd-catalyzed reactions proceed via π-allyl complexes, which at room temperature isomerize via a π–σ–π rearrangement. As a consequence, branched as well as Z- and E-linear starting materials yield the same products, with memory effects being minimal at room temperature [16]. The isomerization processes of (allyl)Ir complexes are usually slow, and accordingly any memory effects are pronounced. The configurational stability of (allyl)Ir intermediates of the Ir-catalyzed allylic substitution was studied by an investigation of substitutions at nonracemic allylic substrates (Scheme 9.5).

The results presented in Scheme 9.5 and the results with (Z)-substrates (cf. Scheme 9.3) clearly demonstrate that the Ir-catalyzed reactions involve intermediary nonsymmetric π- or σ-allyl-Ir-complexes. The intermediary complexes undergo *slow* racemization (or epimerization) via σ–π–σ-rearrangement or sigmatropic 1,3-rearrangement (Scheme 9.6). Evans and Nelson have proposed π/σ- or *enyl*-complexes as intermediates of analogous Rh-catalyzed reactions [13]. Whether the

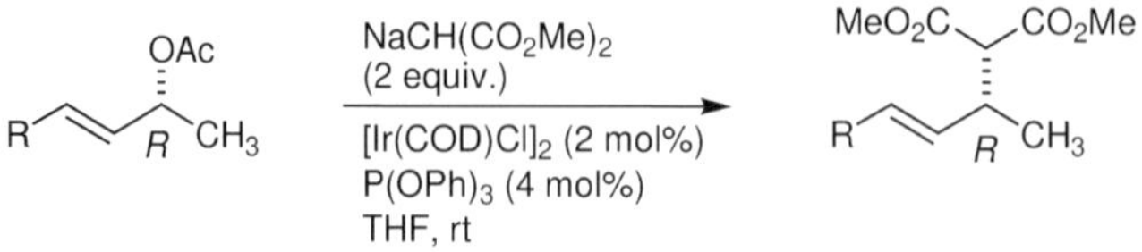

R = H   >99% ee   85% ee (91%)
R = CH$_3$ >99% ee   71% ee (83%)

Scheme 9.5 Stereospecificity of Ir-catalyzed allylic alkylations.

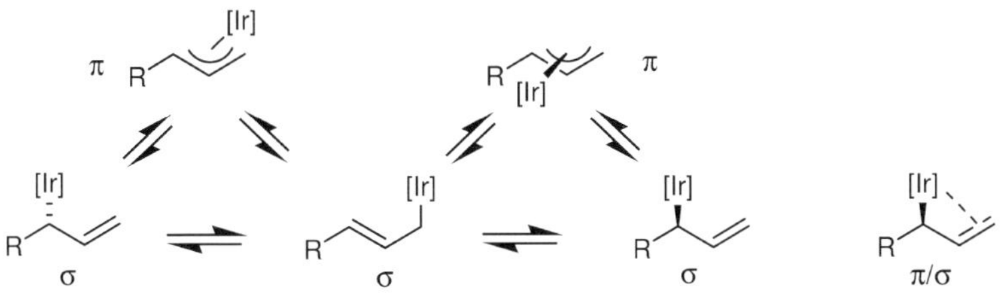

Scheme 9.6 Allyl-complexes of π-, σ- and π/σ-type as possible intermediates of Ir-catalyzed allylic substitutions.

latter are simply nonsymmetric π-complexes or distinct species characterized by separate energy minima is not clear. Density functional theory (DFT) calculations in the author's laboratory have not uncovered an example, which would require this distinction.

9.2.3
Asymmetric Catalysis: The Beginnings with Phosphinooxazolines as Chiral Ligands

Allylic acetates in conjunction with phosphinooxazolines (PHOX) as chiral ligands were used for the first asymmetric Ir-catalyzed allylic substitution (Scheme 9.7) [5]. The reaction was slow, compared to that catalyzed by the $[Ir(COD)Cl]_2/P(OPh)_3$ system or the parent complex $[Ir(COD)Cl]_2$, although both regioselectivity and enantioselectivity were very high. The aminations were generally slow, yet quite interesting results were achieved in intramolecular aminations nevertheless [17, 18]; in particular, a very strong influence of halide salts was found.

The preparation of (π-allyl)(PHOX)Ir^{III} complexes was possible by using standard methods. A typical example is depicted by a crystal structure in Figure 9.1. Here, the configuration around the Ir center is as to be expected on the basis of the *trans* influences of the ligands. The reaction of this complex with dimethyl sodiomalonate proceeds by ligand exchange and addition of the nucleophile at the

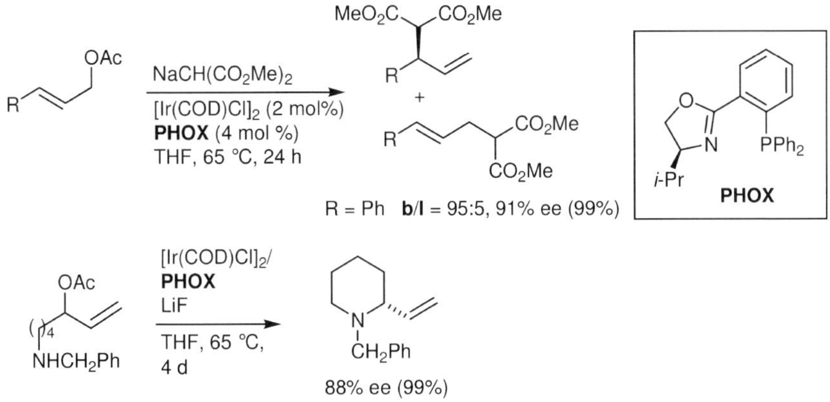

Scheme 9.7 Early asymmetric allylic substitutions.

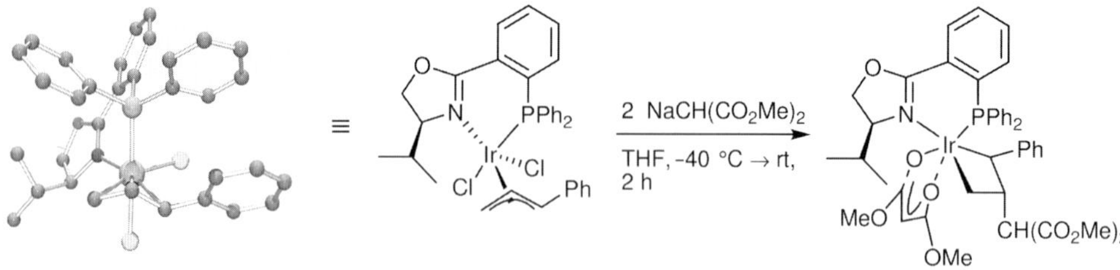

Figure 9.1 Crystal structure of the complex [η^3-(1-phenylallyl)(PHOX)Cl$_2$Ir] and its reaction with dimethyl sodiomalonate.

central rather than the terminal allylic carbon to give an iridacyclobutane [19a]. Although, on first view this reaction may appear unusual, several years ago Bergman and Stryker showed that a reaction with a nucleophile at the central carbon of the allylic moiety is very typical for (π-allyl)Ir^{III} complexes [19b,c].

Unfortunately, further investigation of the PHOX–Ir-catalysts gave disappointing results; in particular, the regioselectivities were low with alkyl-substituted substrates (R = Alkyl).

9.2.4
Phosphoramidites as Ligands for the Ir-Catalyzed Allylic Substitution

9.2.4.1 Survey
Many of the studies on the asymmetric Ir-catalyzed allylic substitution have been carried out with complexes prepared from [Ir(COD)Cl]$_2$ and a phosphoramidite ligand. While numerous phosphoramidites have been investigated, those considered to be the most useful are shown in Figure 9.2.

After a prelude with **Monophos-NMe$_2$** [6a, 20], which gave moderate to good results in allylic alkylations but is inactive in aminations [6b], the ligands **L1** [20], **L2** [9] and **L3** [20] derived from 1-arylethylamines have been most often used. Ligand **L1** is easily available and is usually the first ligand to be tried. The best results have been obtained with **L2**, although enantiomerically enriched 1-(2-methoxyphenyl)ethylamine is required for its synthesis (this compound is not yet commercially available from fine chemicals suppliers [21]). Ligands **L4** [22a,b] and **L5** [22b,c] are simplified versions of **L1–L3** which, nevertheless, have furnished excellent results.

The absolute configurations of many branched products prepared by Ir-catalyzed allylic substitution have been determined. The steric course of all corresponding reactions follows the general rule described in Scheme 9.8.

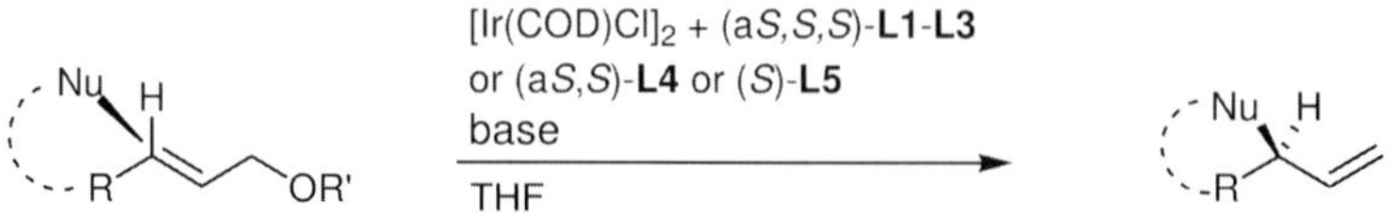

Scheme 9.8 Steric course of all known allylic substitutions catalyzed by Ir complexes prepared from phosphoramidites **L1–L5**.

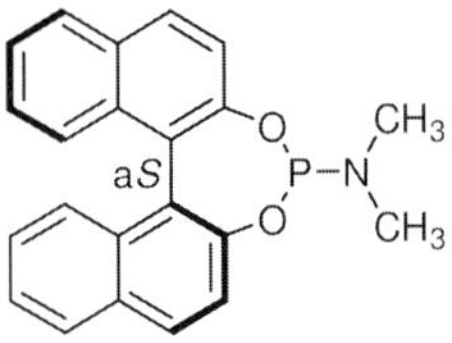

Monophos-NMe₂

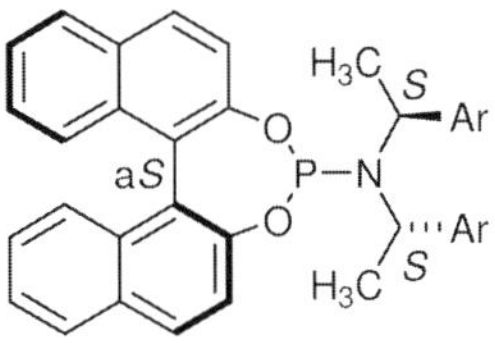

L1 Ar = Ph
L2 Ar = o-(MeO)C₆H₄
L3 Ar = α-naphthyl

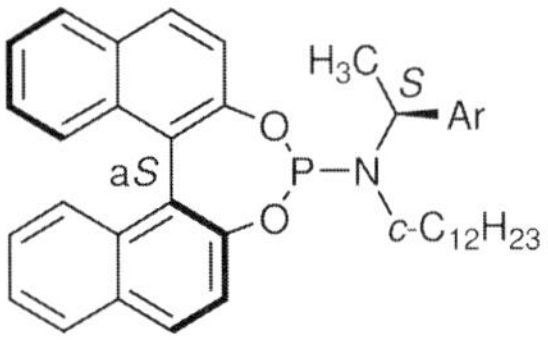

L4a Ar = Ph
L4b Ar = o-(MeO)C₆H₄

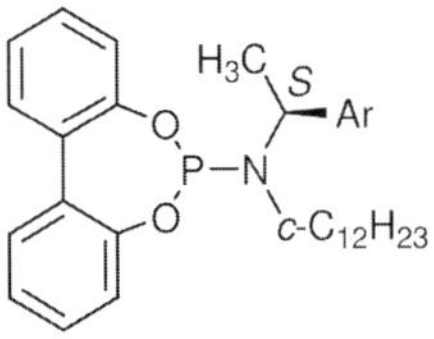

L5a Ar = Ph
L5b Ar = o-(MeO)C₆H₄

Figure 9.2 The chiral phosphoramidites most often employed for Ir-catalyzed allylic substitutions.

9.2.4.2 Catalyst Preparation, Reaction Conditions and Catalytic Cycle

Attention to catalyst preparation and reaction conditions is of crucial importance for the success of an Ir-catalyzed allylic substitution, because many ligands are altered by C—H activation at aryl (see above) or CH₃ groups. For most reactions, tetrahydrofuran (THF) is the preferred solvent; it is important that dry THF (<35 μg H₂O ml^{-1} THF, Karl Fischer titration) is used for catalyst preparation because this step is very sensitive to water. The following procedures have been applied.

(a) Dissolution of [Ir(COD)Cl]₂ and ligand **L*** in 1:2 molar ratio in THF. The application of this procedure is limited to aminations with aliphatic amines [7, 18], and it works particularly well in conjunction with ligand **L2** [9]. A complex [Ir(COD)L*Cl] (**K1**) is initially formed, simply by breaking up the chloro bridges; however, it is most likely that a sufficiently basic nucleophile effects C—H activation *in situ*, as described below (c).

(b) Procedure (a) with additional LiCl (1 equiv. with respect to the allylic carbonate). This procedure was introduced for alkylations with **L1** as ligand [6c, 23], and gave vastly superior results for alkylations upon the use of ligand **L2**, as was shown by Alexakis and Polet [24, 25]. The Alexakis group found that an air-stable, catalytically active Ir-complex can be partially recovered [25a]. According to ^{31}P NMR data recorded in the author's laboratory, the recoverable complex is [Ir(COD)L2Cl] – that is, a complex of type **K1** [26].

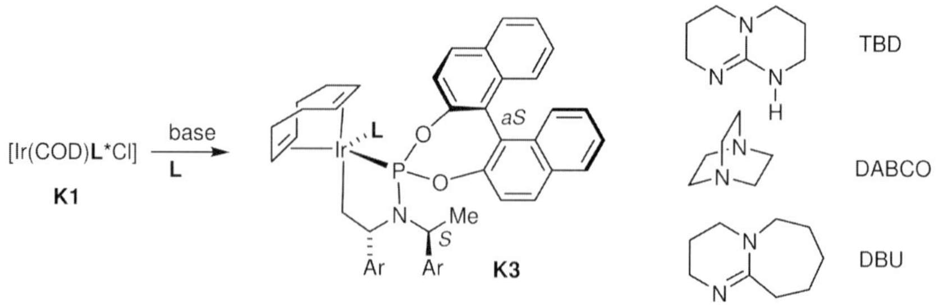

9 Iridium-Catalyzed Asymmetric Allylic Substitutions

Scheme 9.9 C–H activation of (phosphoramidite)Ir complexes.

(c) Treatment of a 1:2 mixture of [Ir(COD)Cl]$_2$ and **L*** with a base (TBD [18]; DABCO [27]; n-propylamine [27]; DBU [28c]). The base induces formation of a complex of type **K3** by C–H activation at the CH$_3$ group (Scheme 9.9) [8], and the reaction can be followed easily by using ^{31}P NMR. The complexes of type **K3** are analogues of the complex **K2** presented above. It is of interest to note that C–H activation is much faster for **L2** than for **L1** [29].

(d) Treatment of a solution of [Ir(COD)Cl]$_2$, **L***, tetrahydrothiophene (THT) and THF with TBD for 2 h at room temperature, followed by addition of the allylic substrate and subsequently of CuI. This procedure yields an excellent catalyst for alkylations, particularly in conjunction with ligand **L1** [30].

(e) Salt-free conditions: Catalyst preparation according to procedure (c) and use of the conjugate acid HNu of an anionic nucleophile Nu$^-$ as pronucleophile [28, 29]. This method is applicable to reactions with allylic carbonates, which generate alkoxide as a strong base capable of deprotonating the pronucleophile HNu. The method was introduced by Tsuji in 1984, and the snappy designation was coined by Plietker [31]. There are two main advantages of salt-free conditions for the Ir-catalyzed reaction with phosphoramidite ligands. First, a strong base introduced for C–H activation, which can cause catalyst destruction, is neutralized. Second, solubility problems are avoided; for example, alkali phthalimides are almost insoluble in THF and cannot be used as pronucleophiles.

Although complexes of type **K3** are coordinatively saturated (18 VE), the ligand L can be exchanged readily, for example by PPh$_3$. Dissociation of L is required in order to generate a reactive species. A catalytic cycle as described in Scheme 9.10 appears likely in analogy to the Pd-catalyzed allylic substitution. Hartwig and Marcović [32] have investigated this proposal for the reaction of cinnamyl methyl carbonate with aniline using **L1** as ligand, and found that the product complex **K6** is the resting state. The allyl complex **K5** was not observed; indeed, the kinetic analysis was compatible with the assumption that the formation of this complex is both reversible and endergonic. The substitution reaction **K5**→**K6** probably proceeds via a late transition state, because the ratio of diastereomeric complexes of type **K6** was almost exactly equal to the ratio of the enantiomeric substitution products.

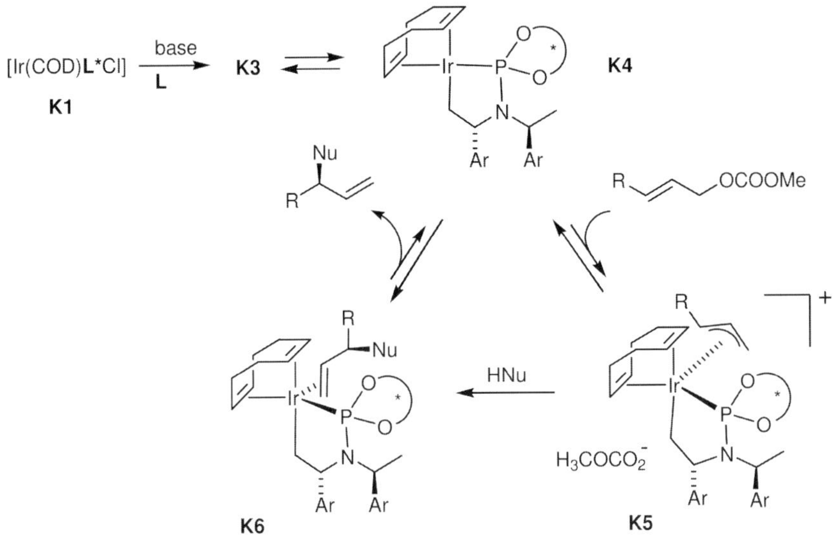

Scheme 9.10 Catalytic cycle of the allylic substitution catalyzed by (phosphoramidite)Ir complexes.

9.2.4.3 Preparation of Phosphoramidites

Phosphoramidites are usually air-stable and can be handled without special precautions. The following routes have been most often applied for their synthesis:

- Treatment of a binaphthol or biphenol with neat PCl_3 yields a chlorophosphite [33], which can be stored at low temperature. The reaction with a lithiated secondary amine produces the phosphoramidite [34]. The yields are good, and the scope of this method is broad.

- In another route, PCl_3 is first reacted with the secondary amine, and the product treated with the binaphthol or biphenol to give the phosphoramidite [35]. This procedure can also be carried out with the hydrochloride of the amine, which is convenient for storage [25]. Methylene chloride or THF are usually applied as solvents. According to the author's experience, this route is well suited in conjunction with electron-rich amines.

9.2.4.4 Variation of the Phosphoramidite Ligands

One strong point of the phosphoramidite ligands is their modular construction. A generalized cyclometallated, catalytically active complex **K4** is shown in Figure 9.3; possible variations of this structure are indicated by the formulas, and discussed in detail below.

- Ligands with axial as well as central units of chirality (**L1–L4**) can belong to the *like* or *unlike* series of diastereoisomers. Diastereoisomers with the (a*S,S,S*)- and the (a*S,R,R*)-configuration, or their enantiomers, have been investigated several times for ligands **L1** [9c, 22b,c] and **L2** [9]. The absolute configuration of the allylation products is determined by the axial configuration. Ligands of the *like*

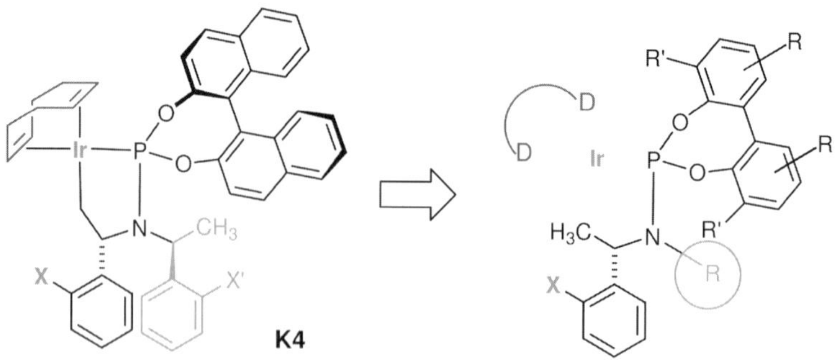

Figure 9.3 Modular make-up of phosphoramidites.

series generally induce higher degrees of selectivity as well as higher reaction rates. In the case of **L1**, the ligand with *unlike* configuration is inactive; this was elegantly demonstrated with a mixture prepared from racemic BINOL; the mixture induced essentially the same enantioselectivity as the pure ligand with *like*-configuration [22c].

- In view of the dominating influence of the axial chirality of **L1–L4**, it is interesting that corresponding ligands with a biphenyl unit also give rise to high levels of selectivity [6c, 22b]. It is likely that atropisomers separated by a low barrier give rise to complexes with catalytic properties, which are similar to those of the BINOL derivatives.

- Models show that space around the Ir center in allyl complexes **K5** is highly crowded. Substituents R′ = CH_3, OCH_3 or aryl give rise to reduced selectivity as well as activity [36, 37].

- Numerous variations in the vicinity of the nitrogen atom have been probed. Overall, the best results were obtained with ligands containing an *ortho*-substituted aryl group, in particular ligand **L2**. Coordination of OCH_3 to Ir has been invoked as reason for the beneficial effect of X = OCH_3; however, similar results have been obtained with a corresponding ligand with X = X′ = CH_3 [25]. It appears plausible that increasing the steric bulk of X facilitates dissociation of ligand **L** of complex **K3**.

- Replacement of one of the *N*-arylethyl groups by a bulky alkyl or cycloalkyl group is possible. The best results were obtained with R = cyclododecyl (cf. the ligands **L4** and **L5**) [22a,b]. Combining different *N*-arylethyl substituents, for example with X = OCH_3 and X′ = H, is also possible [25]. The corresponding ligand, which is easily available from *(S)*-1-phenylethylamine, was found to be a good substitute for **L2**.

9.2.4.5 Further Ligands Used in Ir-Catalyzed Allylic Substitutions

The Ir-catalyzed allylic substitution at arylallyl derivatives can be catalyzed with Ir-complexes derived from very diverse types of chiral ligands, for example

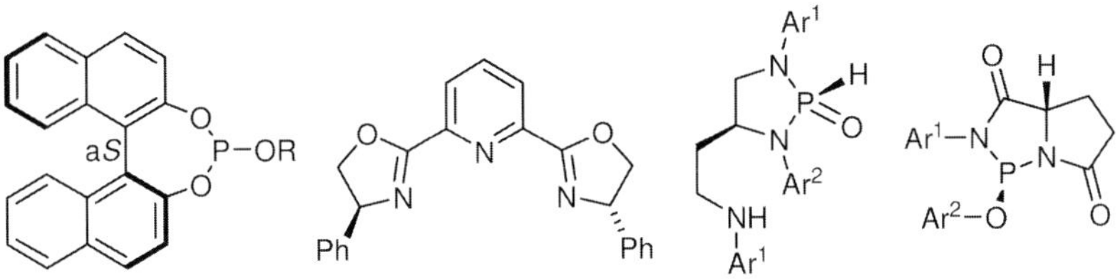

L6a R = Ph
L6b R = $(CH_2)_2SEt$

Ph-Pybox

DIAPHOX

L7

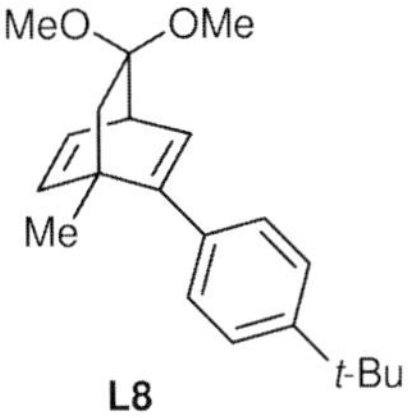

L8

Figure 9.4 Chiral ligands other than phosphoramidites that have been used in Ir-catalyzed allylic substitutions.

phosphites (**L6a** [38], **L6b** [39]), **Pybox** [40] and **DIAPHOX** [41] (Figure 9.4). The Ir-complexes were usually prepared from $[Ir(COD)Cl]_2$; however, COD is not a necessary element of an active catalyst. Thus, chiral (2.2.2)-bicyclooctadienes (e.g. **L8**), derived from (–)-carvone, have been used for kinetic resolutions with phenolates as nucleophiles [42]. A chiral ligand containing a single coordinating double bond was described by the same group (cf. Scheme 9.24). Here, it must be pointed out, that a new complex can only be regarded as competitive if it is able to catalyze the allylic substitution of *alkyl*allyl derivatives with good results.

9.3
C-Nucleophiles

9.3.1
Stabilized Enolates as Nucleophiles

9.3.1.1 Malonates and Related Pronucleophiles

Dimethyl malonate is the standard test pronucleophile for allylic alkylations. During the early studies on the Ir-catalyzed variant, allylic acetates were used as substrates in conjunction with ligand **L1** and procedure (b) (*cf.* Section 9.2.4.2) (Table 9.1) [6c]. Although high enantioselectivities (ca. 90% ee) were obtained immediately, the regioselectivity was low with alkylallyl acetates. Improved results were achieved upon use of ligand **L2**, due to its ability to rapidly undergo C—H activation *in situ*. Further investigations led to the development of procedure (d), which gave preparatively usable results even with **L1** for a wide range of allylic carbonates [30]. However, difficulties were encountered for reactions on a

Table 9.1 Allylic alkylations with dimethyl malonate using (phosphoramidite)–Ir complexes as catalysts.

Procedure (b): [Ir(COD)Cl]₂ (2 mol%), L* (4 mol%), THF, rt, LiCl (1 equiv.)

L1 (X = OAc)	R = Ph	b:l = 91:9	86% ee (98%) [6c]
	R = Me	b:l = 75:25	82% ee (96%) [6c]
	R = i-Pr	b:l = 55:45	94% ee (56%) [6c]
L2 (X = OAc)	R = Ph	b:l = 99:1	97% ee (79%) [25a]
	R = n-Pr	b:l = 87:13	97% ee (87%) [25a]
L2 (X = OCO₂Me)	R = Ph	b:l = 99:1	98% ee (82%) [24]
	R = 2-(MeO)C₆H₄	b:l = >99:1	79% ee (98%)^a [25a]
	R = Cyclohexyl-	b:l = 93:7	98% ee (65%) [24]
	R = n-Pr	b:l = 80:20	96% ee (92%) [25a]
	(boronate)	b:l = 81:19	84% ee (>70%) [43]

Procedure (d): X = OCO₂Me, [Ir(COD)Cl]₂ (2 mol%), L* (4 mol%), THF, rt, TBD (12 mol%), CuI (20 mol%), tetrahydrothiophene (20 mol%)

L1	R = Ph	b:l = 99:1	96% ee (88%) [30]
	R = PhCH₂CH₂	b:l = 81:19	96% ee (92%) [30]
L2	R = Ph	b:l = >99:1	98% ee (92%) [36]
	R = PhCH=CH	b:l = 99:1	98% ee (80%) [36]
	R = PhCH₂CH₂	b:l = 91:9	98% ee (93%) [36]

Procedure (e) (salt-free conditions): X = OCO₂Me, [Ir(COD)Cl]₂ (2 mol%), L* (4 mol%), THF, rt, TBD (8 mol%)

L1	R = Ph	b:l = 99:1	97.5% ee (92%) [29]
L2	R = Ph	b:l = 99:1	>99% ee (96%) [29]
	R = Me	b:l = 98:2	98.5% ee (90%) [29]
	R = CH₂OCPh₃	b:l = 78:22	>99% ee (88%) [29]

a Similar results have been obtained with R¹ = 2-MeC₆H₄ and R¹ = 1-naphthyl; G. Helmchen and D. Beton, unpublished results.

multigram scale because of the poor solubility of dimethyl sodiomalonate in THF. This problem could be solved with the help of procedure (e) – that is, salt-free reaction conditions; in conjunction with **L2**, an enantiomeric excess (ee) of >98% can now be routinely achieved.

In general, it was found that THF was the most suitable solvent [6b] and that carbonates were superior to acetate as a leaving group [23]. Among the ligands, **L2** induced the highest activity as well as regioselectivity and enantioselectivity [24]. Catalyst loadings as low as 0.1 mol% were possible [36].

The following conclusions with respect to the dependence of regioselectivity and enantioselectivity on the substituent R at the allylic moiety can be drawn:

- Allyl carbonates with a sp^2-substituent (alkenyl, aryl) are privileged substrates; regioselectivities of **b/l** ≥ 98:2 are obtained, using ligand **L1** in conjunction with procedures (d) or (e) or ligand **L2** in conjunction with procedures (b), (d) or (e).

- Regioselectivity of >90:10 is typical for alkyl-substituted allylic carbonates with small and rigid substituents (R = Me, cyclohexyl). However, values in the range of 70:30 to 90:10 are encountered with substrates containing flexible or sterically very demanding substituents (R = *n*-Pr, *n*-Octyl, PhCH$_2$CH$_2$, (*t*-Bu)Ph$_2$SiOCH$_2$).

- Enantioselectivity is generally very high, except in the case of allylic substrates with an *o*-substituted aryl group (e.g., R = 2-(MeO)C$_6$H$_4$).

- The scope of the reaction with respect to substituents is broad. An extreme example is Hall's [43] use of sensitive allylic boronates (cf. Table 9.1). Pd-catalysts failed to promote the reaction, while the Ir-catalyzed reaction proceeded with excellent regioselectivity and up to 84% ee.

The details of a number of allylic alkylations of malonates using ligands other than phosphoramidites have also been published. Typically, regioselectivity was high with arylallyl derivatives but low with alkylallyl acetates or carbonates: **PHOX** ligands (cf. Section 9.2.3) [5a], *i*-**Pr-Pybox** [40b], phosphites of type **L6** [38] and **DIAPHOX** ligands (cf. Figure 9.4) [41a].

Substituted malonates and β-keto-esters have also been successfully used as pronucleophiles (Scheme 9.11) [25a, 36]. From β-ketoesters, approximately 1:1 mixtures of epimers are generally formed. Products derived from 2-alkenylmalonates have been subjected to Ru-catalyzed ring-closing metathesis to give cyclopentene derivatives in good yield [25a, 36]. With the ester-amide displayed in Scheme 9.11 as pronucleophile, 1:1 mixtures of epimers were also formed [44]. This pronucleophile serves as the equivalent of the enolate of a methyl ketone, because the methoxycarbonyl group can be removed selectively by saponification/decarboxylation and the resultant Weinreb amide transformed into a ketone. An example is described in Section 9.6.

Cyclizations based on allylic substitutions are often problematic. Pd-catalyzed intramolecular allylic alkylations are accompanied by polymerization and, thus must be run at low concentration. Ir-catalyzed intramolecular reactions do not require high dilution conditions, although the competing noncatalyzed background reaction – which produces the racemic product – must be considered. *Cyclopentane* and *cyclohexane* derivatives (Scheme 9.12) [36] were obtained in good yield from malonic ester derivatives when the anion was prepared at −78 °C in order to suppress the noncatalyzed reaction. Once again, the best results were obtained with **L2** as ligand. Under the same conditions, the vinyl*cyclopropane* was not formed at all and the vinyl*cyclobutane* was produced in poor yield. Good results were again obtained upon application of the salt-free conditions [29].

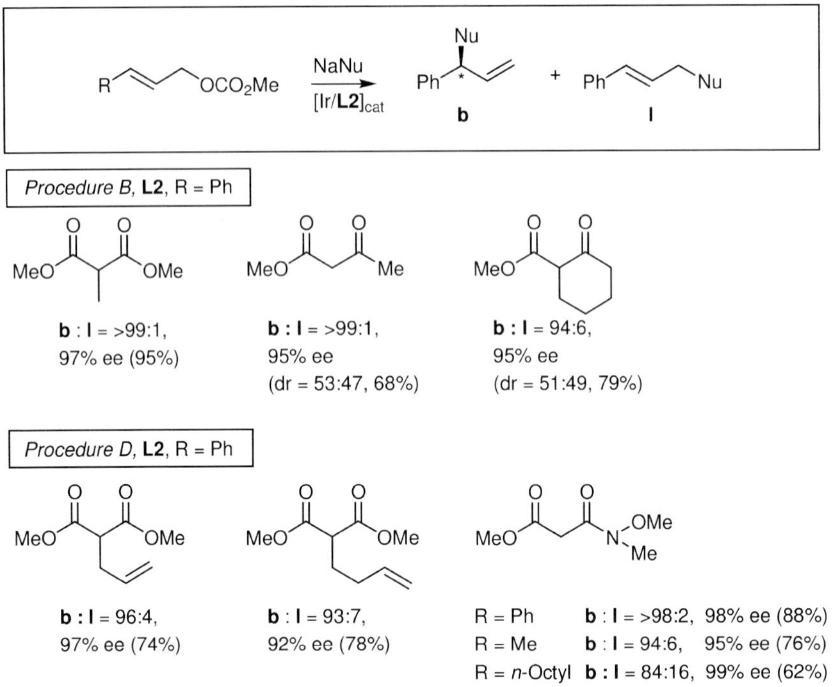

9 Iridium-Catalyzed Asymmetric Allylic Substitutions

$$R\diagdown\diagup OCO_2Me \xrightarrow[\text{[Ir/L2]}_{cat}]{\text{NaNu}} Ph\overset{Nu}{\underset{*}{\diagdown}}\diagup + Ph\diagdown\diagup Nu$$
                                                                b   l

Procedure B, L2, R = Ph

MeO–C(O)–CH(Me)–C(O)–OMe
b : l = >99:1,
97% ee (95%)

MeO–C(O)–CH2–C(O)–Me
b : l = >99:1,
95% ee
(dr = 53:47, 68%)

MeO–C(O)–(2-oxocyclohexyl)
b : l = 94:6,
95% ee
(dr = 51:49, 79%)

Procedure D, L2, R = Ph

MeO–C(O)–CH(allyl)–C(O)–OMe
b : l = 96:4,
97% ee (74%)

MeO–C(O)–CH(homoallyl)–C(O)–OMe
b : l = 93:7,
92% ee (78%)

MeO–C(O)–CH2–C(O)–N(OMe)(Me) (R group)
R = Ph   b : l = >98:2, 98% ee (88%)
R = Me   b : l = 94:6, 95% ee (76%)
R = n-Octyl b : l = 84:16, 99% ee (62%)

Scheme 9.11 Allylic substitutions with 2-substituted malonates and β-keto-esters as pronucleophiles.

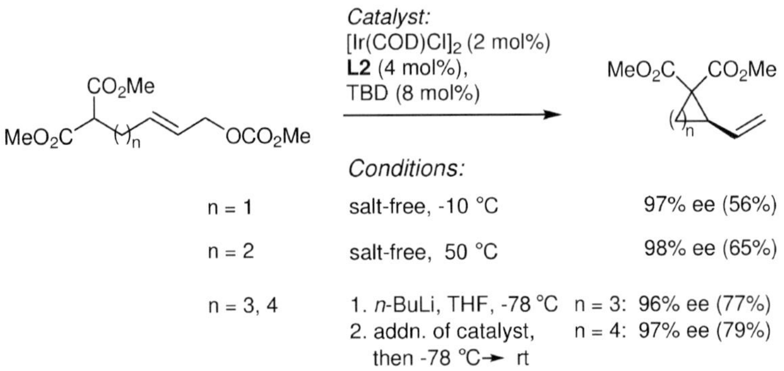

$$MeO_2C\diagdown CH(CO_2Me)-(CH_2)_n-CH=CH-CH_2-OCO_2Me \xrightarrow{\text{Catalyst}} \text{cyclic product with } MeO_2C, CO_2Me$$

Catalyst:
[Ir(COD)Cl]$_2$ (2 mol%)
L2 (4 mol%),
TBD (8 mol%)

Conditions:

n = 1   salt-free, −10 °C   97% ee (56%)
n = 2   salt-free, 50 °C   98% ee (65%)
n = 3, 4 1. n-BuLi, THF, −78 °C   n = 3: 96% ee (77%)
         2. addn. of catalyst,   n = 4: 97% ee (79%)
         then −78 °C → rt

Scheme 9.12 Intramolecular allylic alkylations with malonic ester derivatives.

9.3.2
Aliphatic Nitro Compounds as Pronucleophiles

Aliphatic nitro compounds are intermediates with considerable potential in organic synthesis (*cf.* Section 9.6). Initially, allylic substitutions with nitromethane were investigated, but unfortunately complex mixtures of mono- and dialkylation products were formed. In contrast, good results were obtained with primary and

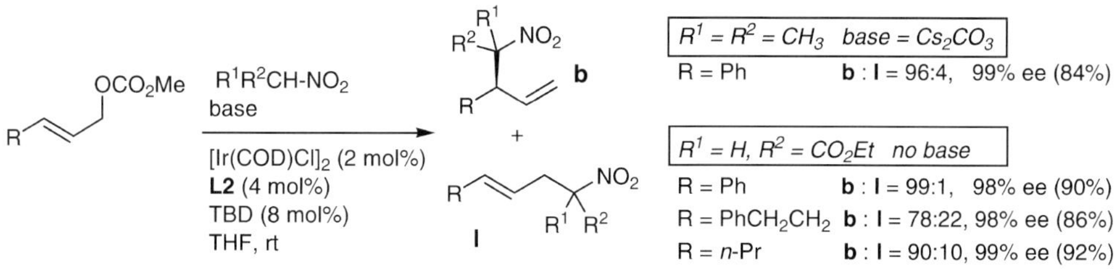

Scheme 9.13 Asymmetric allylic alkylations with aliphatic nitro compounds as pronucleophiles.

secondary nitro compounds as pronucleophiles [45]. Cesium carbonate was used as base for the generation of nitronates; some typical examples are presented in Scheme 9.13 and in Section 9.6.

Commercially available ethyl nitroacetate is an interesting pronucleophile, because it can serve as the synthetic equivalent of either nitromethane or glycine. The ethoxycarbonyl group can also be considered as a protecting group against dialkylation. The allylic alkylation with ethyl nitroacetate did not require an additional base (salt-free conditions). As a consequence of the high acidity of the chirality center α to N, 1:1 mixtures of epimers were formed.

9.3.2.1 A Glycine Equivalent as Pronucleophile

A regio-, diastereo- and enantioselective synthesis of amino acids was reported by Takemoto and coworkers. The glycine equivalent ethyl diphenylimino glycinate was used as pronucleophile (Scheme 9.14), while the ligand was a bidentate chiral phosphite, and 3-arylallyl diethyl phosphates were employed as allylic substrates [39, 46].

Both, diastereoisomeric β-substituted α-amino acid derivatives could be formed selectively, depending on the cation provided by the base. The *(R,S)*-diastereoisomer was the major product with a lithium amide, whereas the

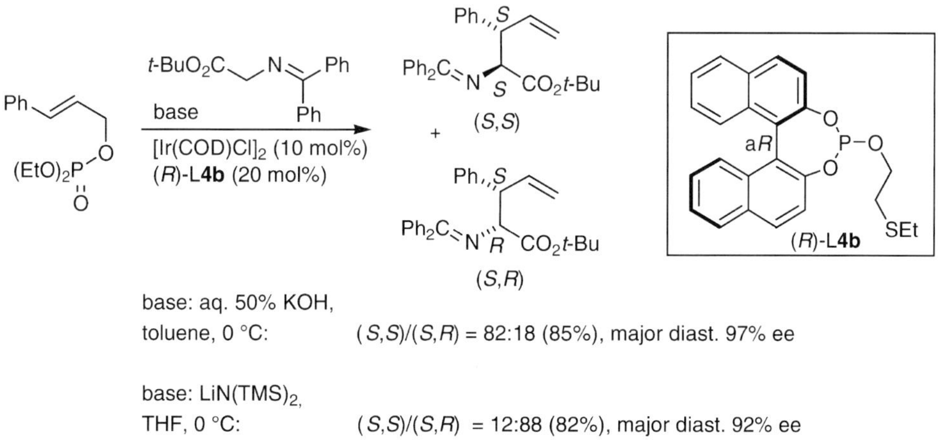

Scheme 9.14 Allylic alkylations with a glycine equivalent.

(S,S)-diastereoisomer was preferentially obtained when KOH was used as base. As an explanation for this stereodichotomic effect, it was proposed that $LiNR_2$ induces the formation of a N,O-chelated *syn*-enolate, and the reaction with KOH leads to an *anti*-enolate. The procedure also allowed α,α-disubstituted amino acids to be prepared.

9.3.3
Allylic Substitutions with Nonstabilized Enolates, Enamines and Organozinc Compounds

Nonstabilized enolates have been studied for many years in Pd-catalyzed allylic substitutions. Generally, they are problematic nucleophiles because of their strongly basic character, which gives rise to enolate equilibration, elimination and cleavage of carbonates [47]. In allylic substitutions, good results were achieved recently with Zn- and Cu-enolates rather than with the traditional lithium enolates. During the past few years, decarboxylative *in situ* generation of enolates – a method introduced by Saegusa and Tsuji [48] – has been successfully developed with Pd-catalysts, and very recently also with Ir-catalysts. Enamines, which are classical enolate equivalents, have also been probed recently as nucleophiles in Ir-catalyzed allylations.

9.3.3.1 Ketone Enolates Derived from Silyl Enol Ethers as Nucleophiles

Highly regioselective and enantioselective reactions of allylic carbonates with enolates generated *in situ* from trimethylsilyl enol ethers were described by Graening and Hartwig (Scheme 9.15) [49]. Upon desilylation with cesium fluoride, regioselectivity was modest and diallylation pronounced, but this problem was overcome with a remarkable fluoride source – a combination of CsF and ZnF_2, which allowed diallylation to be suppressed almost completely. Catalyst activation by base was

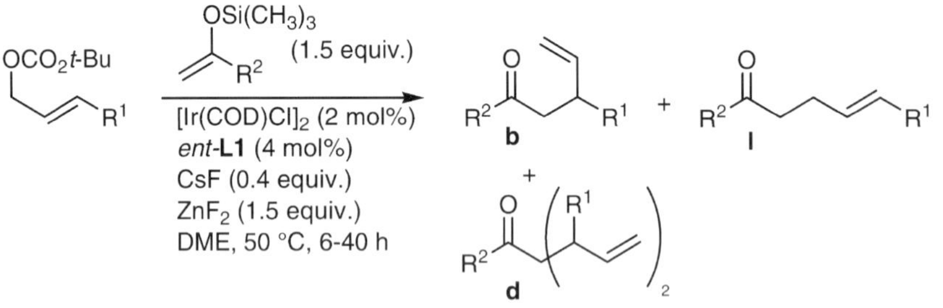

R^1 = Ph,	R^2 = Ph	b : l = 93:7,	94% ee (R) (84%)
R^1 = p-(MeO)C_6H_4,	R^2 = Ph	b : l = 99:1,	96% ee (94%)
R^1 = Ph,	R^2 = $PhCH_2CH_2$	b : l = 99:1,	94% ee (54%)
R^1 = Ph,	R^2 = *i*-Pr	b : l = 95:5,	91% ee (46%)
R^1 = *n*-Pr,	R^2 = Ph	b : l = 87:13,	92% ee (92%)

Scheme 9.15 Alkylations with ketone enolates derived from silyl enol ethers as nucleophiles.

not necessary, and monitoring of the reaction by ^{31}P NMR indicated that C—H activation was occurring *in situ*.

The procedure was applied to a considerable number of substrates. In general aromatic silyl enol ethers gave satisfactory results, while aliphatic silyl enol ethers gave rise to slow reactions and yields in the range 46–54%. The absolute configuration of one of the products was determined and found to be in accordance with the general rule presented in Chapter 2. Only silyl enol ethers of methyl ketones, giving rise to one stereogenic center, have been investigated.

9.3.3.2 Allylation of Enamines

Whilst the method described above appears very elegant, Weix and Hartwig expressed their discontent about the allylations of aliphatic silyl enol ethers and developed an alternative system using enamines as nucleophiles. Once the considerable initial difficulties had been overcome, these authors were able to present a procedure that gave excellent results (Scheme 9.16) [50].

Two features of the above procedure are particularly notable: (i) a combination of the pure complex *ent*-**K3** (Ar = Ph; Scheme 9.9) and [Ir(COD)Cl]$_2$ was used as catalyst; and (ii) as solvent, toluene was superior to THF. Again, only enamines derived from methyl ketones were employed.

R^1 = 4-(MeO)C$_6$H$_4$, R^2 = *i*-Bu	**b** : **l** = 97:3,	96% ee (91%)
R^1 = 4-(CF$_3$)C$_6$H$_4$, R^2 = *i*-Bu	**b** : **l** = 85:15,	94% ee (75%)
R^1 = Me, R^2 = Ph	**b** : **l** = 95:5,	94% ee (68%)
R^1 = *n*-Pr, R^2 = *i*-Bu	**b** : **l** = 89:11,	83% ee (64%)
R^1 = *i*-Pr, R^2 not stated	low conversion	

Scheme 9.16 Allylation of enamines under optimized conditions.

9.3.3.3 Decarboxylative Allylic Alkylation

The first examples of decarboxylative alkylation via Ir-catalyzed reactions of allyl β-ketocarboxylates were presented by You and coworkers (Scheme 9.17) [37]. Typically, good selectivities were obtained with a broad range of arylallylic substrates, while regioselectivities with alkylallylic substrates were slightly lower. It appears likely that improved results will be achieved with ligands other than **L1** (note that CH$_2$Cl$_2$ was used as solvent). A few interesting observations concerning the mechanism of the reaction were made: (i) crossover experiments revealed that reactions proceeded inter- rather then intramolecularly; and (ii) according to absolute

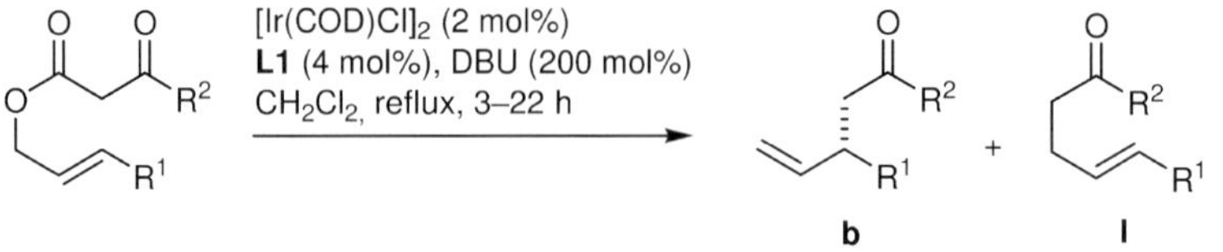

Scheme 9.17 Decarboxylative alkylation of allyl β-ketocarboxylates.

R^1	R^2	Yield (%)	b : l	Ee (%)
Ar	Ar	58–83	98 : 2	91–95
CH_3	Ph	61	94 : 6	90
$n\text{-}C_5H_{11}$	Ph	52	80 : 20	89

configurations, determined for four products with R^1 = aryl, the steric course of the substitutions followed the general rule (cf. Section 9.2.4.1).

9.3.3.4 Reactions with Aryl Zinc Compounds

Allylic substitutions with nonstabilized C-nucleophiles are an important domain of organocopper chemistry [51]. However, on close inspection of the literature, it becomes apparent that regioselectivity in favor of the branched allylic alkylation products is only obtained with *alkyl* copper compounds, while *aryl* copper compounds mainly give the linear alkylation products. This observation was an incentive for Alexakis et al. [52] to probe the reactions of aryl zinc halides in the Ir-catalyzed allylic substitution (Scheme 9.18).

The authors considered diphenylzinc as the reactive intermediate as, apparently, both phenyl groups were transferred during the course of the overall reaction. The additive LiBr has a pronounced beneficial effect. Apart from the relatively low degree of regioselectivity, these substitutions share many features of the corresponding reactions with stabilized carbanions. Absolute configurations of the products were determined and, perhaps astonishingly, were found to follow the general rule previously presented (cf. Section 9.2.4.1). A formal synthesis of the antidepressant sertraline was carried out as an application of the method.

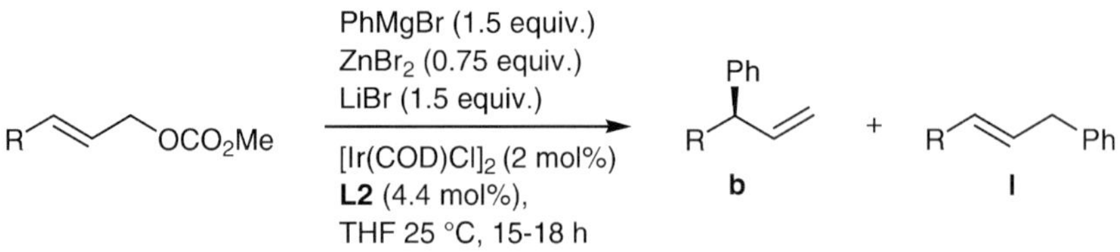

R = 4-(MeO)C_6H_4, b : l = 55:45, 99% ee (83%)
R = 4-(CF_3)C_6H_4, b : l = 45:55, 82% ee (95%)
R = c-Hex, b : l = 69:31, 74% ee (72%)

Scheme 9.18 Reactions of a phenyl zinc reagent with allylic carbonates.

9.4
N-Nucleophiles

9.4.1
Inter- and Intramolecular Reactions with Aliphatic Amines and Ammonia as Nucleophiles

Asymmetric aminations with amines as nucleophiles were first carried out by Hartwig and coworkers, using the combination [Ir(COD)Cl]$_2$/2**L1** without explicit base-activation as catalyst [procedure (a), cf. Section 9.2.3.2] (Table 9.2) [7]. However, given the possibility of effecting C—H activation with n-PrNH$_2$, it appears safe to assume that cyclometallation generally occurs *in situ* when aliphatic amines are employed as nucleophiles. A range of solvents was assessed according to the results obtained for the reaction of cinnamyl carbonate with BnNH$_2$, and THF was found to be the most suitable. Reversibility of the amination was observed when EtOH was the solvent [7].

As might have been expected following the pioneering studies of Takeuchi and colleagues, the monoallylated, branched amine was the major substitution product. As was found for the corresponding alkylations with malonates, the enantiomeric excess and regioselectivity were high for reactions of allylic substrates with R^1 being a sp^2-bound substituent (Table 9.2, entries 1–6). The exception was the case of R^1 = aryl with an electron-withdrawing substituent in the *para* position (entry 7) or with any substituent in the *ortho* position (entry 8). (Similar findings have been obtained with R^1 = 2-MeC$_6$H$_4$ and R^1 = 1-naphthyl; G. Helmchen and D. Beton, unpublished results.) Alkylallyl carbonates yielded products with high enantioselectivity but reduced regioselectivity (entries 9 and 10). It was first shown by Alexakis and coworkers [9] that results were slightly better with ligand **L2** than with ligand **L1**.

Catalyst activation became a necessity when reactions with bulky aliphatic amines and arylamines (cf. Section 9.4.2) as nucleophiles were probed. It was also required for intramolecular aminations [8, 18]. Thus, with Ph$_2$CHNH$_2$, an ammonia equivalent, conversion was only 11% upon application of procedure (a) (Table 9.2, entry 11), while the reaction promoted by the activated catalyst proceeded with high selectivity and yield. Catalyst activation is faster with ligand **L2** than **L1**, and accordingly *in situ* activation occurs more readily for the former (cf. entries 10 and 12). Examples presented in entries 16–20 further demonstrate the advantages of catalyst activation [53] (note that excellent results can be achieved with the simplified ligand **L5a**).

Procedure (d) (cf. Section 9.2.4.2), which is useful for alkylations, was not applicable to aminations because of coordination of Cu^I to the amine. Further screening of the salts of soft cations revealed that the addition of Pb^{II} salts in conjunction with base activation of the precatalyst led to significantly faster reaction rates. The reaction could be accomplished with catalyst loading as low as 0.4 mol% (compare entries 14 and 15).

Table 9.2 Ir-catalyzed asymmetric allylic aminations.

$$R^1\diagdown\!\!\!\diagup\!\!\!\diagdown OCO_2Me \xrightarrow[\text{L* (2 mol\%)}]{R^2NH_2,\ [Ir(COD)Cl]_2\ (1\ mol\%)} R^1\text{-CH(NHR}^2\text{)-CH=CH}_2\ \textbf{b (major)} + R^1\text{-CH=CH-CH}_2\text{NHR}^2\ \textbf{l} + (R^1\text{-CH=CH-CH}_2)_2NR^2\ \textbf{d}$$

Entry	R^1	R^2	L*	Time (h)	Yield of b (%)	b:l:d	ee (%)	Reference
Procedure (a) (no explicit base activation of catalyst):								
1	Ph	Bn	L1	10	84	98:1:1	95	[7]
2	Ph	Bn	L2	n.d.	88	98:2	97	[9]
3	Ph	4-(MeO)C$_6$H$_4$	L1	18	80	99:0:1	94	[7]
4	Ph	n-C$_6$H$_{11}$	L1	9	88	98:2	96	[7]
5	Ph	n-C$_6$H$_{11}$	L2	n.d.	89	98:2	98	[9]
6	PhCH=CH	Bn	L1	24	61	99:1	97	[23]
7	4-(NO$_2$)C$_6$H$_4$	Bn	L1	12	67	83:13:4	86	[7]
8	2-(MeO)C$_6$H$_4$	Bn	L1	16	77	95:4:1	76	[7]
9	n-Pr	Bn	L1	10	66	88:8:4	95	[7]
10	PhCH$_2$CH$_2$	Bn	L2	3	63	84:16	96	A. Dahnz and G. Helmchen (unpublished)
11	Ph	Ph$_2$CH	L1	10	11 (conv.)	–	–	[8]
Procedure (c) (activated catalyst):								
12^a	PhCH$_2$CH$_2$	Bn	L2	0.7	59	84:16	96	A. Dahnz and G. Helmchen (unpublished)
13^b	Ph	Ph$_2$CH	L1	10	85	97:3	98	[8]
14^{a,c}	PhCH$_2$CH$_2$	Bn	L1	72	no r.	–	–	[22a]
15^{a,c,d}	PhCH$_2$CH$_2$	Bn	L1	72	67	81:19	95	[22a]
16^e	4-(NO$_2$)C$_6$H$_4$	Bn	L3	4	82	94:6	96	[53]
17^e	n-Pr	Bn	L1	3	58	94:6	95	[53]
18^e	n-Pr	Bn	L3	0.5	67	94:6	98	[53]
19^e	n-Pr	Bn	L4a	5	76	96:4	96	[22b]
20^e	n-Pr	Bn	L5a	5	72	99:1	94	[22b]
21^e	BnOCH$_2$CH$_2$	Bn	L1	10	94	>99:1	96	[54]

a Catalyst activation with TBD at room temperature.
b Catalyst: 1 mol% of pure K3/[Ir(COD)Cl]$_2$.
c 0.4 mol% of catalyst.
d Addition of 0.4 mol% of Pb(NO$_3$)$_2$ and tetrahydrothiophene.
e Catalyst activation with n-PrNH$_2$ at 50 °C.
n.d. = not determined; no r. = no reaction.

The allylic substitution with NH₃ as nucleophile could yield allylamines without an N-protecting group. Although attempts with Pd-catalysts have failed because of catalyst poisoning, Hartwig and colleagues [55] showed that Ir-catalysis allows this elusive reaction to be realized. Cinnamyl methyl carbonate yielded the disubstitution product exclusively in high yield and with excellent enantioselectivity and diastereoselectivity (Scheme 9.19). Although only one example was reported, it appears likely that similar results will be obtained with other arylallyl carbonates. Preliminary experiments with alkylallyl carbonates produced mixtures of products.

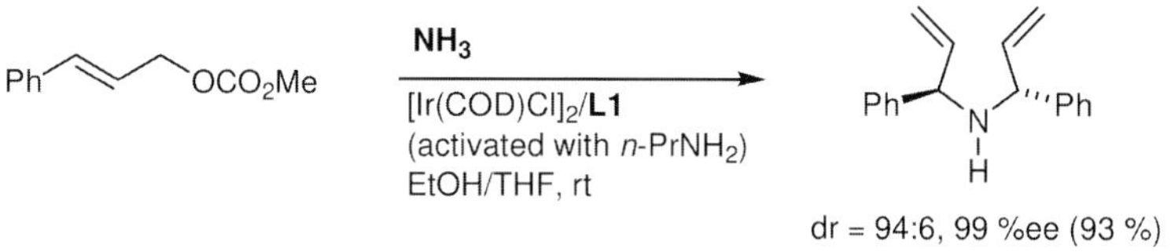

Scheme 9.19 Asymmetric allylic amination with ammonia as nucleophile.

Intramolecular allylic aminations (Scheme 9.20) proceeded with low catalytic efficiency and with ee-values <90% if procedure (a) (cf. Section 9.2.3.2) was used – that is, the catalyst was not activated [18]. The effect of catalyst activation [procedure (c)] was pronounced [18, 22a]; for example, activation with TBD increased the rate of formation of N-benzyl-2-vinylpiperidine by a factor of about 1000. Also notable was the fact that substrate concentration as high as 1 M was possible, thereby demonstrating the high preference of intramolecular over intermolecular substitutions leading to oligomers.

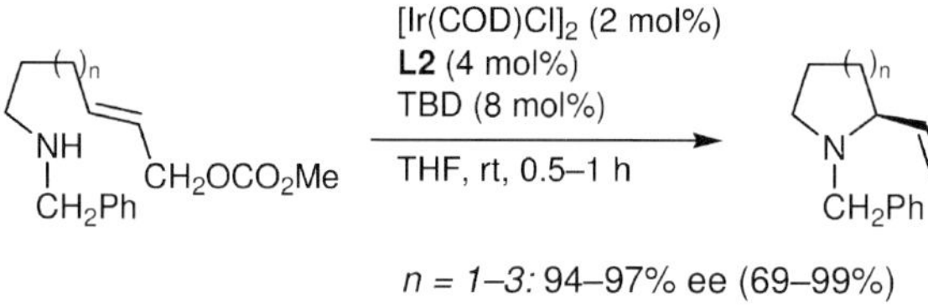

Scheme 9.20 Intramolecular asymmetric allylic aminations.

Trans-divinyl-pyrrolidines and -piperidines were prepared by sequential intermolecular and intramolecular aminations of bis-allylic carbonates (Scheme 9.21) [22a]. Due to double stereoselection, these reactions proceeded with high diastereoselectivity and enantioselectivity.

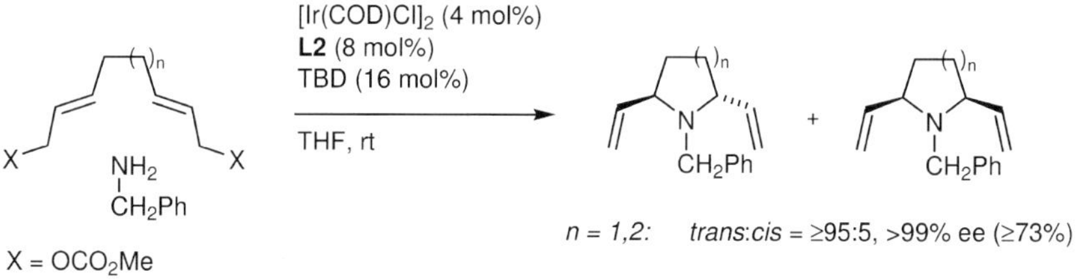

Scheme 9.21 Sequential intermolecular and intramolecular asymmetric allylic aminations.

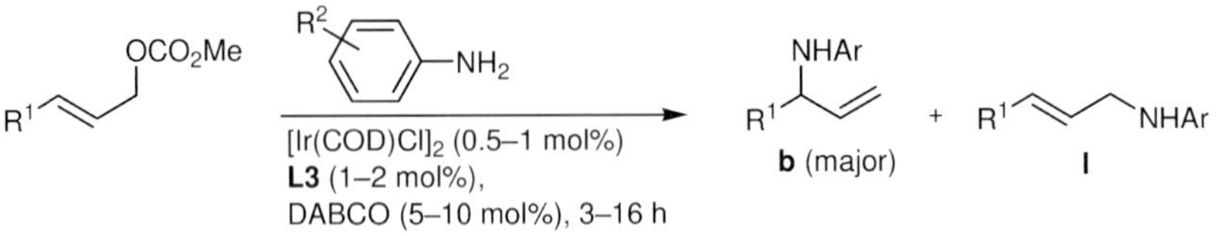

Scheme 9.22

R¹	R²	b:l	ee	yield
Ph	H	>99:1	96% ee (S)	(80%)
	4-OCH₃	98:2	95% ee	(91%)
	4-CF₃	94:6	96% ee	(72%)
n-Pr	H	98:2	95% ee	(87%)
i-Pr	H	97:3	97% ee	(83%)

Scheme 9.22 Arylamines as nucleophiles.

9.4.2
Arylamines as Nucleophiles

Procedure (a) (cf. Section 9.2.3.2) – that is, the use of a nonactivated catalyst – was successful with strongly basic nucleophiles but failed with arylamines as they are unable to effect cyclometallation. Accordingly, a catalyst activation with base was employed [procedure (c)]. Hartwig and coworkers [26] used n-PrNH$_2$ or, preferably, DABCO as activators and, with the exception of the reaction with o-methoxycinnamyl methyl carbonate (74% ee), the ee-value was excellent when the bulky ligand L3 was employed (Scheme 9.22). The regioselectivity was uniformly high, even with alkylallyl carbonates as substrates; this is a distinctive feature of aminations with arylamines in combination with ligand L3. Remarkably good results have also been obtained with ligands L4a and L5a [22b].

9.4.3
Amination of Allylic Alcohols

Ir-catalyzed allylic substitutions employing allylic alcohols as substrates and diethyl malonate as pronucleophile were first reported by Takeuchi and coworkers [11]. Here, the substitution step was found to be preceded by OH activation via transesterification to a malonic ester derivative. The asymmetric alkylation of cinnamic alcohol was similarly accomplished by Helmchen and colleagues, using a PHOX ligand and the procedure described in Section 9.2.3 [19].

Hidden activation (as above) is not possible in an amination, and other means are required. Thus, Hartwig *et al.* have systematically developed a method of amination by activation of the OH group with a Lewis acid (Scheme 9.23) [56]. Two methods of activation were found:

- Good results were obtained for a wide range of substrates with stoichiometric amounts of Nb(OEt)$_5$ as the Lewis acid; the amines used also included a secondary amine, morpholine, which gave results similar to benzylamine.

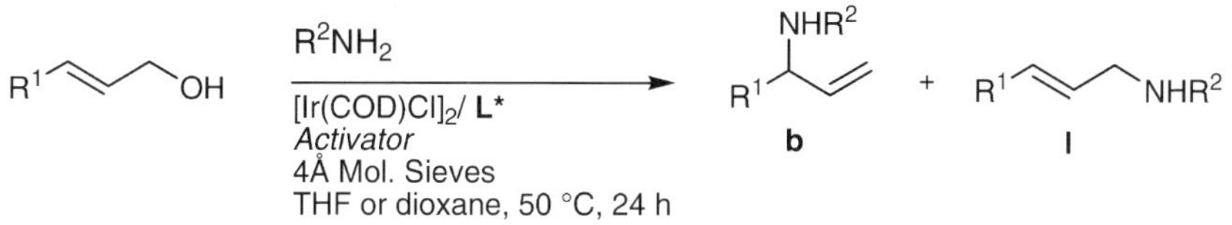

Scheme 9.23 Ir-catalyzed asymmetric aminations of allylic alcohols.

Activator: Nb(OEt)$_5$ (1.2 equiv.), [Ir(COD)Cl]$_2$ (1–2.5 mol%)/ (R)-**L5a** (2–5 mol%), THF

R^1 = Ph,	R^2 = 4-(OMe)C$_6$H$_4$	**b** : **l** = 96:4,	89% ee (84%)
R^1 = Ph,	R^2 = PhCH$_2$	**b** : **l** = 96:4,	93% ee (72%)
R^1 = n-Pr,	R^2 = Ph	**b** : **l** = 92:8,	90% ee (70%)
R^1 = i-Pr,	R^2 = Ph	**b** : **l** = 88:12,	82% ee (45%)

Activator: BPh$_3$ (8 mol%), [Ir(COD)Cl]$_2$ 2.5 mol%/ (R)-**L5a** (5 mol%), dioxane

R^1 = Ph,	R^2 = 4-(OMe)C$_6$H$_4$	**b** : **l** = 94:6,	93 %ee (72 %)
R^1 = 4-(OMe)C$_6$H$_4$,	R^2 = 4-MeC$_6$H$_4$	**b** : **l** = 96:4,	92 %ee (72 %)

- BPh$_3$ was found to act as an activator in catalytic amounts; in this case only results for aryl-substituted substrates were reported. Note that the simple ligand **L5a** (cf. Figure 9.2) was used.

An entirely different approach has been developed by Carreira and coworkers [57], who used sulfamic acid as the ammonia source. In conjunction with dimethyl formamide (DMF), amination occurred via an imidate, which was formed *in situ*. Although only those examples with branched allylic alcohols as substrates were reported, a promising enantiomeric excess was achieved with the novel phosphoramidite **L9** (Scheme 9.24).

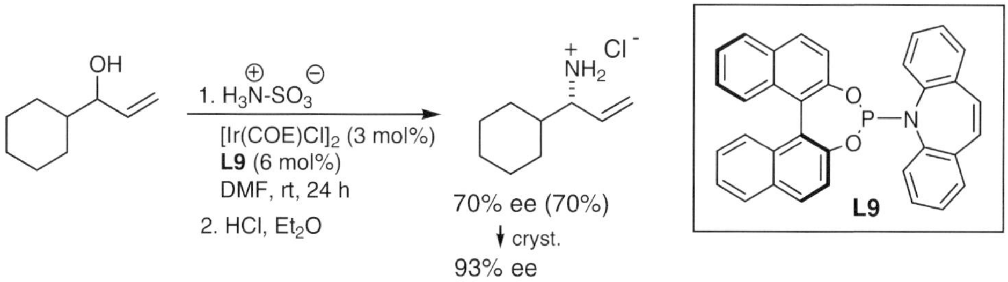

Scheme 9.24 Asymmetric amination of an alcohol according to Carreira et al. [57].

9.4.4
Pronucleophiles Serving as Ammonia Surrogates: N,N-Diacylamines, Trifluoroacetamide and N-Sulfonylamines

Unprotected branched allylamines are interesting chiral building blocks in organic synthesis (cf. Section 9.6). The removal of N-benzyl- and related protecting groups with methods other than catalytic hydrogenation is difficult in the presence of a double bond, but this problem was overcome by the author's group by using a variety of anionic N-nucleophiles. These give rise to allylamines protected by acyl groups (Table 9.3); an example is Boc, which could be readily removed under mild reaction conditions [28]. The variability of the system $MN(Acyl)^1(Acyl)^2$ is very high, and other research groups have added further examples more recently. Unprotected branched allylamines with high enantiomeric purity are now readily available via allylic substitution on a multigram scale.

Table 9.3 Allylic substitutions with N,N-diacylamines.

Entry	Pronucleophile	R	L*	Time (h)	Yield (%)	b:l	ee (%)	Reference
1	phthalimide	Ph	L1	18	66	93:7	96	[28a]
2		Ph	L2	2.5	95	96:4	98	[28a]
3		n-Pr	L2	2.5	82	94:6	96	[28a]
4	NaNBoc$_2$	Ph	L1	18	80	97:3	97.5	[28a]
5	NaNBoc$_2$	Ph	L2	0.7	80	97:3	99	[28a]
6	NaNBoc$_2$	n-Pr	L2	0.7	86	96:4	99	[28a]
7	LiNBoc$_2$	4-(OMe)C$_6$H$_4$	L5b	6	83	93:7	93	[55]
8	LiNBoc$_2$	n-Pr	L5b	24	77	93:7	95	[55]
9[a]	HNBoc$_2$	Ph	L2	0.5	93	98:2	96	[28c]
10[b]	HNBoc$_2$	Ph$_3$CO(CH$_2$)	L2	3	85	92:8	98	[58]
11	HNBoc$_2$	Ph$_3$CO(CH$_2$)$_2$	L2	3	93	96:4	97	[58]
12	HC(O)N(H)Boc	Ph	L1	18	86	97:3	97.5	[28a]
13		Ph	L2	0.7	96	98:2	98.5	[28a]
14		n-Pr	L2	1	98	97:3	96	[28a]
15	Boc-NH-C(O)-CH=CH-Me	Ph	L2[c]	3	79	98:2	>98	[28b]
16		Me	L2	1	74	98:2	>98	[28b]
17		n-Pr	L2	2	78	98:2	99	[28b]
18		CH$_2$OCPh$_3$	L2	4	77	96:4	>99	[28b]
19[a]	Ac-N(H)-Boc	Ph	L2	1	94	>99:1	>99	[28c]
20[a]		H$_3$CCH=CH	L2	0.5	92	98:2	96	[28c]
21[a]		n-Pr	L2	0.5	95	97:3	97	[28c]
22[a]		i-Pr	L2	1.5	85	99:1	>99	[28c]

a DBU (20 mol%) was used for catalyst activation; ethyl carbonates were used as substrates; reaction temperature 55 °C.
b Reaction temperature 50 °C.
c The sodium salt was used as pronucleophile.

Substitutions with *N,N*-diacylamines are best carried out under salt-free conditions in order to minimize the concentration of base in the reaction medium and to circumvent the low solubility of salts in THF. For example, potassium phthalimide could not be reacted in THF because of its insolubility. The reaction under salt-free conditions proceeded smoothly even with **L1** as the ligand (Table 9.3).

The reaction with $HN(Boc)_2$ was found to be slow at room temperature, although complete conversion was obtained with the salts $NaN(Boc)_2$ or $LiN(Boc)_2$ at room temperature or with $HN(Boc)_2$ at a reaction temperature of 50 °C. Enantioselectivity with these pronucleophiles in combination with ligand **L2** is extremely high. The pronucleophile HN(Boc)(CHO) is likely more acidic than $HN(Boc)_2$, and excellent results were obtained at room temperature. Similarly good results were obtained by Han and colleagues [28c] with the additional pronucleophiles listed in Table 9.3.

Potassium trifluoroacetamide was introduced as a pronucleophile by Hartwig *et al.* [55]; some representative reactions are described in Scheme 9.25.

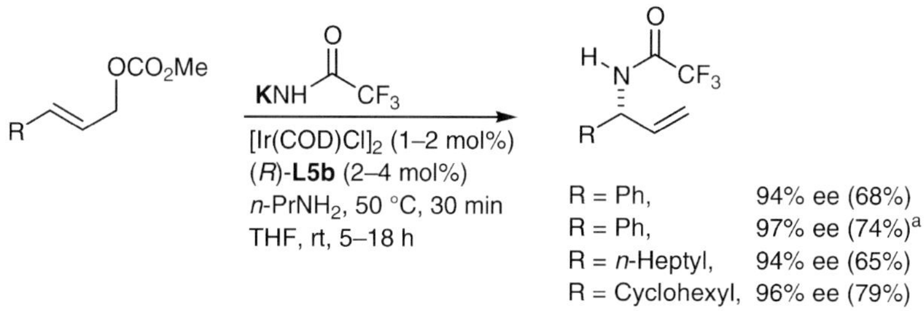

Scheme 9.25 Allylic substitutions with potassium trifluoroacetamide as pronucleophile. [a]Ligand *(R)*-**L2** was used; **b**/**l** ratios were not reported.

Sulfonamides are suitable pronucleophiles, as was first established with *N*-tosylamines [59]. Particularly good results were achieved with **L2** as ligand and activation of the catalyst with TBD (Scheme 9.26); for example, the regioselectivity of **b**/**l** = 98 : 2 and an ee-value of 98% were obtained in the reaction of $LiN(CH_2Ph)p$-Ts with cinnamyl carbonate. With a substrate containing a sp^3-substituent, however,

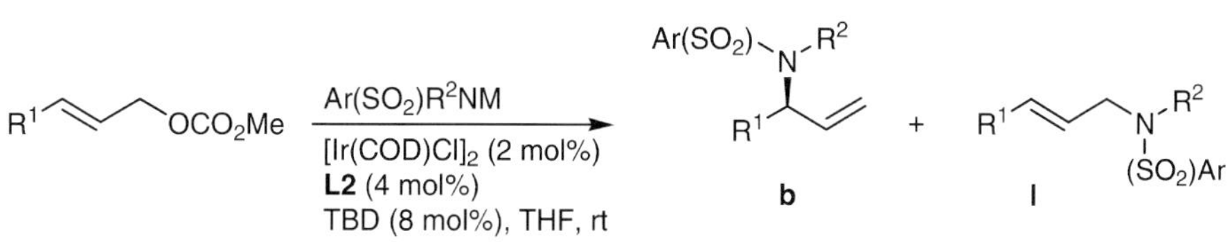

M = Li: R^1 = Ph,	R^2 = Bn	Ar = *p*-Ts	**b** : **l** = 98:2,	98% ee (92%)
R^2 = $PhCH_2CH_2$,	R^2 = Bn	Ar = *p*-Ts	**b** : **l** = 80:20,	95% ee (60%)
M = H: R^1 = Ph,	R^2 = Bn	Ar = *p*-Ns	**b** : **l** = 93:7,	96% ee (91%)
R^1 = Ph,	R^2 = H	Ar = *o*-Ns	**b** : **l** = 95:5,	95% ee (64%)
R^1 = 3-Pyridyl	R^2 = Allyl	Ar = *p*-Ns	**b** : **l** = 93:7,	92% ee (86%)

Scheme 9.26 Sulfonamides as pronucleophiles (data from Refs [28a] and [59]).

neither the yield (60%) nor regioselectivity (b/l = 80:20) were satisfactory. An early attempt at the reaction led to a low selectivity when the phosphoramidite **Monophos-NMe$_2$** was used as ligand, although this was found later to be generally unsuitable for aminations [6b].

N-(Nitrophenyl)sulfonylamines (o- and p-Ns(R)NH) are very useful pronucleophiles, as deprotection of the substitution products reliably yields free amines, unlike N-tosylates. N-Nosylamines are sufficiently acidic to react without additional base – that is, under salt-free conditions, but the yield and regioselectivity were slightly affected by the addition of NEt$_3$. With N-(p-Ns)NHR [59] and o-Ns-NH$_2$ [27a] as nucleophiles, the branched products were generally obtained. An exception was the reaction of N-(o-Ns)NHCH$_2$Ph and cinnamyl methyl carbonate in conjunction with ligands **L2** and **L3**, which gave the linear product as a result of thermodynamic control.

Sequential inter- and intramolecular reactions according to Scheme 9.21 were also possible with p-NsNH$_2$ [59]. In contrast to N-benzyl derivatives, the p-nosyl derivatives could be readily transformed into the nonprotected amines (A. Dahnz and G. Helmchen, unpublished results).

9.4.5
Decarboxylative Allylic Amidation

Prior to the studies of You and colleagues on decarboxylative allylic alkylation (cf. Scheme 9.17), Singh and Han [60] had developed the first asymmetric decarboxylative amidation. These authors used conditions of type (c) as a starting point and, despite careful screening of the ligand, base and solvent, only partial success was achieved. A further probing of additives led to the remarkable finding that the addition of 'proton sponge' increased the reaction rate by a factor of >100 and led to excellent results, some of which are presented in Scheme 9.27.

A mechanistic investigation furnished the following results:

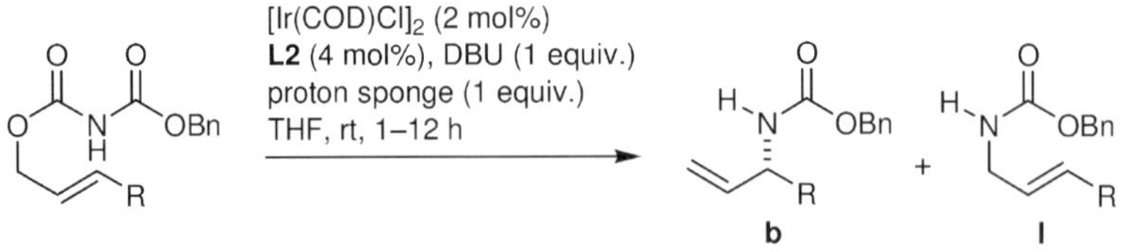

R	Yield (%)	b : l	ee (%)
Ph	90	>99:1	>99
CH=CHCH$_3$	80	98:2	92
n-Pr	90	97:3	96
CH$_2$(CH$_2$)$_2$OSi(t-Bu)Ph$_2$	80	98:2	98

Scheme 9.27 Decarboxylative allylic amidation. Proton sponge = 1,8-Bis-(dimethylamino)-naphthalene.

- A crossover experiment suggests that the reaction follows an intermolecular course.

- The steric course of the reaction follows the general rule – that is, it is the same as that of other Ir-catalyzed allylic substitutions.

- Cinnamyl ethyl carbonate did not react with the lithium salt of benzyl carbamate. This observation led to the suggestion that the reacting nucleophile is a tautomer of the anion that is initially formed rather than its decarboxylation product:

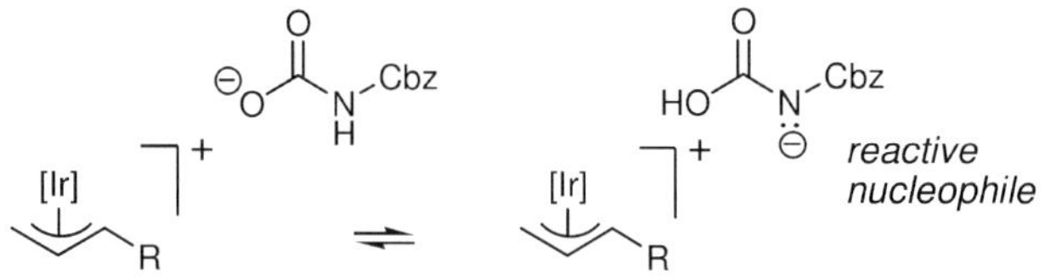

In another report of Singh and Han [61], Ir-catalyzed decarboxylative amidations of benzyl allyl imidodicarboxylates derived from enantiomerically enriched branched allylic alcohols are described. This reaction proceeded with complete stereospecificity – that is, with complete conservation of enantiomeric purity and retention of configuration. This result underlines once again (cf. Section 9.2.2) that the isomerization of intermediary (allyl)Ir complexes is a slow process in comparison with nucleophilic substitution.

9.4.6
Dihydropyrroles and γ-Lactams via Allylic Substitution and Ring-Closing Metathesis

Products from reactions with diacylamines or nosylamines can be very easily deprotected to give primary allylamines. These were used as nucleophiles in allylic substitutions to give secondary amines, which were transformed into unsymmetrically 2,5-disubstituted 2,5-dihydropyrroles (Scheme 9.28) [28a]. Thus, the allylic

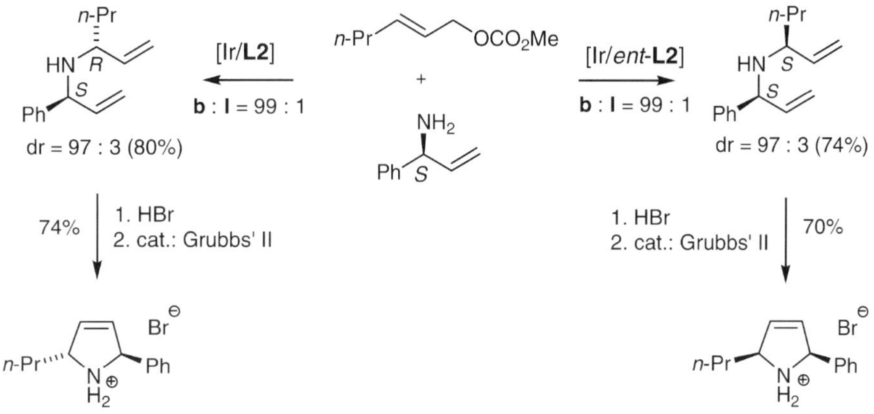

Scheme 9.28 2,5-Disubstituted 2,5-dihydropyrroles via allylic substitution in combination with ring-closing metathesis.

amination of 3-propylallyl methyl carbonate with *(S)*-(1-phenyl-prop-2-enyl)amine, using ligands **L2** or *ent*-**L2**, furnished diastereoisomeric secondary amines with *(S,R)*- and *(S,S)*-configuration, respectively. Excellent regioselectivity and diastereoselectivity were achieved for both reactions; accordingly, the substitution reaction is catalyst- rather than substrate-controlled [62]. The protection of nitrogen by salt formation followed by ring-closing metathesis (RCM), using Grubbs' II catalyst, yielded the *cis*- and *trans*-2,5-disubstituted 2,5-dihydropyrroles in good yields.

N-Boc-*N*-(but-2-enoyl)amine is an excellent pronucleophile for the Ir-catalyzed allylic amination under salt-free conditions (cf. Table 9.3, entries 15–18). The products were subjected to RCM with good results, even upon application of the Grubbs' I catalyst (Scheme 9.29) [27b]. The resultant N-Boc protected α,β-unsaturated γ-lactams are valuable chiral intermediates with applications in natural products synthesis and medicinal chemistry.

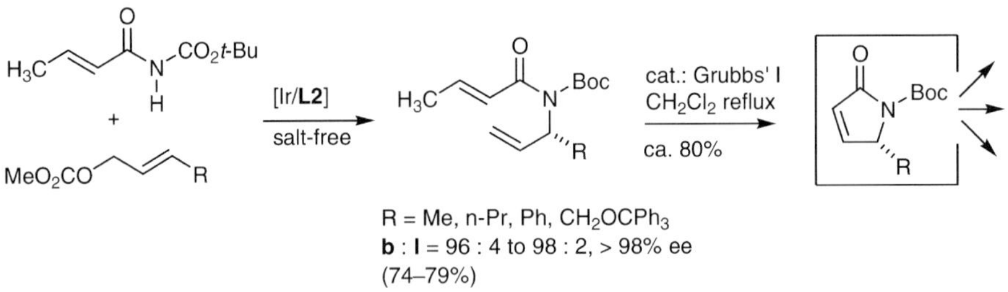

Scheme 9.29 Preparation of α,β-unsaturated γ-lactams via allylic amination/ring-closing metathesis.

Singh and Han [60] have reported the preparation of another dihydropyrrole by using a N-Cbz derivative obtained by decarboxylative amidation (Scheme 9.27, results line 1) as starting material. N-Alkylation with allyl bromide followed by RCM (Grubbs' II catalyst) furnished the dihydropyrrole in excellent yield (95%). Lee *et al.* have similarly transformed the amination product (Table 9.2, entry 21) into a variety of N-heterocycles [54].

9.4.7
Hydroxylamine Derivatives as N-Nucleophiles

Hydroxylamine derivatives are ambident nucleophiles. For example, *N*-benzylhydroxylamine functions as an N-nucleophile in the Ir-catalyzed allylic substitution, while N-Boc-hydroxylamine yields mixtures of the N- and O-substituted products in Ir- as well as Pd-catalyzed allylic substitutions. Accordingly, either O- or N,O-protected hydroxylamine derivatives need to be used as nucleophiles [63].

In our hands, base-activated phosphoramidite–Ir complexes were not suited for the allylation of hydroxylamine and hydrazine derivatives (R. Weihofen and G. Helmchen, unpublished results). However, interesting results were obtained by Takemoto and coworkers with **Pybox** type ligands (Scheme 9.30) [40]. Phosphates

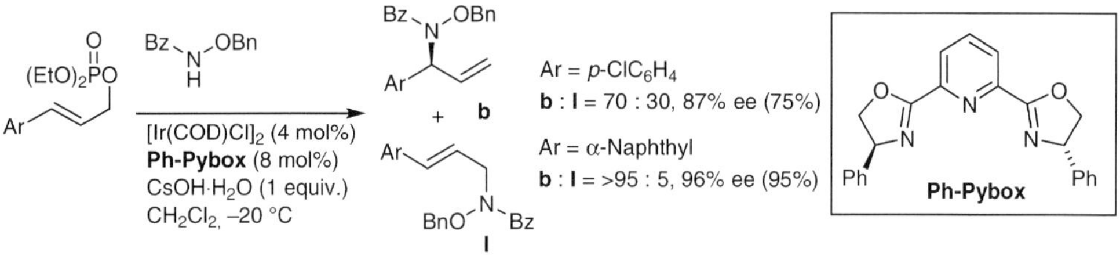

Scheme 9.30 Allylic substitutions with a hydroxylamine derivative as nucleophile.

rather than carbonates (which did not react) were used as substrates, and CH_2Cl_2 was the solvent of choice; a base was usually required as an additive. Good results were obtained with either $Cs(OH)\cdot H_2O$ or $Ba(OH)_2\cdot H_2O$, although so far the procedure appears to be limited to arylallyl substrates.

A few other hydroxylamine derivatives were tested with cinnamyl diethyl phosphate as substrate (Scheme 9.30, Ar = Ph) and $Cs(OH)\cdot H_2O$ as base [39]. The addition of a base was necessary in the case of N,O-dibenzoylhydroxylamine, whereas the reaction with N,O-dibenzylhydroxylamine proceeded without base. The corresponding amide N-benzoylbenzylamine did not undergo the reaction.

9.5
O-Nucleophiles

9.5.1
Phenolates as Nucleophiles

Phenolates as nucleophiles in conjunction with Ir/phosphoramidite catalysts were first investigated by Hartwig and coworkers. The early investigations were carried out with catalysts not activated through the addition of a base (Scheme 9.31) [64]; rather, it is likely that cyclometallation was induced *in situ* by alkoxide generated from the leaving group. The following observations were made with regards to selectivities and yields:

- Alkali phenolates were superior to ammonium phenolates as pronucleophiles.
- Transesterification as a side reaction was observed for a reaction of a sodium phenolate with a methyl carbonate.
- Transesterification was not found with ethyl carbonates or lithium phenolates.
- Generally, arylallyl carbonates gave better results than alkylallyl carbonates.
- The influence of the solvent was investigated and, once again, the best results were obtained with THF.

The deficiencies of the nonactivated catalysts were particularly apparent for reactions with alkylallyl carbonates as substrates. An activated catalyst, formed by use of procedure (c) (cf. Section 9.2.4.2) gave rise to a distinctly improved yield and selectivities (Scheme 9.31) [53].

With regards to the scope of the reaction with respect to the phenolate, donor substituents and halogen were tolerated, and even sterically hindered lithium

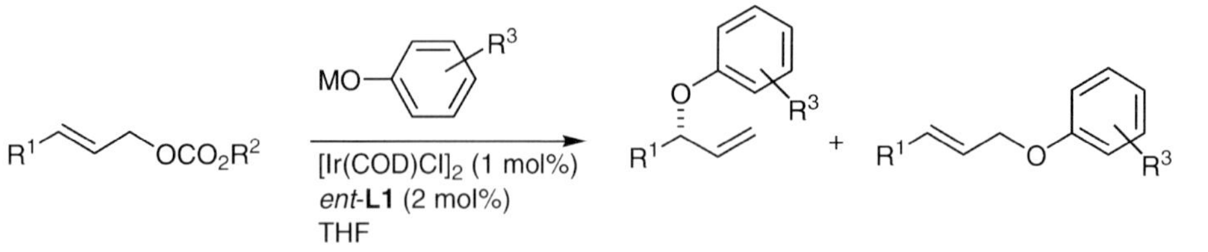

Scheme 9.31

R¹⎯⎯⎯OCO₂R² + MO-C₆H₄-R³ →[[Ir(COD)Cl]₂ (1 mol%), ent-**L1** (2 mol%), THF] branched product + linear product

non-activated cat. (50 °C)

R¹ = Ph, R² = Me, M = Li:
- R³ = 4-(OCH₃)   b : l = 98:2, 97% ee (88%)
- R³ = 3-Ph   b : l = 96:4, 95% ee (76%)
- R³ = 2,4,6-(CH₃)₃ b : l = 93:7, 93% ee (82%)

R¹ = Ph, R² = Et, M = Na:
- R³ = 4-Br   b : l = 96:4, 90% ee (91%)
- R³ = 4-(CF₃)   b : l = 90:10, 80% ee (92%)
- R³ = 4-(NO₂)   no reaction

R¹ = n-Pr, R² = Me, M = Li: R³ = 4-OMe   b : l = 90:10, 85% ee (73%)

activated cat. (rt)

R¹ = n-Pr, R² = Me, M = Li: R³ = 4-OMe   b : l = 93:7, 94% ee (95%)

Scheme 9.31 Allylic substitutions with alkali phenolates as pronucleophiles.

phenolates gave excellent results. In the case of phenolates with electron-withdrawing substituents, good results were obtained when sodium phenolates in combination with ethyl carbonates were used. Phenolates with very strongly electron-withdrawing substituents (i.e. 4-nitro- and 4-cyanophenolates) did not react.

The reaction of lithium phenoxide with cinnamyl methyl carbonate at 50 °C gave excellent selectivities, if the reaction time was less than about 20 h. However, longer reaction times led to a fall in both regioselectivity and enantioselectivity, which indicated that the reaction is reversible.

Interesting results were obtained by Kimura and Uozumi with a series of phosphorodiamidites **L7** as ligands [65]. Here, only cinnamyl methyl carbonate was probed as substrate, and moderate selectivities were achieved; the best results are described in Scheme 9.32. An interesting switch in the configurational course of

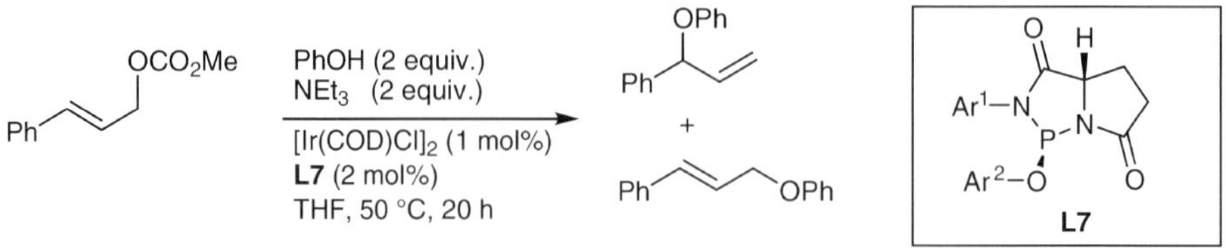

Scheme 9.32

Ph⎯⎯⎯OCO₂Me + PhOH (2 equiv.), NEt₃ (2 equiv.) →[[Ir(COD)Cl]₂ (1 mol%), **L7** (2 mol%), THF, 50 °C, 20 h] Ph-CH(OPh)-CH=CH₂ + Ph-CH=CH-CH₂-OPh

L7: Ar¹-N, Ar²-O, P-N bicyclic phosphorodiamidite

Ar¹ = 2,6-(Me)₂C₆H₃, Ar² = Ph: b : l = 97:3, 70% ee (S) (22%)
Ar¹ = Ph, Ar² = 2,6-(Me)₂C₆H₃: b : l = 73:27, 74% ee (R) (76%)

Scheme 9.32 Application of phosphorodiamidite ligands to allylic etherification.

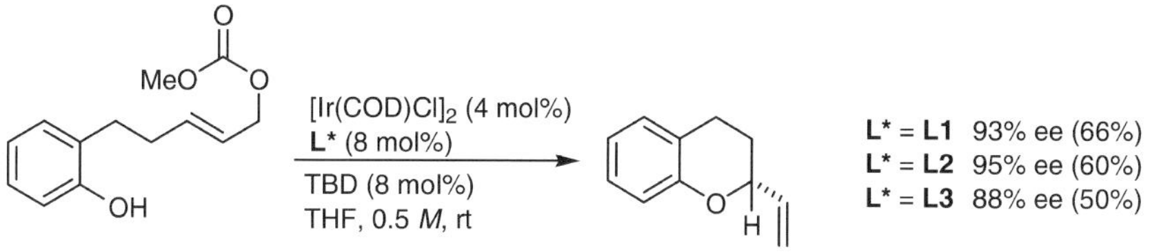
MeO
O
O
OH
[Ir(COD)Cl]2 (4 mol%)
L* (8 mol%)
TBD (8 mol%)
THF, 0.5 M, rt
O
H
L* = L1 93% ee (66%)
L* = L2 95% ee (60%)
L* = L3 88% ee (50%)

Scheme 9.33 Examples of intramolecular allylic etherifications.

the reaction was observed upon permutation of substituents Ar^1 and Ar^2 of **L7**. The authors presented extensive density functional theory (DFT) calculations, although mechanistic details of the reaction were not investigated experimentally. For example, the C–H activation of **L7**/Ir complexes was not considered.

Intramolecular reactions of a phenolate were also reported (Scheme 9.33) [22a]. The preparation of a chromane derivative is described below, where the catalyst was activated with the base TBD. As in the case of intramolecular aminations, these cyclizations could be run at concentrations as high as 0.5–1 *M*.

Phenolates were used as nucleophiles for kinetic resolutions by Carreira and coworkers [42]; these authors used a catalyst which was prepared from $[Ir(COE)_2Cl]_2$ and the chiral bicycle(2.2.2)octadiene **L8** (cf. Section 9.2.5).

9.5.2
Alkoxides as Nucleophiles

Generally, alkoxides are problematic nucleophiles because of their basic character. In metal-catalyzed allylic substitutions, superior results were obtained with Zn-alkoxides (achiral Ir-catalysts) [66] and Cu-alkoxides (achiral Rh-catalyst with chiral substrates) [67]. Shu and Hartwig developed allylic substitutions with alkoxides using Ir/phosphoramidite catalysts [68]; these authors used catalysts obtained from $[Ir(COD)Cl]_2$ and **L1** or **L3** without explicit base activation [procedure (a) in Section 9.2.4.2) (Scheme 9.34).

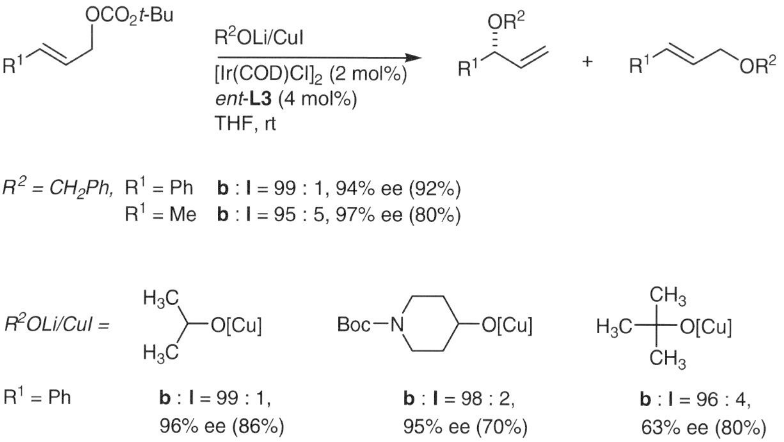
OCO2t-Bu
R1
R2OLi/CuI
[Ir(COD)Cl]2 (2 mol%)
ent-L3 (4 mol%)
THF, rt
OR2
R1
+
R1
OR2
R2 = CH2Ph, R1 = Ph b : l = 99 : 1, 94% ee (92%)
R1 = Me b : l = 95 : 5, 97% ee (80%)
R2OLi/CuI =
H3C
H3C
O[Cu]
Boc
N
O[Cu]
CH3
H3C
O[Cu]
CH3
R1 = Ph
b : l = 99 : 1,
96% ee (86%)
b : l = 98 : 2,
95% ee (70%)
b : l = 96 : 4,
63% ee (80%)

Scheme 9.34 Allylic substitutions with alkoxides as nucleophiles.

In an exploratory study, Cu-alkoxides were found to be superior to Zn-alkoxides, and *tert*-butyl cinnamyl carbonate superior to methyl cinnamyl carbonate, which underwent transesterification. Both, arylallyl and alkylallyl carbonates were successful as substrates.

The optimized reaction conditions were successfully applied to alkoxides derived from primary as well as secondary alcohols. With tertiary alkoxides, the regioselectivities and yields were excellent but the enantioselectivities were comparatively low. Applications of the method are presented in Sections 9.5.5 and 9.6.

9.5.3
Hydroxylamine Derivatives as O-Nucleophiles

Hydroxylamine can act as either a N-nucleophile or O-nucleophile, depending on which of the reactive centers is protected. For all reactions **Ph-Pybox** has been used as ligand, and moderate to high levels of selectivity have been achieved. Hydroxamic acid derivatives and oximes have also been probed as O-nucleophiles [63].

The reactions of hydroxamic acid derivatives required carefully optimized reactions conditions; that is, 3-arylallyl phosphates were used as substrates in conjunction with $PhCF_3$/water 2:1 as solvent (Scheme 9.35) [69]. The reactivity and selectivity were influenced by base, and the best results were obtained with $Ba(OH)_2$.

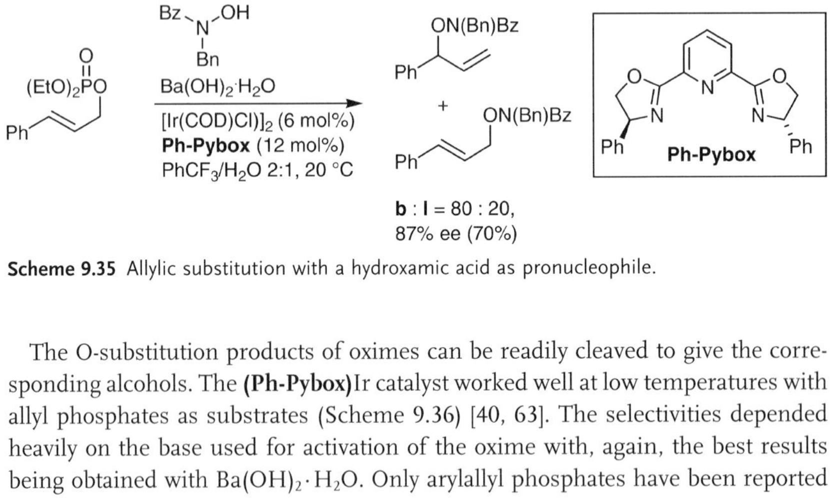

Scheme 9.35 Allylic substitution with a hydroxamic acid as pronucleophile.

The O-substitution products of oximes can be readily cleaved to give the corresponding alcohols. The **(Ph-Pybox)**Ir catalyst worked well at low temperatures with allyl phosphates as substrates (Scheme 9.36) [40, 63]. The selectivities depended heavily on the base used for activation of the oxime with, again, the best results being obtained with $Ba(OH)_2 \cdot H_2O$. Only arylallyl phosphates have been reported as substrates.

9.5.4
Silanolates as Nucleophiles

Carreira and coworkers have established silanolates as useful nucleophiles [70], with allylic substitutions being carried out with an activated catalyst [procedure (c),

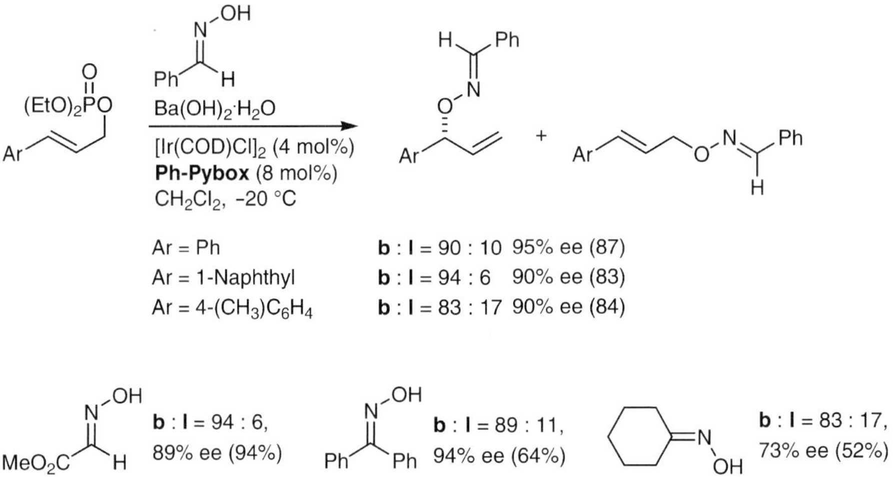

Scheme 9.36 Allylic substitutions with oximes as nucleophiles. In the lower row, Ar = Ph.

using n-propylamine as base; cf. Section 9.2.4.2] (Scheme 9.37). For catalyst preparation, THF was used as a solvent, while the substitution reaction proceeded with optimal results with dichloromethane. Methyl carbonates underwent transesterification as a side reaction, but this was not observed with *tert*-butyl carbonates; allylic acetates yielded linear products exclusively. Several silanolates were tested as nucleophiles, with Et$_3$SiOK proving to be the best suited.

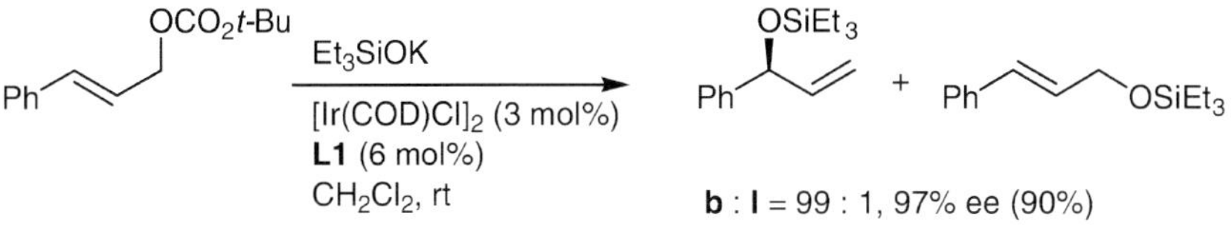

Scheme 9.37 Et$_3$SiOK as a pronucleophile.

The reaction tolerates a wide variety of allylic substrates (Scheme 9.38). Allylic alcohols were obtained from silyl ethers by standard methods.

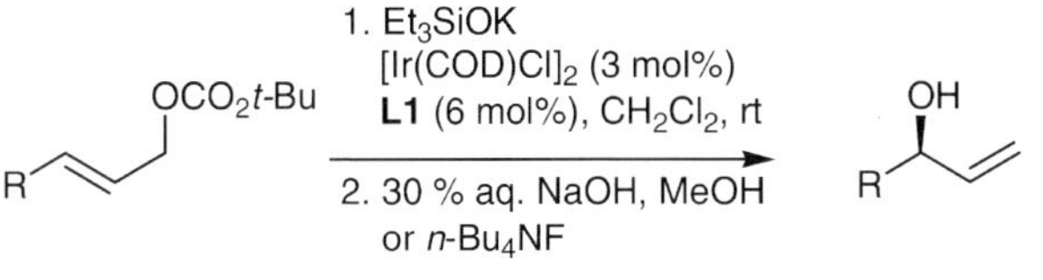

R	Yield (%)	ee (%)
Aryl	64–88	92–98
Heteroaryl	50–67	97–99
Alkenyl	65	97
Alkyl	65	95

Scheme 9.38 Allylic alcohols via Ir-catalyzed allylic etherification.

9.5.5
Dihydrofurans via Allylic Etherification in Combination with RCM

Disubstituted dihydrofurans and dihydropyrans were prepared via allylic etherification [68] in a similar manner to dihydropyrroles (cf. Section 9.4.6). Thus, diastereoisomeric ethers were generated by the reaction of cinnamyl *tert*-butyl carbonate with the copper alkoxide prepared from *(R)*-1-octen-3-ol, depending on which enantiomer of the phosphoramidite ligand was used (Scheme 9.39). Good yields and excellent selectivities were obtained. RCM in a standard manner gave *cis*- and *trans*-dihydrofuran derivatives in good yield, and the same method was used for the preparation of dihydropyrans.

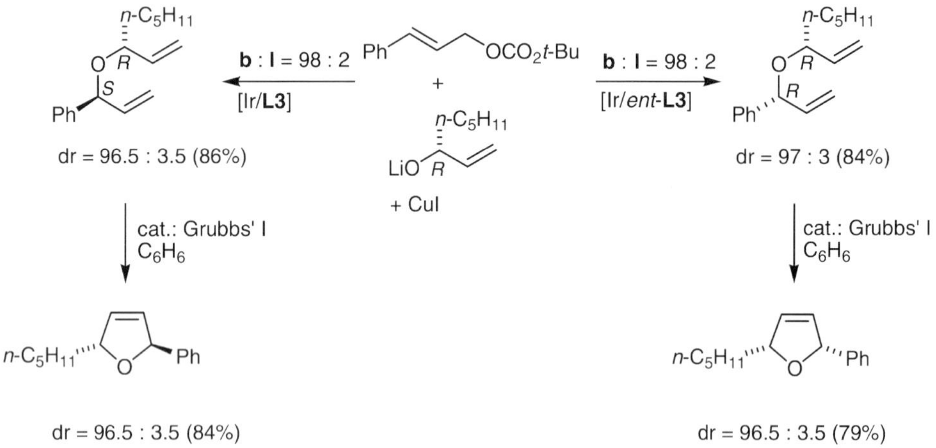

Scheme 9.39 Preparation of dihydrofuran derivatives.

9.6
Synthesis of Biologically Active Compounds via Allylic Substitution

Although the Ir-catalyzed allylic substitution was developed only recently, several applications in the areas of medicinal and natural products chemistry have already been reported. In many syntheses the allylic substitution has been combined with a RCM reaction [71]. Examples not directed at natural products targets have already been described in Sections 9.4 and 9.5. It has also been mentioned that this strategy had previously been used in conjunction with allylic substitutions catalyzed by other transition metals (Figure 9.5). This was pioneered by P. A. Evans and colleagues, who used Rh-catalyzed allylic amination (compound A in Figure 9.5) [72] and etherification (compound B) [73], while Trost and coworkers demonstrated the power of this concept for Pd-catalyzed allylic alkylations (compound C) [74] and Alexakis *et al.* for Cu-catalyzed (compound D) allylic alkylations [75].

A synthesis of the prostaglandin analogue TEI-9826 is presented in Scheme 9.40 as an example of the use of an Ir-catalyzed allylic alkylation [44]. The allylic alkylation with a malonic amide of the Weinreb type as pronucleophile gave the substi-

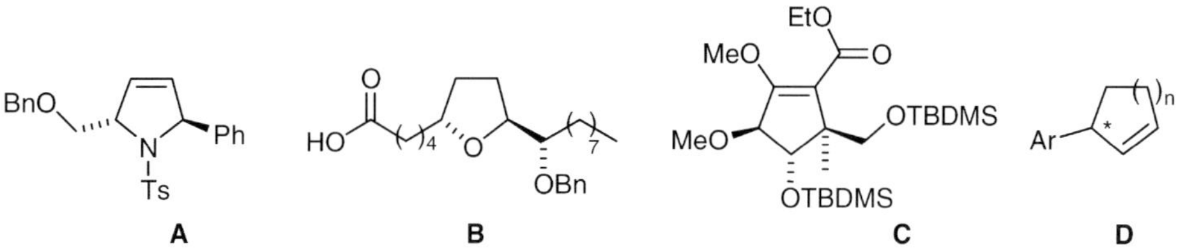

Figure 9.5 Formulas of compounds prepared via combination of an allylic substitution and ring-closing metathesis.

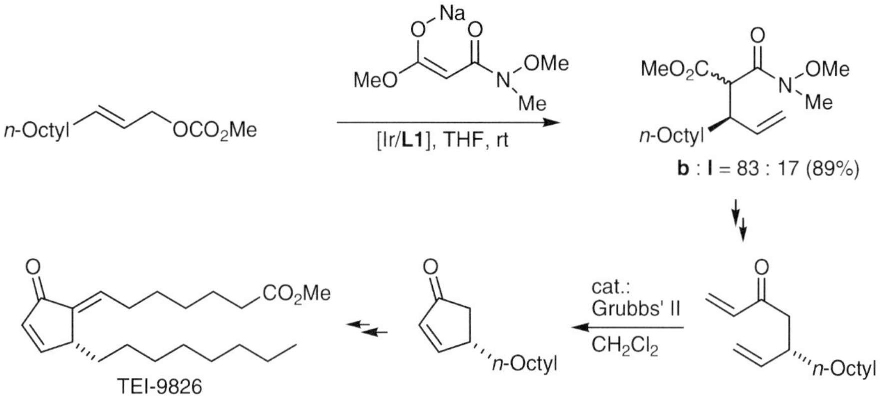

Scheme 9.40 Synthesis of the prostaglandin analogue TEI-9826 via Ir-catalyzed allylic alkylation.

tution product with 96 and 99% ee upon the use of **L1** and **L2**, respectively, as ligand. A few simple steps furnished a dienone, which was subjected to RCM to give 4-(*n*-octyl)-cyclopent-2-enone, without racemization. A final stepwise aldol condensation gave TEI-9826 in excellent overall yield.

Ir-catalyzed alkylation with a nitro compound was applied in a synthesis of *(1S,2R)-trans*-2-phenylcyclopentanamine, a compound with antidepressant activity (Scheme 9.41) [45]. The reaction of cinnamyl methyl carbonate with 4-nitro-1-butene gave the substitution product with 93% ee in 82% yield. A Grubbs' I catalyst sufficed for the subsequent RCM. Further epimerization with NEt$_3$ yielded a *trans*-cyclopentene in 83% yield via the two steps, while additional reduction steps proceeded in 90% yield.

A combination of allylic amination and RCM was used for the synthesis of *(S)*-nicotine (Scheme 9.42) [76]. The Ir-catalyzed amination of methyl 3-(3-pyridyl)-allyl

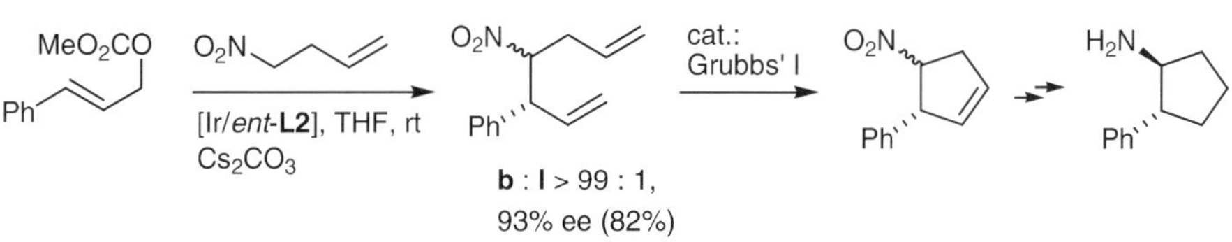

Scheme 9.41 Synthesis of the antidepressant *(1S,2R)-trans*-2-phenylcyclopentanamine.

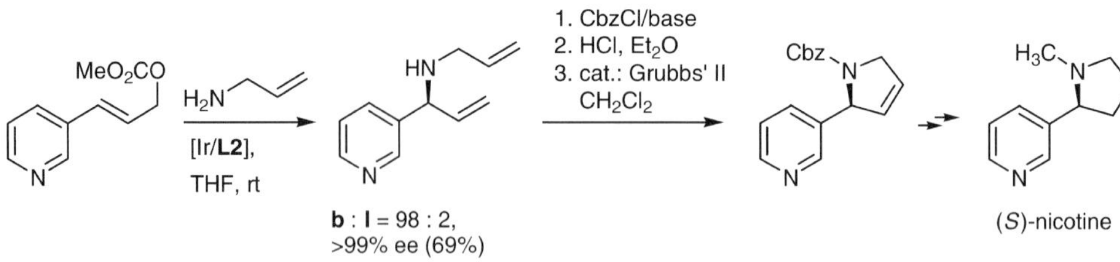

Scheme 9.42 (S)-Nicotine via allylic amination in combination with ring-closing metathesis.

carbonate with allylamine proceeded with excellent regioselectivity and enantioselectivity. Following protection of the nitrogen, RCM with Grubbs' II catalyst gave a 2,5-dihydropyrrole in very good yield, and further steps produced (S)-nicotine in good overall yield.

An Ir-catalyzed etherification in combination with ring rearrangement metathesis (RRM) was used by Blechert and coworkers for an elegant synthesis of the antibiotic centrolobine (Scheme 9.43) [77]. The copper alkoxide of cyclopent-3-en-1-ol was used as the pronucleophile, and the product was obtained in 87% yield with >98% ee. Ring rearrangement metathesis using Grubbs' II catalyst, followed by double-bond isomerization and then a one-pot RCM/catalytic hydrogenation gave centrolobine in good overall yield.

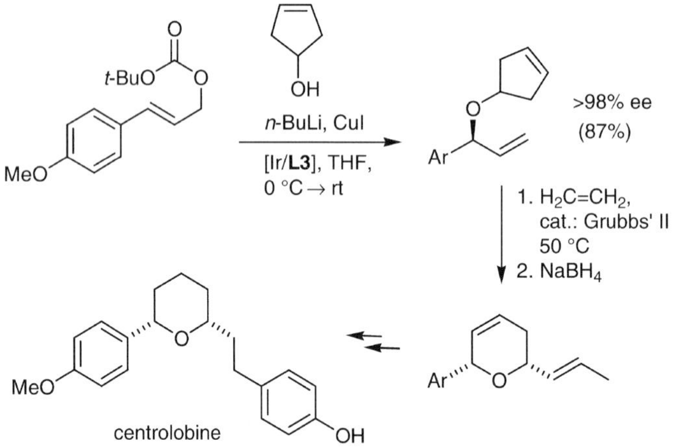

Scheme 9.43 Synthesis of the antibiotic centrolobine.

Further compounds that have been synthesized via allylic alkylation or amination and functionalization of the vinyl group are listed in Scheme 9.44.

9.7
Conclusions

Over the past few years, the asymmetric Ir-catalyzed allylic substitution reaction has been developed into a synthetically significant and useful method. Broadly

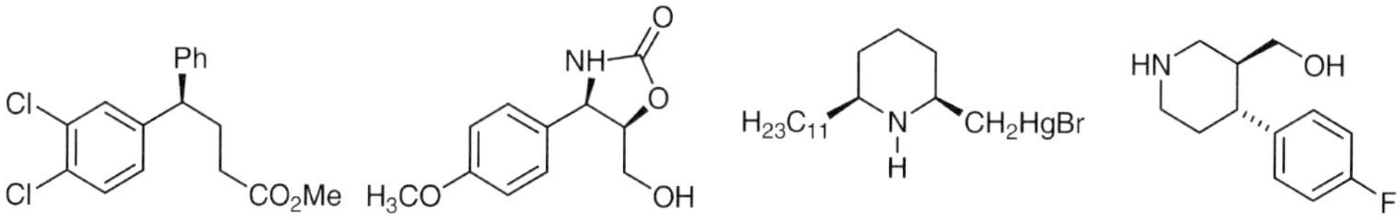

Scheme 9.44 Further targets available via Ir-catalyzed allylic substitution. (From left to right, Refs [28c], [41a], [52] and [60]).

applicable catalysts are available by combining [Ir(COD)Cl]$_2$ and phosphoramidites, the reaction being carried out with a variety of C-, N- and O-nucleophiles to give branched substitution products, in particular chiral allylamines and their derivatives, with excellent regioselectivity and enantioselectivity. The configurational course of all substitutions investigated to date has been identical, despite the highly variable range of substrates. Moreover, the method has been applied to the syntheses of a broad range of biologically active compounds of interest in medicinal chemistry.

Acknowledgments

The investigations of the author's group on Ir-catalyzed allylic substitutions are supported by the Fonds der Chemischen Industrie, the Deutsche Forschungsgemeinschaft (SFB 623), and The Royal Society. Thanks are also offered to Professor K. Ditrich of BASF AG for generous donations of enantiomerically pure amines. The author is also indebted to Amaruka Hazari for linguistic polishing of the manuscript.

References

1 (a) Trost, B.M. and Lee, C. (2000) *Catalytic Asymmetric Synthesis*, 2nd edn (ed. I. Ojima), Wiley-VCH Verlag GmbH, New York, pp. 593–649;
(b) Pfaltz, A. and Lautens, M. (1999) *Comprehensive Asymmetric Catalysis I-III* (eds E.N. Jacobsen, A. Pfaltz and H. Yamamoto), Springer, Berlin, pp. 833–84.

2 (a) Zheng, W.H., Sun, N. and Hou, X.L. (2005) *Organic Letters*, **7**, 5151;
(b) Dai, L.X., Tu, T., You, S.-L., Deng, W.P. and Hou, X.L. (2003) *Accounts of Chemical Research*, **36**, 659;
(c) You, S.L., Zhu, X.Z., Luo, Y.M., Hou, X.L. and Dai, L.X. (2001) *Journal of the American Chemical Society*, **123**, 7471;
(d) Prétôt, R. and Pfaltz, A. (1998) *Angewandte Chemie – International Edition*, **37**, 323;
(e) Hilgraf, R. and Pfaltz, A. (2005) *Advanced Synthesis Catalysis*, **347**, 61;
(f) Pàmies, O., Diéguez, M. and Claver, C. (2005) *Journal of the American Chemical Society*, **127**, 3646;
(g) Trost, B.M. and Toste, F.D. (1998) *Journal of the American Chemical Society*, **120**, 9074;
(h) Hayashi, T., Kawatsura, M. and Uozumi, Y. (1997) *Chemical Communications*, 561;
(i) Hayashi, T., Ohno, A., Lu, S.-J., Matsumoto, Y., Fukuyo, E. and Yanagi, K.

(1994) *Journal of the American Chemical Society*, **116**, 4221.
3 (a) Johns, A.M., Liu, Z. and Hartwig, J.F. (2007) *Angewandte Chemie – International Edition*, **46**, 7259;
(b) Dubovyk, I., Watson, I.D.G. and Yudin, A.K. (2007) *Journal of the American Chemical Society*, **129**, 14172.
4 (a) Takeuchi, R. and Kashio, M. (1997) *Angewandte Chemie – International Edition in English*, **36**, 263; Reviews:
(b) Takeuchi, R. (2002) *Synlett*, 1954;
(c) Takeuchi, R. and Kezuka, S. (2006) *Synthesis*, 3349.
5 Janssen, J.P. and Helmchen, G. (1997) *Tetrahedron Letters*, **38**, 8025.
6 (a) Bartels, B. and Helmchen, G. (1999) *Chemical Communications*, 741;
(b) Bartels, B., García-Yebra, C., Rominger, F. and Helmchen, G. (2002) *European Journal of Inorganic Chemistry*, 2569;
(c) Bartels, B., García-Yebra, C. and Helmchen, G. (2003) *European Journal of Organic Chemistry*, 1097;
(d) For a very detailed account see: Bartels, B. (2001) Regio- und enantioselektive allylische Substitution katalysiert durch Iridiumkomplexe. Dissertation, Universität Heidelberg.
7 Ohmura, T. and Hartwig, J.F. (2002) *Journal of the American Chemical Society*, **124**, 15164.
8 Kiener, C.A., Shu, C., Incarvito, C. and Hartwig, J.F. (2003) *Journal of the American Chemical Society*, **125**, 14272.
9 Tissot-Croset, K., Polet, D. and Alexakis, A. (2004) *Angewandte Chemie – International Edition*, **43**, 2426.
10 Helmchen, G., Dahnz, A., Dübon, P., Schelwies, M. and Weihofen, R. (2007) *Chemical Communications*, 675.
11 Takeuchi, R. and Kashio, M. (1998) *Journal of the American Chemical Society*, **120**, 8647.
12 Ashfeld, B.L., Miller, K.A., Smith, A.J., Tran, K. and Martin, S.F. (2007) *The Journal of Organic Chemistry*, **72**, 9018.
13 Evans, P.A. and Nelson, J.D. (1998) *Journal of the American Chemical Society*, **120**, 5581.
14 (a) Takeuchi, R., Ue, N., Tanabe, K., Yamashita, K. and Shiga, N. (2001) *Journal of the American Chemical Society*, **123**, 9525;
(b) Takeuchi, R. and Shiga, N. (1999) *Organic Letters*, **1**, 265.
15 This complex had been characterized earlier in another context: Bedford, R.B., Castillòn, S., Chaloner, P.A., Claver, C., Fernandez, E., Hitchcock, P.B. and Ruiz, A. (1996) *Organometallics*, **15**, 3990.
16 (a) Lloyd-Jones, G.C., Stephen, S.C., Murray, M., Butts, C.P., Vyskočil, Š. and Kočovský, P. (2000) *Chemistry – A European Journal*, **6**, 4348;
(b) Lloyd-Jones, G.C. and Stephen, S.C. (1998) *Chemistry – A European Journal*, **4**, 2539.
17 Koch, O. (2000) Enantioselektive Synthese von Hetero- und Carbocyclen mit Übergangsmetall-katalysiertem Schlüsselschritt. Dissertation, Unversität Heidelberg.
18 Welter, C., Koch, O., Lipowsky, G. and Helmchen, G. (2004) *Chemical Communications*, 896.
19 (a) García-Yebra, C., Janssen, J.P., Rominger, F. and Helmchen, G. (2004) *Organometallics*, **23**, 5459;
(b) McGhee, W.D. and Bergman, R.G. (1985) *Journal of the American Chemical Society*, **107**, 3388;
(c) Wakefield, J.B. and Stryker, J.M.J. (1991) *Journal of the American Chemical Society*, **113**, 7057.
20 Feringa, B.L. (2000) *Accounts of Chemical Research*, **33**, 346.
21 This compound is available from BASF AG, Ludwigshafen, in their ChiPros program. Distributor Apha Aesar (Item No. H26984).
22 (a) Welter, C., Dahnz, A., Brunner, B., Streiff, S., Dübon, P. and Helmchen, G. (2005) *Organic Letters*, **7**, 1239;
(b) Leitner, A., Shekhar, S., Pouy, M.J. and Hartwig, J.F. (2005) *Journal of the American Chemical Society*, **127**, 15506;
(c) Leitner, A., Shu, C. and Hartwig, J.F. (2004) *Proceedings of the National Academy of Sciences of the United States of America*, **101**, 5830.
23 Lipowsky, G. and Helmchen, G. (2004) *Chemical Communications*, 116.
24 Alexakis, A. and Polet, D. (2004) *Organic Letters*, **6**, 3529.

25 For experimental details see: (a)Polet, D., Alexakis, A., Tissot-Croset, K., Corminboeuf, C. and Ditrich, K. (2006) *Chemistry – A European Journal*, **12**, 3596;
(b) Polet, D. and Alexakis, A. (2005) *Organic Letters*, **7**, 1621, Supporting Information.
26 Spiess, S., Welter, C., Franck, G., Taquet, J.-P. and Helmchen, G. (2008) *Angewandte Chemie – International Edition*.
27 Shu, C., Leitner, A. and Hartwig, J.F. (2004) *Angewandte Chemie – International Edition*, **43**, 4797.
28 (a) Weihofen, R., Tverskoy, O. and Helmchen, G. (2006) *Angewandte Chemie – International Edition*, **45**, 5546;
(b) Spiess, S., Berthold, C., Weihofen, R. and Helmchen, G. (2007) *Organic and Biomolecular Chemistry*, **5**, 2357;
(c) Singh, O.V. and Han, H. (2007) *Tetrahedron Letters*, **48**, 7094.
29 Gnamm, C., Förster, S., Miller, N., Brödner, K. and Helmchen, G. (2007) *Synlett*, 790.
30 Lipowsky, G., Miller, N. and Helmchen, G. (2004) *Angewandte Chemie – International Edition*, **43**, 4595.
31 (a) Tsuji, J., Minami, I. and Shimizu, I. (1984) *Tetrahedron Letters*, **25**, 5157;
(b) Plietker, B. (2006) *Angewandte Chemie – International Edition*, **45**, 1469.
32 Marković, D. and Hartwig, J.F. (2007) *Journal of the American Chemical Society*, **129**, 11680.
33 Matsumura, K., Saito, T., Sayo, N., Kumobayashi, H. and Takaya, H. (1994) European Patent 0614902 A1.
34 van Zijl, A.W., Arnold, L.A., Minnaard, A.J. and Feringa, B.L. (2004) *Advanced Synthesis Catalysis*, **346**, 413, Supporting Information.
35 Tissot-Croset, K., Polet, D., Gille, S., Hawner, C. and Alexakis, A. (2004) *Synthesis*, 2586.
36 Streiff, S., Welter, C., Schelwies, M., Lipowsky, G., Miller, N. and Helmchen, G. (2005) *Chemical Communications*, 2957.
37 He, H., Zheng, X.-J., Li, Y., Dai, L.-X. and You, S.-L. (2007) *Organic Letters*, **9**, 4339.
38 (a) Fuji, K., Kinoshita, N., Tanaka, K. and Kawabata, T. (1999) *Chemical Communications*, 2289;
(b) Kinoshita, N., Marx, K.H., Tanaka, K., Tsubaki, K., Kawabata, T., Yoshikai, N., Nakamura, E. and Fuji, K. (2004) *The Journal of Organic Chemistry*, **69**, 7960.
39 Kanayama, T., Yoshida, K., Miyabe, H. and Takemoto, Y. (2003) *Angewandte Chemie – International Edition*, **42**, 2054.
40 (a) Miyabe, H., Matsumura, A., Moriyama, K. and Takemoto, Y. (2004) *Organic Letters*, **6**, 4631;
(b) Lavastre, O. and Morken, J.P. (1999) *Angewandte Chemie – International Edition*, **38**, 3163.
41 (a) Nemoto, T., Sakamoto, T., Fukuyama, T. and Hamada, Y. (2007) *Tetrahedron Letters*, **48**, 4977;
(b) Nemoto, T., Sakamoto, T., Matsumoto, T. and Hamada, Y. (2006) *Tetrahedron Letters*, **47**, 8737.
42 Fischer, C., Defieber, C., Suzuki, T. and Carreira, E.M. (2004) *Journal of the American Chemical Society*, **126**, 1628.
43 Peng, F. and Hall, D.G. (2007) *Tetrahedron Letters*, **48**, 3305.
44 Schelwies, M., Dübon, P. and Helmchen, G. (2006) *Angewandte Chemie – International Edition*, **45**, 2466.
45 Dahnz, A. and Helmchen, G. (2006) *Synlett*, 697.
46 Kanayama, T., Yoshida, K., Miyabe, H., Kimachi, T. and Takemoto, Y. (2003) *The Journal of Organic Chemistry*, **68**, 6197.
47 Reviewss: (a) Braun, M. and Meier, T. (2006) *Synlett*, 661;
(b) Kazmaier, U. (2003) *Current Organic Chemistry*, **7**, 317.
48 Tunge, J.A. and Burger, E.C. (2005) *European Journal of Organic Chemistry*, 1715.
49 Graening, T. and Hartwig, J.F. (2005) *Journal of the American Chemical Society*, **127**, 17192.
50 Weix, D.J. and Hartwig, J.F. (2007) *Journal of the American Chemical Society*, **129**, 7720.
51 (a) Alexakis, A., Malan, C., Lea, L., Tissot-Crosset, K., Polet, D. and Falciola, C. (2006) *Chimia*, **60**, 124;
(b) López, F., Minaard, A.J. and Feringa, B.L. (2007) *Accounts of Chemical Research*, **40**, 179.

52 Alexakis, A., Hajjaji, El, S., Polet, D. and Rathgeb, X. (2007) *Organic Letters*, **9**, 3393.

53 Leitner, A., Shu, C. and Hartwig, J.F. (2005) *Organic Letters*, **7**, 1093.

54 Lee, J.H., Shin, S., Kang, J. and Lee, S. (2007) *The Journal of Organic Chemistry*, **72**, 7443.

55 Pouy, M.J., Leitner, A., Weix, D.J., Ueno, S. and Hartwig, J.F. (2007) *Organic Letters*, **9**, 3949.

56 Yamashita, Y., Gopalarathnam, A. and Hartwig, J.F. (2007) *Journal of the American Chemical Society*, **129**, 7508.

57 Defieber, C., Ariger, M.A., Moriel, P. and Carreira, E.M. (2007) *Angewandte Chemie – International Edition*, **46**, 3139.

58 Gnamm, C., Franck, G., Miller, N., Stork, T., Brödner, K. and Helmchen, G. *Synthesis* (accepted for publication).

59 Weihofen, R., Dahnz, A., Tverskoy, O. and Helmchen, G. (2005) *Chemical Communications*, 3541.

60 Singh, O.V. and Han, H. (2007) *Journal of the American Chemical Society*, **129**, 774.

61 Singh, O.V. and Han, H. (2007) *Organic Letters*, **10**, 4801.

62 Very recently, we found that these substitutions are accompanied by the formation of urethanes, which are produced via reactions with CO_2; Gärtner, M. (2007) Enantioselektive Synthese von Xenovenin durch Iridium-katalysierte allylische Aminierung. Diploma Thesis, Heidelberg.

63 Miyabe, H. and Takemoto, Y. (2005) *Synlett*, 1641.

64 López, F., Ohmura, T. and Hartwig, J.F. (2003) *Journal of the American Chemical Society*, **125**, 3426.

65 Kimura, M. and Uozumi, Y. (2007) *The Journal of Organic Chemistry*, **72**, 707.

66 Roberts, J.P. and Lee, C. (2005) *Organic Letters*, **7**, 2679.

67 Evans, P.A. and Leahy, D.K. (2002) *Journal of the American Chemical Society*, **124**, 7882.

68 Shu, C. and Hartwig, J.F. (2004) *Angewandte Chemie – International Edition*, **43**, 4794.

69 Miyabe, H., Yoshida, K. Takemoto, M. and Yamauchi, Y. (2005) *The Journal of Organic Chemistry*, **70**, 2148.

70 Lyothier, I., Defieber, C. and Carreira, E.M. (2006) *Angewandte Chemie – International Edition*, **45**, 6204.

71 Grubbs, R.H. (ed.) (2003) *Handbook of Metathesis*, Wiley-VCH Verlag GmbH, New York.

72 Review: Leahy, D.K., Evans, P.A. and Evans, P.A. (2005) *Modern Rhodium-Catalyzed Organic Reactions*, John Wiley & Sons, Inc., New York, p. 191.

73 (a) Evans, P.A. (2003) *Chemstracts*, **16**, 567; (b) Evans, P.A. and Robinson, J.E. (1999) *Organic Letters*, **1**, 1929; (c) Evans, P.A., Leahy, D.K., Andrews, W.J. and Uraguchi, D. (2004) *Angewandte Chemie – International Edition*, **43**, 4788.

74 Trost, B.M. and Jiang, C. (2003) *Organic Letters*, **5**, 1563.

75 Alexakis, A. and Polet, D. (2002) *Organic Letters*, **4**, 4147.

76 Welter, C., Moreno, R.M., Streiff, S. and Helmchen, G. (2005) *Organic and Biomolecular Chemistry*, **3**, 3266.

77 Böhrsch, V. and Blechert, S. (2006) *Chemical Communications*, 1968.

10
Iridium-Catalyzed Coupling Reactions

Yasutaka Ishii, Yasushi Obora and Satoshi Sakaguchi

10.1
Introduction

Organoiridium complexes have been frequently used as model compounds to understand the elementary steps in transition-metal-catalyzed reactions [1]. On the other hand, since [Ir(cod)(PCy$_3$)(py)]PF$_6$ was first used as a catalyst for the hydrogenation of olefins by Crabtree and coworkers in 1977 [2], Ir(I) complexes have been widely employed as efficient catalysts for alkene hydrogenation. In 1983, Stork and Kahne reported the control of stereochemistry in the catalytic hydrogenation of alkenes directing by the hydroxy function [3], while a few years later Pfaltz and colleagues disclosed details of the iridium complex-catalyzed enantioselective hydrogenations of tri- and tetrasubstituted simple alkenes [4]. Recent developments in iridium chemistry have shown that Ir complexes can be used as efficient catalysts for hydrogen transfer and related reactions [5], as well as for Claisen rearrangements [6]. In addition, the iridium-catalyzed hydrogenation of alkynes [7], hydrosilylation of alkenes [8], hydroboration of alkenes and alkynes [9], isomerization [10] and cycloisomerization [11] have also been reported.

This chapter reviews an overview of the iridium-complex-catalyzed crosscoupling reactions to form carbon–carbon and carbon–heteroatom bonds [12].

10.2
Iridium-Catalyzed Dimerization and Cyclotrimerization of Alkynes

The dimerization of alkynes is a useful method for forming compounds such as enynes from simple alkynes [13]. The iridium-catalyzed dimerization of 1-alkynes was first reported by Crabtree, and afforded *(Z)*-head-to-head enynes using [Ir(biph)(PMe$_3$)Cl] (biph = biphenyl-2,2′-diyl) as a catalyst [14]. Thereafter, an iridium complex generated *in situ* from [Ir(cod)Cl]$_2$ and a phosphine ligand catalyzed the dimerization of 1-alkynes **1** to give *(E)*-head-to-head enyne **2**, *(Z)*-head-to-head enyne **3**, or 1,2,3-butatriene derivatives **4** in the presence of triethylamine

Iridium Complexes in Organic Synthesis. Edited by Luis A. Oro and Carmen Claver
Copyright © 2009 WILEY-VCH Verlag GmbH & Co. KGaA, Weinheim
ISBN: 978-3-527-31996-1

(Equation 10.1). The selectivity was markedly influenced by the ligand used; for example, the PPh_3 ligand gave the *(E)*-enyne while the $P(n\text{-}Pr)_3$ ligand gave the *(Z)*-enyne [15].

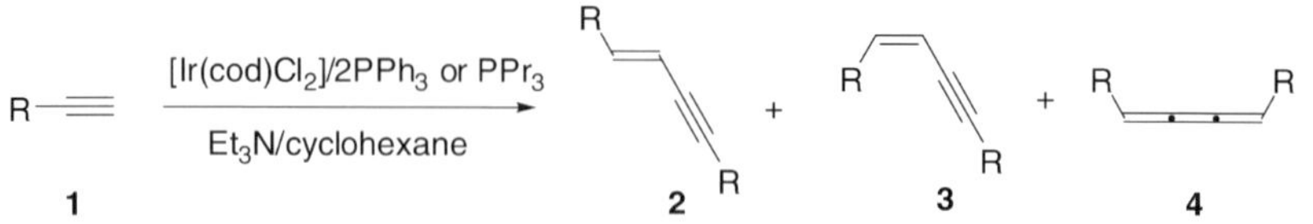
[Ir(cod)Cl2]/2PPh3 or PPr3
Et3N/cyclohexane
R
1
2
3
4

(10.1)

The cross-dimerization of various electron-rich 1-alkynes **5** with electron-deficient internal alkynes such as methyl phenylpropiolate **6** was promoted by an $[IrCl(cod)]_2$ combined with bidentate phosphine ligands such as *(rac)*-BINAP (Equation 10.2) [16]. This reaction produces a 1 : 1 adduct **7** in high regioselectivity and stereoselectivity.

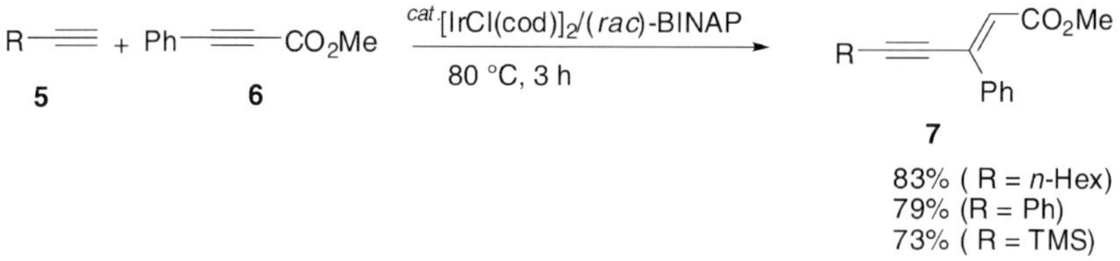
cat.[IrCl(cod)]2/(rac)-BINAP
80 °C, 3 h
R
Ph
CO2Me
5
6
7
83% (R = n-Hex)
79% (R = Ph)
73% (R = TMS)

(10.2)

A plausible reaction pathway was suggested by an experiment using labels, where the reaction proceeded via the pathway shown in Scheme 10.1. First, the oxidative addition of terminal alkyne **8** to an iridium complex leads to an alkynyl–

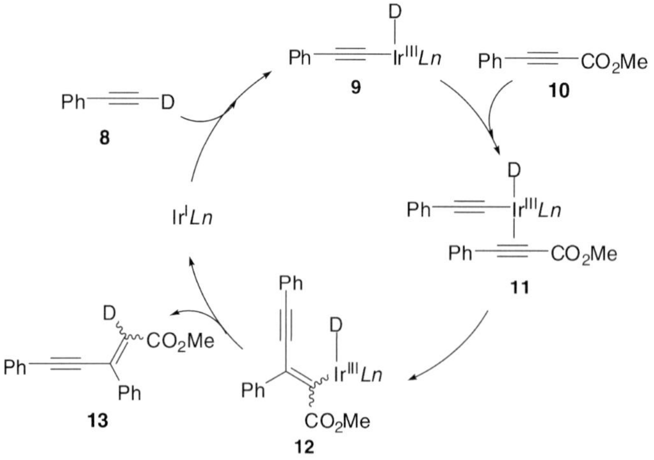
Ph
D
CO2Me
IrIIILn
IrILn
8
9
10
11
12
13

Scheme 10.1

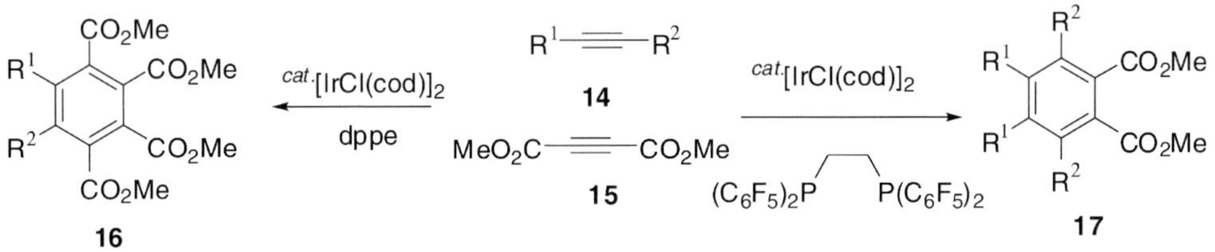

10.3 Iridium-Catalyzed, Three-Component Coupling Reactions of Aldehydes, Amines and Alkynes | 253

Scheme 10.2

iridium complex **9**, after which the coordination of an internal alkyne **10** to **9**, followed by insertion, gives a vinyliridium complex **12**. Reductive elimination of **12** produces the enyne **13**, and the complex **9** may be regenerated.

An iridium complex also catalyzes the cyclotrimerization of alkynes. The regioselective cyclotrimerization of phenylacetylene was induced by an [IrH(cod)(dppe)] catalyst to give 1,2,4-triphenylbenzene in quantitative yield [17]. The cross-cyclotrimerization of terminal and internal alkynes with dimethyl acetylene dicarboxylate (DMAD) was achieved in the presence of [IrCl(cod)]$_2$ combined with diphosphine ligands to give (2+2+2) adducts (Scheme 10.2) [4, 6]. The ligand used in the cyclotrimerization had a significant effect on the reaction. Thus, when dppe was used as the ligand, a 2:1 adduct of DMAD and alkyne was obtained, whereas when 1,2-bis(dipentafluorophenylphosphino)ethane was used as the ligand a 1:2 adduct of DMAD and alkyne was obtained [18].

The [Ir(cod)Cl]$_2$/diphosphine is an efficient catalyst for the (2+2+2) cycloaddition of α,ω-diynes **18** with alkyne **19** to give indane **20** and tetraline derivatives (Equation 10.3) [19].

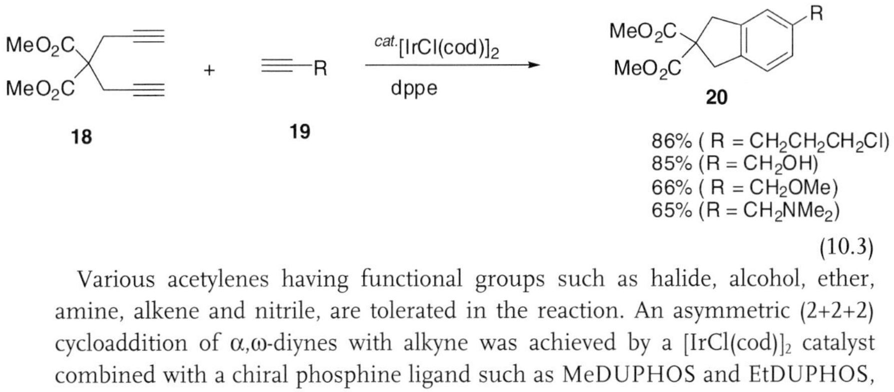

86% (R = CH$_2$CH$_2$CH$_2$Cl)
85% (R = CH$_2$OH)
66% (R = CH$_2$OMe)
65% (R = CH$_2$NMe$_2$)

(10.3)

Various acetylenes having functional groups such as halide, alcohol, ether, amine, alkene and nitrile, are tolerated in the reaction. An asymmetric (2+2+2) cycloaddition of α,ω-diynes with alkyne was achieved by a [IrCl(cod)]$_2$ catalyst combined with a chiral phosphine ligand such as MeDUPHOS and EtDUPHOS, and gave axially chiral aromatic compounds [20].

10.3
Iridium-Catalyzed, Three-Component Coupling Reactions of Aldehydes, Amines and Alkynes

[IrCl(cod)]$_2$ leads to three-component coupling products of aldehydes **21**, primary amines **22** and 1-alkynes **23** to afford **24** via an activation of the C—H bond adjacent to the nitrogen atom of imines (Equation 10.4) [21]. For example, the reaction of butyraldehyde, butylamine and 1-octyne in the presence of [IrCl(cod)]$_2$ gave N-butylidene(2-hexyl-1-propylallyl)amine (**24**) in 72% yield. The iridium-catalyzed reaction was not affected by the water generated during imine formation from amine and aldehyde.

$$R^1CHO + R^2NH_2 + HC\equiv CR^3 \xrightarrow[\text{THF, 60 °C, 15 h}]{cat. [IrCl(cod)]_2} \text{24} \quad (10.4)$$

21   **22**   **23**   **24**

72% (R^1, R^2, R^3 = ^{n}Pr, ^{n}Pr, ^{n}Hex)
74% (R^1, R^2, R = ^{n}Pr, ^{n}Pr, ^{n}Bu)
73% (R^1, R^2, R = ^{i}Pr, ^{n}Pr, ^{n}Hex)
62% (R^1, R^2, R = ^{t}Bu, ^{n}Pr, ^{n}Hex)

The coupling product consisted of a single geometrical isomer, and the reaction of N-butylidene-1,1-dideuteriobutylamine (**25**) under the same catalytic conditions gave a coupling product (**27**) in which a deuterium is incorporated to the terminal triple bond of the imine (Equation 10.5). Therefore, the reaction would proceed through the pathway shown in Scheme 10.3. Initially, an Ir^I complex **28** coordinates to imine generated *in situ* from aldehyde and amine. Oxidative addition of the C—H bond adjacent to the nitrogen atom in the imine to the Ir^I complex would afford an Ir^{III} complex, **30**. The coordination of alkyne, followed by insertion to the complex **30**, may give an iridium complex **32**, on which subsequent reductive elimination forms the coupling product.

$$\text{25} + \text{26} \xrightarrow[\text{THF, 60 °C}]{cat. [IrCl(cod)]_2} \text{27}$$

(10.5)

When the secondary amine **33** was used instead of a primary amine, a different type of three-component coupling reaction took place with aldehyde **34** and 1-alkynes **35** to afford the corresponding allylamines (**36** and **37**) [22]. In this reaction, Ir–hydride generated by amine (**33**) with Ir^I, would be a key intermediate. The reaction may proceed by addition of the Ir–hydride to the enamine derived from amine **33** and alkyne **34**, followed by insertion of aldehyde and dehydration to give the coupling product (**36** and **37**).

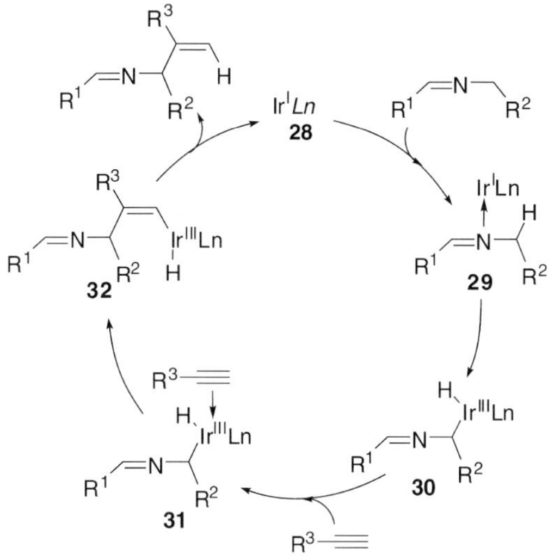

10.3 Iridium-Catalyzed, Three-Component Coupling Reactions of Aldehydes, Amines and Alkynes

Scheme 10.3

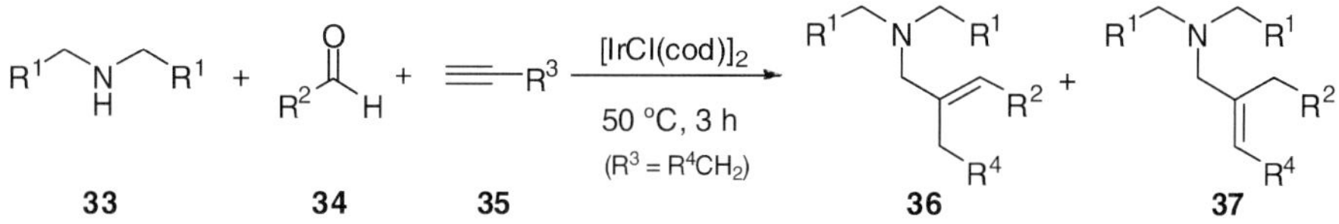

Furthermore, when trimethylsilylacetylene **40** was used as an alkyne in the [IrCl(cod)]$_2$-catalyzed reaction, propargylic amines (where the alkyne was added to the double bond of imine) were obtained (Equation 10.7) [21, 23]. It is probable that the reaction proceeds through oxidative addition of the terminal C—H bond of alkyne to the Ir^I complex, followed by the insertion of imine to the resulting Ir–H complex. The crosscoupling reaction of trimethylsilyl (TMS)-acetylene with aldimines took place by [IrCl(cod)]$_2$, leading to the corresponding adducts (Equation 10.8) [24].

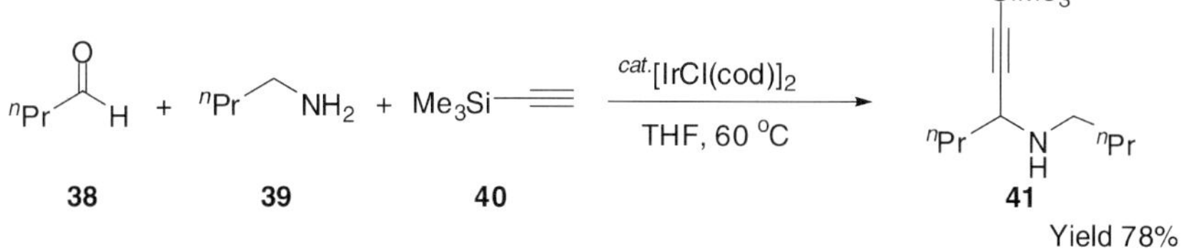

Equation (10.6): 92% (78:22) (R^1 = nPr, R^2 = nPr, R^3 = nHex); 78% (74:26) (R^1 = Et, R^2 = Et, R^3 = nHex)

Equation (10.7): Yield 78%

$$\underset{42}{\underset{R}{\overset{N^{\cdot R^2}}{\bigwedge}}_H} + \underset{43}{Me_3Si\text{---}\!\!\equiv\!\!\text{---}} \xrightarrow[THF,\ 23\ °C]{cat.[IrCl(cod)]_2} \underset{44}{\underset{R^1}{\overset{SiMe_3}{\underset{\|}{\underset{\|}{\bigwedge}}}\!\!\!\underset{H}{\overset{}{N}}\!\!\!R^2}}$$

76% (R^1 = Ph, R^2 = CH_2Ph)
85% (R^1 = Ph, R^2 = 4-MeO-C_6H_5)

(10.8)

In addition, the three-component coupling reaction of aliphatic aldehydes **45**, aliphatic amines **46** and ethyl diazoacetate (EDA) **47** was achieved under the influence of an [IrCl(cod)]$_2$ catalyst to afford the aziridine derivatives **48**, under mild conditions (Equation 10.9) [25].

$$\underset{45}{R^1\overset{O}{\underset{}{\bigcup}}H} + \underset{46}{R^2\text{-}NH_2} + \underset{\underset{47}{EDA}}{N_2CHCOOEt} \xrightarrow[THF,\ -10\ °C,\ 3\ h]{cat.[IrCl(cod)]_2\ or\ IrCl_3} \underset{48}{\underset{R^1}{\overset{R^2}{\underset{}{\bigtriangleup}}}\!\!COOEt}$$

71% R^1, R^2 = ^nPr, ^nBu
76% R^1, R^2 = ^nPr, ^iBu
83% R^1, R^2 = ^nPr, ^{sec}Bu
83% R^1, R^2 = ^nPr, ^tBu

(10.9)

The present Ir-catalyzed, three-component coupling reaction was thought to progress via the formation of an imine, which then reacts with EDA. The use of a bulky amine such as *tert*-butylamine gave the corresponding *cis*-aziridine in high stereoselectivity. Thus, the reaction must proceed through coordination of the Ir complex to an imine, followed by a nucleophilic attack of EDA from the direction to reduce the steric repulsion between the ester moiety of the incoming EDA and *tert*-butyl group of the imine to form the corresponding *cis*-product.

10.4
Head-to-Tail Dimerization of Acrylates

The dimerization of functional alkenes such as acrylates and acrylonitrile represents an attractive route to obtain bifunctional compounds such as dicarboxylates and diamine, respectively. The head-to-tail dimerization of acrylates and vinyl ketones was catalyzed by an iridium hydride complex generated *in situ* from [IrCl(cod)]$_2$ and alcohols in the presence of P(OMe)$_3$ and Na$_2$CO$_3$ [26]. The reaction of butyl acrylate **51** in the presence of [IrCl(cod)]$_2$ in 1-butanol led to a head-to-tail dimer, 2-methyl-2-pentenedioic acid dibutyl ester (53%), along with butyl propionate (35%) which is formed by hydrogen transfer from 1-butanol. In order to avoid

10.4 Head-to-Tail Dimerization of Acrylates

the formation of butyl propionate, 1-butanol was removed from the catalytic solution after treatment of a mixture of [IrCl(cod)$_2$], P(OMe)$_3$ and Na$_2$CO$_3$ in 1-butanol at 100 °C for 1 h, after which butyl acrylate **51** in toluene was added and reacted at 100 °C for 5 h to give a dimer **52** in 86% yield (Equation 10.10). Under these conditions, methyl vinyl ketone **53** was also dimerized to form a head-to-tail dimer, 3-methy-3-heptene-2,6-dione **54** in 79% yield (Equation 10.10). The present reaction was promoted in methanol, ethanol and 2-butanol, whereas no reaction occurred in *tert*-butyl alcohol (which has no α-hydrogen). These results suggest that an iridium–hydride complex generated *in situ* from [IrCl(cod)]$_2$ and alcohols in the presence of P(OMe)$_3$ and Na$_2$CO$_3$ promotes the head-to-tail dimerization of acrylates.

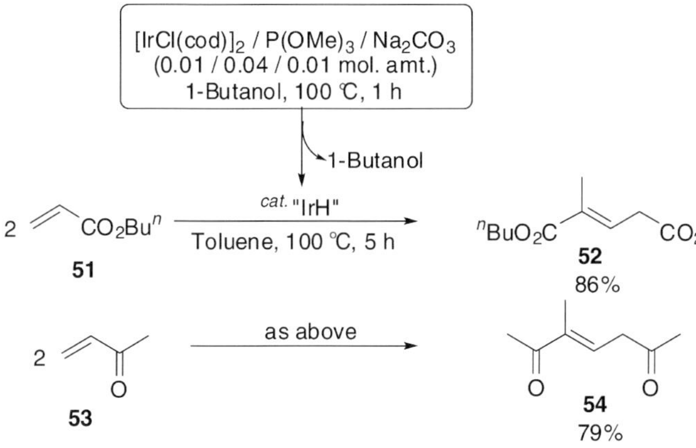

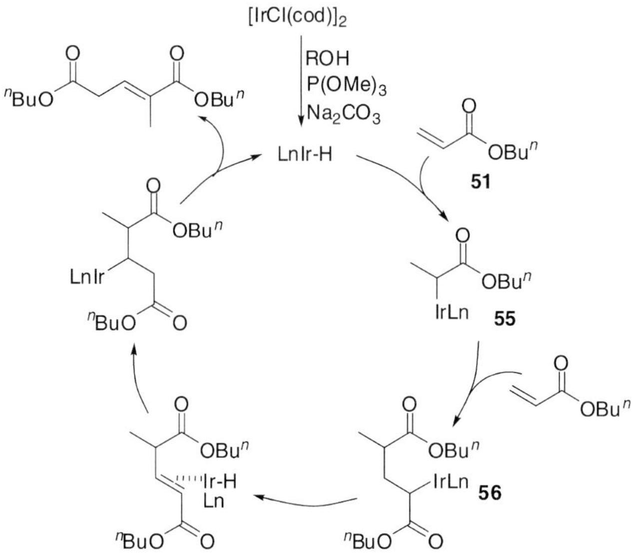

Scheme 10.4

The reaction may proceed as follows (see Scheme 10.4). An acrylate **51** coordinates to the iridium–hydride complex generated *in situ*, and then inserts into the Ir–H bond to form a σ–Ir complex **55**; the coordination and insertion of another acrylate to **55** leads to an iridium complex, **56**. A β-hydride elimination of the iridium–hydride from the intermediate **56**, followed by isomerization of the double bond to a more stable internal alkene, results in a head-to-tail dimer.

10.5
A Novel Synthesis of Vinyl Ethers via an Unusual Exchange Reaction

Vinyl ethers are important raw materials in the production of glutaraldehyde, as well as of vinyl polymer materials which contain oxygen and are expected to degrade easily in Nature. The [IrCl(cod)]$_2$ catalyzes an efficient exchange reaction between vinyl acetate **57** and alcohols or phenols **58**, leading to the corresponding vinyl ethers **59** (Equation 10.11) [27]. Usually, the acid-catalyzed exchange reaction between alcohols and vinyl acetate results in alkyl acetates **60**, and also to vinyl alcohol **61** which is readily isomerized to acetaldehyde **62**.

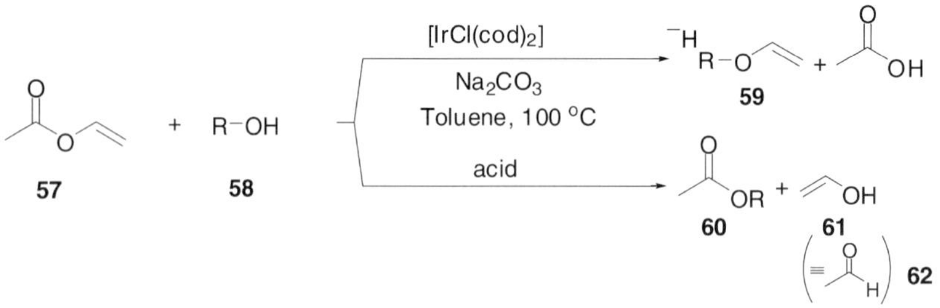

(10.11)

Thus, a wide variety of vinyl ethers could be synthesized using this method (Scheme 10.5). This catalytic vinylation system was found to be applicable to the synthesis of vinyl ethers from secondary and tertiary alcohols. No deuterium was introduced into the resulting phenyl vinyl ether when phenol-*d* was allowed to

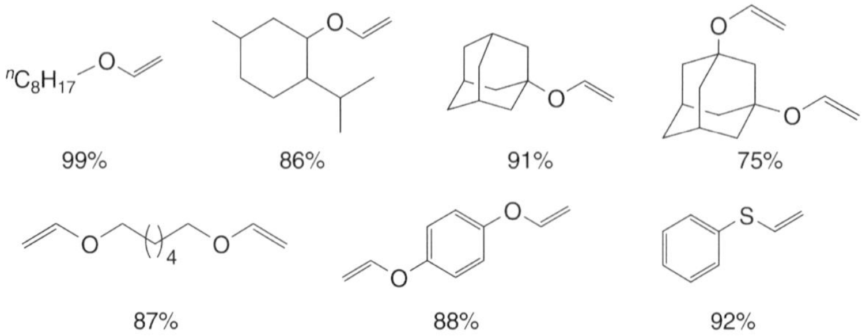

Scheme 10.5

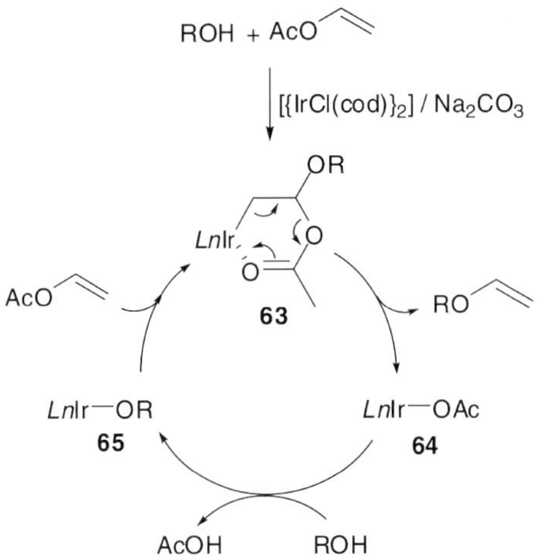

10.5 A Novel Synthesis of Vinyl Ethers via an Unusual Exchange Reaction

Scheme 10.6

react with vinyl acetate under these conditions. This may suggest that the following reaction proceeds through an intermediate **63**, which resulted from the reaction of [IrCl(cod)]$_2$ with vinyl acetate and alcohol under the influence of Na$_2$CO$_3$ (Scheme 10.6). The release of an alkyl vinyl ether from the intermediate **63** gives rise to an iridium acetoxy complex **64**, which then reacts with alcohol, leading to an iridium alkoxy complex **65**; coordination of the vinyl acetate to complex **65**, followed by insertion, then regenerates **63**.

The transfer vinylation between carboxylic acids **64** and vinyl acetate **65** was also achieved under the influence of [IrCl(cod)]$_2$ and NaOAc; in this way a variety of carboxylic acids were converted into the corresponding vinyl esters **66**, with excellent yields (Equation 10.12 and Scheme 10.7) [28].

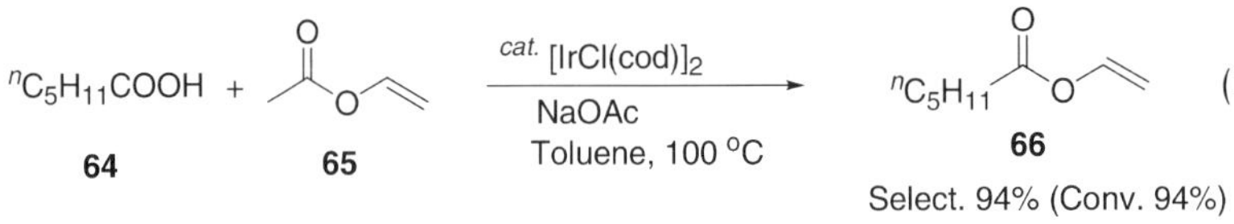

$$^nC_5H_{11}COOH + \text{AcO-CH=CH}_2 \xrightarrow[\text{Toluene, 100 °C}]{cat. \text{ [IrCl(cod)]}_2 \text{ / NaOAc}} {}^nC_5H_{11}\text{COO-CH=CH}_2 \quad (10.12)$$

64   **65**   **66**

Select. 94% (Conv. 94%)

^{n}Bu(Et)CHCOOCH=CH$_2$ CH$_3$CH=CHCOOCH=CH$_2$ PhCOOCH=CH$_2$ ^{t}BuCOOCH=CH$_2$ PhCH=CHCOOCH=CH$_2$

94% (89%)   88% (94%)   91% (77%)   89% (88%)   98% (80%)

Select. (Conv.)

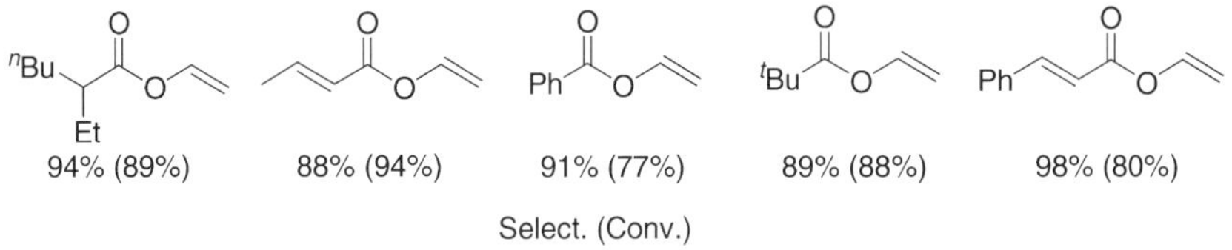

Scheme 10.7

Furthermore, the reaction of allyl alcohols **67** and vinyl or isopropenyl acetates **68** was reported to afford γ,δ-unsaturated carbonyl compounds **70** (Equation 10.13) [29].

10 Iridium-Catalyzed Coupling Reactions

$$\text{Ph} \diagup \text{OH} \ (\mathbf{67}) + \text{AcO-vinyl} \ (\mathbf{68}) \xrightarrow[\text{Mesitylene, 100 °C, 3 h}]{\text{cat. [IrCl(cod)]}_2 / \text{Na}_2\text{CO}_3} [\ \mathbf{69}\ ] \xrightarrow{140\ °\text{C}} \mathbf{70}\ (83\%)$$

(10.13)

The reaction was achieved through transfer vinylation of **67** with **68** by action of the [IrCl(cod)]$_2$ complex to afford allyl homoallyl ethers **69**, followed by a Claisen rearrangement of the ether **70**. The Claisen rearrangement of allyl homoallyl ethers to γ,δ-unsaturated aldehydes has been reported previously [6].

This method can be extended for the synthesis of allyl alkyl ethers from alcohols with allyl acetate. Thus, the iridium cationic complex [Ir(cod)$_2$]$^+$BF$_4^-$, catalyzes the allylation of alcohols **71** with allyl acetate **72** to afford allyl ethers **73** (Equation 10.14) [30].

$$\text{R-OH}\ (\mathbf{71}) + \text{AcO-allyl}\ (\mathbf{72}) \xrightarrow[\text{toluene, 100 °C}]{\text{cat. [Ir(cod)}_2\text{]}^+\text{BF}_4^-} \text{R-O-allyl}\ (\mathbf{73})$$

98% (R = nBu)
90% (R = 4-CH$_3$C$_6$H$_4$)
93% (R = 4-CH$_3$OC$_6$H$_4$)

(10.14)

The reaction would proceed through formation of the π-allyl–iridium complex, followed by a nucleophilic attack of the alcohol.

The iridium-catalyzed transformation between carboxylic acid and vinyl acetate [28] or allyl acetate [30] was also promoted to afford vinyl or allyl carboxylates in good yields.

The [Ir(cod)$_2$]$^+$BF$_4^-$ complex catalyzed the reaction of alkyl and aromatic amines with allyl acetate, leading to the corresponding allyl amines in fair to good yields [28] (see also Section 10.6).

10.6
Iridium-Catalyzed Allylic Substitution

For further details of this reaction, the reader is referred to Chapter 9. The catalytic allylation with nucleophiles via the formation of π-allyl metal intermediates has produced synthetically useful compounds, with the palladium-catalyzed reactions being known as Tsuji–Trost reactions [31]. The reactivity of π-allyl-iridium complexes has been widely studied [32]; for example, in 1997, Takeuchi identified a [IrCl(cod)]$_2$ catalyst which, when combined with P(OPh)$_3$, promoted the allylic alkylation of allylic esters **74** with sodium diethyl malonate **75** to give branched

10.6 Iridium-Catalyzed Allylic Substitution

alkylated compounds (**76** and **77**) with high regioselectivity (Equation 10.15) [33]. Unlike the Pd-catalyzed allylic alkylation [31], the iridium-catalyzed reaction gave a branched adduct **72**, and the triphenylphosphite ligand was found to be essential for high branched selectivity. In the iridium-catalyzed reaction, both linear and branched allyl acetates reacted with **75** to afford a branched product selectively [33].

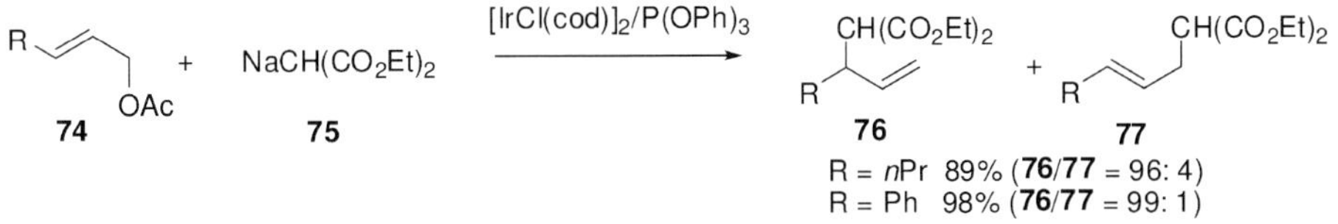

$$R\text{-CH=CH-CH}_2\text{OAc} + \text{NaCH(CO}_2\text{Et)}_2 \xrightarrow{[\text{IrCl(cod)}]_2/\text{P(OPh)}_3} \text{76} + \text{77}$$

74, **75**, **76**, **77**

R = nPr 89% (**76/77** = 96:4)
R = Ph 98% (**76/77** = 99:1)

(10.15)

The asymmetric allylic alkylation of allylic esters was achieved with a high enantioselectivity (up to 95% enantiomeric excess; ee) by using $[\text{IrCl(cod)}]_2$ in the presence of chiral phosphinooxazolines **78** [34], phosphoramidites **79** [35] or phosphates **80** [36] as ligands (Equation 10.16). Furthermore, an asymmetric allylic alkylation of allyl phosphates with (diphenylimino)glycinates afforded chiral β-substituted α-amino acids in high enantioselectivity (up to 97% ee) [37]. In this reaction system, chiral bidentate phosphates bearing an (ethylthio)ethyl group promoted the allylic alkylation.

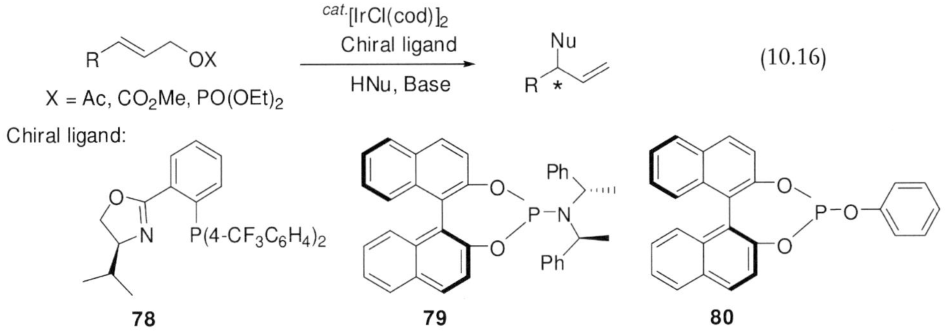

$$R\text{-CH=CH-CH}_2\text{OX} \xrightarrow[\text{HNu, Base}]{cat.[\text{IrCl(cod)}]_2 \text{ Chiral ligand}} R^*\text{-CH(Nu)-CH=CH}_2$$

X = Ac, CO$_2$Me, PO(OEt)$_2$

(10.16)

Chiral ligand:

78: phosphinooxazoline with P(4-CF$_3$C$_6$H$_4$)$_2$
79: BINOL-phosphoramidite with N(CHPhMe)$_2$
80: BINOL-phosphate with OPh

Allyl carbonate **81** reacts with primary and secondary amines **82** to afford branched allylic amination products **83** under the influence of a $[\text{IrCl(cod)}]_2/\text{P(OPh)}_3$ catalyst (Equation 10.17) [38]. In this reaction, alcohols such as ethanol and methanol proved to be the best solvents, while both intermolecular and intramolecular enantioselective allylic aminations were reported by using chiral phosphoramidites as ligands [39]. The allylic amination of (Z)-2-nonenyl carbonate **85** with **82** in the presence of $[\text{IrCl(cod)}]_2$ and P(OPh)$_3$ afforded Z-linear allylic amines (Equation 10.18) [38]. In this reaction, the Z-geometry of **85** was completely retained.

10.7
Alkylation of Ketones with Alcohols

The α-alkylation of enolates derived from ketones with alkyl halides is a very important and frequently used method for forming new carbon–carbon bonds in organic synthesis [40]. Yet, if the α-alkylation of enolates derived from ketones with alkyl halides can be replaced by the direct reaction of ketones with alcohols, this method would provide a very useful waste-free, 'green' route to α-alkylation, producing no side products other than water.

[IrCl(cod)]$_2$, in the presence of PPh$_3$ and KOH, catalyzed the α-alkylation of ketones with alcohols [41]. As an example, the reaction of 2-octanone **87** with 1-butanol **88** was catalyzed by the iridium complex to give 6-dodecanone **89** in 80% yield (Equation 10.19). The alkylation proceeded with complete regioselectivity at the less-hindered side of 2-octanone, and the reaction was promoted by a catalytic quantity of KOH (10 mol%) in the absence of both a hydrogen acceptor and a solvent.

This method provides a very convenient route to aliphatic ketones, to which a carbonyl function can be introduced in the desired position by selecting the ketones and alcohols employed.

The reaction is thought to proceed via the following pathway:

(i) hydrogen transfer from alcohol **88** to an Ir complex, giving aldehyde **90** and an Ir–hydride complex
(ii) a base-catalyzed aldol condensation between the resulting aldehyde **90** and ketone **to** give α,β-unsaturated ketone **91**
(iii) selective hydrogenation of **91** by an Ir–hydride complex generated during the course of the reaction, to form the α-alkylated ketone **92** (Scheme 10.8).

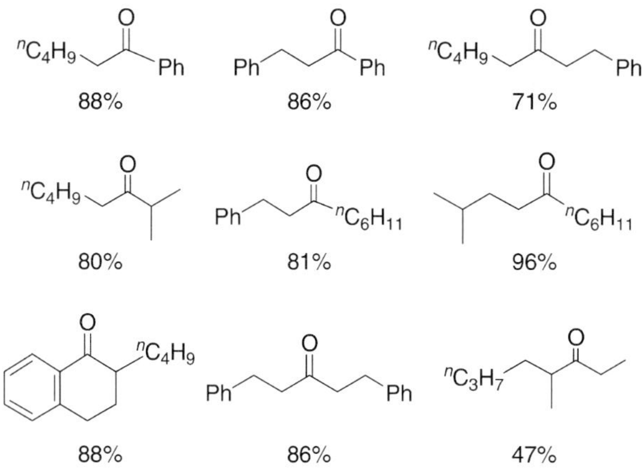

10.7 Alkylation of Ketones with Alcohols

[Structures with yields: 88%, 86%, 71%, 80%, 81%, 96%, 88%, 86%, 47%]

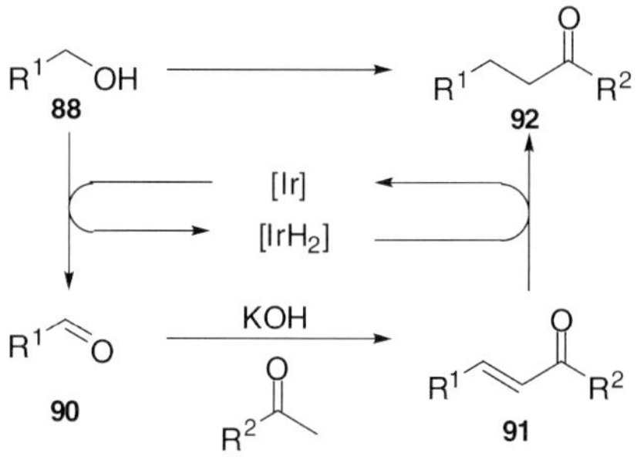

[Scheme 10.8: catalytic cycle showing R^{1}CH$_2$OH (88) → R^{1}CHO (90) via [Ir]/[IrH$_2$]; condensation with R^{2}COCH$_3$ and KOH gives enone 91; reduction gives ketone 92]

Scheme 10.8

The iridium catalytic system can also be applied to the α-alkylation of active methylene compounds. The alkylation of cyanoacetates **93** with primary alcohols **94** was achieved by using [IrCl(coe)$_2$]$_2$ and PPh$_3$ to afford saturated α-alkylated products **95** (Equation 10.20) [42]. Here, the alkylation reaction was efficiently accomplished, without the need for any base.

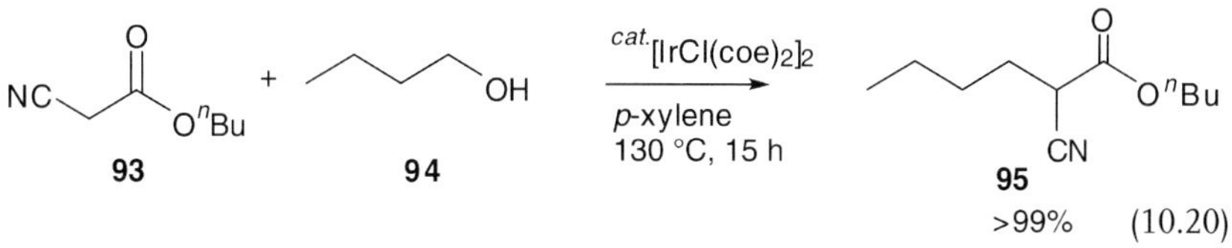

[Equation 10.20: NC-CH$_2$-CO$_2^n$Bu (93) + n-propanol (94) → α-butylated cyanoacetate (95), >99%, with cat. [IrCl(coe)$_2$]$_2$, p-xylene, 130 °C, 15 h]

(10.20)

The [Cp*IrCl$_2$]$_2$-catalyzed alkylation of arylacetonitrile **96** with primary alcohols has also been reported (Equation 10.21) [43]. Here, KOH proved to be the most successful base and the reaction proceeded in the absence of any solvent.

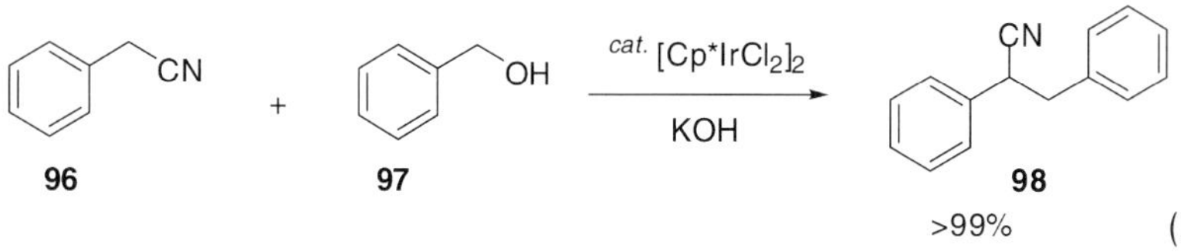

[Equation 10.21: PhCH$_2$CN (96) + PhCH$_2$OH (97) → Ph-CH(CN)-CH$_2$-Ph (98), >99%, with cat. [Cp*IrCl$_2$]$_2$, KOH]

(10.21)

The [Cp*IrCl$_2$]$_2$ and [IrCl(cod)]$_2$ catalysts have also been used in the alkylation of barbituric acid [43] and nitroalkanes [44] with primary alcohols.

The [IrCl(cod)]$_2$ and [Cp*IrCl$_2$]$_2$ complexes provided an efficient catalysis of the Guerbet reaction of primary alcohols to afford β-alkylated dimer alcohols [45]. As an example, the reaction of 1-butanol **99** in the presence of [Cp*IrCl$_2$]$_2$ (1 mol%), *t*-BuOK (40 mol%) and 1,7-octadiene (10 mol%) produced 2-ethyl-1-hexanol **100** in 93% yield (Equation 10.22). In this reaction, the addition of base and a small amount of hydrogen acceptor (e.g. 1,7-octadiene) were needed. A variety of primary alcohols have been shown to undergo the Guerbet reaction under the influence of Ir complexes to give the corresponding dimer alcohols, in good yields. This method provides an alternative, direct route to β-alkylated primary alcohols, which are usually prepared by the aldol condensation of aldehydes, followed by hydrogenation.

$$(10.22)$$

The reaction of secondary alcohols **101** with primary alcohols **99** using [Cp*IrCl$_2$]$_2$ and a base such as NaO^{t}Bu was reported to afford the corresponding β-alkylation products **102** (Equation 10.23) [46].

$$(10.23)$$

Another reported example of the carbon–carbon bond-forming reaction from alcohols is that of the indirect Wittig reaction, which utilizes an [IrCl(cod)]$_2$/dppp/ CsCO$_3$ catalyst system and leads to the production of alkanes (Equation 10.24) [47].

$$(10.24)$$

10.8
N-Alkylation of Amines

The [Cp*IrCl$_2$]$_2$/K$_2$CO$_3$ system serves as an efficient catalyst for the N-alkylation of amines **106** with alcohols **107** (Equation 10.25) [48]. Alternatively, [IrCl(cod)]$_2$

combined with bis(diphenylphosphino)ferrocene (dppf) also shows the effective catalyst for the N-alkylation of primary amines [49]; in this reaction both primary and secondary alcohols can be used. This N-alkylation of amines may proceed in a similar step-wise manner to the α-alkylation of aldehydes, as shown above: (i) dehydrogenation of alcohols to aldehydes; (ii) the formation of imines from aldehydes and amines, and (iii) the hydrogenation of imines to amines.

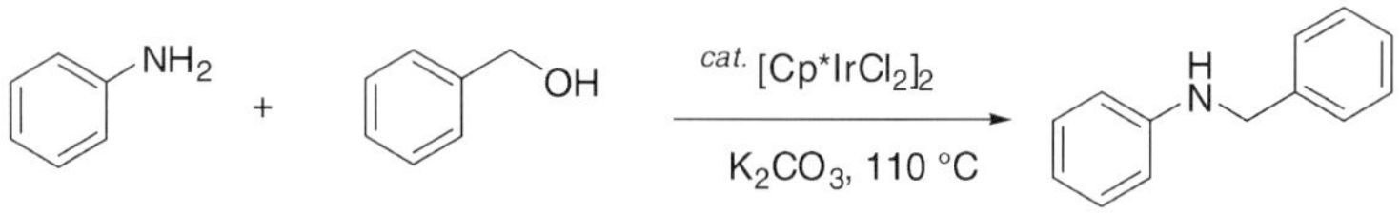

$$\underset{106}{\text{PhNH}_2} + \underset{107}{\text{PhCH}_2\text{OH}} \xrightarrow[\text{K}_2\text{CO}_3,\ 110\ °C]{\text{cat. }[\text{Cp*IrCl}_2]_2} \underset{\underset{88\%}{108}}{\text{PhNHCH}_2\text{Ph}} \quad (10.25)$$

The iridium-catalyzed reaction of primary amines with diols gave cyclic amines. The reaction of amine **109** with diol **110** in the presence of [Cp*IrCl$_2$]$_2$/NaHCO$_3$ catalyst gave heterocyclization product **111** (Equation 10.26) [50].

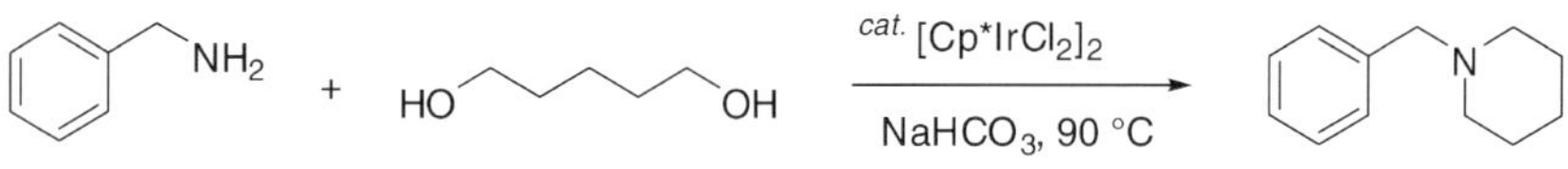

$$\underset{109}{\text{PhCH}_2\text{NH}_2} + \underset{110}{\text{HO(CH}_2)_5\text{OH}} \xrightarrow[\text{NaHCO}_3,\ 90\ °C]{\text{cat. }[\text{Cp*IrCl}_2]_2} \underset{\underset{91\%}{111}}{\text{PhCH}_2\text{-N-piperidine}} \quad (10.26)$$

Intramolecular cyclization of amino alcohols **112** took place with [Cp*IrCl$_2$]$_2$/K$_2$CO$_3$ to give indoles **113** (Equation 10.27) [51].

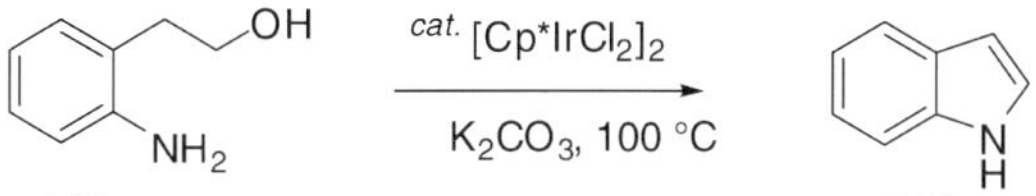

$$\underset{112}{\text{2-(2-aminophenyl)ethanol}} \xrightarrow[\text{K}_2\text{CO}_3,\ 100\ °C]{\text{cat. }[\text{Cp*IrCl}_2]_2} \underset{\underset{90\%}{113}}{\text{indole}} \quad (10.27)$$

Furthermore, the N-alkylation of 2-aminobenzyl alcohol **114** with ketones **115** in the presence of [IrCl(cod)]$_2$ and KOH gave quinoline derivatives **116** (Equation 10.28) [52]. The reaction may be initiated by the formation of ketimine from **114** and **115**, and the ketimine thus formed is oxidized by Ir catalyst and the **114** which serves as a hydrogen acceptor giving the corresponding aldehyde, which is eventually converted into quinoline **116** through intramolecular aldol-type condensation.

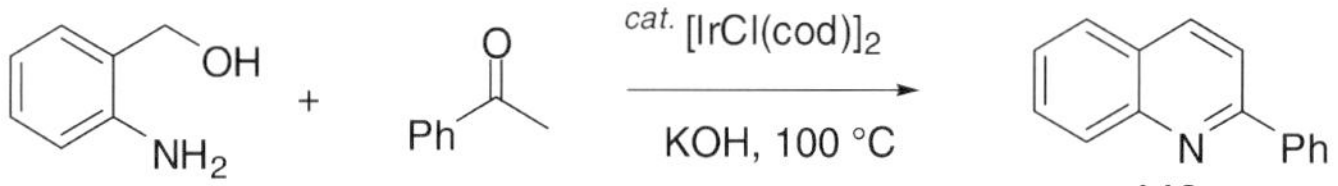

$$\underset{114}{\text{2-aminobenzyl alcohol}} + \underset{115}{\text{PhCOCH}_3} \xrightarrow[\text{KOH,}\ 100\ °C]{\text{cat. }[\text{IrCl(cod)}]_2} \underset{\underset{90\%}{116}}{\text{2-phenylquinoline}} \quad (10.28)$$

Another type of N-alkylation was achieved by the [IrCl(cod)]$_2$-catalyzed reductive alkylation of secondary amine with aldehyde and silane (Equation 10.29) [53]. For example, the treatment of dibutylamine **117** with butyraldehyde **118** and Et$_3$SiH **119** (a 1:1:1 molar ratio amine, aldehyde and silane) or polymethylhydrosiloxane (PMHS) in 1,4-dioxane at 75 °C under the influence of a catalytic amount of [IrCl(cod)]$_2$, gave tributylamine **120**.

$$(10.29)$$

10.9
Oxidative Dimerization of Primary Alcohols to Esters

Primary alcohols **121** undergo an efficient oxidative dimerization by [IrCl(coe)$_2$]$_2$ under air, without any solvent, to form esters **122** in fair to good yields (Equation 10.30) [54]. The reaction is initiated by the *in situ* generation of an Ir–hydride complex via hydrogen transfer from alcohols to afford aldehydes, followed by the dehydrogenation of hemiacetals derived from alcohols and aldehydes by action of the Ir–complex to afford esters.

$$(10.30)$$

10.10
Iridium-Catalyzed Addition of Water and Alcohols to Terminal Alkynes

The addition of water and alcohols to nonactivated terminal alkynes was promoted by [Ir(cod)$_2$]$^+$BF$_4^-$ combined with Lewis acids and phosphate [55]. Thus, terminal alkynes such as 1-octyne **122** reacted with butanol **123** to afford 2-octanone dibutylketal **124** (Equation 10.31).

$$(10.31)$$

The iridium-catalyzed addition of water could be applied to the reaction of α,ω-diynes. Thus, 1,7-octadiyne **125** was converted to 1-(2-methylcyclopent-1-enyl)ethanone **127** (Equation 10.32). The formation of **127** was explained by assuming intramolecular aldol condensation of the resulting 2,7-octadione by Lewis acid.

$$\equiv\text{—(CH}_2)_4\text{—}\equiv \;+\; \text{BuOH} \;+\; \text{H}_2\text{O} \;\xrightarrow[\text{70 °C}]{\substack{[\text{Ir(cod)}_2]^+\text{BF}_4^- \\ \text{P(O}^i\text{Pr})_3 \\ \text{ZrCl}_4}}\; \text{127} \quad (10.32)$$

125, **123**, **126**, **127** 84%

10.11
Iridium-Catalyzed Direct Arylation of Aromatic C—H Bonds

Arylation of benzene **128** with aryl iodides **129** via direct C—H bond activation was achieved in the presence of [Cp*IrHCl]$_2$ and KO^{t}Bu to afford corresponding biaryl **130** (Equation 10.33) [56]. The phenyl radical would participate as an intermediate in this reaction.

$$\text{128} \;+\; \text{129 (4-I-C}_6\text{H}_4\text{-OMe)} \;\xrightarrow[\text{KO}^t\text{Bu}]{cat.\;[\text{Cp*IrHCl}]_2}\; \text{130 (Ph-C}_6\text{H}_4\text{-OMe)} \quad (10.33)$$

66% **130**

10.12
Iridium-Catalyzed Anti-Markovnikov Olefin Arylation

The iridium(III)-complex, [Ir(μ-acac-O,O,C^3)(acac-O,O)(acac-C^3)]$_2$, mediates the activation of unactivated aromatic C—H bond with unactivated alkenes to form anti-Markovnikov products [57]. The reaction of benzene **131** with propene **132** (0.78 MPa of propylene, 1.96 MPa of N$_2$) leads to the formation of n-propylbenzene **133** in 61% selectivities (turnover number (TON) = 13; turnover frequency (TOF) = 0.0110 s^{-1}) (Equation 10.34). The reaction of benzene with ethane at 180 °C for 3 h gave ethylbenzene (TON = 455; TOF = 0.0421 s^{-1}). The anti-Markovnikov selectivity was also proven for the reaction with 1-hexane and isobutene, giving 1-phenylhexane (69% selectivity) and isobutylbenzene (82% selectivity), respectively.

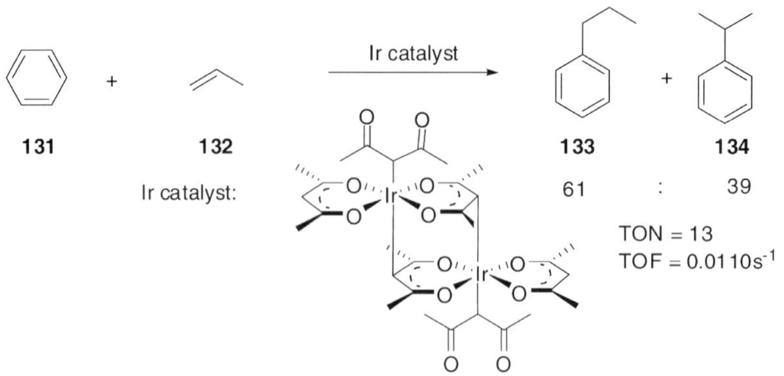

(10.34)

[Ir(CH$_3$)(Py)(trop-O,O)$_2$] (trop=O,O-κ^2-O,O-tropolonato) is also active catalyst for the C—H activation of arenes [58]. Mechanistic investigations for the Ir-catalyzed reaction have been studied [59].

10.13
Iridium-Catalyzed Silylation and Borylation of Aromatic C—H Bonds

The iridium complex [Ir(OMe)(cod)$_2$] with 4,4'-di-*tert*-butyl-2,2'-bipyridine (dtbpy) or 2,9-diisopropyl-1,10-phenanthroline (dipphen) as ligand shows a catalytic activity for aromatic C—H silylation of aromatic compounds by disilane [60]. The reaction of 1,2-dimethylbenzene **135** with 1,2-di-*tert*-butyl-1,1,2,2,-tetrafluorodisilane **136** in the presence of [Ir(OMe)(cod)$_2$] and dtbpy gives 4-silyl-1,2-dimethylbenzene **137** in 99% yield (Equation 10.35).

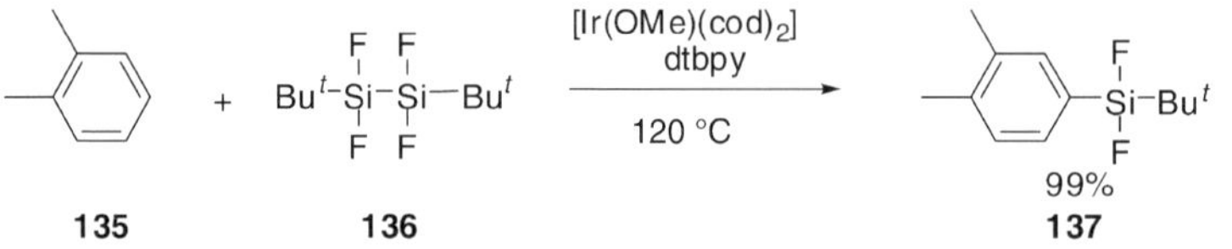

(10.35)

The direct borylation of arenes was catalyzed by iridium complexes [61–63]. Iridium complex generated from [IrCl(cod)]$_2$ and 2,2'-bipyridine (bpy) showed the high catalytic activity of the reaction of bis(pinacolato)diboron (B$_2$Pin$_2$) **138** with benzene **139** to afford phenylborane **140** (Equation 10.36) [61]. Various arenes and heteroarenes are allowed to react with B$_2$Pin$_2$ and pinacolborane (HBpin) in the presence of [IrCl(cod)]$_2$/bipyridne or [Ir(OMe)(cod)]$_2$/bipyridine to produce corresponding aryl- and heteroarylboron compounds [62]. The reaction is considered to proceed via the formation of a tris(boryl)iridium(III) species and its oxidative addition to an aromatic C—H bond.

10.14
Miscellaneous Reactions Catalyzed by Iridium Complexes

The cationic iridium complex [Ir(cod)(PPh$_3$)$_2$]OTf, when activated by H$_2$, catalyzes the aldol reaction of aldehydes **141** or acetal with silyl enol ethers **142** to afford **143** (Equation 10.37) [63]. The same Ir complex catalyzes the coupling of α,β-enones with silyl enol ethers to give 1,5-dicarbonyl compounds [64]. Furthermore, the alkylation of propargylic esters **144** with silyl enol ethers **145** catalyzed by [Ir(cod)[P(OPh)$_3$]$_2$]OTf gives alkylated products **146** in high yields (Equation 10.38) [65]. An iridium-catalyzed enantioselective reductive aldol reaction has also been reported [66].

The [IrCl(cod)]$_2$-catalyzed reductive coupling of acrylates and imines provides *trans*-β-lactams with high diastereoselectivity (Equation 10.39) [67]. With regards to the reaction mechanism, *in situ* generated Ir–hydride reacts with acrylate **148** to produce an Ir enolate, which then reacts with the **147** to afford the β-amido ester **149**.

It has been reported that a cationic iridium such as [Ir(cod)$_2$]BARF (BARF = {3,5-(CF$_3$)$_2$C$_6$H$_3$}$_4$B), when combined with 1,1′-bis(diphenylphosphino)ferrocene (DPPF), catalyzed a hydrogen-mediated reductive carbon–carbon bond formation [68]. Thus, the reaction of alkynes **150** with α-ketoesters **151** produces β,γ-unsaturated-α-hydroxy ketones **152** (Equation 10.40).

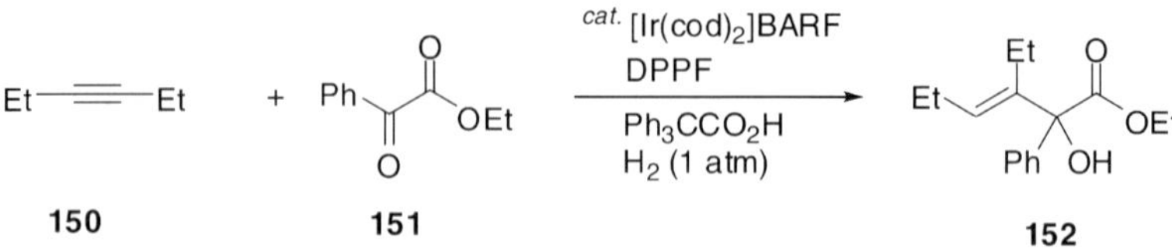

$$\text{Et}\!-\!\!\equiv\!\!-\text{Et} \;+\; \text{Ph-CO-CO-OEt} \xrightarrow[\text{Ph}_3\text{CCO}_2\text{H},\ \text{H}_2\ (1\ \text{atm})]{\text{cat. }[\text{Ir(cod)}_2]\text{BARF},\ \text{DPPF}} \text{Et-CH=C(Et)-C(Ph)(OH)-CO-OEt}$$

150   **151**   **152** 93%   (10.40)

The [Ir(cod)$_2$]BARF complex also showed high catalytic activity in the hydrogenative coupling of alkyne with aldimines to lead to reductive coupling products, allyl amines [69].

The combination of [IrCl(cod)Cl]$_2$ complex with P(t-Bu)$_3$ efficiently catalyzes aromatic homologation using internal alkyne [70]. For example, the reaction of benzoyl chloride **153** with 4-octyne **154** afforded 1,2,3,4-tetrapropylnaphthalene **155** (Equation 10.41). The reaction with 2-thenoyl and 2-naphthoyl chlorides also affords benzothiophene and anthracene, respectively, in high yields. The reaction would proceed as follows (Scheme 10.9): (i) oxidative addition of aroyl chloride

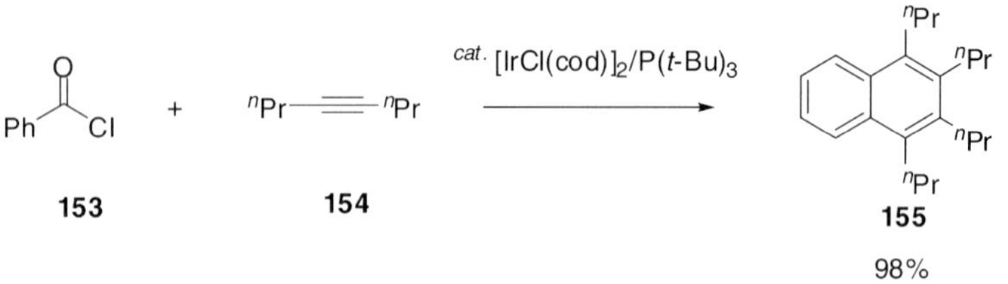

$$\text{Ph-CO-Cl} \;+\; {}^n\text{Pr}\!-\!\!\equiv\!\!-{}^n\text{Pr} \xrightarrow{\text{cat. }[\text{IrCl(cod)}]_2/\text{P}(t\text{-Bu})_3} \text{1,2,3,4-tetra-}{}^n\text{Pr-naphthalene}$$

153   **154**   **155** 98%   (10.41)

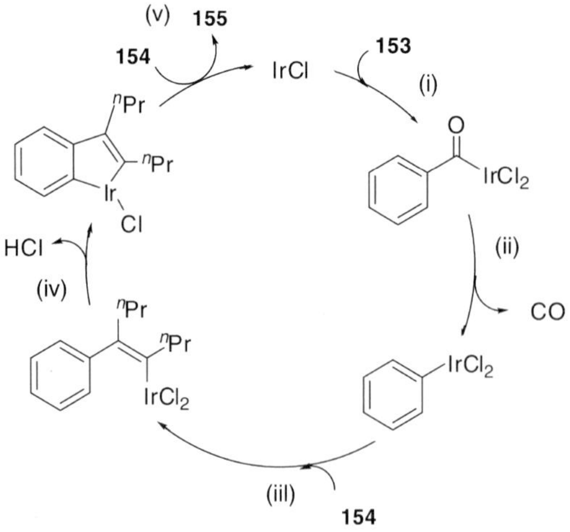

Scheme 10.9

to Ir(I) to form aroyliridium intermediate; (ii) decarbonylation of the aroyliridium; (iii) insertion of alkyne into the Ar–Ir bond; (iv) *ortho*-iridation followed by insertion of the second alkyne; and (v) reductive elimination to form the product.

Naphthalene derivatives **158** were also prepared by the oxidative coupling of benzoic acids **156** with internal alkynes such as diphenylacetylene **157** in the presence of [Cp*IrCl$_2$]$_2$ complex combined with Ag$_2$CO$_3$ as oxidant (Equation 10.42) [71].

$$\text{156} + \text{157} \xrightarrow[\text{o-xylene, 160 °C}]{[\text{Cp*IrCl}_2]_2, \text{Ag}_2\text{CO}_3} \text{158} \quad 88\% \quad (10.42)$$

References

1. (a) Peruzzini, M., Bianchini, C. and Gonsalvi, L. (2007) *Comprehensive Organometallic Chemistry III*, Vol. 7 (ed. R.H. Crabtree), Elsevier, Oxford, pp. 267–425;
(b) Vaska, L. (1968) *Accounts of Chemical Research*, **1**, 335.

2. (a) Crabtree, R.H. and Morris, G.E. (1977) *Journal of Organometallic Chemistry*, **135**, 135;
(b) Crabtree, R.H., Felkin, H. and Morris, G.E. (1977) *Journal of Organometallic Chemistry*, **141**, 205;
(c) Crabtree, R.H. (1979) *Accounts of Chemical Research*, **12**, 331.

3. Stork, G. and Kahne, D.E. (1983) *Journal of the American Chemical Society*, **105**, 1072.

4. (a) Lightfoot, A., Schnider, P. and Pfaltz, A. (1998) *Angewandte Chemie – International Edition in English*, **37**, 2897;
(b) Kainz, S., Brinkmann, A., Leitner, W. and Pfaltz, A. (1999) *Journal of the American Chemical Society*, **121**, 6421;
(c) Vázquez-Serrano, L.D., Owens, B.T. and Buriak, J.M. (2002) *Chemical Communications*, 2518;
(d) Perry, M.C., Cui, X., Powell, M.T., Hou, D.-R., Reibenspies, J.H. and Burgess, K. (2003) *Journal of the American Chemical Society*, **125**, 113;
(e) Wang, W.-B., Lu, S.-M., Yang, P.-Y., Han, X.-W. and Zhou, Y.-G. (2003) *Journal of the American Chemical Society*, **125**, 10536;
(f) Pfaltz, A., Blankenstein, J., Hilgraf, R., Hormann, E., McIntyre, S., Steven, M., Menges, F., Schonleber, M., Smidt, S.P., Wustenberg, B. and Zimmermann, N. (2003) *Advanced Synthesis Catalysis*, **345**, 33.

5. (a) Sakaguchi, S., Yamaga, T. and Ishii, Y. (2001) *The Journal of Organic Chemistry*, **66**, 4710;
(b) Prtra, D.G.I., Kamer, P.C.J., Spek, A.L., Schemaker, H.E. and Van Leeuwen, P.W.N.M. (2000) *The Journal of Organic Chemistry*, **65**, 3010;

(c) Ogo, S., Makihara, N., Kaneko, Y. and Watanabe, Y. (2001) *Organometallics*, **20**, 4903;
(d) Penicaud, V., Maillet, C., Janvier, P., Pipelier, M. and Bujoli, B. (1999) *European Journal of Organic Chemistry*, 1745;
(e) Albrecht, M., Miecznikowski, J.R., Samuel, A., Faller, J.W. and Crabtree, R.H. (2002) *Organometallics*, **21**, 3596;
(f) Fujita, K., Kitatsuji, C., Furukawa, S. and Yamaguchi, R. (2004) *Tetrahedron Letters*, **45**, 3215;
(g) Mashima, K., Abe, T. and Tani, K. (1998) *Chemistry Letters*, 1199;
(h) Murata, K., Ikariya, T. and Noyori, R. (1999) *The Journal of Organic Chemistry*, **64**, 2186;
(i) Suzuki, T., Morita, K., Tsuchida, M. and Hiroi, K. (2002) *Organic Letters*, **4**, 2361;
(j) Fujita, K., Yamamoto, K. and Yamaguchi, R. (2002) *Organic Letters*, **4**, 2691;
(k) Abuta, T., Ogo, S., Watanabe, Y. and Fukuzumi, S. (2003) *Journal of the American Chemical Society*, **125**, 4149;
(l) Takaya, H., Naota, T. and Murahashi, S.-I. (1998) *Journal of the American Chemical Society*, **120**, 4244;
(m) Murahashi, S.-I. and Takaya, H. (2000) *Accounts of Chemical Research*, **33**, 225;
(n) Fujita, K., Tanino, N. and Yamaguchi, R. (2007) *Organic Letters*, **9**, 109;
(o) Hanasaka, F., Fujita, K. and Yamaguchi, R. (2006) *Organometallics*, **25**, 4643;
(p) Hanasaka, F., Fujita, K. and Yamaguchi, R. (2005) *Organometallics*, **24**, 3422;
(q) Fujita, K. and Yamaguchi, R. (2005) *Synlett*, 560;
(r) Cami-Kobeci, G. and Williams, J.M.J. (2004) *Chemical Communications*, 1072;
(s) Cami-Kobeci, G., Slatford, P.A., Whittlesey, M.K. and Williams, J.M.J. (2005) *Bioorganic and Medicinal Chemistry*, **15**, 535;
(t) Tani, K., Iseki, A. and Yamagata, T. (1999) *Chemical Communications*, 1821;
(u) Inoue, S., Nomura, K., Hashiguchi, S., Noyori, R. and Izawa, Y. (1997) *Chemistry Letters*, 957;
(v) Nishibayashi, Y., Singh, J.D., Arikawa, Y., Uemura, S. and Hidai, M. (1997) *Journal of Organometallic Chemistry*, **531**, 13.

6 Higashino, T., Sakaguchi, S. and Ishii, Y. (2000) *Organic Letters*, **2**, 4193.

7 Kim, Y., Kim, S., Kim, S.-J., Lee, M.K., Kim, M., Lee, H. and Chin, C.S. (2004) *Chemical Communications*, 1692.

8 (a) Esteruelas, M.A., Oliván, M., Oro, L.A. and Tolosa, J.I. (1995) *Journal of Organometallic Chemistry*, **487**, 143;
(b) Field, L.D. and Ward, A.J. (2003) *Journal of Organometallic Chemistry*, **681**, 91.

9 (a) Yamamoto, Y., Furukawa, R., Umemoto, T. and Miyaura, N. (2004) *Tetrahedron*, **60**, 10695;
(b) Ohmura, T., Yamamoto, Y. and Miyaura, N. (2000) *Journal of the American Chemical Society*, **122**, 4990.

10 (a) Matsuda, I., Kato, T., Sato, S. and Izumi, Y. (1986) *Tetrahedron Letters*, **27**, 5747;
(b) Ohmura, T., Shirai, Y., Yamamoto, Y. and Miyaura, N. (1998) *Chemical Communications*, 1337;
(c) Ohmura, T., Yamamoto, Y. and Miyaura, N. (1999) *Organometallics*, **18**, 413;
(d) Yamamoto, Y., Miyairi, T., Ohmura, T. and Miyaura, N. (1999) *The Journal of Organic Chemistry*, **64**, 296.

11 (a) Chatani, N., Inoue, H., Morimoto, T., Muto, T. and Murai, S. (2001) *The Journal of Organic Chemistry*, **66**, 4433;
(b) Yamamoto, Y., Hayashi, H., Saigoku, T. and Nishiyama, H. (2005) *Journal of the American Chemical Society*, **127**, 10804;
(c) Kezuka, S., Okado, T., Niou, E. and Takeuchi, R. (2005) *Organic Letters*, **7**, 1711.

12 Recent reviews: (a) Takeuchi, R. and Kezuka, S. (2006) *Synthesis*, 3349;
(b) Ishii, Y. and Sakaguchi, S. (2004) *Bulletin of the Chemical Society of Japan*, **77**, 909;
(c) Fujita, K. and Yamaguchi, R. (2005) *Synlett*, 560;
(d) Takeuchi, R. (2002) *Synlett*, 1954.

13 (a) Trost, B.M. (1995) *Angewandte Chemie – International Edition*, **34**, 259;
(b) Trost, B.M. (1991) *Science*, **254**, 1471.

14 Jun, C.-H. and Crabtree, R.H. (1992) *Tetrahedron Letters*, **33**, 7119.
15 Ohmura, T., Yarozuya, S., Yamamoto, Y. and Miyaura, N. (2000) *Organometallics*, **19**, 365.
16 Hirabayashi, T., Sakaguchi, S. and Ishii, Y. (2005) *Advanced Synthesis Catalysis*, **347**, 872.
17 (a) Fabbian, M., Marsich, N. and Farnetti, E. (2004) *Inorganica Chimica Acta*, **357**, 2881;
(b) Farnetti, E. and Marsich, N. (2004) *Journal of Organometallic Chemistry*, **689**, 14.
18 Takeuchi, R. and Nakaya, Y. (2003) *Organic Letters*, **5**, 3659.
19 (a) Takeuchi, R., Tanaka, S. and Nakaya, Y. (2001) *Tetrahedron Letters*, **42**, 2991;
(b) Kezuka, S., Tanaka, S., Ohe, T., Nakaya, Y. and Takeuchi, R. (2006) *The Journal of Organic Chemistry*, **71**, 543;
(c) Dufková, L., Císaøová, I., Štepnièka, P. and Kotra, M. (2003) *European Journal of Organic Chemistry*, 2882.
20 Shibata, T., Fuiimoto, T., Yokota, K. and Takagi, K. (2004) *Journal of the American Chemical Society*, **126**, 8382.
21 Sakaguchi, S., Kubo, T. and Ishii, Y. (2001) *Angewandte Chemie – International Edition in English*, **40**, 2534.
22 Sakaguchi, S., Mizuta, T. and Ishii, Y. (2006) *Organic Letters*, **8**, 2459.
23 Sakaguchi, S., Mizuta, T., Furuwan, M., Kubo, T. and Ishii, Y. (2004) *Chemical Communications*, 1638.
24 (a) Fisher, C. and Carreria, E.M. (2001) *Organic Letters*, **3**, 4319;
(b) Fisher, C. and Carreria, E.M. (2004) *Synthesis*, 1497.
25 Kubo, T., Sakaguchi, S. and Ishii, Y. (2000) *Chemical Communications*, 625.
26 Nakagawa, H., Sakaguchi, S. and Ishii, Y. (2003) *Chemical Communications*, 502.
27 (a) Okimoto, Y., Sakaguchi, S. and Ishii, Y. (2002) *Journal of the American Chemical Society*, **124**, 1590;
(b) Hirabayashi, T., Sakaguchi, S., Ishii, Y., Davies, P.W. and Fürstner, A. (2005) *Organic Synthesis*, **82**, 55.
28 Nakagawa, H., Okimoto, Y., Sakaguchi, S. and Ishii, Y. (2003) *Tetrahedron Letters*, **44**, 103.
29 Nakagawa, H., Hirabayashi, T., Sakaguchi, S. and Ishii, Y. (2004) *The Journal of Organic Chemistry*, **69**, 3474.
30 Morita, M., Sakaguchi, S. and Ishii, Y. (2006) *The Journal of Organic Chemistry*, **71**, 6285.
31 Tsuji, J. (2004) *Palladium Reagents and Catalysts: New Perspective for the 21st Century*, John Wiley & Sons, Inc., Chichester, UK, p. 431.
32 (a) Chini, P. and Martinengo, S. (1967) *Inorganic Chemistry*, **6**, 837;
(b) Tjaden, E.B. and Stryker, J.M. (1990) *Journal of the American Chemical Society*, **112**, 6420.
33 (a) Takeuchi, R. and Kashio, M. (1997) *Angewandte Chemie – International Edition*, **36**, 263;
(b) Takeuchi, R. and Kashio, M. (1998) *Journal of the American Chemical Society*, **120**, 8647;
(c) Takeuchi, R. (2000) *Polyhedron*, **19**, 557.
34 (a) Janssen, J.P. and Helmchen, G. (1997) *Tetrahedron Letters*, **38**, 8025;
(b) Garcia-Yebrá, C., Janssen, J.P., Rominger, F. and Helmchen, G. (2004) *Organometallics*, **23**, 5459.
35 (a) Bartels, B. and Helmchen, G. (1999) *Chemical Communications*, 741;
(b) Bartels, B., Garcia-Yebrá, C., Rominger, F. and Helmchen, G. (2002) *European Journal of Inorganic Chemistry*, 2569;
(c) Bartels, B., Garcia-Yebrá, C. and Helmchen, G. (2003) *European Journal of Organic Chemistry*, 1097;
(d) Lipowsky, G., Miller, N. and Helmchen, G. (2004) *Angewandte Chemie – International Edition*, **43**, 4595;
(e) Alexakis, A. and Polet, D. (2004) *Organic Letters*, **6**, 3529;
(f) Polet, D. and Alexalis, A. (2005) *Organic Letters*, **7**, 1621;
(g) Polet, D., Alexakis, A., Tissot-Croset, K., Corminboeuf, C. and Ditrich, K. (2006) *Chemistry – A European Journal*, **12**, 3596.
36 (a) Fuji, K., Kinoshita, N., Tanaka, K. and Kawabata, T. (1999) *Chemical Communications*, 2289;
(b) Kinoshita, N., Marx, K.H., Tanaka, K., Tsubaki, K., Kawabata, T., Yoshikai, N., Nakamura, E. and Fuji, K. (2004) *The Journal of Organic Chemistry*, **69**, 7960.
37 (a) Kanayama, T., Yoshida, K., Miyabe, H. and Takemoto, Y. (2003) *Angewandte Chemie – International Edition*, **42**, 2054;

(b) Kanayama, T., Yoshida, K., Miyabe, H., Kimachi, T. and Takemoto, Y. (2003) *The Journal of Organic Chemistry*, **68**, 6197.
38 (a) Takeuchi, R., Ue, N., Tanabe, K., Yamashita, K. and Shiga, N. (2001) *Journal of the American Chemical Society*, **123**, 9525;
(b) Takeuchi, R. and Shiga, N. (1999) *Organic Letters*, **1**, 265.
39 (a) Ohmura, T. and Hartwig, J.F. (2002) *Journal of the American Chemical Society*, **124**, 15164;
(b) Shu, C., Leitner, A. and Hartwig, J.F. (2004) *Angewandte Chemie – International Edition*, **43**, 4797;
(c) Tissot-Croset, K., Polet, D. and Alexakis, A. (2004) *Angewandte Chemie – International Edition*, **43**, 2426;
(d) Miyabe, H., Matsumura, A., Moriyama, K. and Takemoto, Y. (2004) *Organic Letters*, **6**, 4631;
(e) Lipowsky, G. and Helmchen, G. (2004) *Chemical Communications*, 116;
(f) Welter, C., Koch, O., Lipowsky, G. and Helmchen, G. (2004) *Chemical Communications*, 896;
(g) Miyabe, H., Yoshida, K., Kobayashi, Y., Matsumura, A. and Takemoto, Y. (2003) *Synlett*, 1031;
(h) Welter, C., Dahnz, A., Brunner, B., Steiff, S., Dubon, P. and Helmchen, G. (2005) *Organic Letters*, **7**, 1239.
40 Carine, D. (1989) *Carbon-Carbon Bond Formation*, Vol. 1 (ed. R.L. Augustine), Marcel Dekker Inc., New York, pp. 85–352.
41 Taguchi, K., Nakagawa, H., Hirabayashi, T., Sakaguchi, S. and Ishii, Y. (2004) *Journal of the American Chemical Society*, **126**, 72.
42 Morita, M., Obora, Y. and Ishii, Y. (2007) *Chemical Communications*, 2850.
43 (a) Löfberg, C., Grigg, R., Whittaker, M.A., Keep, A. and Derrick, A. (2006) *The Journal of Organic Chemistry*, **71**, 8023;
(b) Löfberg, C., Grigg, R., Keep, A., Derrick, A., Sridharan, V. and Kilner, C. (2006) *Chemical Communications*, 5000.
44 Balck, P.J., Cami-Kobeci, G., Edwards, M.G., Slatford, P.A., Whittlesey, K. and Williams, J.M.J. (2006) *Organic and Biomolecular Chemistry*, **1**, 116.
45 Matsu-ura, T., Sakaguchi, S., Obora, Y. and Ishii, Y. (2006) *The Journal of Organic Chemistry*, **71**, 8306.
46 Fujita, K., Asai, C., Yamaguchi, T., Hanasaka, F. and Yamaguchi, R. (2005) *Organic Letters*, **7**, 4017.
47 Edwards, M.G. and Williams, J.M.J. (2002) *Angewandte Chemie – International Edition*, **41**, 4740.
48 Fujita, K., Li, Z., Ozeki, N. and Yamaguchi, R. (2003) *Tetrahedron Letters*, **44**, 2687.
49 Cami-Kobeci, G., Slatford, P.A., Whittlesey, M.K. and Williams, J.M.J. (2005) *Bioorganic and Medicinal Chemistry Letters*, **15**, 535.
50 (a) Fujita, K., Fujii, T. and Yamaguchi, R. (2004) *Organic Letters*, **6**, 3525;
(b) Fujita, K., Enoki, Y. and Yamaguchi, R. (2006) *Organic Syntheses*, **83**, 217;
(c) Eary, C.T. and Clausen, D. (2006) *Tetrahedron Letters*, **47**, 6899.
51 Fujita, K., Yamamoto, K. and Yamaguchi, R. (2002) *Organic Letters*, **4**, 2691.
52 Taguchi, K., Sakaguchi, S. and Ishii, Y. (2005) *Tetrahedron Letters*, **46**, 4539.
53 Mizuta, T., Sakaguchi, S. and Ishii, Y. (2005) *The Journal of Organic Chemistry*, **70**, 2195.
54 Izumi, A., Obora, Y., Sakaguchi, S. and Ishii, Y. (2006) *Tetrahedron Letters*, **47**, 9199.
55 Hirabayashi, T., Okimoto, Y., Saito, A., Morita, M., Sakaguchi, S. and Ishii, Y. (2006) *Tetrahedron*, **62**, 2231.
56 Fujita, K., Nonogawa, M. and Yamaguchi, R. (2004) *Chemical Communications*, 1926.
57 (a) Matsumoto, T., Taube, D.J., Periana, R.A., Taube, H. and Yoshida, H. (2000) *Journal of the American Chemical Society*, **122**, 7414;
(b) Matsumoto, T., Periana, R.A., Taube, D.J. and Yoshida, H. (2002) *Journal of Molecular Catalysis A – Chemical*, **180**, 1;
(c) Periana, R.A., Liu, X.Y. and Bhalla, G. (2002) *Chemical Communications*, 3000;
(d) Matsumoto, T., Periana, R.A., Taube, D.J. and Yoshida, H. (2002) *Journal of Catalysis*, **206**, 272.

58 (a) Bhalla, G. and Periana, R.A. (2005) *Angewandte Chemie – International Edition*, **44**, 1540;
(b) Bhalla, G., Oxgaard, J., Goddard, W. A. III, and Periana, R.A. (2005) *Organometallics*, **24**, 3229.

59 (a) Bhalla, G., Liu, X.-Y., Oxgaard, J., Goddard, W.A. and Periana, R.A. (2005) *Journal of the American Chemical Society*, **127**, 11372;
(b) Oxgaard, J., Bhalla, G., Periana, R.A. and Goddard, W.A. (2006) *Organometallics*, **25**, 1618.

60 (a) Ishiyama, T., Sato, K., Nishio, Y. and Miyaura, N. (2003) *Angewandte Chemie – International Edition*, **42**, 5346;
(b) Saiki, T., Nishio, Y., Ishiyama, T. and Miyaura, N. (2006) *Organometallics*, **25**, 6068.

61 Ishiyama, T., Takagi, J., Ishida, K., Miyaura, N., Anastasi, N.R. and Hartwig, J.F. (2002) *Journal of the American Chemical Society*, **124**, 390.

62 (a) Takagi, J., Sato, K., Hartwig, J.F., Ishiyama, T. and Miyaura, N. (2002) *Tetrahedron Letters*, **43**, 5649;
(b) Ishiyama, T., Takagi, J., Hartwig, J.F. and Miyaura, N. (2002) *Angewandte Chemie – International Edition*, **41**, 3056;
(c) Tse, M.K., Cho, J.-Y. and Smith, M.R., III (2001) *Organic Letters*, **3**, 2831;
(d) Ishiyama, T., Takagi, J., Yonekawa, Y., Hartwig, J.F. and Miyaura, N. (2003) *Advanced Synthesis Catalysis*, **345**, 1103;
(e) Tzschucke, C.C., Murphy, J.M. and Hartwig, J.F. (2007) *Organic Letters*, **9**, 761;
(f) Boller, T.M., Murphy, J.M., Hapke, M., Ishiyama, T., Miyaura, N. and Hartwig, J. F. (2005) *Journal of the American Chemical Society*, **127**, 14263;
(g) Ishiyama, T. and Miyaura, N. (2006) *Pure and Applied Chemistry*, **78**, 1369.

63 Matsuda, I., Hesegawa, Y., Makino, T. and Itoh, K. (2000) *Tetrahedron Letters*, **41**, 1405.

64 Matsuda, I., Makino, T., Hasegawa, Y. and Itoh, K. (2000) *Tetrahedron Letters*, **41**, 1409.

65 Matsuda, I., Komori, K. and Itoh, K. (2002) *Journal of the American Chemical Society*, **124**, 9072.

66 Zhao, C.-X., Duffey, M.O., Taylor, S.J. and Morken, J.P. (2001) *Organic Letters*, **3**, 1829.

67 Towns, J.A., Evans, M.A., Queffelec, J., Taylor, S.J. and Morken, J.P. (2002) *Organic Letters*, **4**, 2537.

68 Ngai, M., Barchuk, A. and Krische, M.J. (2007) *Journal of the American Chemical Society*, **129**, 280.

69 (a) Ngai, M., Barchuk, A. and Krische, M.J. (2007) *Journal of the American Chemical Society*, **129**, 12644;
(b) Barchuk, A. and Krische, M.J. (2007) *Journal of the American Chemical Society*, **129**, 8432.

70 Yasukawa, T., Satoh, T., Miura, M. and Nomura, M. (2002) *Journal of the American Chemical Society*, **124**, 12680.

71 Ueura, K., Satoh, T. and Miura, M. (2007) *The Journal of Organic Chemistry*, **72**, 5362.

11
Iridium-Catalyzed Cycloadditions

Takanori Shibata

11.1
Introduction

The transition-metal-catalyzed cycloaddition of unsaturated motifs, such as alkynes and alkenes, is an atom-economical and reliable protocol for the construction of carbocyclic and heterocyclic skeletons [1]. The process was first introduced in Reppe's report in 1948, where a Ni-mediated cyclo-oligomerization of acetylene gave several cyclic compounds including benzene as a [2+2+2] cycloadduct [2]. During the 1970s, the development of Co-catalyzed cyclotrimerization of alkynes [3] and Co-mediated carbonylative alkyne–alkene coupling as [2+2+1] cycloaddition (Pauson–Khand reaction) [4] gave recognition to the significance of transition-metal-catalyzed cycloaddition as a synthetic tool.

Lewis acid-catalyzed cycloaddition is also a powerful synthetic method, and various types of cycloaddition have been reported. In particular, enantioselective variants using chiral Lewis acids have been comprehensively studied; some of these were used as key reactions for natural product syntheses [5]. However, they generally require one or more heteroatoms in the substrates, such as enones or enoates, to which (chiral) Lewis acids can coordinate. In contrast, in the case of transition-metal-catalyzed cycloadditions, the metals coordinate directly to the π-electron and activate unsaturated motifs, which means that the heteroatom(s) are unnecessary. Moreover, the direct coordination to the reaction site can realize highly enantioselective reaction using chiral transition-metal complexes.

In fact, a variety of transition metals have been used as efficient catalysts in a range of cycloaddition reactions. Among these, the Co, Ni, Ru, Rh and Pd complexes have been the major players, whilst the Ir complexes have played only a minor role. Nonetheless, several Ir-catalyzed cycloadditions have been recently reported, which cannot be realized by other metal complexes. This chapter summarizes Ir complex-catalyzed cycloadditions, which include several types of cycloisomerization and cyclization.

Iridium Complexes in Organic Synthesis. Edited by Luis A. Oro and Carmen Claver
Copyright © 2009 WILEY-VCH Verlag GmbH & Co. KGaA, Weinheim
ISBN: 978-3-527-31996-1

11.2
[2+2+2] Cycloaddition

In 1967, when Yamazaki reported the Co-mediated cycloaddition of diphenylacetylene, this opened up a new route for transition-metal-catalyzed [2+2+2] cycloaddition in organic synthesis [6]. Vollhardt's development of the Co-catalyzed reaction of alkynes, and its application in natural product synthesis, upgraded [2+2+2] cycloaddition as a synthetic protocol [3]. Since then, [2+2+2] cycloaddition of alkynes has been comprehensively investigated, and a variety of transition-metal catalysts – including Ni, Rh, Pd and Ru – complexes – have been reported [7]. In the case of the Ir-complex, Collman first reported in 1968 that oxidative coupling of [IrCl(N$_2$)(PPh$_3$)$_2$] with dimethyl acetylenedicarboxylate (DMAD) gave iridacyclopentadiene, which could function as a catalyst for the cyclotrimerization of alkynes [8]; to date, however, no practical examples of Ir-catalyzed [2+2+2] cycloadditions have been reported [9].

Takeuchi has comprehensively studied the practical use of iridium complexes in catalytic reactions [10], and has paid particular attention to their catalytic activity for [2+2+2] cycloaddition. In fact, Takeuchi reported an Ir-catalyzed intermolecular cycloaddition of diynes with monoalkynes [11]; here, the choice of phosphorus ligand for the metal was very important, and the bidentate ligand DPPE (1,2-bis(diphenylphosphino)ethane) proved to be the best. The reaction of a symmetrical diyne with unsubstituted alkyne termini and 1-hexyne proceeded at room temperature, and a bicyclic product was obtained in good yield (Scheme 11.1). Alkynes with oxygen and nitrogen functionalities were also good coupling partners, but at a higher reaction temperature.

When an unsymmetrical diyne is used, two regioisomers of *ortho* and *meta* isomers are formed. Then, by choosing the bidentate phosphorus ligand (DPPE or DPPF (1,1'-bis(diphenylphosphino)ferrocene), the ratio of their formation could be controlled from 1:4 to 7:1 (Scheme 11.2) [12].

The Ir-catalyzed [2+2+2] cycloaddition of diynes and monoalkyne advocates a new synthetic approach to silafluorenes [13]. The Ir–PPh$_3$-complex-catalyzed reaction of benzene and silicon-tethered 1,6-diynes with disubstituted alkynes gave tetra-substituted silafluorenes (Scheme 11.3). The consecutive [2+2+2] cycloaddition of two types of tetrayne provided a ladder-type silafluorene and a spirosilabifluorene (Schemes 11.4 and 11.5).

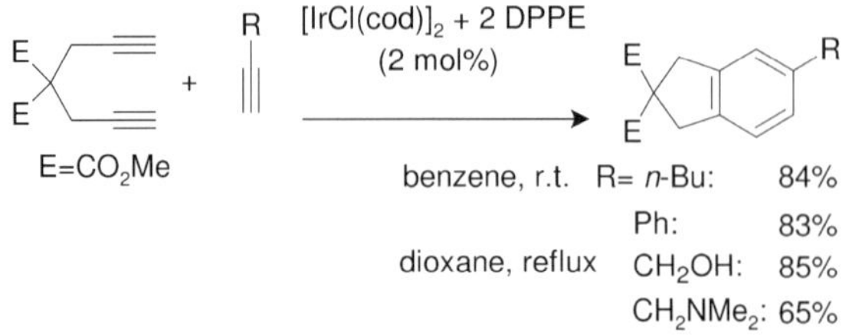

E=CO$_2$Me

[IrCl(cod)]$_2$ + 2 DPPE (2 mol%)

benzene, r.t.   R= *n*-Bu: 84%
                Ph: 83%
dioxane, reflux CH$_2$OH: 85%
                CH$_2$NMe$_2$: 65%

Scheme 11.1

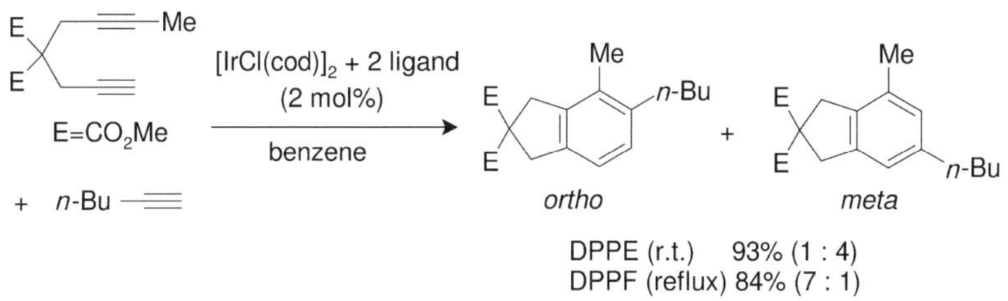

11.2 [2+2+2] Cycloaddition

Scheme 11.2

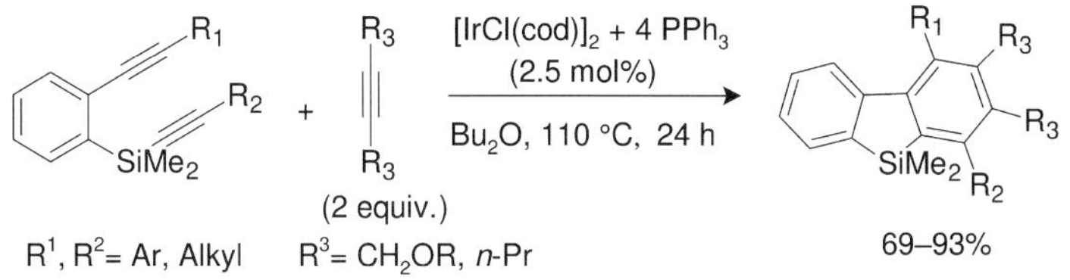

Scheme 11.3

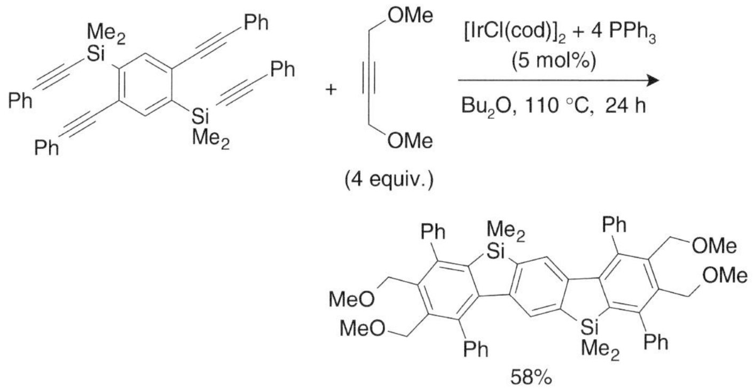

Scheme 11.4

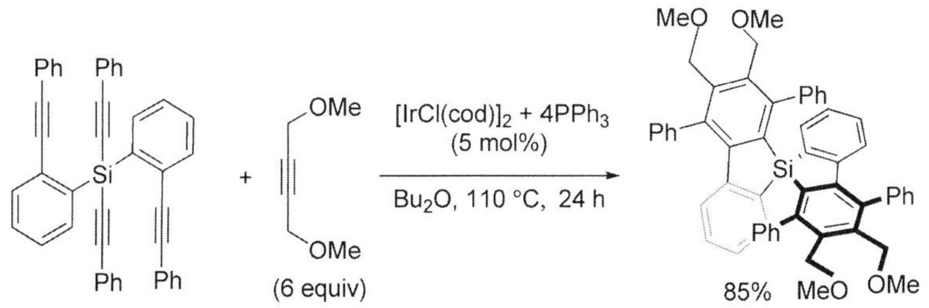

Scheme 11.5

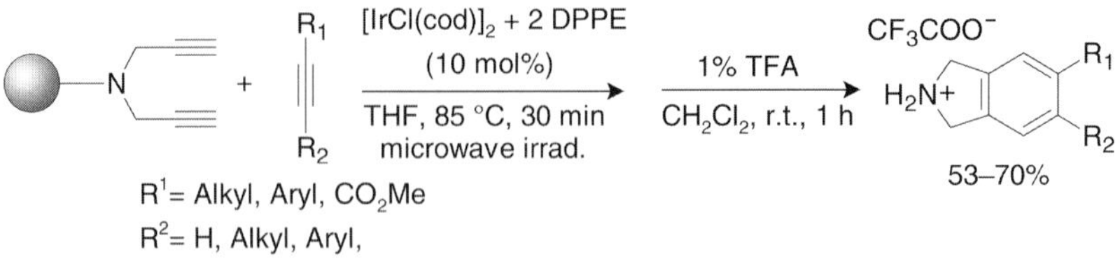

Scheme 11.6

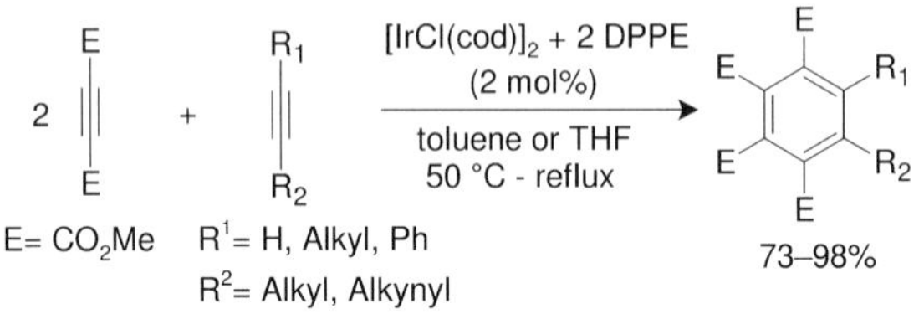

Scheme 11.7

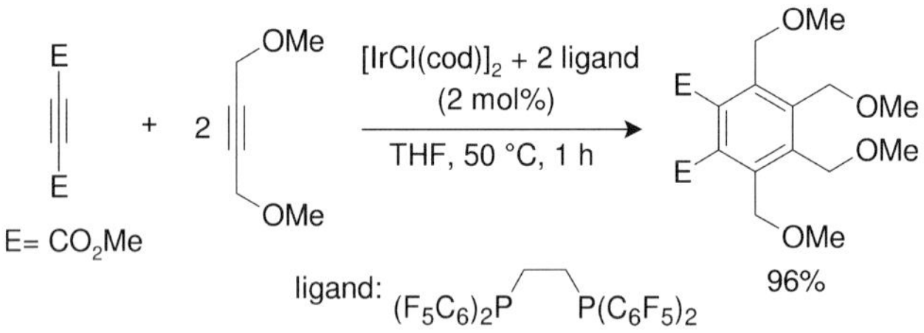

Scheme 11.8

Solid-phase [2+2+2] cycloaddition was also catalyzed by the Ir complex (Scheme 11.6) [14]. Here, the reaction of a resin-bound dipropargylamine and various monoalkynes proceeded under heating by microwave-irradiation, and isoindoline derivatives were obtained in moderate yield after trifluoroacetic acid (TFA)-cleavage of the resin-bound intermediates.

Ir-complexes also demonstrate catalytic activity in the intermolecular [2+2+2] cycloaddition of three monoalkynes, when Takeuchi examined the mixed cyclotrimerization of two alkynes. In this situation the choice of ligand was shown to determine the excellent chemoselectivity; for example, when the Ir–dppe complex was used, 2:1 cycloadducts of DMAD and mono- or disubstituted alkynes were obtained, but when Ir–F–dppe (1,2-bis[bis(pentafluorophenyl)phosphino]ethane) one was used, a 1:2 cycloadduct of DMAD and 1,4-dimethoxybut-2-yne was obtained (Schemes 11.7 and 11.8) [15]. A regioselective cyclotrimerization of three alkynes was achieved by [IrH(cod)(dppm)] (bis(diphenylphosphino)methane), and 1,2,4-triarylbenzenes were obtained exclusively (Scheme 11.9) [16].

The Ir–dppe complex also functioned well as a catalyst in the [2+2+2] cycloaddition of a 1,6-enyne and monoalkynes (Scheme 11.10). For instance, when 1,6-enyne with an asymmetric carbon was submitted to the reaction with 3-hexyne, a bicyclic cyclohexa-1,3-diene was obtained in diastereoselective manner (Scheme

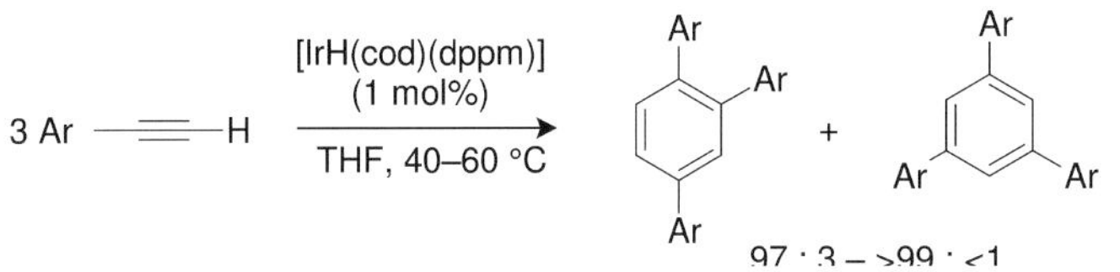

Scheme 11.9

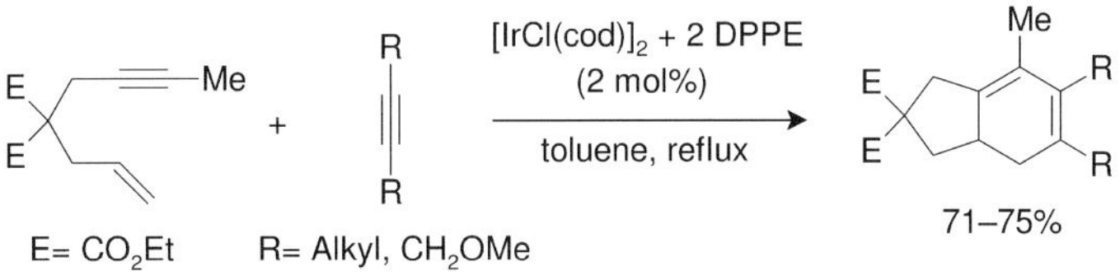

$$3 \text{ Ar}{-}{\equiv}{-}\text{H} \xrightarrow[\text{THF, 40–60 °C}]{[\text{IrH(cod)(dppm)}] \text{ (1 mol\%)}} \text{1,2,4-Ar}_3\text{C}_6\text{H}_3 + \text{1,3,5-Ar}_3\text{C}_6\text{H}_3$$

97 : 3 – >99 : <1

Scheme 11.10

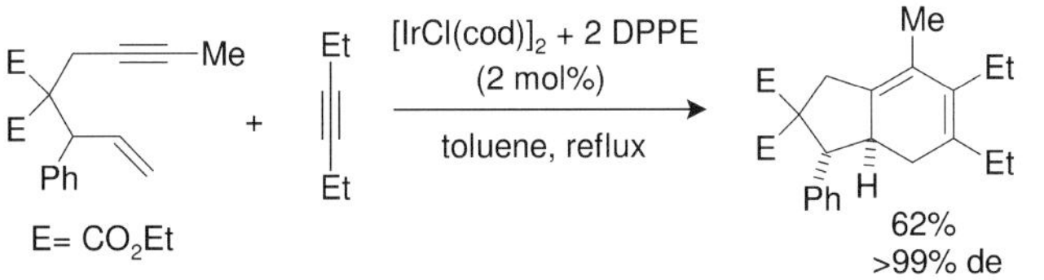

E = CO$_2$Et, R = Alkyl, CH$_2$OMe

[IrCl(cod)]$_2$ + 2 DPPE (2 mol%), toluene, reflux

71–75%

Scheme 11.11

E = CO$_2$Et

[IrCl(cod)]$_2$ + 2 DPPE (2 mol%), toluene, reflux

62%, >99% de

11.11) [17]. The enantioselective variants of the corresponding [2+2+2] cycloaddition were already achieved by chiral Rh catalyst; however, the coupling partners of enynes were limited to monoalkynes with functionality, such as ester or alcohol [18].

11.3
Enantioselective [2+2+2] Cycloaddition

Two precedent examples had been reported of the enantioselective [2+2+2] cycloaddition of alkynes. In one case, an enantioposition-selective intermolecular reaction of a triyne with acetylene generated an asymmetric carbon at the benzylic position of a formed benzene ring [19]. In the other case, an intramolecular reaction of a triyne induced helical chirality [20]. Both reactions were developed by chiral Ni catalysts.

Shibata, in contrast, disclosed Ir-catalyzed [2+2+2] cycloaddition of diynes with monoalkynes, where axial chiralities were generated along with benzene ring formation [21]. Here, the intermolecular reaction of 1,6-diynes, possessing naphthyl groups on their termini, and but-2-yne-1,4-diol gave C_2-symmetrical teraryl compounds with two axial chiralities between naphthyl groups and the formed benzene ring. The Ir–MeDUPHOS (1,2-bis(2,5-dimethylphospholano)benzene) catalyst realized almost perfect enantioselectivity (Scheme 11.12).

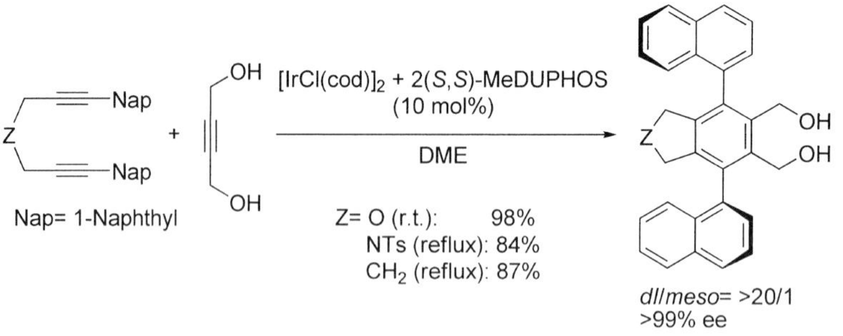

Scheme 11.12

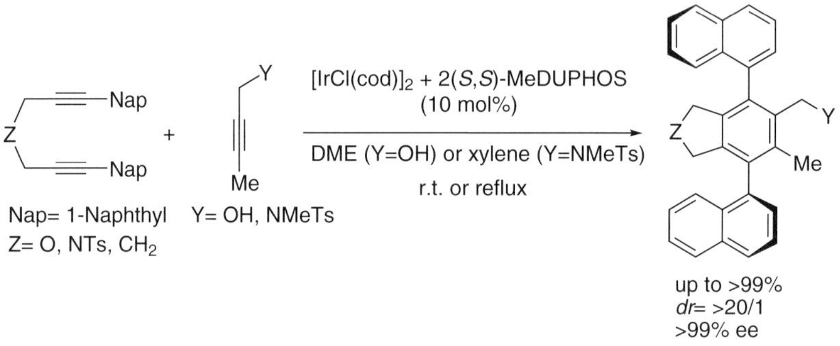

Scheme 11.13

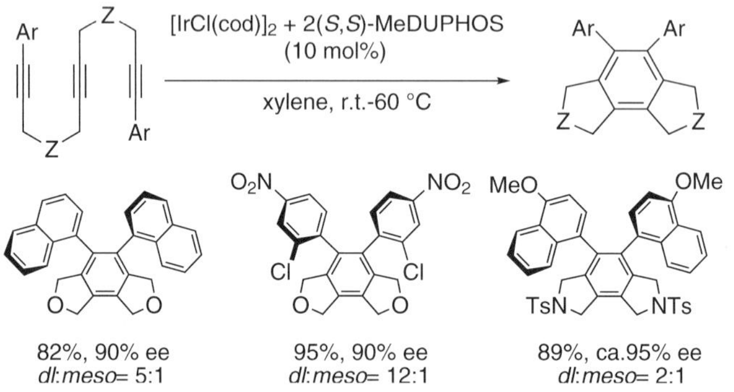

Scheme 11.14

From unsymmetrical propargyl alcohol and propargyl amine derivatives as coupling partners of diynes, the corresponding axially chiral alcohols and amines were obtained in almost perfect diastereo- and enantioselectivities (Scheme 11.13).

The Ir–MeDUPHOS catalyst also functioned efficiently in an intramolecular reaction, where triynes, which possessed *ortho*-substituted aryl groups on their termini, were transformed into *ortho*-diarylbenzene derivatives, which have adjacent two axial chiralities (Scheme 11.14) [22].

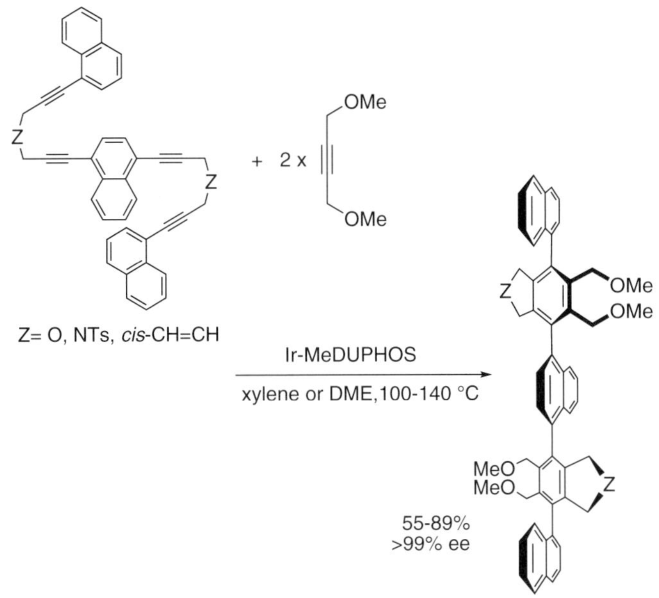

Scheme 11.15

Shibata further examined consecutive reactions of poly-ynes. When tetraynes (where two 1,6-diyne moieties were connected with a naphthalene spacer) were submitted, quinquearyl compounds with four consecutive axial chiralities were obtained (Scheme 11.15). Even an octayne proved to be a good substrate, and a noviaryl compound with eight consecutive axial chiralities was obtained in almost perfect enantioselectivity (Scheme 11.16) [23].

In the intermolecular reaction of tetraynes, where two 1,6-diyne moieties were directly connected, with monoalkynes, CHIRAPHOS (2,3-bis(diphenylphosphino) butane) was the choice of chiral ligand, and axial chirality was enantiomerically generated between the formed benzene rings (Scheme 11.17). Hexaynes with a 1,3-diyne moiety also underwent an intramolecular [2+2+2] cycloaddition, and the Ir–xylylBINAP (2,2′-bis[di(3,5-xylyl)phosphino]-1,1′-binaphthyl) catalyst induced an excellent enantiomeric excess (ee) (Scheme 11.18) [24].

Independently, the same concept of enantioselective [2+2+2] cycloaddition for the construction of axial chirality was achieved with chiral Co and Rh catalysts [25].

11.4
[2+2+1] Cycloaddition

Khand and Pauson reported a Co-mediated intermolecular [2+2+1] cycloaddition of an alkyne, an alkene and carbon monoxide (the Pauson–Khand reaction) [4, 26] wherein an alkyne–$Co_2(CO)_6$ complex, which had been prepared from $Co_2(CO)_8$

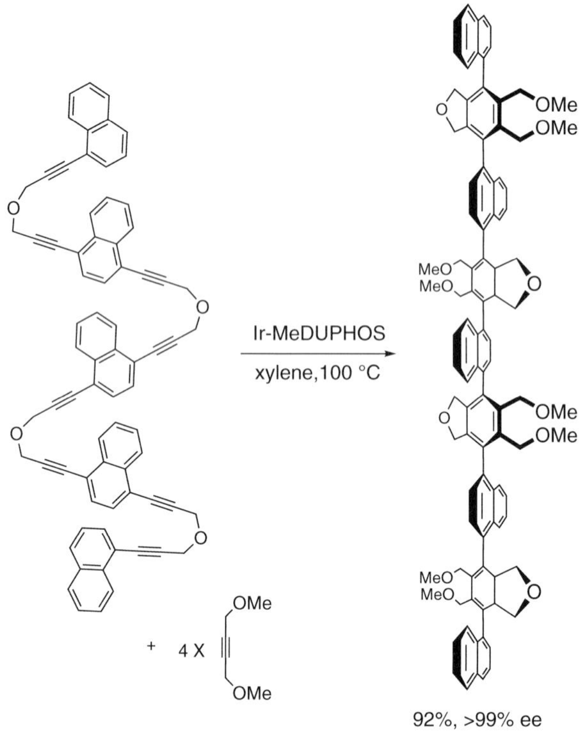

Scheme 11.16

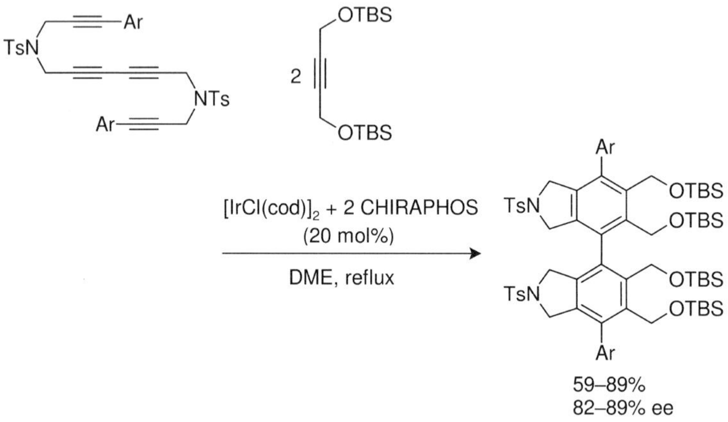

Scheme 11.17

and an alkyne, reacted with an alkene to give a synthetically useful cyclopentenone as a mixture of regioisomers. The intramolecular reaction of enynes, which gives bicyclic cyclopentenones, has been used as a key reaction of natural product synthesis. As a catalytic reaction, Jeong reported an intramolecular Pauson–Khand reaction of enynes using a mixture of $Co_2(CO)_8$ and $P(OPh)_3$ [27]. In addition to

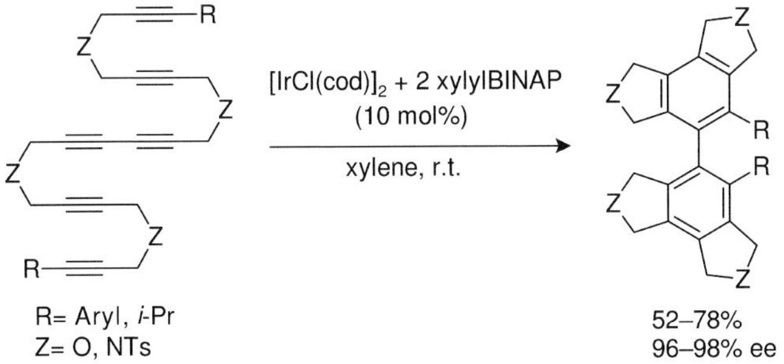

Scheme 11.18

R= Aryl, i-Pr
Z= O, NTs

[IrCl(cod)]$_2$ + 2 xylylBINAP (10 mol%)
xylene, r.t.

52–78%
96–98% ee

Co complexes, Buchwald reported the Ti-catalyzed carbonylative coupling of enynes – the so-called 'Pauson–Khand-type' reaction [28] – and realized the first such catalytic and enantioselective reaction using a chiral Ti complex [29]. Here, a variety of enynes were transformed into bicyclic cyclopentenones with good to high ee-values; however, several steps were required to prepare the chiral Ti catalyst, while the low-valent complex proved to be so unstable that it had to be treated under oxygen-free conditions in a glove box.

By contrast, in 2000 Shibata reported the Ir-catalyzed enantioselective Pauson–Khand-type reaction of enynes [30a]. The chiral Ir catalyst was readily prepared *in situ* from [IrCl(cod)]$_2$ and tolBINAP (2,2'-bis(di-*p*-tolylphosphino)-1,1'-binaphthyl), both of which are commercially available and air-stable, and the reaction proceeded under an atmospheric pressure of carbon monoxide. The Ir-catalyzed carbonylative coupling had a wide generality in enynes with various tethers (Z), substituents on the alkyne terminus (R^1) and the olefinic moiety (R^2). In the case of less-reactive enynes, a lower partial pressure of carbon monoxide achieved a higher yield and ee-value (Table 11.1) [30b].

The Ir–tolBINAP catalyst also functions well in the desymmetrization of dienynes, where a highly enantioselective and diastereoselective Pauson–Khand-type reaction proceeded to give vinyl-substituted bicyclic cyclopentenones with two chiral centers (Scheme 11.19) [31].

Subsequently, it was found that aldehydes could be used as a CO source rather than the toxic CO gas. However, the choice of aldehydes proved to be very important; for example, when Shibata used cinnamaldehyde and Chan used decanal, highly enantioselective Ir-catalyzed Pauson–Khand-type reactions were achieved independently [30b, 32] (Scheme 11.20). Whilst Shibata realized at an early stage that the Rh–tolBINAP complex-catalyzed enantioselective Pauson–Khand-type reaction served as a CO source [33], it was apparent that the Ir-catalysts could induce a greater enantioselectivity.

Allenyne represents an interesting substrate for the intramolecular Pauson–Khand(-type) reaction, where an allene moiety acts as an ene component. Here, there are two possible reaction pathways (Scheme 11.21): (i) the reaction of an external π-bond of allene moiety gives a bicyclic dienone (type **A**); or (ii) the reaction of an internal π-bond gives a bicyclic cyclopentenone with an alkylidene substituent (type **B**).

Table 11.1 Ir-catalyzed enantioselective Pauson–Khand-type reaction of enynes.

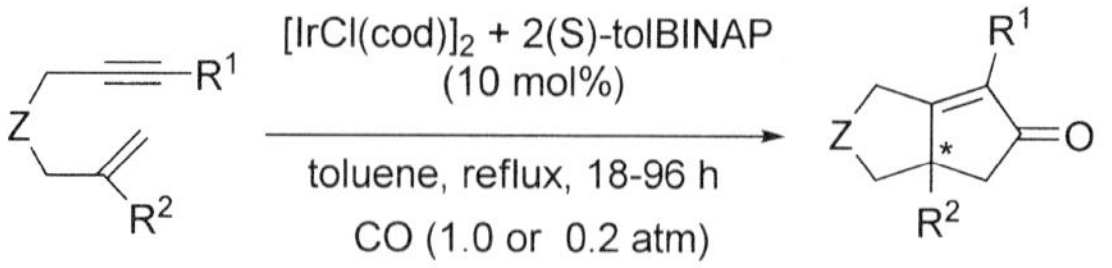

Entry	Enyne	Cyclopentenone	CO (atm)	Yield (%)	ee (%)
1	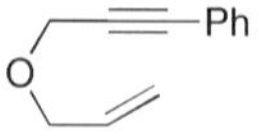 Z=O, R¹=Ph, R²=H	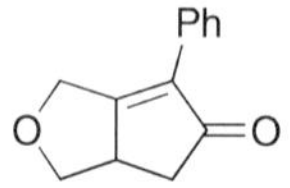 Ph-substituted bicyclic enone	1.0	83	93
2	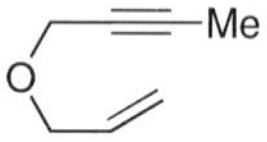 Z=O, R¹=Me, R²=H	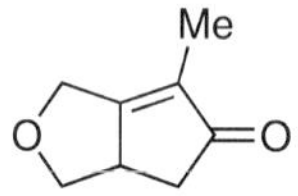 Me-substituted bicyclic enone	1.0	61	98
3	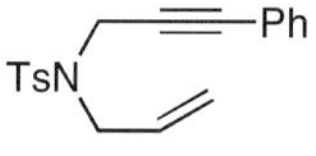 Z=TsN, R¹=Ph, R²=H	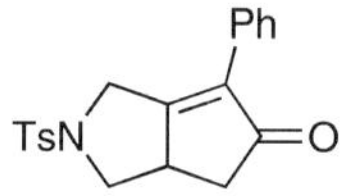 Ph-substituted bicyclic enone	1.0	85	95
4	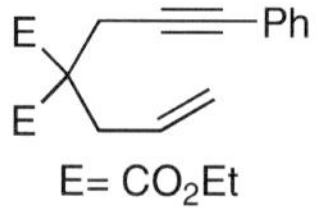 Z=C(CO₂Et)₂, R¹=Ph, R²=H	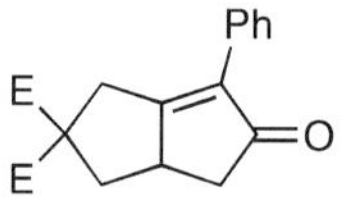 Ph-substituted bicyclic enone	1.0 / 0.2	74 / 87	84 / 86
5	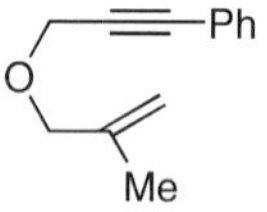 Z=O, R¹=Ph, R²=Me	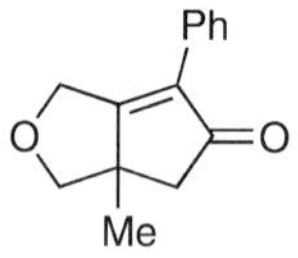 Ph, Me-substituted bicyclic enone	1.0 / 0.2	30 / 86	88 / 93
6	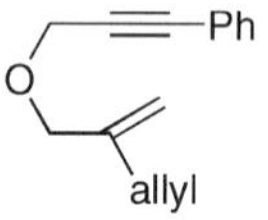 Z=O, R¹=Ph, R²=allyl	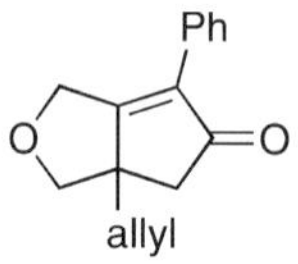 Ph, allyl-substituted bicyclic enone	0.2 / 1.0	22 / 62	86 / 94

Conditions: [IrCl(cod)]$_2$ + 2(S)-tolBINAP (10 mol%), toluene, reflux, 18–96 h, CO (1.0 or 0.2 atm).

Scheme 11.19

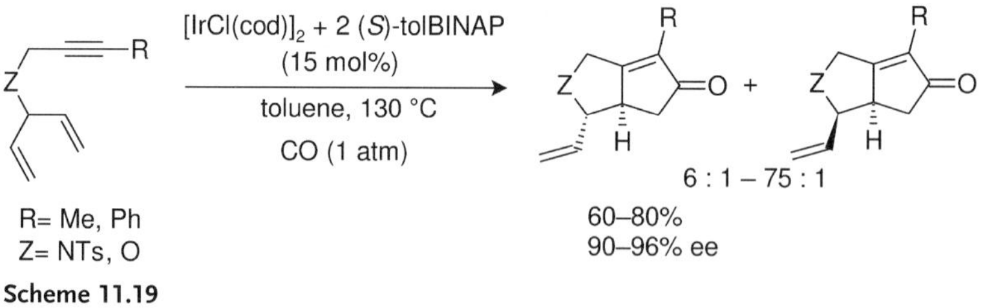

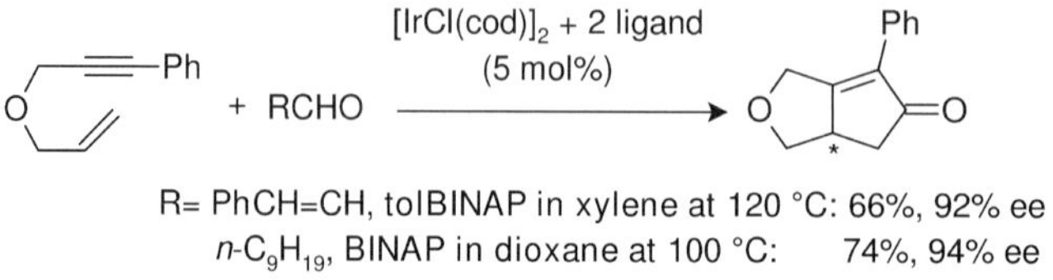

[IrCl(cod)]$_2$ + 2 (S)-tolBINAP (15 mol%), toluene, 130 °C, CO (1 atm)

R = Me, Ph
Z = NTs, O

6 : 1 – 75 : 1
60–80%
90–96% ee

Scheme 11.20

Enyne + RCHO → cyclopentenone product
[IrCl(cod)]$_2$ + 2 ligand (5 mol%)

R = PhCH=CH, tolBINAP in xylene at 120 °C: 66%, 92% ee
n-C$_9$H$_{19}$, BINAP in dioxane at 100 °C: 74%, 94% ee

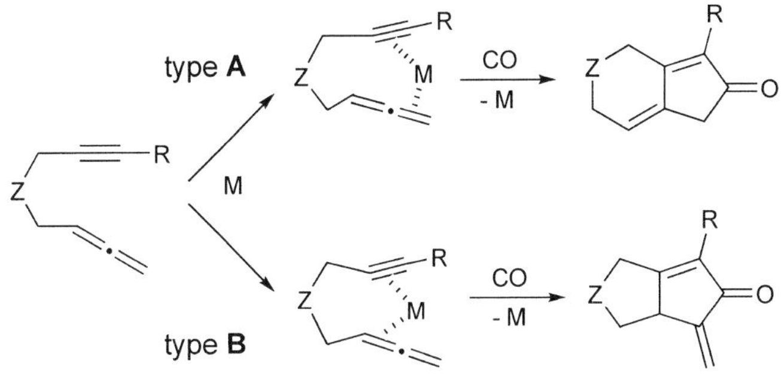

11.4 [2+2+1] Cycloaddition

Scheme 11.21

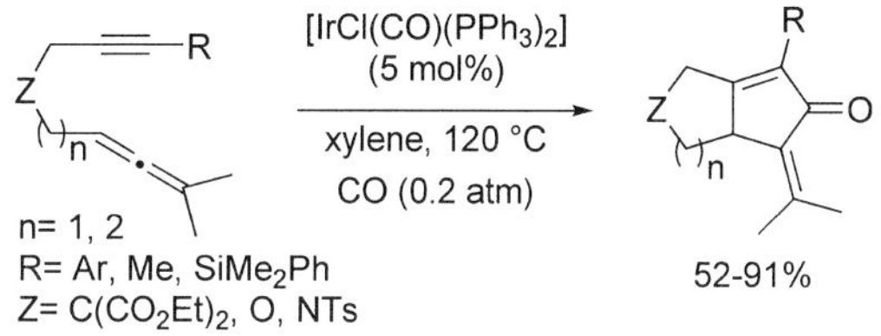

Scheme 11.22

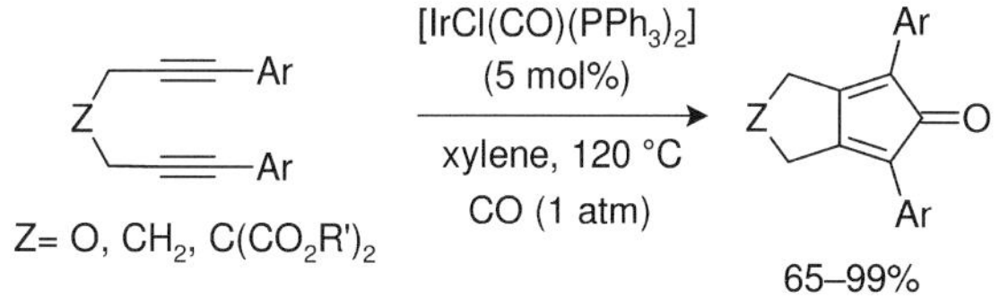

n = 1, 2
R = Ar, Me, SiMe$_2$Ph
Z = C(CO$_2$Et)$_2$, O, NTs

Scheme 11.23

Z = O, CH$_2$, C(CO$_2$R')$_2$

Although all of the reported Rh-catalyzed reactions of allenynes were of type **A**, an Ir catalyst resulted in a different regioselectivity. That is, when allenynes with two substituents on the allene terminus were used under a low partial pressure of CO, the type **B** reaction proceeded exclusively such that bicyclic cyclopentenones with an alkylidene substituent were obtained (Scheme 11.22) [34]. However, when [RhCl(CO)(PPh$_3$)$_2$] was used as a catalyst under the same reaction conditions in place of [IrCl(CO)(PPh$_3$)$_2$], the type **A** reaction was predominant. These results imply that the metal centers of the catalysts control the regioselectivity of two olefinic moieties of allene to some extent.

Vaska's complex ([IrCl(CO)(PPh$_3$)$_2$]) also catalyzed the carbonylative coupling of diynes, which provided bicyclic cyclopentadienones (Scheme 11.23) [35]. Due to the instability of the products, the substrates are limited to symmetrical diynes with aromatic groups on their termini; nonetheless, this reaction still serves as the catalytic and practical procedure for the synthesis of cyclopentadienones, which are anti-aromatic with a 4π system and serve as active synthetic intermediates.

11.5
[4+2] and [5+1] Cycloadditions

Beside [2+2+2] cycloaddition, [4+2] and [5+1] cycloadditions represent other approaches for the construction of six-membered ring systems. In particular, the intermolecular and intramolecular [4+2] cycloadditions of diene and alkyne have been extensively studied, and a variety of transition-metal complexes – including those of Fe, Ni and Rh – have been reported as efficient catalysts. The first enantioselective reaction was achieved with a chiral Rh complex, although the substrates were limited to dienynes with a substituent on the diene terminus [36]. Later, Shibata and coworkers developed an intramolecular and enantioselective [4+2] cycloaddition using an Ir–BDPP (1,3-bis(diphenylphosphino)pentane) complex (Scheme 11.24) [37], where dienynes with an unsubstituted diene terminus were transformed into bicyclic cyclohexa-1,4-diene with up to 98% ee.

Vaska's complex catalyzed the transformation of allenylcyclopropane into 2-alkenylidenecyclohex-3-enone under conditions of pressurized CO (Scheme 11.25) [38]. In this reaction, the π-coordination to internal olefinic moiety of the allene brings the metal closer to the cyclopropane ring. Release of the cyclopropane ring strain then facilitates the oxidative addition of vinylcyclopropane moiety along with C–C bond cleavage, such that metallacyclohexene is obtained; a subsequent carbonyl insertion and reductive elimination then provides the product. Hence, the reaction can be recognized as a [5+1] cycloaddition of vinylcyclopropane and CO.

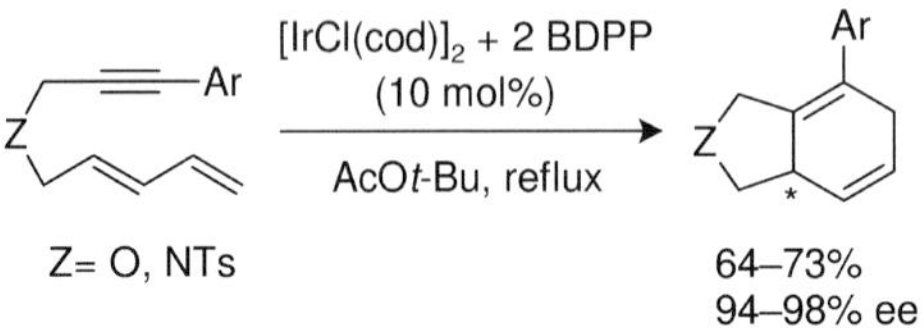

Scheme 11.24

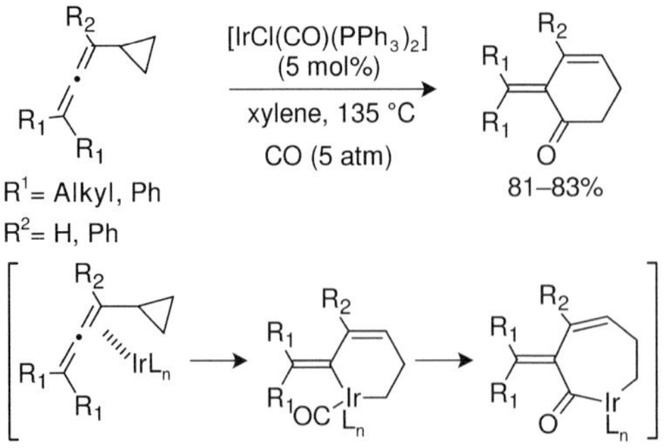

Scheme 11.25

11.6
Cycloisomerization

Cycloisomerization represents another approach for the construction of cyclic compounds from acyclic substrates, with iridium complexes functioning as efficient catalysts. The reaction of enynes has been widely studied; for example, Chatani et al. reported the transformation of 1,6-enynes into 1-vinylcyclopentenes using [IrCl(CO)$_3$]$_n$ (Scheme 11.26) [39]. In contrast, when 1,6-enynes were submitted in the presence of [IrCl(cod)]$_2$ and AcOH, cyclopentanes with two *exo*-olefin moieties were obtained (Scheme 11.27) [39]. Interestingly, however, when the Ir–DPPF complex was used, the geometry of olefinic moiety in the product was opposite (Scheme 11.28) [17]. The Ir-catalyzed cycloisomerization was efficiently utilized in a tandem reaction along with a Cu(I)-catalyzed three-component coupling, Diels–Alder reaction, and dehydrogenation for the synthesis of polycyclic pyrroles [40].

In the presence of [IrCl(cod)]$_2$, heteroatom-tethered enynes with (*E*)-olefinic moieties were transformed into cyclic 1,4-dienes. The ene-type reaction was achieved previously with a Rh catalyst, but only enynes with (*Z*)-olefinic moieties were used. It is worthy of note here that the cycloisomerization showed a clear acceleration when an ionic liquid was used as the solvent (Scheme 11.29) [41].

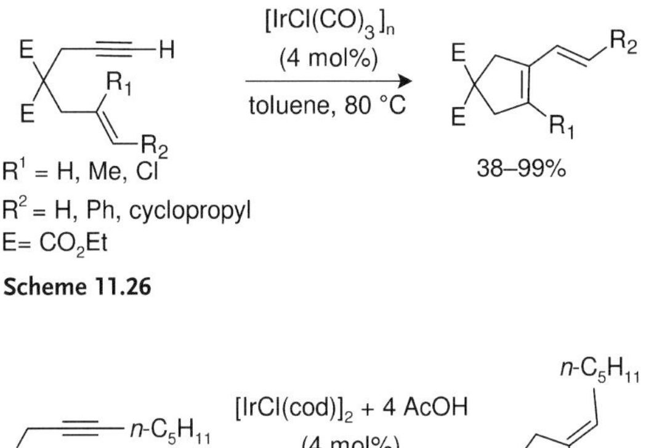

R^1 = H, Me, Cl
R^2 = H, Ph, cyclopropyl
E = CO$_2$Et

Scheme 11.26

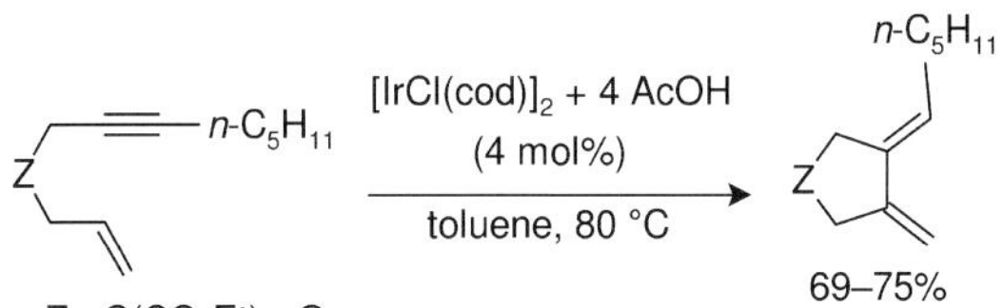

Z = C(CO$_2$Et)$_2$, O

Scheme 11.27

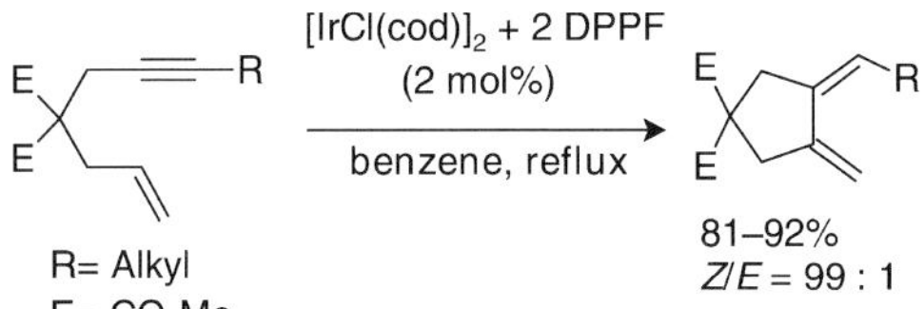

R = Alkyl
E = CO$_2$Me

81–92%
Z/E = 99 : 1

Scheme 11.28

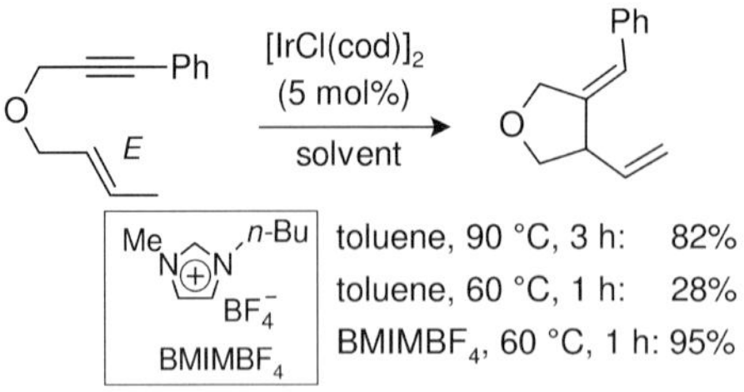

Scheme 11.29

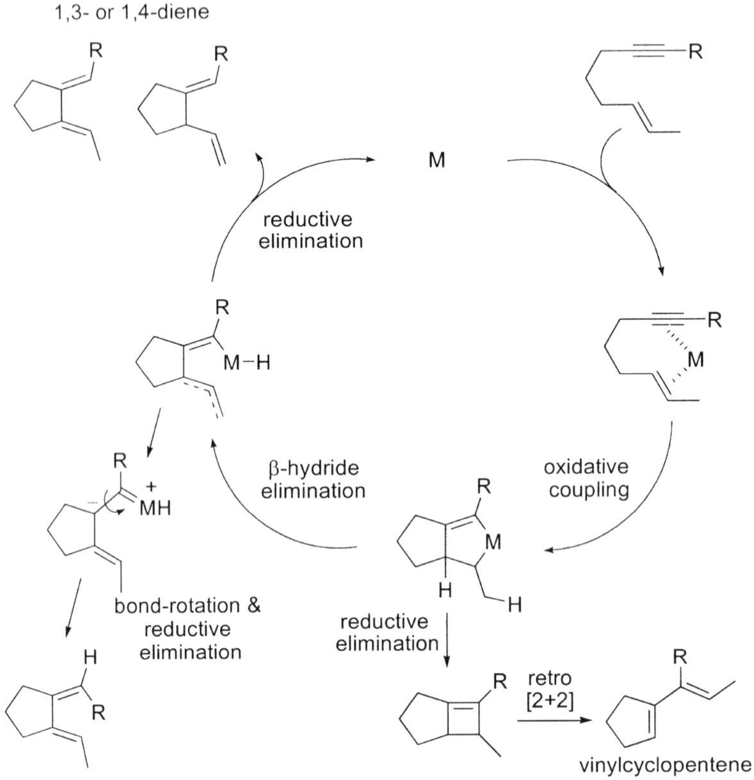

Conditions	Yield
toluene, 90 °C, 3 h:	82%
toluene, 60 °C, 1 h:	28%
BMIMBF$_4$, 60 °C, 1 h:	95%

Scheme 11.30

The proposed mechanism of the above cycloisomerizations are depicted in Scheme 11.30. The oxidative coupling of a metal to an enyne yields a bicyclic metallacyclopentene, which is a common intermediate. The reductive elimination and subsequent retro-[2+2] cycloaddition gave vinylcyclopentene derivatives, while the two patterns of β-elimination and subsequent reductive elimination gave cyclic 1,3- and 1,4-dienes, respectively. The existence of a carbene complex intermediate might explain the isomerization of the olefinic moiety.

In the presence of a cationic iridium catalyst, nitrogen-tethered enynes, possessing 1,1-disubstituted ene moiety, were transformed into cyclohexene fused with a

Scheme 11.31

cyclopropane ring (Scheme 11.31) [42]. Although details of the same type of cycloisomerization of enynes using Pt- and Au-catalysts had been published [43], the chiral Ir complex realized the first enantioselective cycloisomerization.

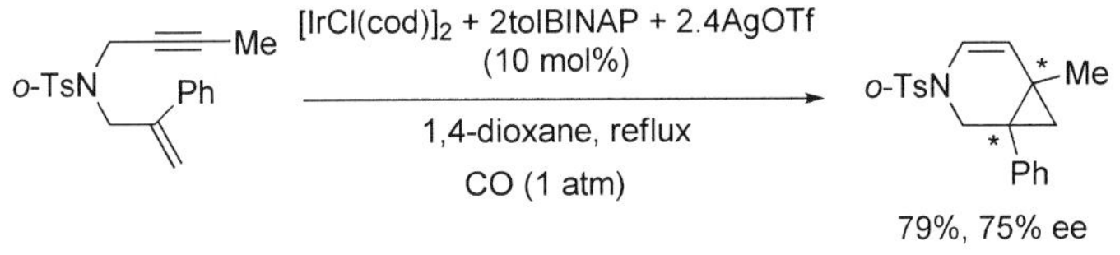

11.7 Ir(III)-Catalyzed Cyclizations

In the previous sections, a number of Ir(I)-catalyzed cycloadditions were described where the oxidative coupling of unsaturated motifs (e.g. alkynes and alkenes) provides metallacycles as common intermediates. At this point, mention will be made of three examples of Ir(III)-catalyzed cyclizations.

- Ir(III) complexes act like Lewis acid: Carmona and coworkers pioneered investigations into an enantioselective Diels–Alder reaction using chiral Cp*Ir(III) complexes (Cp* = pentamethylcyclopentadienyl). Following the screening of various chiral ligands in the reaction of cyclopentadiene with methacrolein [44], Carmona's group achieved high enantioselectivity (90% ee) using PROPHOS (1,2-bis(diphenylphosphino)propane) (Scheme 11.32) [44d]. The same chiral Ir(III) catalyst also functioned well in an enantioselective [3+2] cycloaddition of nitrones and α,β-unsaturated aldehydes, with chiral isoxazolinecarbaldehydes being obtained with excellent ee-values (Scheme 11.33) [45].

R^1 = H, *p*-F, Br, CO_2Me, CN, *o*-OMe
R^2 = H, *p*-F, CO_2Me, CN, *o*-OMe, Br
R^3= Alkyl

Scheme 11.32

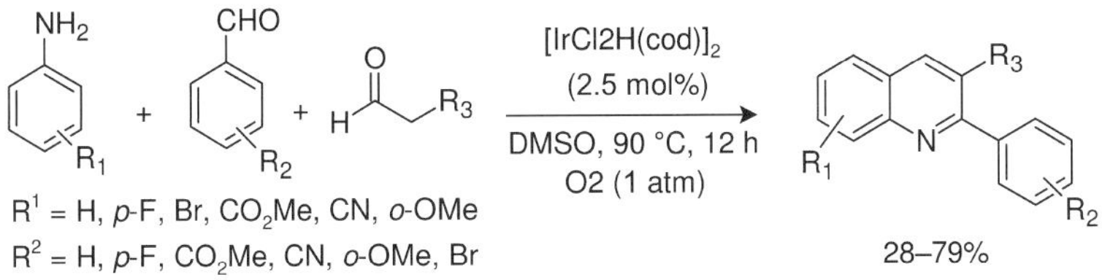

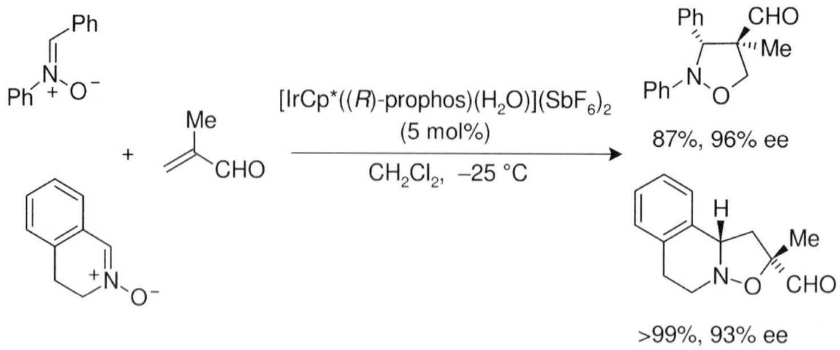

Scheme 11.33

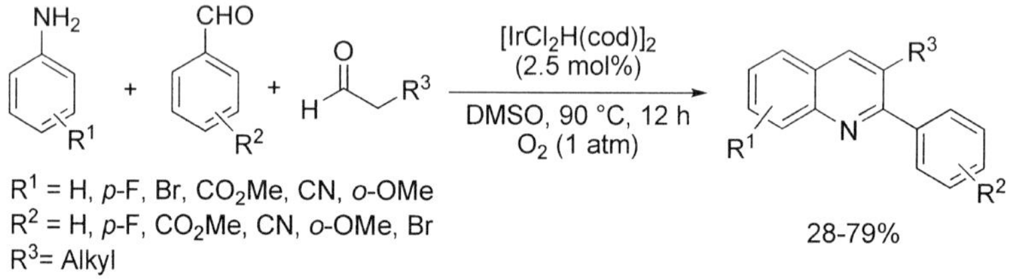

Scheme 11.34

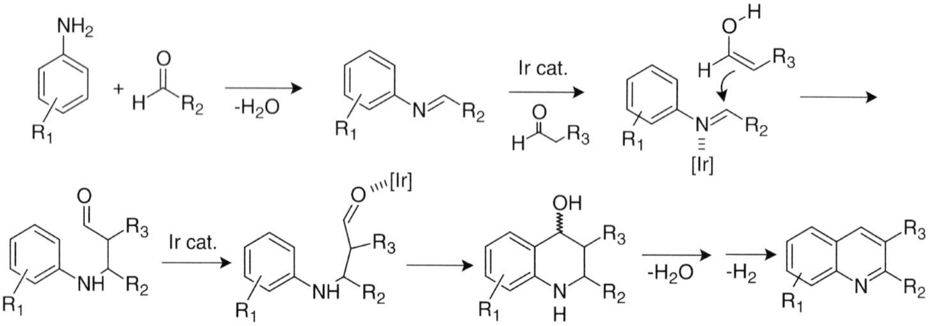

R^1 = H, p-F, Br, CO$_2$Me, CN, o-OMe
R^2 = H, p-F, CO$_2$Me, CN, o-OMe, Br
R^3= Alkyl

Scheme 11.35

- [IrCl$_2$H(cod)]$_2$ catalyzed the synthesis of substituted quinolines, where the reaction of aniline derivatives, aromatic and alkyl aldehydes efficiently proceeds under an oxygen atmosphere (Scheme 11.34) [46]. The plausible mechanism consists of a Mannich reaction, a Friedel–Craft-type aromatic substitution, dehydration, and dehydrogenation. This can be recognized as a formal [4+2] cycloaddition of N-aryl imine and enol (Scheme 11.35).

- The Ir(III) complex also functioned as a catalyst in a tandem Nazarov cyclization–Michael addition. The reaction of monocyclic α-alkylidene-β-keto-γ,δ-unsaturated ester with nitroalkene gave bicyclic cyclopentenones which possessed an alkyl side chain, with high yield and diastereoselectivity (Scheme 11.36) [47].

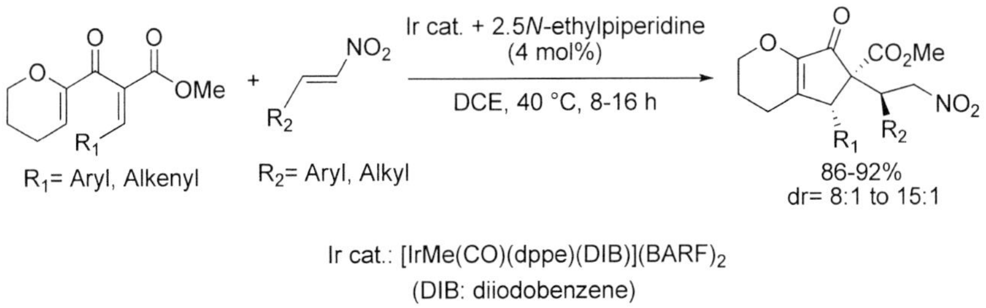

[Scheme 11.36 reaction diagram]

R₁= Aryl, Alkenyl   R₂= Aryl, Alkyl

Ir cat. + 2.5 N-ethylpiperidine (4 mol%)
DCE, 40 °C, 8-16 h

86-92%
dr= 8:1 to 15:1

Ir cat.: [IrMe(CO)(dppe)(DIB)](BARF)$_2$
(DIB: diiodobenzene)

Scheme 11.36

11.8
Miscellaneous Cycloadditions

In this section I refer to a number of cycloadditions which cannot be categorized into the above-described types. For example, Murai and colleagues reported the four-component coupling of alkyne, hydrazone, hydrosilane and CO. Here, the [Ir$_4$(CO)$_{12}$]-catalyzed reaction proceeded under pressurized CO conditions at high temperature, whereby a seven-membered nitrogen heterocycle was obtained (Scheme 11.37) [48].

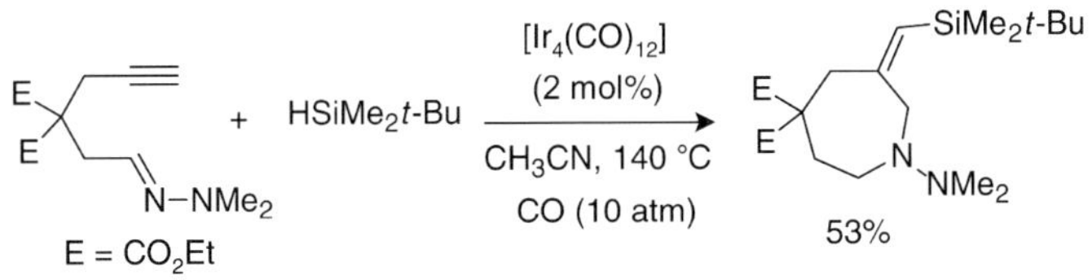

[Scheme 11.37 reaction diagram]

E = CO$_2$Et

[Ir$_4$(CO)$_{12}$] (2 mol%)
CH$_3$CN, 140 °C
CO (10 atm)

53%

Scheme 11.37

When using an Ir–P(t-Bu)$_3$ complex as catalyst, aroyl chloride reacted with two disubstituted alkynes to give polysubstituted naphthalenes (Scheme 11.38). The reaction included decarbonylation and two-time alkyne insertions along with C–H bond cleavage of the *ortho*-position of the chloroformyl group (Scheme 11.39) [49a]. When using [IrCp*Cl$_2$]$_2$ as a catalyst and Ag$_2$CO$_3$ as an oxidant, benzoic acid also reacted with two disubstituted alkynes, accompanied by decarboxylation (Scheme 11.40) [49b]. These reactions may be recognized as the formal [2+2+2] cycloaddition of benzene and two alkynes.

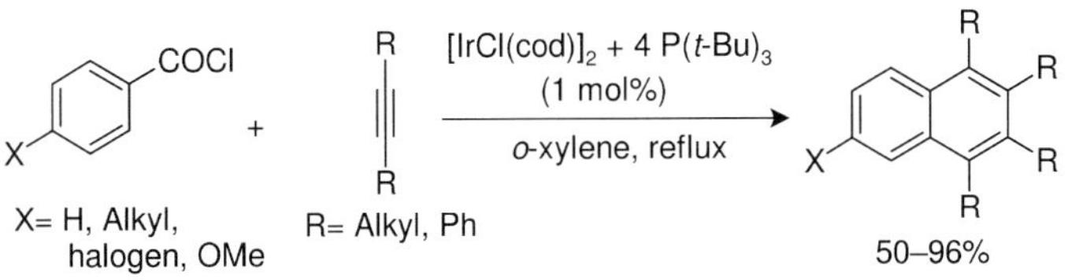

[Scheme 11.38 reaction diagram]

X= H, Alkyl, halogen, OMe   R= Alkyl, Ph

[IrCl(cod)]$_2$ + 4 P(t-Bu)$_3$ (1 mol%)
o-xylene, reflux

50–96%

Scheme 11.38

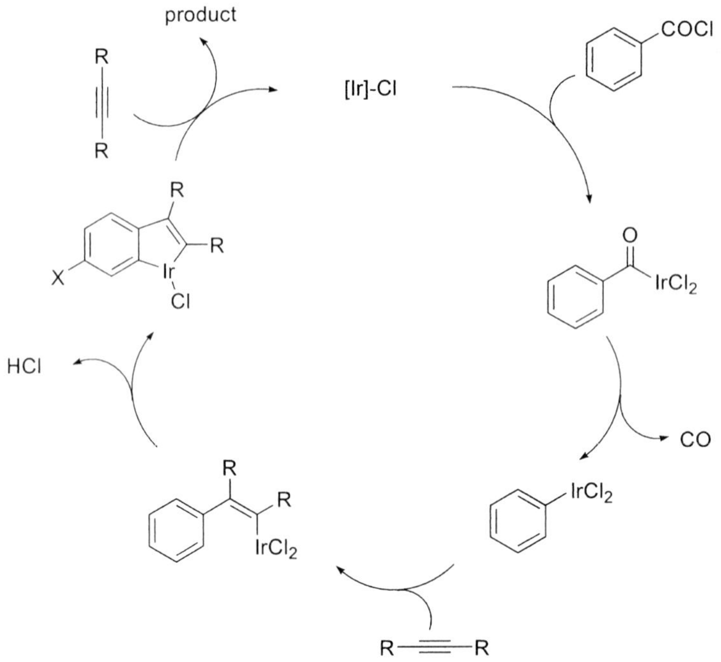

Scheme 11.39

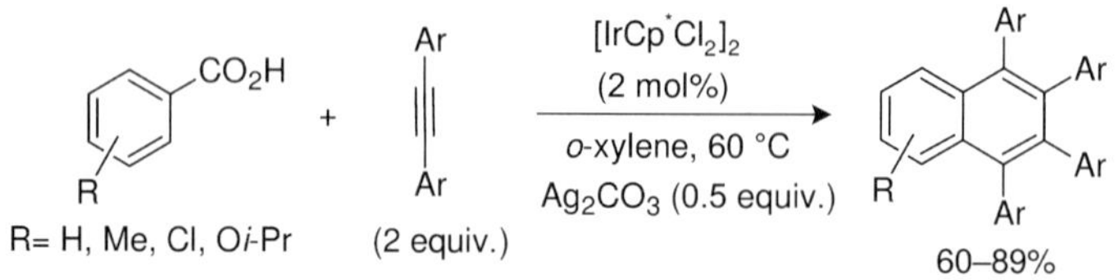

Scheme 11.40

[Ir(OH)(cod)]$_2$ catalyzed a formal [3+2] cycloaddition of 2-formylphenylboronic acid and 1,3-dienes (Scheme 11.41) [50]. The transmetallation of boronic acid with iridium would yield aryliridium, where the carbonyl group coordinates to the metal. An electrophilic attack of the diene terminus to formyl carbon would then

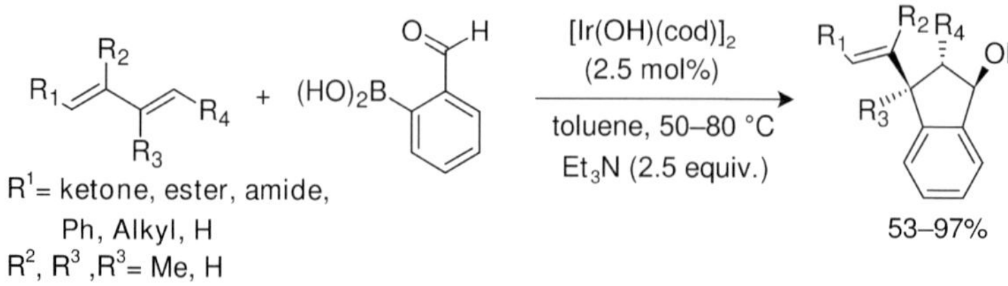

Scheme 11.41

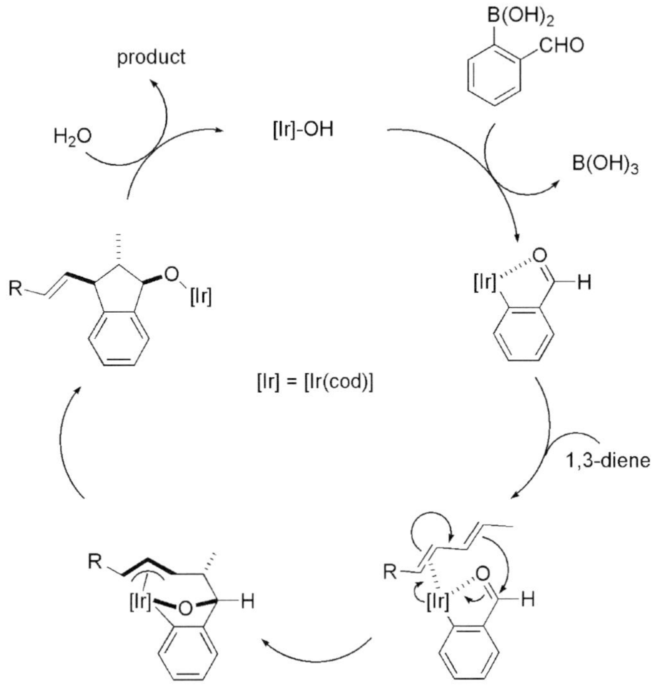

Scheme 11.42

provide the π-allyliridium complex along with C–C bond formation. The subsequent reductive elimination and hydrolysis would provide indanol derivatives (Scheme 11.42).

11.9
Conclusions

In this chapter, the details of several types of Ir–complex-catalyzed cycloadditions have been summarized. Although, compared to other late transition-metal complexes – such as those of Pd, Ni, Ru and Rh – the examples are few in number, some notable Ir-catalyzed cyclizations have recently been reported which cannot be achieved when utilizing other metal catalysts. Until now it has not yet been possible to identify any distinct explanation for the unique reactivity of iridium, and in particular its different reactivity compared to rhodium, which is located just above iridium in the Periodic Table of the elements. Nonetheless, many further developments of efficient and practical Ir-catalyzed cycloadditions are to be expected in the near future.

References

1 (a) Ojima, I., Tzamarioudaki, M., Li, Z. and Donovan, R.J. (1996) *Chemical Reviews*, **96**, 635–62;
(b) Aubert, C., Buisine, O. and Malacria, M. (2002) *Chemical Reviews*, **102**, 813–34;
(c) Nakamura, I. and Yamamoto, Y. (2004) *Chemical Reviews*, **104**, 2127–98.

2 Reppe, W., Schlichting, O., Klager, K. and Toepel, T. (1948) *Justus Liebig's Annalen der Chemie*, **560**, 1–92.

3 (a) Aalbersberg, W.G.L., Barkovich, A.J., Funk, R.L., Hillard, III, R.L. and Vollhardt, K.P.C. (1975) *Journal of the American Chemical Society*, **97**, 5600–2; Reviews:
(b) Vollhardt, K.P.C. (1977) *Accounts of Chemical Research*, **10**, 1–8;
(c) Vollhardt, K.P.C. (1984) *Angewandte Chemie*, **96**, 525–41.

4 Khand, I.U., Knox, G.R., Pauson, P.L., Watts, W.E. and Foreman, M.I. (1973) *Journal of the Chemical Society – Perkin Transactions*, **1**, 977–81.

5 Recent selected reviews: (a) Carmona, D., Lamata, M.P. and Oro, L.A. (2000) *Coordination Chemistry Reviews*, **200–202**, 717–72;
(b) Hayashi, Y. (2001) *Cycloaddition in Organic Synthesis* (eds S. Kobayashi and K.A. Jørgensen), Wiley-VCH Verlag GmbH, Weinheim, pp. 5–55;
(c) Corey, E.J. (2002) *Angewandte Chemie – International Edition*, **41**, 1650–67.

6 (a) Yamazaki, H. and Hagihara, N. (1967) *Journal of Organometallic Chemistry*, **7**, P 22–3;
(b) Wakatsuki, Y., Kuramitsu, T. and Yamazaki, H. (1974) *Tetrahedron Letters*, 4549–52.

7 Recent reviews: (a) Yamamoto, Y. (2005) *Current Organic Chemistry*, **9**, 503–9;
(b) Kotha, S., Brahmachary, E. and Lahiri, K. (2005) *European Journal of Organic Chemistry*, 4741–67;
(c) Chopade, P.R. and Louie, J. (2006) *Advanced Synthesis & Catalysis*, **348**, 2307–27.

8 Collman, J.P., Kang, J.W., Little, W.F. and Sullivan, M.F. (1968) *Inorganic Chemistry*, **7**, 1298–303.

9 (a) Gardner, S.A., Andrews, P.S. and Rausch, M.D. (1973) *Inorganic Chemistry*, **12**, 2396–402;
(b) Baddley, W.H. and Tupper, G.B. (1974) *Journal of Organometallic Chemistry*, **67**, C16–18.

10 (a) Takeuchi, R. (2002) *Synlett*, 1954–65;
(b) Takeuchi, R. and Kezuka, S. (2006) *Synlett*, 3349–66.

11 Takeuchi, R., Tanaka, S. and Nakaya, Y. (2001) *Tetrahedron Letters*, **42**, 2991–4.

12 Kezauka, S., Tanaka, S., Ohe, T., Nakaya, Y. and Takeuchi, R. (2006) *Journal of Organic Chemistry*, **71**, 543–52.

13 Matsuda, T., Kadowaki, S., Goya, T. and Murakami, M. (2007) *Organic Letters*, **9**, 133–6.

14 Shanmugasundaram, M., Aguirre, A.L., Leyva, M., Quan, B. and Martinez, L.E. (2007) *Tetrahedron Letters*, **48**, 7698–701.

15 Takeuchi, R. and Nakaya, Y. (2003) *Organic Letters*, **5**, 3659–62.

16 Farnetti, E. and Marsich, N. (2004) *Journal of Organometallic Chemistry*, **689**, 14–17.

17 Kezuka, S., Okado, T., Niou, E. and Takeuchi, R. (2005) *Organic Letters*, **7**, 1711–14.

18 (a) Evans, P.A., Lai, K.W. and Sawyer, J.R. (2005) *Journal of the American Chemical Society*, **127**, 12466–7;
(b) Shibata, T., Arai, Y. and Tahara, Y. (2005) *Organic Letters*, **7**, 4955–7.

19 Sato, Y., Nishimata, T. and Mori, M. (1994) *Journal of Organic Chemistry*, **59**, 6133–5.

20 Stará, I.G., Starý, I., Kollárovič, A., Teplý, F., Vyskočil, Š. and Šaman, D. (1999) *Tetrahedron Letters*, **40**, 1993–6.

21 (a) Shibata, T., Fujimoto, T., Yokota, K. and Takagi, K. (2004) *Journal of the American Chemical Society*, **126**, 8382–3;
(b) Shibata, T., Arai, Y., Takami, K., Tsuchikama, K., Fujimoto, T., Takebayashi, S. and Takagi, K. (2006) *Advanced Synthesis & Catalysis*, **348**, 2475–83.

22 Shibata, T., Tsuchikama, K. and Otsuka, M. (2006) *Tetrahedron: Asymmetry*, **17**, 614–19.

23 Shibata, T. and Tsuchikama, K. (2005) *Chemical Communications*, 6017–19.

24 Shibata, T., Yoshida, S., Arai, Y., Otsuka, M. and Endo, K. (2008) *Tetrahedron*, **64**, 821–30.
25 (a) Gutnov, A., Heller, B., Fischer, C., Drexler, H.-J., Spannenberg, A., Sundermann, B. and Sundermann, C. (2004) *Angewandte Chemie – International Edition*, **43**, 3795–7;
(b) Tanaka, K., Nishida, G., Wada, A. and Noguchi, K. (2004) *Angewandte Chemie – International Edition*, **43**, 6510–12.
26 Recent reviews of Pauson–Khand-type reaction: (a) Hanson, B.E. (2002) *Comments on Inorganic Chemistry*, **23**, 289–318;
(b) Blanco-Urgoiti, J., Añorbe, L., Pérez-Serrano, L., Domínguez, G. and Pérez-Castells, J. (2004) *Chemical Society Reviews*, **33**, 32–42;
(c) Gibson, S.E. and Mainolfi, N. (2005) *Angewandte Chemie – International Edition*, **44**, 3022–37;
(d) Shibata, T. (2006) *Advanced Synthesis & Catalysis*, **348**, 2328–36.
27 Jeong, N., Hwang, S.H., Lee, Y. and Chung, Y.K. (1994) *Journal of the American Chemical Society*, **116**, 3159–60.
28 (a) Hicks, F.A., Kablaoui, N.M. and Buchwald, S.L. (1996) *Journal of the American Chemical Society*, **118**, 9450–1;
(b) Hicks, F.A., Kablaoui, N.M. and Buchwald, S.L. (1999) *Journal of the American Chemical Society*, **121**, 5881–98.
29 (a) Hicks, F.A. and Buchwald, S.L. (1996) *Journal of the American Chemical Society*, **118**, 11688–9;
(b) Hicks, F.A. and Buchwald, S.L. (1999) *Journal of the American Chemical Society*, **121**, 7026–33.
30 (a) Shibata, T. and Takagi, K. (2000) *Journal of the American Chemical Society*, **122**, 9852–3;
(b) Shibata, T., Toshida, N., Yamazaki, M., Maekawa, S. and Takagi, K. (2005) *Tetrahedron*, **61**, 9974–9.
31 Jeong, N., Kim, D.H. and Choi, J.H. (2004) *Chemical Communications*, 1134–5.
32 Kwong, F.Y., Lee, H.W., Lam, W.H., Qiu, L. and Chan, A.S.C. (2006) *Tetrahedron: Asymmetry*, **17**, 1238–52.
33 (a) Shibata, T., Toshida, N. and Takagi, K. (2002) *Organic Letters*, **4**, 1619–21;
(b) Shibata, T., Toshida, N. and Takagi, K. (2002) *Journal of Organic Chemistry*, **67**, 7446–50.
34 Shibata, T., Kadowaki, S., Hirase, M. and Takagi, K. (2003) *Synlett*, 573–5.
35 Shibata, T., Yamashita, K., Ishida, H. and Takagi, K. (2001) *Organic Letters*, **3**, 1217–19.
36 (a) McKinstry, L. and Livinghouse, T. (1994) *Tetrahedron*, **50**, 6145–54;
(b) O'Mahony, D.J.R., Belanger, D.B. and Livinghouse, T. (1998) *Synlett*, 443–5;
(c) Gilbertson, S.T., Hoge, G.S. and Genov, D.G. (1998) *Journal of Organic Chemistry*, **63**, 10077–80; (d) Heath, H., Wolfe, B., Livinghouse, T. and Bae, S.K. (2001) *Synthesis*, 2341–7.
37 Shibata, T., Takasaku, K., Takesue, Y., Hirata, N. and Takagi, K. (2002) *Synlett*, 1681–2.
38 Murakami, M., Itami, K., Ubukata, M., Tsuji, I. and Ito, Y. (1998) *Journal of Organic Chemistry*, **63**, 4–5.
39 Chatani, N., Inoue, H., Morimoto, T., Muto, T. and Murai, S. (2001) *Journal of Organic Chemistry*, **66**, 4433–6.
40 Yamamoto, Y., Hayashi, H., Saigoku, T. and Nishiyama, H. (2005) *Journal of the American Chemical Society*, **127**, 10804–5.
41 Shibata, T., Yamasaki, M., Kadowaki, S. and Takagi, K. (2004) *Synlett*, 2812–14.
42 Shibata, T., Kobayashi, Y., Maekawa, S., Toshida, N. and Takagi, K. (2005) *Tetrahedron*, **61**, 9018–24.
43 Review: Bruneau, C. (2005) *Angewandte Chemie – International Edition*, **44**, 2328–34.
44 (a) Carmona, D., Lahoz, F.J., Elipe, S., Oro, L.A., Lamata, M.P., Viguri, F., Mir, C., Cativiela, C. and López-Ram de Víu, M.P. (1998) *Organometallics*, **17**, 2986–95;
(b) Carmona, D., Lahoz, F.J., Elipe, S., Oro, L.A., Lamata, M.P., Viguri, F., Sánchez, F., Martínez, S., Cativiela, C. and López-Ram de Víu, M.P. (2002) *Organometallics*, **21**, 5100–14;
(c) Carmona, D., Lamata, M.P., Viguri, F., Rodríguez, R., Lahoz, F.J., Dobrinovitch, I.T. and Oro, L.A. (2007) *Dalton Transactions*, 1911–21;
(d) Carmona, D., Lamata, M.P., Viguri, F., Rodríguez, R., Barba, C., Lahoz, F.J.,

Martín, M.L., Oro, L.A. and Salvatella, L. (2007) *Organometallics*, **26**, 6493–6.

45 (a) Carmona, D., Lamata, M.P., Viguri, F., Rodríguez, R., Oro, L.A., Lahoz, F.J., Balana, A.I., Tejero, T. and Merino, P. (2005) *Journal of the American Chemical Society*, **127**, 13386–98;
(b) Carmona, D., Lamata, M.P., Viguri, F., Rodríguez, R., Fischer, T., Lahoz, F.J., Dobrinovitch, I.T. and Oro, L.A. (2007) *Advanced Synthesis & Catalysis*, **349**, 1751–8;
(c) Carmona, D., Lamata, M.P., Viguri, F., Rodríguez, R., Lahoz, F.J. and Oro, L.A. (2007) *Chemistry – A European Journal*, **13**, 9746–56.

46 (a) Igarashi, T., Inada, T., Sekioka, T., Nakajima, T. and Shimizu, I. (2005) *Chemistry Letters*, **34**, 106–7; (b) Nakajima, T., Inada, T., Igarashi, T., Sekioka, T. and Shimizu, I. (2006) *Bulletin of the Chemical Society of Japan*, **79**, 1941–9.

47 Janka, M., He, W., Haedicke, I.E., Fronczek, F.R., Frontier, A.J. and Eisenberg, R. (2006) *Journal of the American Chemical Society*, **128**, 5312–13.

48 Chatani, N., Yamaguchi, S., Fukumoto, Y. and Murai, S. (1995) *Organometallics*, **14**, 4418–20.

49 (a) Yasukawa, T., Satoh, T., Miura, M. and Nomura, M. (2002) *Journal of the American Chemical Society*, **124**, 12680–1;
(b) Ueura, K., Satoh, T. and Miura, M. (2007) *Journal of Organic Chemistry*, **72**, 5362–7.

50 Nishimura, T., Yasuhara, Y. and Hayashi, T. (2007) *Journal of the American Chemical Society*, **129**, 7506–7.

12
Pincer-Type Iridium Complexes for Organic Transformations
Martin Albrecht and David Morales-Morales

12.1
Introduction

Pincer ligand scaffolds have been shown to possess an extraordinarily useful potential for directing the properties and reactivity patterns of coordinated metal (M) centers. This is illustrated not only by the sheer number of publications on this topic, but also by the various applications that have been disclosed by using pincer-type ligands, such as highly efficient sensors, switches and catalysts [1].

Pincer-type ligands are characterized by a monoanionic terdentate bonding mode comprising typically an ECE skeleton, where C denotes a central carbanion (typically an aryl anion), where E represents a neutral, two-electron donor such as an amine (NCN), imine (NCN), phosphine (PCP), phosphinite (PCP), sulfide (SCS), and so on (Figure 12.1). Most frequently encountered are *meta*-xylyl-derived systems (i.e. Y = CH_2), although the importance of resorcinol-based ligands (Y = O) has grown considerably during the past few years. In general, pincer-type ligands adopt a meridional, terdentate coordination mode. As a consequence of these bonding features, the M—C bond in pincer complexes is considerably shielded and displays stability properties that are significantly enhanced as compared with bidentate or monodentate coordinated aryl ligands. Moreover, the electron density at the coordinated metal center can be modulated and eventually tailored by a number of ligand modifications which do not affect the general bonding mode, including the incorporation of electroactive groups on the aryl ring (R′) as well as on the heteroatom E (R). Given this high modularity and the general catalytic activity of iridium, it is not surprising that a number of highly active iridium–pincer catalyst systems have been developed during the past few years. The specific impact of the pincer ligand appeared to be of prime importance for the activation of unreactive bonds in particular. In some cases, the chelating bonding mode has been proven useful to stabilize crucial intermediates, thus providing significant mechanistic insights. In other cases, such bonding was revealed to be essential for increasing the catalytic activity, as the terdentate meridional coordination mode arranges the frontier orbitals of the metal energetically, as well as in terms of

Iridium Complexes in Organic Synthesis. Edited by Luis A. Oro and Carmen Claver
Copyright © 2009 WILEY-VCH Verlag GmbH & Co. KGaA, Weinheim
ISBN: 978-3-527-31996-1

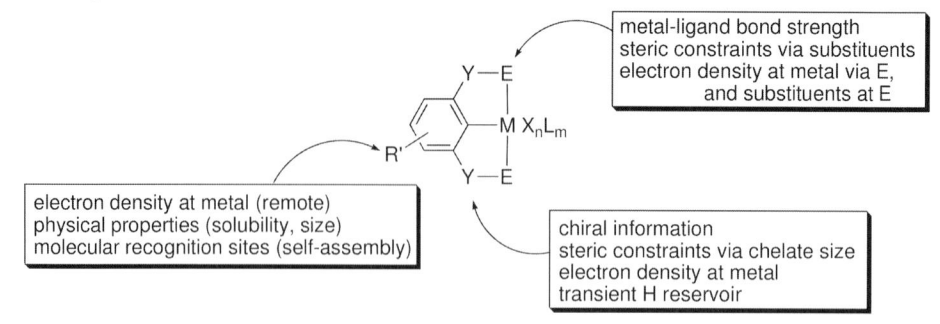

Figure 12.1 General representation of an ECE pincer ligand coordinated to a metal fragment (MX_nL_m), including the different functions that can be regulated by appropriate ligand modifications.

Boxes (clockwise from top right):
- metal-ligand bond strength; steric constraints via substituents; electron density at metal via E, and substituents at E
- chiral information; steric constraints via chelate size; electron density at metal; transient H reservoir
- electron density at metal (remote); physical properties (solubility, size); molecular recognition sites (self-assembly)

Figure 12.2 Bonding of a neutral PNP pincer ligand to an IrX_nL_m fragment.

accessibility. Most remarkably, iridium–pincer complexes have allowed for a relatively mild and selective activation of hexane and other unfunctionalized alkanes under homogeneous conditions [2]. This outstanding reactivity has recently been expanded to the first homogeneous alkane metathesis process [3], which may have significant relevance for fuel refinement, and perhaps eventually also for fine chemical production.

Thus, this chapter aims at compiling iridium-catalyzed processes that exploit the unique properties of pincer ligands. For the sake of completeness, complexes containing pincer-like ligands – such as formally neutral PNP ligands – have also been included here. These ligands comprise a central N-coordinated pyridine rather than an anionic aryl unit (see Figure 12.2). Within the chapter, truly catalytic processes will be discussed, as well as mechanistic studies that provide considerable insight into the key bond activation steps, although these have not yet led to a closed catalytic cycle.

12.2
Iridium PCP-Catalyzed Activation of C(sp^3)—H Bonds in Unfunctionalized Alkanes

12.2.1
Scope of the Reaction

For many years the activation of unfunctionalized alkanes has been the 'Holy Grail' of organic synthesis and, indeed, it has only been during the past few years that catalysts have evolved which allow an alkane C—H bond to be selectively

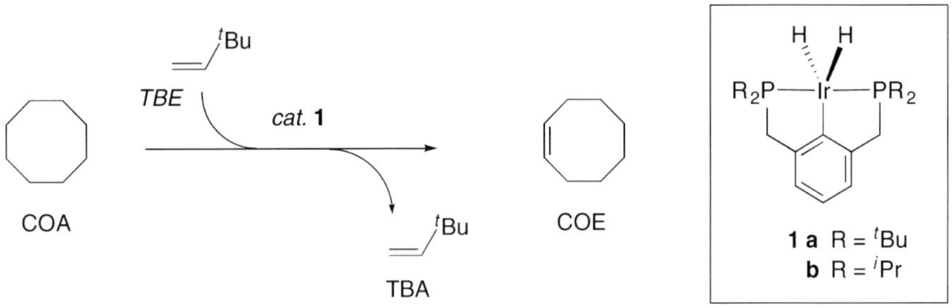

12.2 Iridium PCP-Catalyzed Activation of C(sp³)—H Bonds in Unfunctionalized Alkanes

Scheme 12.1

addressed under relatively mild conditions (<250 °C, ambient pressure) [4]. Two main concepts appeared to be particularly useful in this respect: (i) σ-bond metathesis via agostic interactions; and (ii) oxidative C—H addition to a metal center [5]. While σ-bond metathesis relies on low-valent, metal centers – typically in d^0 electronic configuration – an oxidative addition protocol clearly requires electron-rich metal centers such as d^8 platinum group metals, and coordination sites that are available, at least transiently. It is in this latter area where iridium pincer chemistry has made major contributions by providing some of the most efficient catalysts for alkane C—H bond activation.

The iridium(I) PCP pincer complexes **1** exhibit remarkable activity in the catalytic dehydrogenation of unfunctionalized alkanes (Scheme 12.1). The H_2, which is formally produced during this process, may be transferred to either *tert*-butylethylene (TBE) or norbornene (NBE) as a sacrificial hydrogen acceptor. For example, complex **1a** converts cyclooctane (COA) to cyclooctene (COE) in the presence of TBE, which in turn is reduced to *tert*-butylethane (TBA; *neo*-hexane) [6].

The results of some initial experiments indicated that the catalytic activity of **1** is strongly temperature-dependent. For instance, at 100 °C, the rates are relatively low but there is an appreciable turnover [turnover frequency (TOF) = $20.5\,h^{-1}$]. Increasing the reaction temperature to 200 °C increases the TOF to $720\,h^{-1}$; hence, the thermal robustness of the iridium catalyst is pivotal for optimal catalyst performance. The tridentate pincer-type ligands provide a particularly stable platform for metal confinement through covalent Ir—C bonding combined with the terdentate chelating coordination.

A variety of alkanes can be dehydrogenated with catalysts such as **1**. Whilst COA dehydrogenation usually stops at the COE stage, cyclohexane dehydrogenation affords benzene in good yields [7]. Substituted cyclohexanes are dehydrogenated at both exocyclic and endocyclic positions; for example, ethylcyclohexane dehydrogenation produces styrene as the major product (Equation 12.1).

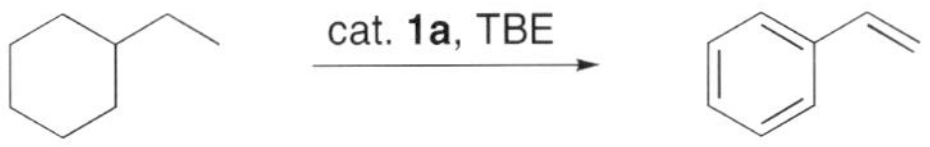

$$\text{ethylcyclohexane} \xrightarrow{\text{cat. } \mathbf{1a},\ \text{TBE}} \text{styrene} \quad (12.1)$$

Linear alkanes have been successfully dehydrogenated under similar conditions. Initially, selective activation of the terminal position has been observed, leading

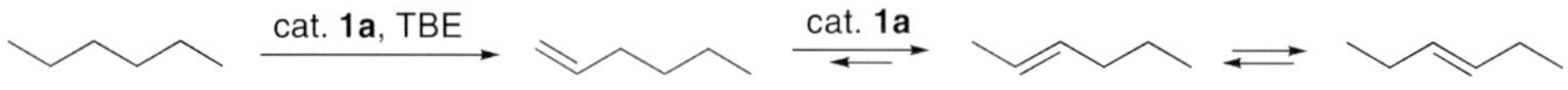

Scheme 12.2

to terminal olefins. However, double bond migration gradually increases the fraction of internal olefins, thus providing ultimately an isomeric mixture of alkenes (Scheme 12.2) [8]. Careful control of the reaction time, as well as modification of the hydrogen acceptor, have been suggested as potential means of preventing undesired isomerization processes and thus increasing the yield of terminal olefins.

It is noteworthy that, while the dehydrogenation of ethylcyclohexane proceeds smoothly, the dehydrogenation of ethylbenzene is inefficient and essentially ceases after about 50 turnovers. This reduced activity may suggest that terminal alkyl moieties are better substrates than disubstituted (cyclic) alkane residues. However, time-dependent analysis of the reaction mixture revealed a considerable build-up of ethylcyclohexene and ethylbenzene, indicating that the reduction of the ethyl substituent is relatively slow. High arene concentrations seem, however, to inhibit the catalysis. Similarly, heteroatoms slow down the catalyst activity and heterocyclic compounds such as tetrahydrofuran (THF) appeared to be poor substrates. Structural studies have demonstrated that even very weakly coordinating species such as N_2 effectively compete with alkane C—H bonds for coordination to the iridium center, thus interfering with one of the key steps in the catalytic cycle. The presence of dinitrogen led to catalyst inhibition and afforded the crystallographically characterized complex **2**, which featured an unusually stable bridging $\mu-\eta^1:\eta^1$-coordination of N_2 to two Ir—PCP fragments (Equation 12.2) [9].

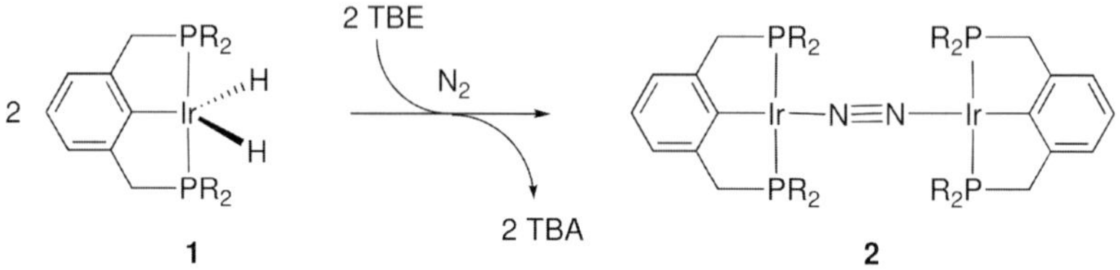

(12.2)

Accordingly, catalytic dehydrogenation with complexes such as **1** must be performed under an atmosphere of argon that is devoid of traces of N_2. Similarly, alkenes exert an inhibitory effect, and high concentrations of either the product olefin or the hydrogen acceptor TBE inhibit catalyst activity.

12.2.2
Mechanistic Considerations

Based on the observations outlined above and further experimental and theoretical investigations, a detailed mechanism has been proposed for the Ir—PCP-catalyzed transfer dehydrogenation of alkanes (Scheme 12.3) [2]. Similar to the transfer

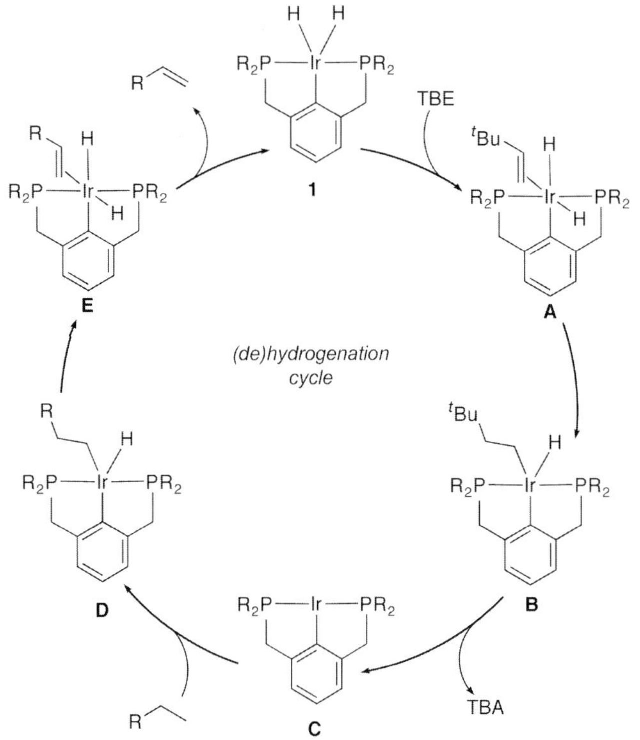

Scheme 12.3

(de)hydrogenation of ketones via a dihydride mechanism, the catalytic cycle of alkane dehydrogenation can be split into reductive and oxidative parts, comprising the transfer hydrogenation of the sacrificial acceptor (**1** → **C**) and dehydrogenation of the substrate (**C** → **1**), respectively. The reductive half-cycle is initiated by TBE coordination to **1** and subsequent migratory insertion to give the (*neo*-hexyl)(hydride) complex **B**. Reductive elimination of TBA then produces the coordinatively unsaturated complex **C**, which fulfils the requirements for catalytic oxidative C—H bond addition – that is, a coordinatively unsaturated, electron-rich metal center in a conformationally stable PCP-ligand environment. The oxidative addition of a substrate alkane produces the (alkyl)(hydride) complex **D**, which may undergo β-hydrogen elimination to afford the six-coordinate intermediate **E**. Upon dissociation of the dehydrogenated alkene, the dihydride complex **1** is regenerated.

Given the structural similarity of intermediates **A** and **E**, and also of **B** and **D**, it seems obvious that each reaction step in the cycle is potentially reversible. The equilibrium is shifted, however, into the desired direction by the steric shielding of the sacrificial hydrogen acceptor, which makes the reductive elimination of TBA an essentially irreversible process.

In order to rationalize the olefin isomerization reaction, a second cycle must be considered (Scheme 12.4) comprising, as a first step, the reverse of product

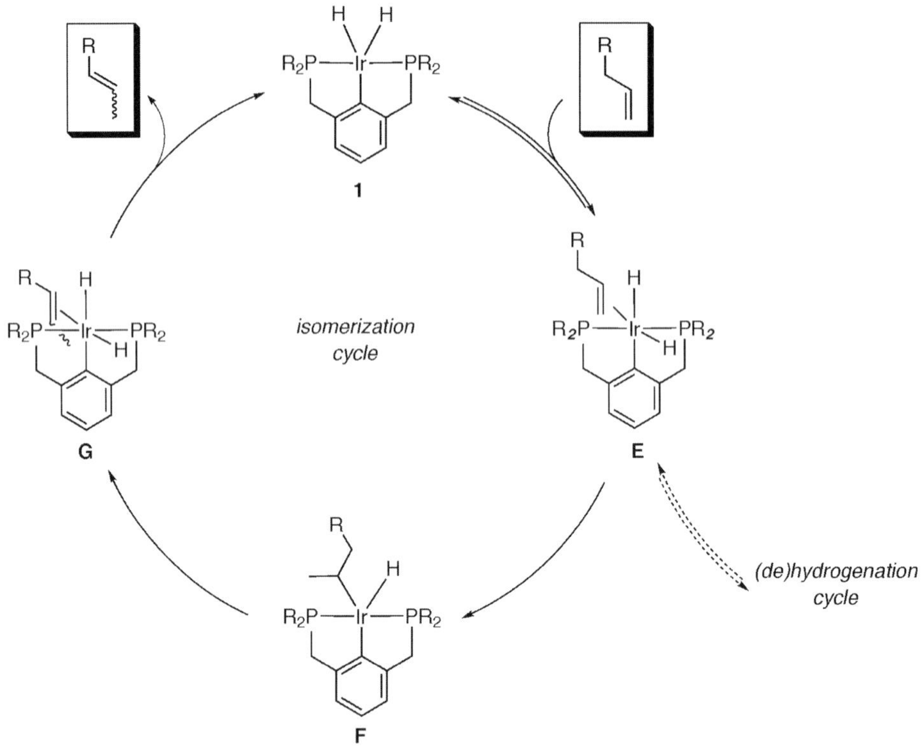

Scheme 12.4

release – that is, the coordination of a product molecule to form intermediate **E** again. Alternatively, a small portion of product may engage in this cycle directly before being released from the metal coordination sphere. Migratory insertion in **E** may then take place either via metal coordination of the terminal carbon, thus forming again the (alkyl)(hydride) **D** and hence entering the (de)hydrogenation cycle (Scheme 12.3). Alternatively, the β-carbon may coordinate to the metal center, thus forming intermediate **F** comprising a secondary alkyl ligand. Subsequent β-hydrogen elimination can again occur at the terminal position or, thermodynamically more favorably, at the γ-carbon, thus producing the corresponding 2-alkene via the olefin complex **G** [8].

Kinetic analyses and deuterium-labeling experiments have demonstrated that, remarkably, the reductive elimination of TBA and the formation of intermediate **C** is the rate-determining step in the (de)hydrogenation cycle. Accordingly, hydrogenation of the acceptor appears to be slower than dehydrogenation of the alkane substrate. This contrasts with the fact that catalytic olefin hydrogenation is well-established in transition-metal-mediated chemistry [10].

The existence of the coordinatively unsaturated complex **C** also rationalizes the sensitivity of the catalyst towards weakly coordinating species such as N_2 and heterocycles, as well as towards substrates containing more labile C—H bonds, such as arenes. In this light it is not surprising that detailed kinetic investigations revealed catalyst deceleration also due to olefin coordination. Optimal catalyst operation therefore relies on a delicate balance of product olefin and hydrogen acceptor concentrations [11]. On a practical basis, large product:substrate ratios are readily avoided by using the substrate alkane as solvent. Similarly, high initial TBE concentrations are unfavorable and lead to undesired side reactions such as $C(sp^2)$—H activation and formation of the (vinyl)(hydride) complex **3** (Equation 12.3) [12].

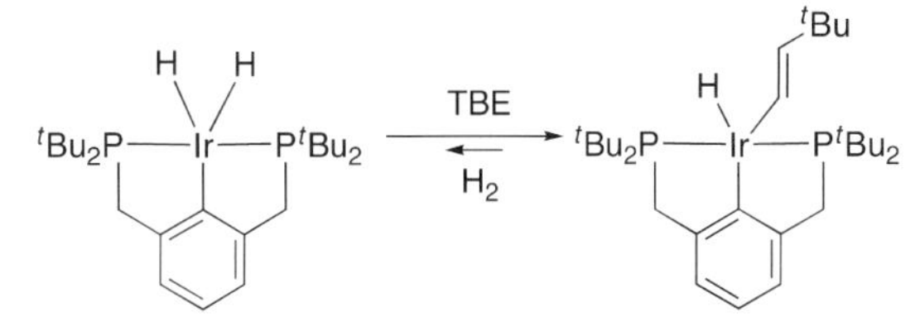

$$(12.3)$$

Complex **3** undergoes rapid exchange with free TBE on the nuclear magnetic resonance (NMR) time scale, even at ambient temperature. The reversible loss of TBE has been suggested to involve the transient formation of the unsaturated species **C** [11]. However, the equilibrium of this reaction is largely on the (vinyl)(hydride) complex side, thus deactivating a major portion of the catalyst. Such deactivation may be suppressed by keeping the hydrogen acceptor concentration low throughout the reaction; this may be achieved, for example, by the continuous addition of TBE rather than providing all TBE at the onset of the reaction.

In this context, it is worth noting that complexes such as **1** and the thermally even more robust anthracene-derived complex **4** catalyze the dehydrogenation of cycloalkanes such as cyclodecane and cyclododecane also in the absence of TBE (Scheme 12.5) [13]. Such acceptorless dehydrogenation circumvents the difficulties associated with controlling the TBE concentration, although the TOFs are typically lower. The liberated H_2 has been efficiently removed by passing a stream of argon over the reaction mixture. Again, high reaction temperatures are

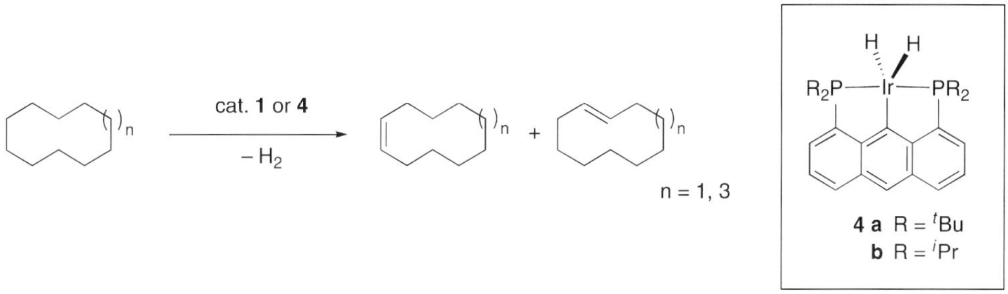

Scheme 12.5

crucial for compensating the large positive enthalpy of dehydrogenation (ca. 125–150 kJ mol^{-1}, endothermic). For this purpose, complex **4** appeared to be superior and can be used up to 250 °C, without noticeable catalyst decomposition.

The results of both, mechanistic and computational studies, have suggested that the acceptorless dehydrogenation is initiated by rate-determining dissociation of H$_2$ from the Ir(H)$_2$PCP system, thus forming the unsaturated intermediate **C** [14]. Associative pathways have been calculated to include free-energy barriers that are substantially higher. These findings are supported by recent studies on the nature of the dihydride ligand in the precursor complex. In the catalytically very active iridium complex **5**, which contains a resorcinol-based PCP ligand, an equilibrium between an iridium(III)–dihydride species and the iridium(I)–dihydrogen complex **6** comprising an elongated H—H bond has been established by detailed NMR spectroscopic analyses (Equation 12.4) [15].

$$\text{(12.4)}$$

5

a R' = H
b R' = Me
c R' = OMe
d R' = F
e R' = C$_6$F$_5$
f R' = 3,5-(CF$_3$)$_2$-C$_6$H$_3$

6

The dissociation of dihydrogen from **6** is supposed to provide a smooth access to the catalytically active species **C**. Measurement of the relaxation time, T_1, indicates that the equilibrium between complexes **5** and **6** is heavily dependent on the solvent, the temperature, and the substitution pattern at the pincer ligand. For example, complex **5f** comprises strongly electron-withdrawing substituents exists predominantly in the dihydride form in chlorinated solvents at room temperature. In pentane – and presumably also in the alkanes used as solvents for the dehydrogenation – this complex is best described by an equilibrium between an elongated dihydrogen complex of iridium(I) (**6f**) and an iridium(III)–dihydride complex **5f**. Based on the pronounced temperature-dependence of the coupling constants and the T_1 measurements, the equilibrium between the two distinct structures is rapidly established.

The effect of the substituent R' on the pincer ligand becomes predictable when considering the generally accepted notion that increasing the metal-to-ligand π-backdonation into the σ* HH orbital elongates the H—H bond, and in the extreme case actually leads to H—H bond cleavage and concomitant oxidative addition [16]. Thus, increasing the ability of the PCP ligand to donate electron density into the σ* orbital of the bound H$_2$ ligand facilitates oxidative addition and hence shifts the equilibrium of Equation 12.4 unfavorably to the dihydride species **5**. For high catalytic activity, of course, the opposite effect is desirable.

12.2.3
Catalyst Optimization

Given the mechanistic considerations referred to in Section 12.2.2, the catalyst activity is evidently delicately dependent on the donor properties of the PCP ligand. A too-strong donation decelerates the rate-limiting dissociation of TBA or H$_2$ due to the increased stability of the oxidized iridium(III) (alkyl)(hydride) or dihydride species, respectively, whereas a too-low donation disfavors oxidative addition of the alkane to the unsaturated iridium(I) species **C**. The phosphinite PCP pincer iridium complexes **5**, having a reduced donor ability as compared to phosphines in **1** and **4**, appears to meet these requirements particularly well [17]. In addition, fine-tuning of the donor properties has been achieved by incorporating different substituents R′, thus taking advantage of the unique bonding features of pincer ligands. In the presence of TBE as hydrogen acceptor, turnover numbers (TONs) as high as 2200 have been observed for COA dehydrogenation (see Scheme 12.1). This is markedly higher than the TONs typically observed for phosphine-based Ir—PCP systems (ca. 250 under identical conditions). Electron-withdrawing substituents such as C$_6$F$_5$ (**5e**) and 3,5-(CF$_3$)$_2$-C$_6$H$_3$ (**5f**) considerably improve the catalytic activity (Table 12.1). Notably, the MeO-substituted Ir—PCP complex **5c** seems to operate differently, and remains active over a longer time range than the other phosphinite complexes, thus reaching final TONs that are comparable to those of **5f**.

The reduced donor ability of the phosphinite complexes such as **5e** and **5f** has an impact beyond the catalyst activation stipulated above. Apparently, the decreased tendency to undergo oxidative addition reactions also disfavors catalyst deactivation via oxidative olefin addition. Accordingly, (vinyl)(hydride) complexes such as **3** are less relevant. Simultaneously, product oxidative addition is restricted and, as

Table 12.1 Catalytic activity of Ir—PCP complexes in COA dehydrogenation.

Catalyst	TON (8 min)	TON (40 h)	TOF$_{ini}^a$ (s^{-1})	COE:COD^b
1a	156	227	0.3	100:0
5a	922	1583	1.9	84:16
5b	811	1484	1.7	86:14
5c	806	1904	1.7	81:19
5d	840	1530	1.8	84:16
5e	1150	2041	2.4	78:22
5f	1162	2070	2.4	76:24

a TOF$_{ini}$ determined after 8 min of reaction.
b COE:COD ratio after 40 h of reaction.

a consequence, olefinic substrates can also be dehydrogenated. This is illustrated by the increasing ratio of 1,3-cyclooctadiene (COD) formed in the dehydrogenation of COA with **5f**.

The dehydrogenation of COE has been observed already at a low alkene concentration. Mechanistic investigations revealed that the reverse reaction – that is, the hydrogenation of COE with the iridium–dihydride complexes **5** – proceeds very rapidly even at temperatures as low as −70 °C [18]. These results suggest that the reductive elimination of alkane from an (alkyl)(hydride) species is also rapid. Likewise, TBA elimination from the phosphinite PCP analogue of intermediate **B** is expected to be fast, thus providing a rational for lowering the rate-determining step in catalytic cycles using phosphine PCP-ligated iridium centers.

Phosphinite pincer iridium systems have also been shown to have a lower tendency to oxidatively add TBE to give (vinyl)(hydride) complexes similar to **3** [18]. While this has been identified as one of the major catalyst deactivation processes in phosphine pincer iridium catalysis, apparently with complexes such as **5**, only olefin coordination can occur. However, this is a considerably weaker bonding and is less detrimental to catalyst activity. Based on steric arguments, product olefin coordination (e.g. COE) is favored over TBE coordination, and therefore at a high TON and high product concentrations the phosphinite catalysts **5** are markedly less active than the phosphine analogues **1**.

Further catalyst improvement has been achieved by incorporating a methoxy-substituent into the aryl framework of the phosphine PCP ligand. When using complex **7** comprising less bulky P^iPr_2 donor groups, TONs of up to 3000 have been achieved for the acceptorless dehydrogenation of cyclodecane [19]. In transfer dehydrogenation with TBE as hydrogen acceptor – that is, under conditions similar to those used in Table 12.1 – the TON of complex **7** was slightly slower than that of the phosphinites, although it is obviously intrinsically difficult to perform catalytic reactions in different laboratories under mutually identical conditions. Complex **7** is also efficient in the transfer dehydrogenation of *n*-alkanes, although product mixtures and isomer distributions are typically complex. The high activity of complex **7** has been attributed to a favorable interplay of steric and electronic effects. Indeed, the analogous catalyst comprising P^tBu_2 donor groups is considerably less active. Theoretical evaluations suggest that the apparently unfavorable electron-donating effect of the methoxy substituent at the aryl ring may be rationalized by an advantageous mesomeric stabilization of the aryl ligand, which may adjust the π acceptor properties of the ligand so as to favor oxidative $C(sp^3)$—H bond activation.

Further pincer ligand modification at the carbanionic coordination site provided the most active homogeneous alkane dehydrogenation catalyst known thus far. Replacing the aryl anion by a ferrocenyl unit (**8**; Figure 12.3) improved the catalytic activity by a factor of 2 and allowed for a decrease in the reaction temperature for dehydrogenation [20]. At 180 °C, TONs as high as 3300 have been obtained with complex **8**, while the phosphinite catalyst **5a** provided a TON of 1840 under these conditions. The ruthenocene-derived catalyst **9** (Figure 12.3; TON 2570) was found to be slightly less active than the ferrocenyl analogue. The origin of the enhanced

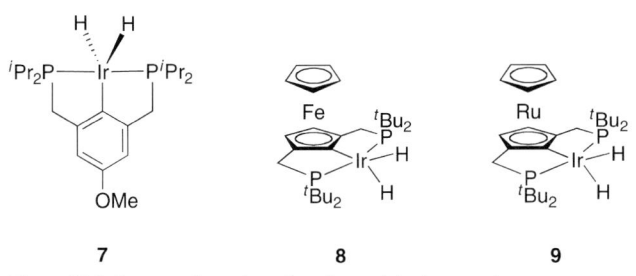

Figure 12.3 Improved catalysts for alkane dehydrogenation.

catalytic activity of these metallocene pincer iridium complexes is not yet fully understood, although several different factors have been suggested as contributing. For example, the five-membered nature of the aromatic ring in **8** leads to a decrease of the P—Ir—P angle as compared to related phenyl-derived systems, thus facilitating accessibility to the iridium center. In addition, the ferrocene backbone in **8** interacts with one face of the iridium coordination environment, thus inducing a mutually tilted conformation of the two cyclopentadienyl rings. This tilting is less pronounced when larger ruthenium(II) centers are involved. Clearly, the π-accepting orbitals of the iridium-coordinated cyclopentadienyl ring are also arranged differently due to bonding to the [M(cp)]$^+$ fragment, which may have effects that are even stronger than those discussed for the methoxy-substituted PCP complex **7**.

12.2.4
Application of Alkane Functionalization

12.2.4.1 Alkane Metathesis

The high activity of iridium PCP pincer complexes in transfer dehydrogenation has been applied in a very elegant approach to devise the first homogeneous alkane metathesis process (Equation 12.5) [3].

$$R\frown R' + R\frown R' \longrightarrow R\frown R + R'\frown R' \quad (12.5)$$

Catalytic metathesis has been accomplished via a tandem combination of catalytic alkane dehydrogenation, olefin metathesis and subsequent olefin hydrogenation (Scheme 12.6). Two factors are crucial for this transformation:

- The transfer (de)hydrogenation is fully reversible – that is, the iridium pincer complexes catalyze the alkane dehydrogenation as well as the alkene hydrogenation.

- The reaction conditions applied for transfer dehydrogenation using complex **5a** are compatible with the conditions required for olefin metathesis with Schrock's molybdenum alkylidene complex **10** [21].

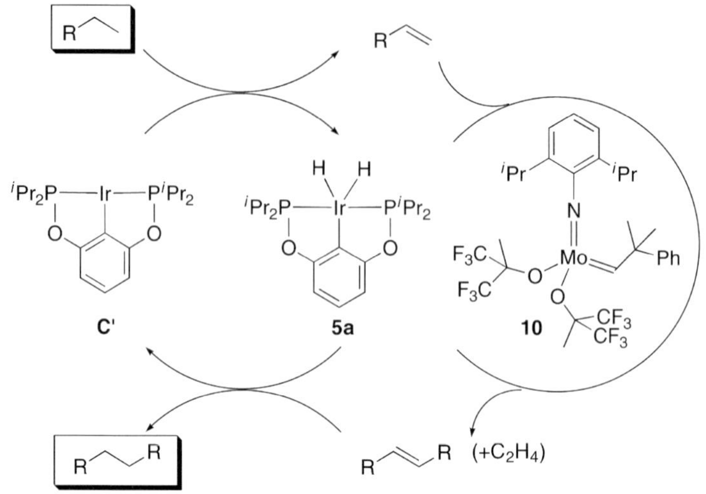

Scheme 12.6

The system successfully mediates the transformation of *n*-hexane into high-molecular-weight hydrocarbons, and it is also effective for long alkane chain metathesis. The major product fraction from hexane is linear decane, although the full range of linear alkanes has been obtained. This is in part due to multiple reactions, since products can serve as new substrates, and also due to the fact that both complexes **5a** and **10** catalyze olefin isomerization (double-bond migration). The isomerization of 1-hexene into 2-hexene will lead, for example, to the extrusion of propene rather than ethene in the metathesis step, eventually leading to the C_9 rather than the C_{10} alkane. Although isomerization is less pronounced when using catalyst **1a**, the benefits in selectivity are thwarted by reduced TOFs. Such alkane metathesis reactions have been suggested as an alternative to the Fischer–Tropsch synthesis for the production of alkanes in the C_{10} to C_{20} range, the major component of diesel engine fuel.

Recently, this tandem catalytic conversion was successfully applied also to cyclic alkanes [22]. In the presence of the molybdenum catalyst **10**, the formed cycloalkene undergoes ring-opening metathesis and cyclo-oligomerization before being reduced to the ring-expanded cycloalkane. Accordingly, COA has been converted to cyclohexadecane as the major product (Scheme 12.7). Larger ring systems containing multiples of eight carbon atoms constituted the predominant product fraction, indicating that multiple ring-opening metathesis reactions may take place, while olefin isomerization is less important.

12.2.4.2 Polymer Functionalization

Alkane dehydrogenation has been demonstrated as a suitable method for the functionalization of polyolefins such as atactic poly(1-hexene) under homogeneous conditions (Equation 12.6) [23].

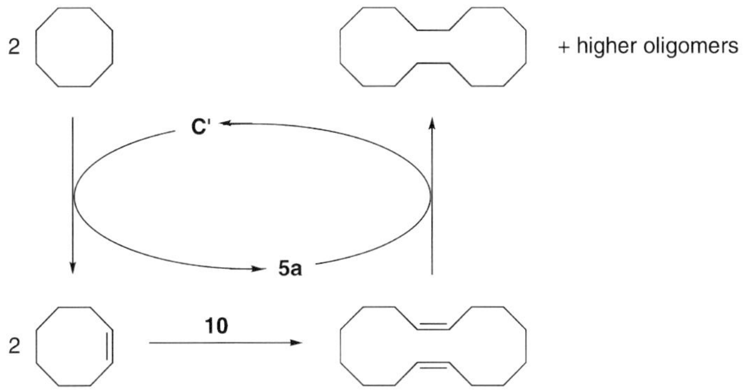

Scheme 12.7

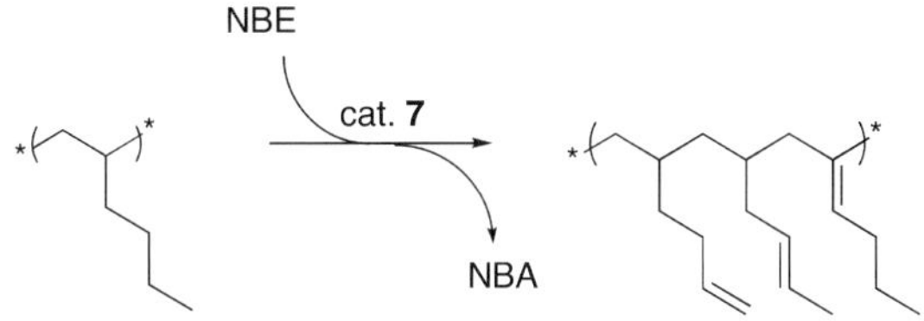

Time-dependent monitoring of the dehydrogenation reaction mediated by complex **7** revealed that dehydrogenation occurs virtually exclusively at the side chains, leaving the polymer backbone unaffected. A kinetic selectivity for the terminal position has been established, which is in line with observations on low-molecular-weight alkane dehydrogenation (see Section 12.2.1). Double-bond migration takes place rapidly, leading predominantly to the formation of disubstituted and trisubstituted olefins and, after prolonged reaction times, also to conjugated olefins. When using polyethylene as substrate, dehydrogenation occurs in moderate yields. At a high catalyst loading (3 mole equiv. per monomer repeat unit), up to 4.4% of C—C bonds of the polymer backbone have been hydrogenated. Here, ligand substitution in the pincer framework had strong effects. The Ir—PCP analogue of **7** containing $^tBu_2P$ donor groups was seen to be considerably less active in dehydrogenating polyhexane side chains, and failed to react with polyethylene.

12.3
Arene C(sp²)—H and Alkyne C(sp¹)—H Bond Activation

Although neither C_{aryl}—H nor $C_{alkynyl}$—H bonds have been functionalized in a catalytic process by iridium pincer complexes, a substantial mechanistic understanding has been accumulated that may lead to catalytic reactions in the near future. For this reason—and also due to the fact that some of these activation protocols are different from standard processes—a short discussion of the achievements to date is included at this point.

12.3.1
Activation of C(sp^2)—H Bonds

The transition-metal-mediated activation of C$_{aryl}$—H bonds is very common, and represents one of the oldest and most frequently used strategies to metallate arenes. Cycloiridation is typically supported by an initial heteroatom coordination, followed by *ortho*-directed C—H bond activation, and is generally assumed to occur with appropriately heteroatom-substituted benzene derivatives [24]. At first sight, the formation of complex **13** from a reaction of the iridium pincer complex **1a** with nitrobenzene therefore appears trivial (Scheme 12.8) [25]. Complex **13** corresponds to the kinetic product with the aryl and hydride ligands in mutual *cis* positions, and represents also the thermodynamically most stable configuration with the hydride as the strongest *trans* ligand in *cis* position to both aryl ligands.

Low-temperature experiments have indicated, however, that the iridium-mediated C—H bond activation is not heteroatom assisted, and instead, direct C—H bond activation occurs. This process is under steric control and leads predominantly to the *para*-activated complex **11** and in minor portions also to the analogous product from *meta* C—H bond activation. Subsequent migration of the metal center across the aryl skeleton is probably triggered by weak N—O···H interactions between the nitro group and the metal-bound hydride, thus yielding first the octahedral complex **12**. Upon heating, a further rearrangement takes place, perhaps involving cleavage of the relatively weak Ir—O bond in **12** to give finally the thermodynamically most stable product **13**.

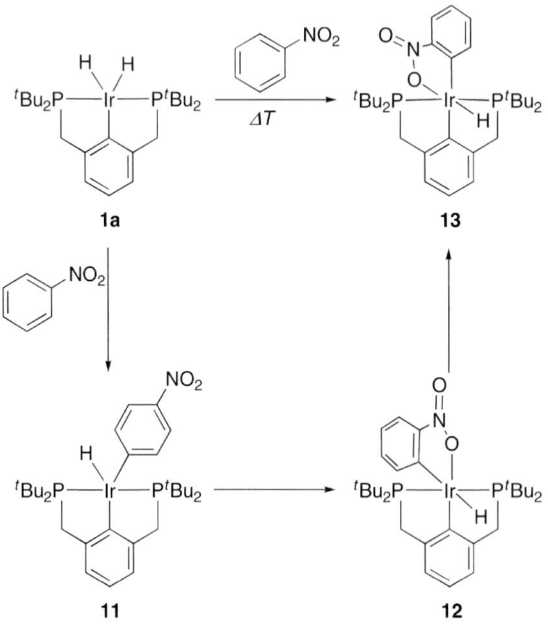

Scheme 12.8

In contrast to these results, C_{aryl}—H bond activation of related Ir—PNP pincer complexes appeared to be a heteroatom-directed process. The iridium(I) complex **14** induces *ortho* C—H bond activation of a variety of haloarenes and anisole to afford the corresponding iridium(III) complexes **15** (Equation 12.7) [26].

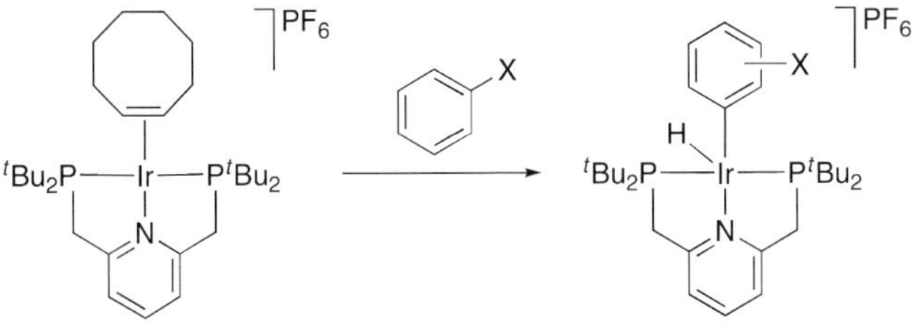

$$\text{(12.7)}$$

14

a X = H
b X = F
c X = Cl
d X = Br
e X = OMe

15

Presumably, due to steric congestion at the iridium center, competitive *meta* and *para* functionalization of the arene has been observed. The selectivity is highest for anisole containing the strongest donor in the series and decreases in the sequence Br > Cl > F. This trend is consistent with both, the coordination ability of the heteroatom as well as its steric requirement. Remarkably, no C—X bond oxidative addition was observed, clearly distinguishing iridium chemistry from crosscoupling catalysis with Group 10 metals. The results of deuterium-labeling experiments indicate that complex **15a** undergoes reversible C_{aryl}—H activation and, depending on the conditions, allows for deuterium labeling of the metal-bound hydride as well as of the phenyl ligand. In contrast, the heteroatom-substituted *ortho*-metallated phenyl ligands exhibit a higher barrier to arene exchange.

Similar heteroatom-assisted C—H bond activation also occurs with aliphatic substrates. For example, complex **14** activates the terminal C—H bond of 2-butanone and affords already at temperatures as low as 60 °C the cyclometallated complex **16** (Equation 12.8) [27].

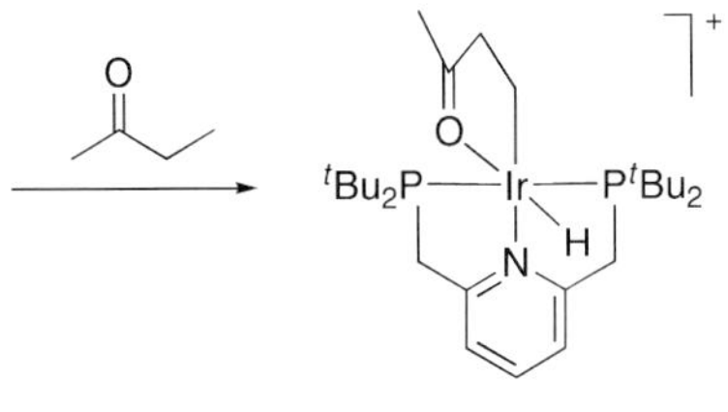

$$\text{(12.8)}$$

14

16

The benzylic protons of the PNP ligands in **14** are acidic and have been deprotonated at low temperature to give complexes **17** comprising a dearomatized,

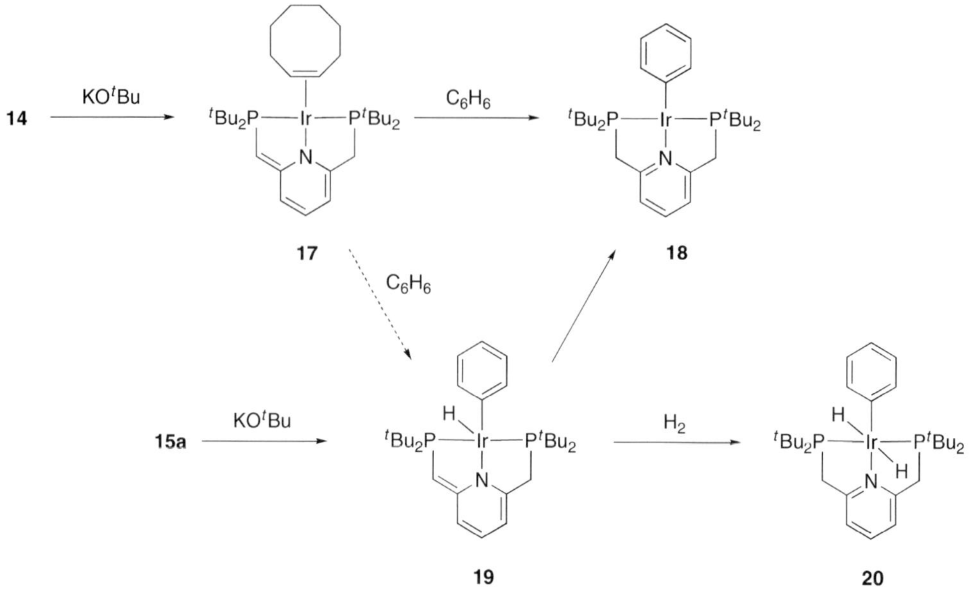

Scheme 12.9

monoanionic PNP amide pincer ligand (Scheme 12.9). This complex is highly active in functionalizing arene C—H bonds, and leads in the presence of benzene to the formation of the phenyl complex **18** [28]. While this activation formally does not alter the iridium oxidation state, an oxidative addition–reductive elimination sequence involving the (phenyl)(hydride) iridium(III) species **19** has been proposed as intermediate. Support for this reaction trajectory is derived from analyses of the reactivity pattern of isolated complex **19** obtained from deprotonation of the iridium complex **15a**. Notably, iridium reduction from complex **19** involves rearomatization of the pyridine ligand via a hydrogen migration from the metal to the benzylic carbon. Mechanistically, this migration probably proceeds through a hydride dissociation rather than via a sigmatropic rearrangement. Deuterium-labeling studies did not reveal any isotope exchange, but rather indicated an intramolecular process.

The de- and re-aromatization of the pyridine moiety of the pincer ligand appears to be crucial for this process. This is also the key underlying feature in the oxidative addition of H_2 by complex **19** in an apparent iridium(III) oxidation state, which results in the formation of the dihydride complex **20** (Scheme 12.9). Similarly, the addition of CO to the iridium(I) complex **18** formally results in oxidation of the metal center and provides the iridium(III) complex **21** (Equation 12.9).

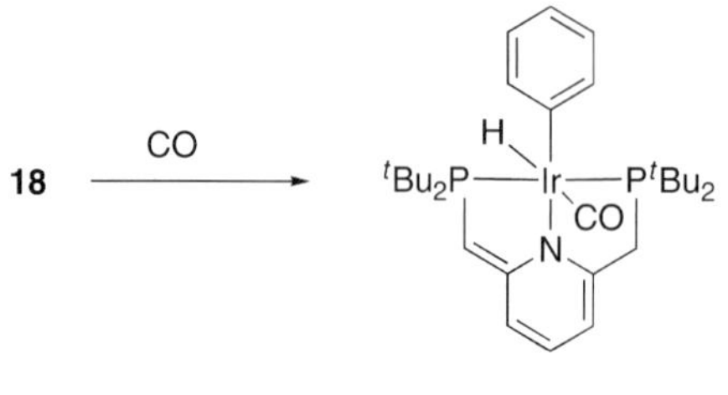

(12.9)

Evidently, reversible migration of the benzylic protons to the metal coordination sphere must be taken into account when designing catalytic processes with such PNP pincer ligands, in combination with iridium and other metal centers [29].

12.3.2
Activation of C(sp^1)—H Bonds

The activation of acetylenic C—H bonds by iridium–PCP complexes has been the topic of a recent study [30]. The iridium complex **22** has been prepared by subjecting the iridium PCP complex **1a** to norbornene, thus giving 1 equiv. of hydrogenated norbornane as a side product. The addition of 2 molar equivalents of phenylacetylene results in an activation of the C$_{alkynyl}$—H bonds, and gives the alkenyne complex **23** in high yields (Scheme 12.10). A detailed analysis of the reaction profile of this C—C coupling reaction revealed that, initially, an oxidative addition of one acetylene takes place to give the (alkynyl)(hydride) iridium(III) system **24**, A second acetylene substrate is subsequently inserted into the iridium–hydride bond. While steric arguments may favor a 1,2-insertion and the formation of a β-substituted vinyl ligand (as in **25**), kinetic studies clearly demonstrated that 2,1-insertion and the formation of an α-substituted vinyl ligand is preferred. This 2,1-insertion appeared to be reversible, and provided complex **26** as a mixture of two rotamers (**26a** and **26b**). Despite the steric congestion in the metal coordination sphere, reductive elimination of the α-vinyl ligand from complex **26** has not been observed. In contrast, the sterically less-crowded complex **25** has been found to undergo facile and clean C—C bond elimination to yield the alkenyne complex **23**. The exclusive *trans* selectivity of the enyne coupling product can be rationalized by a side-on insertion of acetylene into the Ir—H bond. This selective product formation makes an eventual involvement of a vinylidene intermediate unlikely.

Theoretical analyses have suggested that an appropriate orbital orientation of the vinyl ligand with respect to the alkynyl system is crucial for reductive C—C bond formation. This orientation requires rotation about the C$_{vinyl}$—Ir bond, a process that is severely hampered for α-substituted vinyl ligands such as in **26**. Apparently, the driving force for reductive elimination and C—C bond making in these Ir—PCP systems is electronic rather than steric in nature.

Attempts to turn this acetylene dimerization reaction into a catalytic polymerization process have failed thus far. In the presence of excess phenylacetylene, the iridium(I) complex **23** activates another C$_{alkynyl}$—H bond and transforms, after a hydrogen shift, to the stable (vinyl)(alkynyl) iridium(III) system **27** (Equation 12.10).

(12.10)

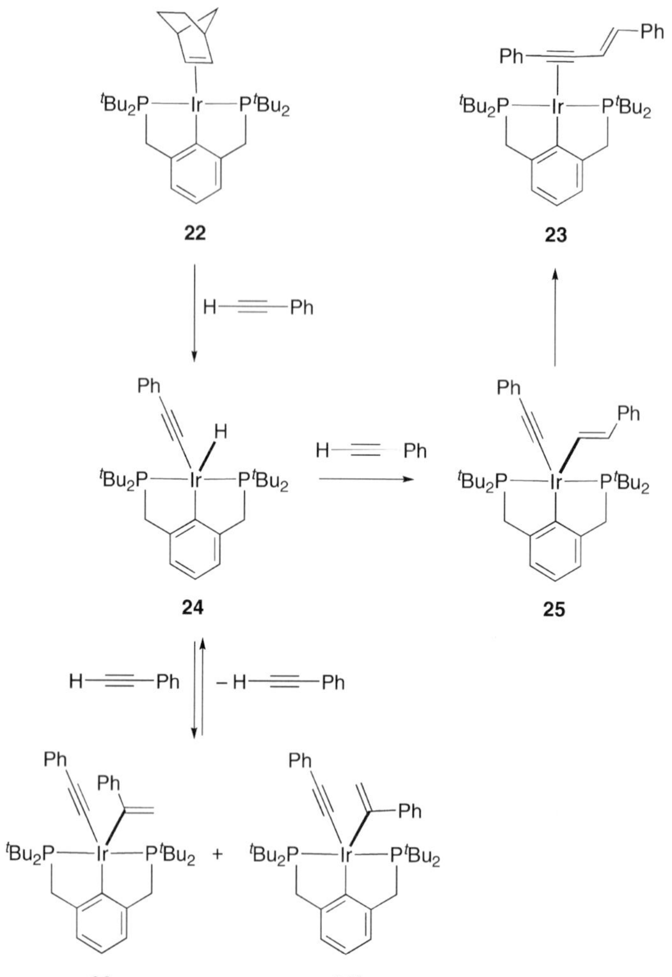

Scheme 12.10

No further reaction has been observed, presumably because the vinyl ligand in **27** again contains a substituent in the α position, which prohibits rotation and mutual orbital alignment for subsequent reductive elimination. The catalytic dimerization of acetylenes is also prevented, due presumably to the strong bonding of acetylenes to iridium(I). While in complex **22**, the olefinic ligand dissociates readily, thus providing an active species for engaging in oxidative C—H addition, the acetylenic ligand in **23** remains in the metal coordination sphere. Hence, formation of the coordinatively unsaturated iridium(I) intermediate **C** is precluded and oxidative acetylene addition provides complex **27** rather than complex **24**, as required for a catalytic dimerization.

12.4
C—E Bond Activation

12.4.1
Activation of Carbon–Halogen Bonds

Abstraction of one of the metal-bound hydrides from complex **5a** provides the cationic iridium(III) complex **28**, which is an efficient precatalyst for alkyl halide reduction in the presence of Et$_3$SiH (Equation 12.11) [31].

$$\text{R–X} + \text{Et}_3\text{Si–H} \xrightarrow{0.5 \text{ mol\% } \mathbf{28}} \text{R–H} + \text{Et}_3\text{Si–X} \qquad (12.11)$$

A variety of alkyl halides have been reduced at room temperature, including benzyl halides, primary, secondary and tertiary alkyl halides. The reaction times depend on the halide, and vary between 20 min (benzyl bromide, 0.5 mol% **28**) up to several days (iodopentane, fluoropentane). The reactivity of alkyl halides decreases in the order R—Br > R—Cl > R—I when reductions are performed in separate flasks. Several mechanistic details of the reaction have been uncovered by *in situ* monitoring of the reaction by NMR spectroscopy. The precatalyst **28** appeared to be activated by a rapid reduction of the coordinated acetone to ^{i}PrO—SiEt$_3$ and concomitant coordination of an alkyl halide (**H**; Scheme 12.11). This complex represents a resting state that is in equilibrium with a σ-silane

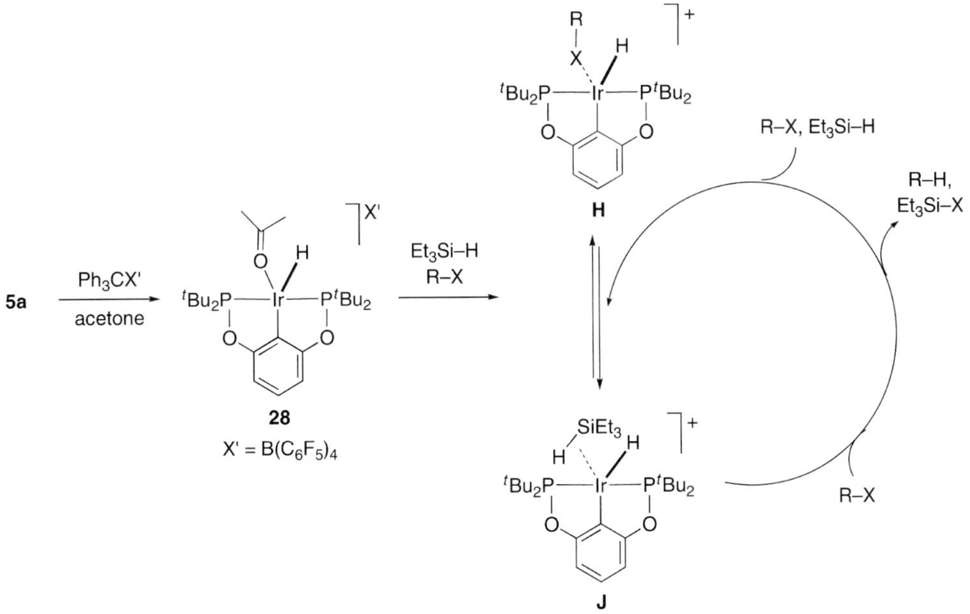

Scheme 12.11

complex (**J**) as the active species. The location of this equilibrium therefore constitutes a pivotal parameter for high catalytic activity. Alkyl iodides, as relatively good donor sites, shift the equilibrium strongly to the resting state **H**, thus leading to low activity, whereas bromides and chlorides favor formation of the reactive σ-silane complex. In line with this mechanism, competition experiments between two different alkyl halides follow the expected reactivity order R—I > R—Br > R—Cl. For example, the reduction of 1-iodoheptane in the presence of 250 equiv. 1-chlorohexane showed a 1200:1 selectivity for iodoalkane reduction. These results indicate that highly chemoselective dehalogenation can be achieved using the catalyst precursor **28**.

12.4.2
Activation of Carbon–Oxygen Bonds

The use of CO_2 as a reagent for synthetic purposes would be highly desirable, due not only to the vast availability of this gas but also its environmental concerns. The stoichiometric activation of CO_2 has been achieved with the iridium–PCP complex **29** comprising an alkyl rather than an aryl skeleton (Scheme 12.12) [32]. The addition of CO_2 to the dihydride complex results in C=O insertion into the iridium–hydride bond, and affords the formate complex **30**. However, this complex is not stable and disproportionates spontaneously into the virtually insoluble bicarbonate complex **31** and the carbonyl dihydride **32**. Such disproportionation is suppressed when the iridium metal center is replaced by rhodium [33], which is generally assumed to have a lower hydride affinity than iridium.

12.4.3
Activation of Carbon–Carbon Bonds

Iridium pincer complexes – and to a larger extent also the corresponding rhodium complexes – have provided useful insights into the mechanism of C_{aryl}—C_{alkyl} bond

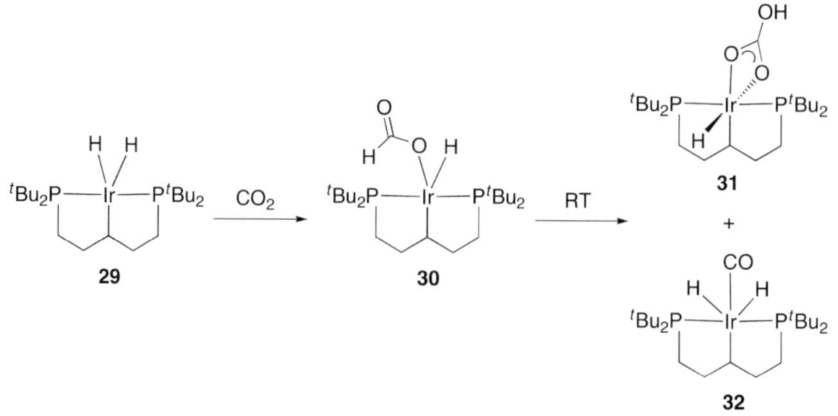

Scheme 12.12

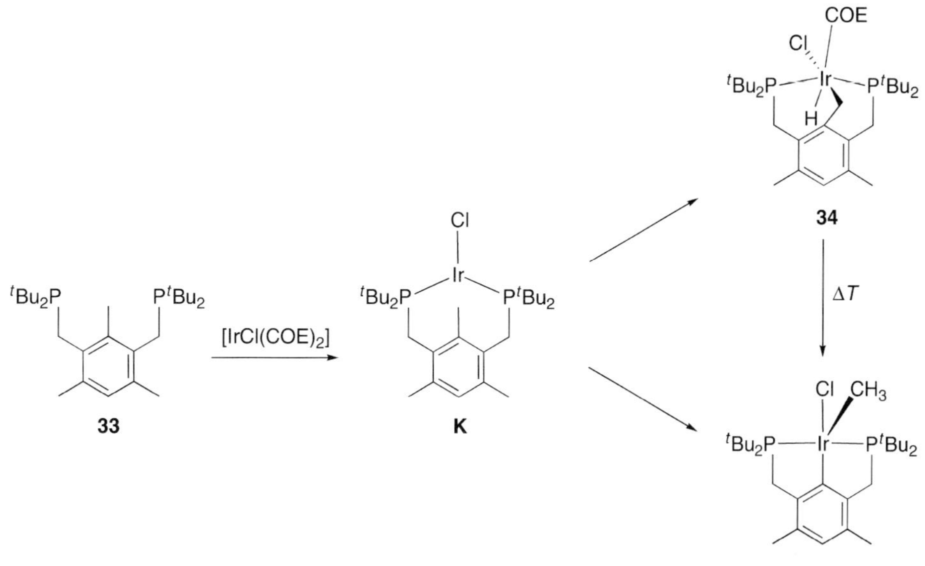

Scheme 12.13

activation [34]. Metallation of the PCP ligand precursor **33** with an appropriate iridium(I) salt results in competitive C_{alkyl}—H and C_{aryl}—C_{alkyl} bond activation and affords complexes **34** and **35**, respectively (Scheme 12.13) [35]. A time-dependent analysis of the reaction indicated a constant ratio of the two products, which was in agreement with two independent and concurrent processes for the formation of complexes **34** and **35**. As both complexes are formed irreversibly, the product distribution must be under kinetic control. Moreover, the product ratio appeared to be temperature-insensitive in the 20–60 °C range, which points to similar activation parameters for both C—H and C—C bond activation. The absence of any intermediates implies that the rate-determining step occurs very early in the reaction coordinate, presumably at the formation of **K** as the common intermediate for both C_{alkyl}—H and C_{aryl}—C_{alkyl} bond cleavage. Notably, heating of complex **34** to 100 °C results in the exclusive formation of complex **35**, perhaps via an initial reductive elimination and intermediate formation of **K**. Independent of the exact mechanism, this experiment indicates that in this case, C—C bond formation is thermodynamically more favorable than C—H bond activation.

12.5
Ammonia Borane Dehydrogenation

Chemical hydrogen storage – where hydrogen is stored in a chemical compound and released via a reversible chemical reaction – represents a promising strategy [36]. Ammonia borane (H_3NBH_3) and related aminoborane compounds have emerged as particularly attractive candidates for hydrogen-storage materials due to their high percentage by weight of available hydrogen, and the potential reversibility of their hydrogen release reactions. However, catalysts are required to effect the release of hydrogen from aminoborane compounds at efficient rates (Equation 12.12).

$$n\, H_3NBH_3 \xrightleftharpoons{\text{catalyst}} [H_2NBH_2]_n + n\, H_2 \qquad (12.12)$$

Given the excellent reactivity that iridium phosphinite PCP pincer complexes display in alkane dehydrogenation, complex **1a** was employed in the dehydrogenation reaction of H_3NBH_3, and showed an exceptionally high catalytic activity [37]. Thus, when catalyst **1a** was added to a THF solution of ammonia borane, under an argon atmosphere, a vigorous evolution of hydrogen gas was observed. Accompanied by this reaction, the initially red-colored solution of **1a** faded to pale yellow and a white solid precipitate appeared, indicative of the formation of $[H_2NBH_2]_n$. After complete consumption of the substrate, gas evolution was immediately resumed upon the addition of more ammonia borane. Complex **1a** has been reported to be the best catalyst to date for the dehydrogenation of ammonia borane. The catalytic system was tested in the presence and absence of elemental mercury and exhibited the same rate in both cases, which suggested that the iridium catalyst remained homogeneous throughout the entire dehydrogenation process. Control reactions with $H_3B \cdot THF$ aimed at investigating the fate of the catalyst, led to the isolation of complex **36**, a BH_3 adduct of **1a** (Scheme 12.14). The crystal structure of this adduct was determined and suggested to be the dormant form of the iridium–PCP pincer phosphinite catalyst **1a** and its hydrogenated congener, **L**.

The result of theoretical investigations have suggested that cleavage of a B—H bond occurs to initiate ammonia borane dehydrogenation [38]. Alternatively, the oxidative N—H addition of ammonia to the dehydrogenated intermediate **C** may constitute a feasible reaction pathway due, in particular, to the fact that ammonia and aniline oxidative addition to **1a** and related iridium–PCP systems has been reported experimentally [39].

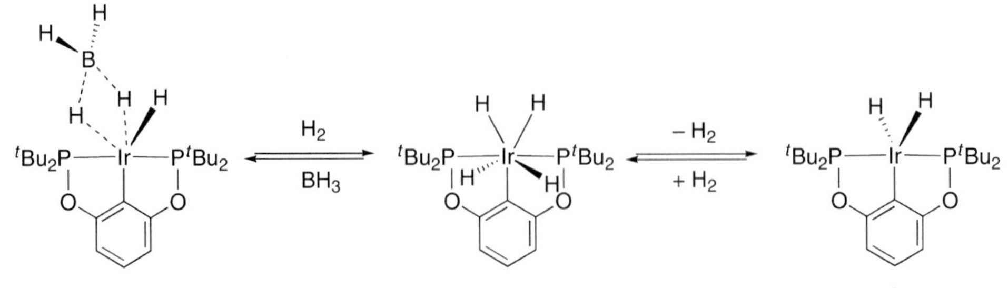

Scheme 12.14

12.6
Conclusions

During the past decade, iridium–PCP pincer complexes have undergone steady development, due mainly to their potential catalytic applications. Initial developments based on the study of simple processes such as C—H activation have led to some of the most commercially promising homogeneous alkane dehydrogenation processes. These advances – together with knowledge of the limitations of these systems – have led to further investigations affording ingenious alternatives to overcome the caveats of the processes, such as internal isomerization and inhibition of catalytic activity by the product. The consequent tandem alkane dehydrogenation–olefin metathesis systems which have been developed, although still imperfect, invite the continuation of research studies in order to further improve the reactions. Motivated not only by these results but also by the robustness of the Ir—PCP pincer system, a number of research groups have turned their attention to other difficult-to-activate bonds. The positive results obtained so far from these experiments resemble very much the early results obtained with the activation of alkane C—H bonds. Hence, it is quite safe to predict that the development of further applications of Ir—PCP pincer derivatives in other important organic transformations will flourish during the next few years, and will most likely lead to highly efficient reactions for the creation of high-value products such as pharmaceuticals and cosmetics.

Acknowledgments

M.A. would like to thank Gerard van Koten for providing a stimulating introduction into the chemistry of pincer complexes, and also the Alfred Werner Foundation for an Assistant Professorship. Sally Brooker and the University of Otago are acknowledged for their generous hospitality during the editing of this chapter. M.A.'s studies have been supported financially by the Swiss National Science Foundation, COST action D40, and ERA-net chemistry. D.M-M would like to thank the support and enthusiasm of former and current group members and colleagues. The research studies conducted by D.M-M's group is supported financially by CONACYT (F58692) and DGAPA-UNAM (IN227008).

References

1 (a) Morales-Morales, D. and Jensen, C. (eds) (2007) *The Chemistry of Pincer Compounds*, Elsevier, Amsterdam; (b) van Koten, G. and Albrecht, M. (2001) *Angewandte Chemie-International Edition*, **40**, 3750; (c) van der Boom, M.E. and Milstein, D. (2003) *Chemical Reviews*, **103**, 1759; (d) Milstein, D. (2003) *Pure and Applied Chemistry*, **75**, 445; (e) Morales-Morales, D. (2008) *Mini Reviews in Organic Chemistry*, **5**, 141.

2 Jensen, C.M. (1999) *Chemical Communications*, 2443.

3 Goldman, A.S., Roy, A.H., Huang, Z., Ahuja, R., Schinski, W. and

Brookhart, M. (2006) *Science*, **312**, 257.
4 (a) Murai, S. (ed.) (1999) *Topics in Organometallic Chemistry*, 3;
(b) Goldberg, K.I. and Goldman, A.S. (eds) (2004) *Activation and Functionalization of C—H Bonds*, ACS Symposium Series 885, ACS, Washington.
5 (a) Labinger, J.A. and Bercaw, J.E. (2002) *Nature*, **417**, 507;
(b) Crabtree, R.H. (1995) *Chemical Reviews*, **95**, 987.
6 Gupta, M., Hagen, C., Flesher, R.J., Kaska, W.C. and Jensen, C.M. (1996) *Chemical Communications*, 2083.
7 (a) Gupta, M., Hagen, C., Kaska, W.C., Cramer, R.E. and Jensen, C.M. (1997) *Journal of the American Chemical Society*, **119**, 840;
(b) Gupta, M., Kaska, W.C. and Jensen, C.M. (1997) *Chemical Communications*, 461.
8 Liu, F., Pak, E.B., Singh, B., Jensen, C.M. and Goldman, A.S. (1999) *Journal of the American Chemical Society*, **121**, 4086.
9 (a) Lee, D.W., Kaska, W.C. and Jensen, C.M. (1998) *Organometallics*, **17**, 1;
(b) Gosh, R., Kanzelberger, M., Emge, T.J., Hall, G.S. and Goldman, A.S. (2006) *Organometallics*, **25**, 5668.
10 de Vries, J. and Elsevier, C.J. (eds) (2007) *Handbook of Homogeneous Hydrogenation*, Wiley-VCH Verlag GmbH, Weinheim.
11 Renkema, K.B., Kissin, Y.V. and Goldman, A.S. (2003) *Journal of the American Chemical Society*, **125**, 7770.
12 Kanzelberger, M., Singh, B., Czerw, M., Krogh-Jesper, K. and Goldman, A.S. (2000) *Journal of the American Chemical Society*, **122**, 11017.
13 (a) Xu, W.-W., Rosini, G.P., Gupta, M., Jensen, C.M., Kaska, W.C., Krogh-Jespersen, K. and Goldman, A.S. (1997) *Chemical Communications*, 2273;
(b) Haenel, M.W., Oevers, S., Angermund, K., Kaska, W.C., Fan, H.-J. and Hall, M.B. (2001) *Angewandte Chemie-International Edition*, **40**, 3596.
14 Krogh-Jespersen, K., Czerw, M., Summa, N., Renkema, K.B., Achord, P.D. and Goldman, A.S. (2002) *Journal of the American Chemical Society*, **124**, 11404.
15 Göttker-Schnetmann, I., Heinekey, D.M. and Brookhart, M. (2006) *Journal of the American Chemical Society*, **128**, 17114.
16 Kubas, G.J. (2001) *Journal of Organometallic Chemistry*, **635**, 37.
17 (a) Göttker-Schnetmann, I., White, P. and Brookhart, M. (2004) *Journal of the American Chemical Society*, **126**, 1804;
(b) Morales-Morales, D., Redón, R., Yung, C. and Jensen, C.M. (2004) *Inorganica Chimica Acta*, **357**, 2953.
18 (a) Göttker-Schnetmann, I. and Brookhart, M. (2004) *Journal of the American Chemical Society*, **126**, 9330;
(b) Göttker-Schnetmann, I., White, P.S. and Brookhart, M. (2004) *Organometallics*, **23**, 1766.
19 Zhu, K., Achord, P.D., Zhang, X., Krogh-Jespersen, K. and Goldman, A.S. (2004) *Journal of the American Chemical Society*, **126**, 13044.
20 Kuklin, S.A., Sheloumov, A.M., Dolgushin, F.M., Ezernitskaya, M.G., Peregudov, A.S., Petrovskii, P.V. and Koridze, A.A. (2006) *Organometallics*, **25**, 5466.
21 Schrock, R.R. (1990) *Accounts of Chemical Research*, **23**, 158.
22 Ahuja, R., Kundu, S., Goldman, A.S., Brookhart, M., Vicente, B.C. and Scott, S.L. (2008) *Chemical Communications*, 253.
23 Ray, A., Zhu, K., Kissin, Y.V., Cherian, A.E., Coates, G.W. and Goldman, A.S. (2005) *Chemical Communications*, 3388.
24 (a) Moulton, C.J. and Shaw, B.L. (1976) *Journal of the Chemical Society–Dalton Transactions*, 1020;
(b) Nemeh, S., Jensen, C., Binamira-Soriaga, E. and Kaska, W.C. (1983) *Organometallics*, **2**, 1442.
25 Zhang, X., Kanzelberger, M., Emge, T.J. and Goldman, A.S. (2004) *Journal of the American Chemical Society*, **126**, 13192.
26 Ben-Ari, E., Cohen, R., Gandelman, M., Shimon, L.J.W., Martin, J.M.L. and Milstein, D. (2006) *Organometallics*, **25**, 3190.
27 Feller, M., Karton, A., Leitus, G., Martin, J.M.L. and Milstein, D. (2006) *Journal of the American Chemical Society*, **128**, 12400.
28 Ben-Ari, E., Leitus, G., Shimon, L.J.W. and Milstein, D. (2006) *Journal of the American Chemical Society*, **128**, 15390.

29 Gunanathan, C., Ben-David, Y. and Milstein, D. (2007) *Science*, **317**, 790.
30 Ghosh, R., Zhang, X., Achord, P., Emge, T.J., Krogh-Jespersen, K. and Goldman, A.S. (2007) *Journal of the American Chemical Society*, **129**, 853.
31 Yang, J. and Brookhart, M. (2007) *Journal of the American Chemical Society*, **129**, 12657.
32 McLoughlin, M.A., Keder, N.L., Harrison, W.T.A., Flesher, R.J., Mayer, H.A. and Kaska, W.C. (1999) *Inorganic Chemistry*, **38**, 3223.
33 Vigalok, A., Ben-David, Y. and Milstein, D. (1996) *Organometallics*, **15**, 1839.
34 Gozin, M., Weisman, A., Ben-David, Y. and Milstein, D. (1993) *Nature*, **364**, 699.
35 Rybtchinski, B., Vigalok, A., Ben-David, Y. and Milstein, D. (1996) *Journal of the American Chemical Society*, **118**, 12406.
36 (a) U.S. DOE (2006) *Hydrogen, and Fuel Cells & Infrastructure Technologies Program* (http://www1.eere.energy.gov/hydrogenfuelcells/storage);
(b) The American Physical Society (2004), *The Hydrogen Initiative* (http://www.aps.org/public_affairs/index.cfm).
37 Denney, M.C., Pons, V., Hebden, T.J., Heinekey, D.M. and Goldberg, K.I. (2006) *Journal of the American Chemical Society*, **128**, 12048.
38 Paul, A. and Musgrave, C.B. (2007) *Angewandte Chemie-International Edition*, **46**, 8153.
39 (a) Kanzelberger, M., Zhang, X., Emge, T.J., Goldman, A.S., Zhao, J., Incarvito, C. and Hartwig, J.F. (2002) *Journal of the American Chemical Society*, **125**, 13644;
(b) Zhao, J., Goldman, A.S. and Hartwig, J.F. (2005) *Science*, **307**, 1080;
(c) Sykes, A.C., White, P. and Brookhart, M. (2006) *Organometallics*, **25**, 1664.

13
Iridium-Mediated Alkane Dehydrogenation
David Morales-Morales

13.1
Introduction

C—H activation is one of the most profitable, studied – and as yet unsolved – problems in chemistry. Since the first experiments performed to identify this phenomenon during the early 1960s by Kleinman and Dubeck [1] and by Chatt and Davison [2], countless studies have been carried out by many research groups worldwide. The main interest in this chemistry is the great potential that a clear understanding of the process may bring to basic chemistry, allowing the organic transformation of simple, cheap and easily attainable starting materials into high-value products [3]. There is, however, an even more valuable motivation for studying the C—H activation process, namely to allow the conversion of small molecules such as methane into potentially clean energy sources, such as methanol [4]. It is for exactly these reasons – and the well-known intrinsic difficulty of activating this ubiquitous bond – that the chemistry community has denominated this process as one of the 'Holy Grails' of chemistry [5].

Among the historic developments of studies on C—H activation, several methods and conditions have been used, from thermal studies to photochemical activation, with all of which have been mediated by derivatives of metals ranging from transition metals to lanthanides and actinides. Hence, given the importance that the chemistry surrounding the C—H activation process entails, several excellent reviews and monographs have already been published [6]. Unfortunately, in most of these cases the chemistry described has been devoted entirely to descriptions of the activation of a particular C—H bond in a given process, and consequently some division has arisen as to how the C—H activation process does in fact proceed [7].

From the outset, iridium compounds have played an important role in the better understanding of the C—H activation process, and consequently in the development of efficient alkane dehydrogenation reactions [8]. Hence, in this chapter we will review the participation of iridium complexes in the optimization of chemical processes for C—H activation which, today, have led to some highly promising

Iridium Complexes in Organic Synthesis. Edited by Luis A. Oro and Carmen Claver
Copyright © 2009 WILEY-VCH Verlag GmbH & Co. KGaA, Weinheim
ISBN: 978-3-527-31996-1

reactions for industrially scalable alkane dehydrogenation under mild conditions. Given the major relevance of iridium pincer compounds during the past decade in metal-mediated organic transformations – and especially in alkane dehydrogenation – this theme will be mentioned only briefly at this point; a more extensive coverage can be found in Chapter 12.

13.1.1
The Beginning

Whilst C—H bonds represent the most ubiquitous chemical linkage in Nature, they are at the same time some of the most difficult bonds to cleave, although they are not completely inert [5b,c]. For example, in 1963 Kleinman and Dubeck reported the possibility of C—H bond cleavage in azobenzene by the Cp_2Ni complex (Scheme 13.1) [1]. The structure originally proposed by Kleinman and Dubeck considered the nickel center to be coordinated η2 to the N=N π-bond (**2**) (Scheme 13.2).

Later in 1965, Chatt and Davidson [2] reported the first example of cyclometallation of an sp^3 C—H bond in $[Ru(dmpe)_2]$ (**3**); dmpe = dimethyl **phos**phinoethane. These authors found not only that this complex spontaneously cyclometallates at the phosphorus methyl groups to produce complex $[Ru(H)(CH_2P(Me)CH_2CH_2PMe_2)(dmpe)]$ (**4**; see Scheme 13.4) (a later examination by Cotton and coworkers [9] of this compound provided crystallographic evidence that the cyclometalated form of $[Ru(dmpe)_2]$ is in fact a dimer (**5**) of the type shown in Scheme 13.3), but also that the system reacts with free naphthalene via the oxidative addition of a C—H bond to the zero-valent ruthenium center to produce complex [cis-Ru(H)(2-naphthyl)(dmpe)$_2$] (**6**). This species was in equilibrium with the π-coordinated naphthalene ruthenium complex $[Ru(naphthalene)(dmpe)_2]$ (**7**) (Scheme 13.4).

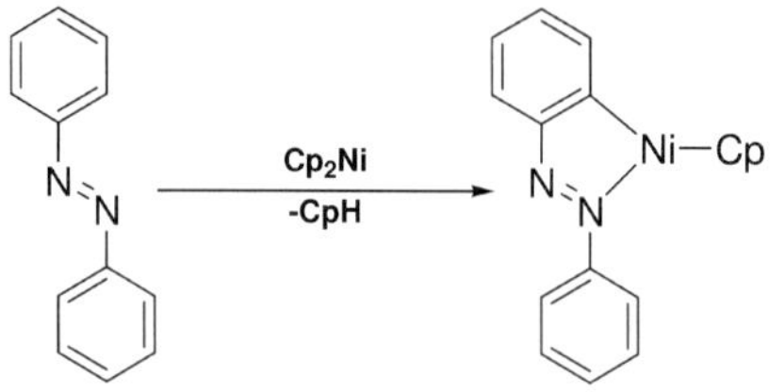

Scheme 13.1

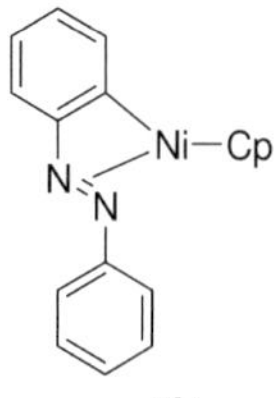

Scheme 13.2

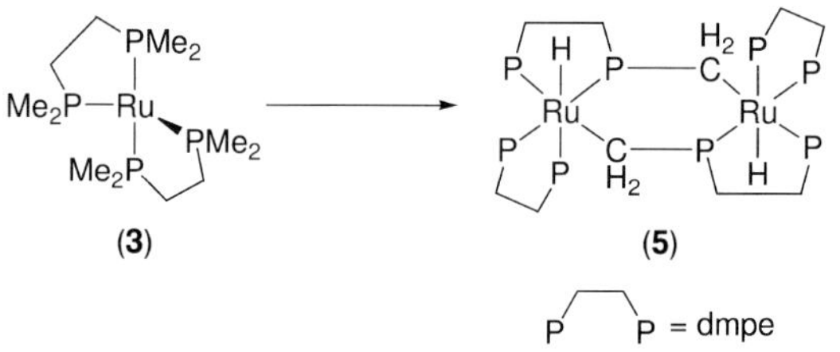

Scheme 13.3

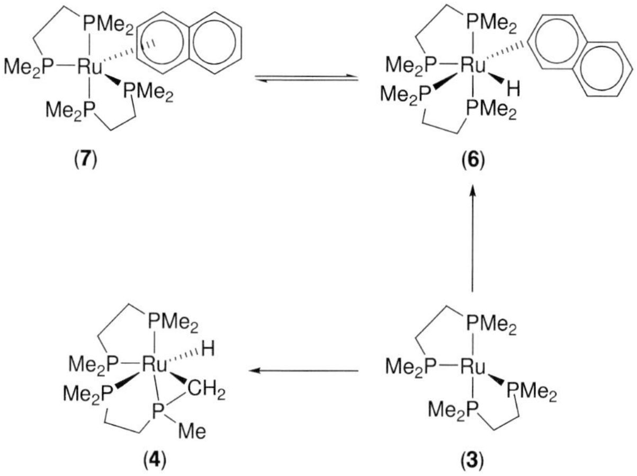

Scheme 13.4

This was the very first explicit connection to be made between the intermolecular and intramolecular varieties of C—H bond breaking by transition-metal complexes.

13.2
Alkane C—H Activation with Ir Derivatives

One of the most direct ways of attaining fundamental information about the C—H activation process is to study the direct stoichiometric reactions between metal centers and carbon–hydrogen bonds. However, this simple process was not directly observed with simple alkanes until 1981. Thus, individual efforts by the Bergman, Graham and Jones research groups [10] during the early 1980s produced the first examples of the oxidative addition of unactivated C—H bonds to transition-metal centers that resulted in the direct observation of alkyl hydride products [10]. In these instances, the photolysis of complexes Cp*(PMe$_3$)IrH$_2$ (**8**) [10a,b], Cp*Ir(CO)$_2$ (**9**) [10c] or Cp*(PMe$_3$)RhH$_2$ (**10**) [10d,e] (Cp* = C$_5$Me$_5$) in a hydrocarbon solvent resulted in the loss of dihydrogen (complexes **8** and **10**) or CO in the case of complex **9** and C—H activation of the solvent (Scheme 13.5).

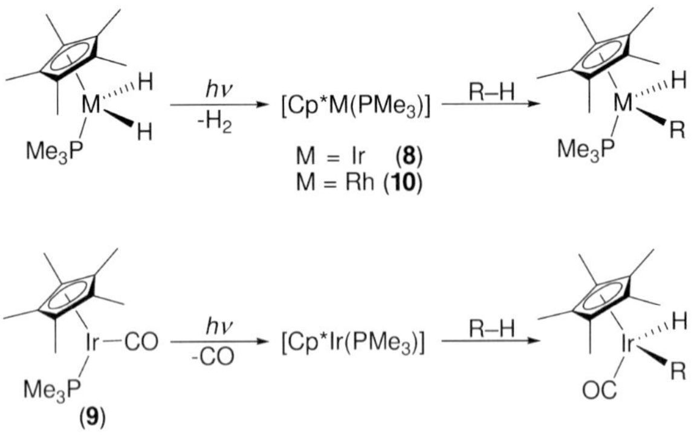

Scheme 13.5

13.3
Alkane Dehydrogenation with Ir Complexes

In addition to these basic processes that involve the insertion of a metal center into a C—H bond, the conversion of alkanes to olefins by soluble transition-metal complexes represents not only a C—H activation process but also a formal functionalization of the alkane – in this particular case through the removal of dihydrogen. Olefins constitute the most important feedstock of the chemical industry, and the dehydrogenation of alkanes is the most direct and potentially most economical route for olefin production. While heterogeneous catalysts have long been used for the dehydrogenation of ethane and, less successfully for propane and isobutene [11], no systems capable of dehydrogenating higher alkanes to olefins with good chemoselectivity have been found. This process was first noted by Crabtree and colleagues [12] almost three decades ago, and occurred stoichiometrically with complex $[IrH_2(PPh_3)_2(acetone)_2]^+ BF_4^-$ (11) and cyclic alkanes in noncoordinating solvents such dichloromethane (DCM) or 1,2-dichloroethane to produce either the arene (12) or olefin derivatives (13) (Scheme 13.6). These complexes were especially interesting not only because they were the first examples of olefin–hydrido complexes, but also because they appeared to be the active intermediates in the reverse process (hydrogenation) reported earlier by Crabtree and coworkers [13].

The results of further studies showed that the addition of *tert*-butylethylene (tbe) to this system acted as a 'hydrogen acceptor', thus driving the endothermic loss of H_2 and resulting in the catalytic generation of dehydrogenation products [14].

Crabtree and coworkers proposed that the initial step in these processes was the oxidative addition of an alkane C—H bond to the metal. Here, the role of tbe was in part kinetic – to dehydrogenate and so to activate complex 11, and in part thermodynamic – to provide an additional driving force for the overall process. The thought behind this idea was the belief that any alkyl hydride intermediate formed by oxidative addition would be unstable and would have to be trapped by subsequent β-elimination to afford the corresponding olefin. The resultant metal hydride would then be dehydrogenated by the tbe, and the cycle would continue until a stable product had been formed. *Tert*-butylethylene is one of the very few olefins

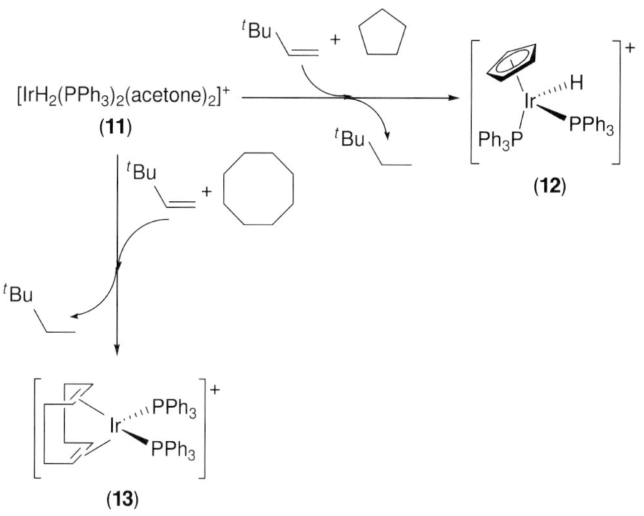

13.3 Alkane Dehydrogenation with Ir Complexes

Scheme 13.6

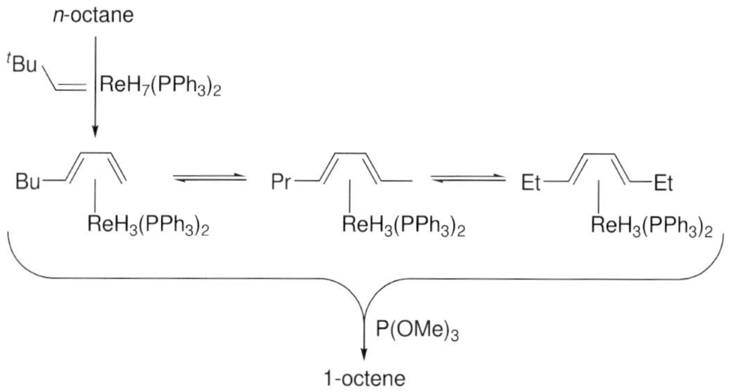

Scheme 13.7

that is an effective hydrogen acceptor, this being most likely due to its lack of allylic C—H bonds which, if cleaved by the metal, would lead to an allyl complex that would block the active sites on the metal center. *Tert*-butylethylene is also sufficiently bulky, and so does not bind strongly to the metal center; moreover, when compared to other common olefins it hydrogenates rapidly to produce *tert*-butylethane (tba).

Felkin and coworkers [15] were able to use the same sacrificial hydrogen acceptor tbe with $ReH_7(PPh_3)_2$, and succeeded in dehydrogenating linear alkanes to diene complexes. For *n*-octane, a series of interconverting isomers was formed. Of note here was the fact that an addition of $P(OMe)_3$ led to the specific release of 1-octene, the least stable but most desirable alkene isomer (Scheme 13.7). This led in turn to the first true homogeneous catalyst for the transfer dehydrogenation of alkanes to alkenes, affording up to nine turnovers of cyclooctene at 80 °C and 1.6

turnovers even at 30 °C. The homogeneity of the catalyst was confirmed, and a mechanism proposed in which the olefin dissociates after two C—H bond-breaking steps. The H_2 abstracted from the alkane, and now bound to the metal, could then be removed by tbe and the cycle could continue.

In the case of iridium, complex $[IrH_2(PPh_3)_2(acetone)_2]^+ BF_4^-$ (11) was the first to carry out catalytically the dehydrogenation of cycloalkanes [13, 14]. However, it was later realized that the halocarbons used as solvents reacted with 11 to produce the stable species $[HL_2Ir(\mu\text{-}Cl)_2(\mu\text{-}X)IrL_2H]BF_4$ (X = Cl (14) or H (15)) [16] (Scheme 13.8), and that elimination of the solvent by running the reactions in neat alkane not only improved yields but also permitted the activation of other previously unreactive cycloalkanes, such as methyl- and ethyl-cyclopentane. However, it was also noted that the system in some cases was not catalytic, due mainly to decomposition of the catalyst at the temperatures employed [16].

Variations on complex 11 led the same authors to use neutral complexes of the type $[IrH_2(CF_3CO_2)(PR_3)_2]$ R = C_6H_4-4-F (16) and cyclohexyl (17) [17]. Once again, whilst both compounds provided thermal alkane dehydrogenation, it was of note that almost simultaneously and independently Felkin and collaborators [18] observed thermal dehydrogenation from $[IrH_5(P^iPr_3)_2]$ (18) in the presence of $MeCO_2H$. In this case, the authors considered the species $[IrH_2(MeCO_2)(PPh_3)_2]$ (19) to be involved in the process. Complex 17 also provided a novel photochemical catalytic dehydrogenation at room temperature, even in the absence of tbe; perhaps more importantly, no decomposition products were detected even after 7 days of irradiation (254 nm) at room temperature (Scheme 13.9) [17].

At this point it became clear that, although a wide variety of complexes could be employed for both transfer dehydrogenation and photochemical dehydrogena-

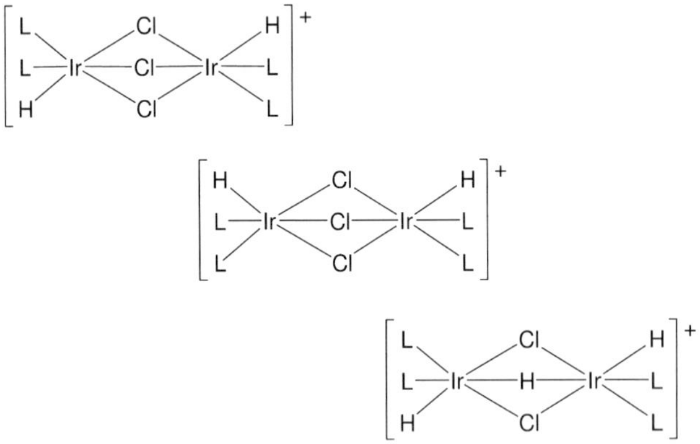

Scheme 13.8

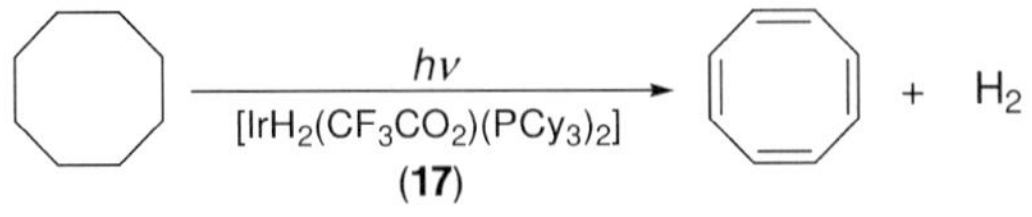

Scheme 13.9

$$RCH_2CH_3 \longrightarrow RCH=CH_2 + H_2 \qquad (a)$$

$$RCH_2CH_3 + {}^tBuCH=CH_2 \longrightarrow RCH=CH_2 + {}^tBuCH_2CH_3 \qquad (b)$$

$$RCH_2CH_3 \xrightarrow{h\nu} RCH=CH_2 + H_2 \qquad (c)$$

Scheme 13.10

tion, one of the main issues to be solved was indeed the thermal stability of the catalyst.

One way of overcoming this problem was to use even more active species that would enable a given system to afford higher turnover rates and yields, thus compensating in some way the potential decomposition of the catalytic species. Hence, much research effort was turned towards studying and using rhenium-based catalysts, as this metal was known to be more active than its counterparts, the iridium-based systems. Important contributions by Nomura and Saito [19], by Tanaka and coworkers [20] and by Goldman and colleagues [21] to the application of Rh-Vaska's analogues and Wilkinson-type catalysts led to some fundamental advances in the quest for efficient dehydrogenation catalysts, although these were not completely satisfactory.

A revisit to the well-known dehydrogenation methods allowed it to be established that alkane dehydrogenation (Scheme 13.10a), as a normally endoergonic process, had to be driven in some manner; consequently, hydrogen acceptor or photochemical methods were used. Although hydrogen transfer (Scheme 13.10b) itself does not lead to a net dehydrogenation process, neither is the efficiency of photochemical processes (Scheme 13.10c) high, this being due to the fact that the photon energy greatly exceeds the reaction endothermicity and any build-up of light-absorbing impurities will stop the photo-process.

In 1990, Fujii and Saito [22] made the important observation that refluxing the alkane substrate allowed the dehydrogenation to proceed spontaneously while using Wilkinson's complex as catalyst (Scheme 13.10a). The principle was simple – under reflux conditions, the dehydrogenation reaction would overcome the equilibrium restriction by separating the product hydrogen not only from the catalyst solution but also from the gas phase in contact with the solution. This was feasible because these conditions cause the vapor phase to consist exclusively of solvent molecules. Consequently, molecular hydrogen could be spontaneously removed from the catalyst solution immediately after its generation.

In the case of iridium, the above-mentioned procedure was soon employed by Aoki and Crabtree [23] who, by using the compound $[IrH_2(CF_3CO_2)(PCy_3)_2]$ (**17**), showed that the system could dehydrogenate alkanes in the absence of a hydrogen acceptor, under reflux conditions proposed by Fujii and Saito [22], with sole evolution of the alkane and hydrogen. These authors were also able to determine that the dehydrogenation rate was independent of the reflux rate, although reflux conditions were required in order for the reaction to proceed. For example, when the reaction temperature was reduced to 140 °C under 1 atm pressure, both the reflux and dehydrogenation processes stopped. However, if the pressure was then

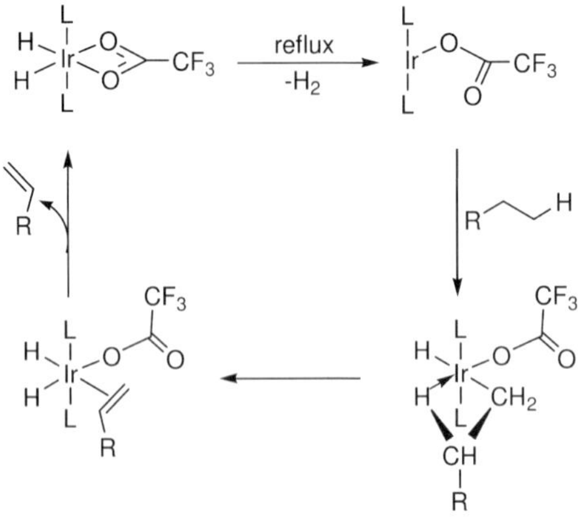

Scheme 13.11

reduced so that the system began to reflux at 140 °C, the dehydrogenation resumed, albeit at a slightly lower rate. These observations were consistent with the proposal that the role of the reflux is to remove the H_2 and to displace the equilibrium of reaction (a) in Scheme 13.9 in favor of the alkene. When the metallic mercury test [24] was performed on this system, the results suggested the need for an homogeneous origin in order for the chemistry to proceed. As in previous cases, the main caveat of the system was the thermal instability and, therefore, the deactivation of the catalyst, which was ascribed mainly to P—C bond cleavage of the phosphine [17b, 25]; however, deactivation of the catalyst also occurred by product inhibition. A mechanistic proposal of the entire dehydrogenation reaction was proposed (Scheme 13.11).

Similar experiments using other catalysts were also performed [23]. Hence, complexes [IrH$_2$(RCO$_2$)(PCy$_3$)$_2$] R = C$_2$F$_5$ (**20**) and PhCH$_2$ (**21**) were tested in identical dehydrogenation experiments. Of particular interest were the results attained using compound **20**, which was shown not only to be active but also to be more robust than catalyst **17**. These results were important because they showed that phosphine degradation could, in principle, be prevented by modifying the non-phosphine ligands included in the catalyst.

The consideration that reflux of the substrate limited the temperature range over which the reaction could be carried out, led to further improvements of the system by two different approaches: (i) to consider using an inert, volatile cosolvent as a refluxing medium; and (ii) by bubbling an inert gas through the reaction mixture. In the first case the use of perfluorodecalin (PFD) [23] resulted in an efficient process, affording almost a doubling of turnovers compared to reflux in neat alkane, such as cyclo-octane (COA). The bubbling of an inert gas (e.g. Ar) was also effective, although when COA was employed as substrate the procedure led to an excessive loss of COA by evaporation. When the relatively nonvolatile cyclodecane was used, however, the process was effective. Although both procedures presented with limitations, they were significant in that alkanes with boiling points unsuitable for the usual reflux process could now be studied.

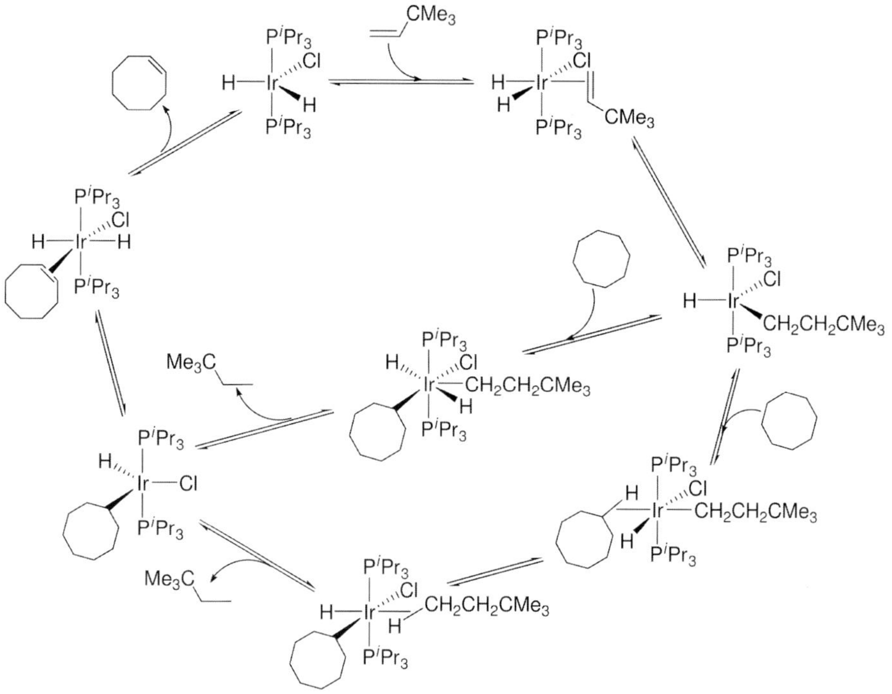

13.4 Alkane Dehydrogenation Catalyzed by Ir Pincer Complexes

Scheme 13.12

Until now, for most of the systems described here it has been accepted that alkane activation occurred through oxidative addition to the 14-electron intermediate complexes. Yet, Belli and Jensen [26] showed, for the first time, evidence for an alternative reaction path for the catalytic dehydrogenation of COA with complex [IrClH$_2$(P^{i}Pr$_3$)$_2$] (**22**) which invoked an Ir(V) species. Catalytic and labeling experiments led these authors to propose an active mechanism (Scheme 13.12), on the basis of which they concluded that the dehydrogenation of COA by compound **22** did not involve an intermediate 14-electron complex [17–21], but rather the association of COA to an intermediate alkyl–hydride complex (Scheme 13.12).

13.4 Alkane Dehydrogenation Catalyzed by Ir Pincer Complexes

13.4.1 Ir-PCP Pincer Compounds

Based on these results, it became clear that there was indeed a need for new, more robust complexes capable of withstanding the higher temperatures at which it was believed the alkane dehydrogenation process would occur more efficiently. Consequently, the chemistry of pincer compounds appeared on the scene. In 1976,

Moulton and Shaw reported [27] for the first time the synthesis of the complex IrHCl{C$_6$H$_3$-2,6-(CH$_2$P^{t}Bu$_2$)$_2$} (**23**). This compound was found to exhibit a high thermal stability, and sublimed without visible decomposition at temperatures as high as 180 °C. These results, independently, led to the research groups of Jensen [28], Goldman [29] and Leitner [30] using derivatives of this complex for its potential application as catalysts in the dehydrogenation of alkanes. Thus, Jensen and coworkers [28] reported using the dihydride–rhodium complex RhH$_2${C$_6$H$_3$-2,6-(CH$_2$P^{t}Bu$_2$)$_2$} (**24**) in the dehydrogenation of COA at 150 °C, using *tert*-butylethylene as the sacrificial hydrogen acceptor. This compound, when tested at reaction temperatures as high as 200 °C, was shown to be stable for periods of weeks, although the dehydrogenation reaction using this species only afforded 1.8 turnovers at 200 °C. Further experiments led Jensen and coworkers [31] to determine that it was in fact the iridium derivative IrH$_2${C$_6$H$_3$-2,6-(CH$_2$P^{t}Bu$_2$)$_2$} (**25**) and not its rhodium counterpart (**24**) the best dehydrogenation catalyst. Under optimized conditions, Jensen and coworkers were able to attain a turnover number (TON) of 720 at a reaction temperature of 200 °C, even though this catalyst exhibited excellent activity at temperatures as low as 150 °C (TON = 82). Unfortunately, complex **25** is quickly deactivated by the product of the reaction (i.e. product inhibition occurred) [31].

Recently, complex IrH$_2${C$_6$H$_3$-2,6-(CH$_2$P^{t}Bu$_2$)$_2$} (**25**) has been used in the activation of C—H bonds of several substrates (Scheme 13.13) [32], including alkane polymers and oligomers (Scheme 13.14).

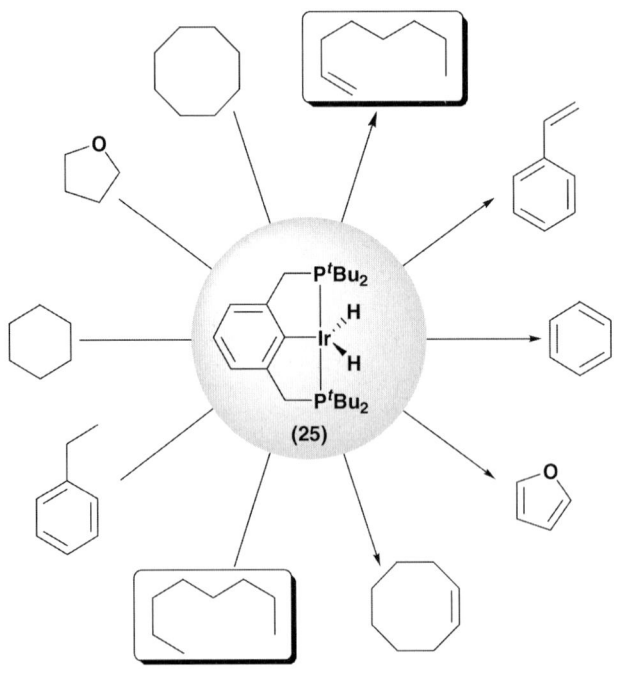

Scheme 13.13

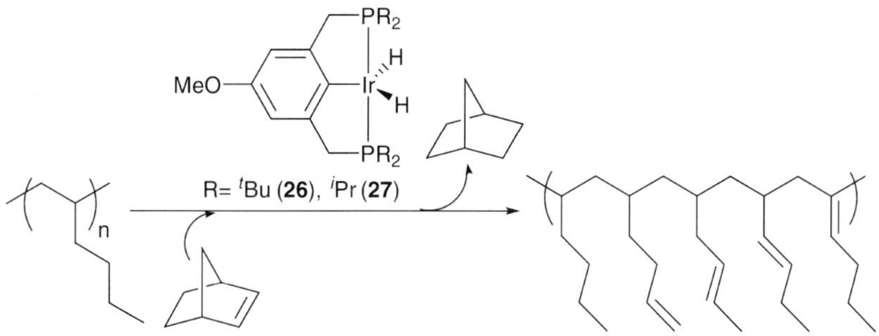

Scheme 13.14

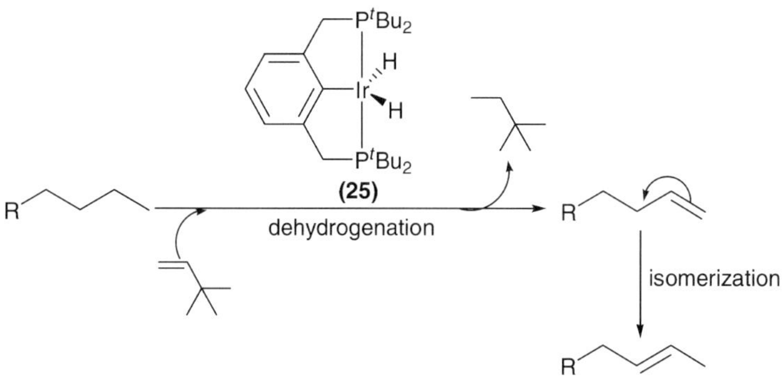

Scheme 13.15

Among these examples, probably the most notable case is the dehydrogenation of linear alkanes to their corresponding terminal alkenes (α-olefins), this being the kinetically favored process over the production of the internal alkenes. However, the same complex slowly catalyzes an isomerization of the terminal alkene to internal alkenes, as the latter are the thermodynamic products (Scheme 13.15) [33].

Further modifications in the PCP pincer ligand, such as changing substituents at the P moiety and using different hydrogen acceptors, led Liu and Goldman [34] to increase the efficiency of the catalytic system to a maximum of 68% selectivity for the terminal alkene in the catalytic dehydrogenation of n-octane, with TONs on the order of 143. Other improvements to the system were an elimination of the need for a sacrificial hydrogen acceptor – an achievement that was merited by joint efforts of the Jensen and Goldman research groups, who achieved up to 1000 turnovers under optimized acceptorless conditions [35]. Studies using theoretical calculations [36] led to the postulation of a tentative reaction mechanism through which complexes $IrH_2\{C_6H_3\text{-}2,6\text{-}(CH_2P^tBu_2)_2\}$ (**25**) and $IrH_2\{C_6H_3\text{-}2,6\text{-}(CH_2P^iPr_2)_2\}$ (**28**) carried out the alkane dehydrogenation both under hydrogen acceptor and acceptorless conditions (Scheme 13.16) [32].

Recently, Kaska and coworkers [37] reported a rigid PCP pincer system based on an anthracene backbone (**29**) (Scheme 13.17). The iridium derivative of this

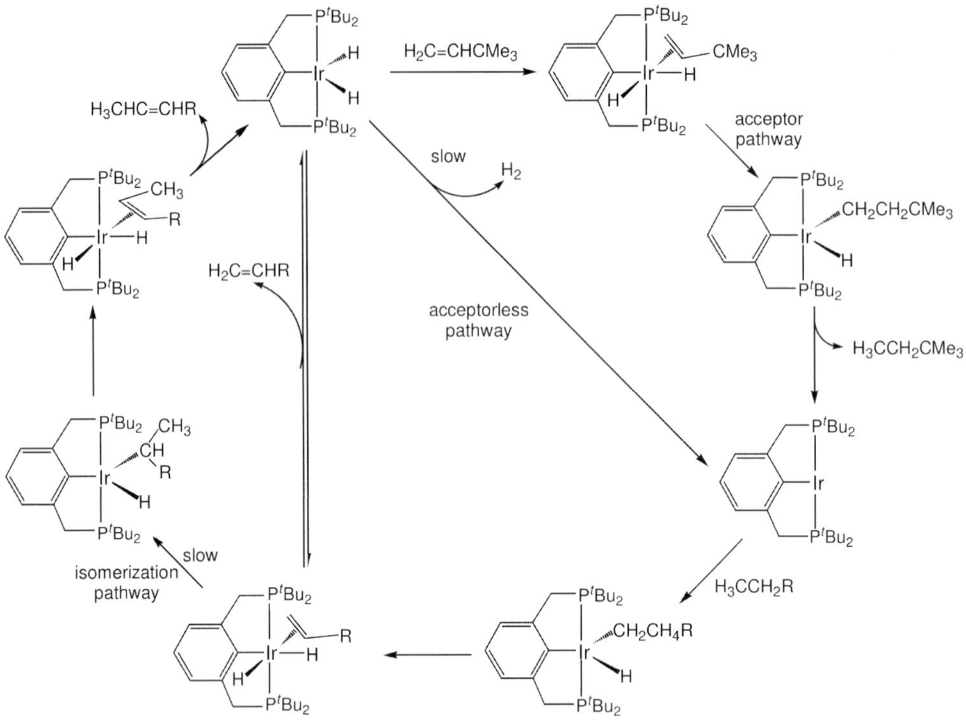

Scheme 13.16

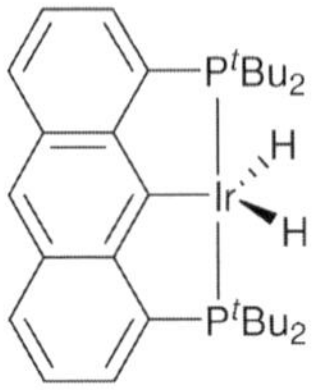

(29) **Scheme 13.17**

ligand has demonstrated catalytic activity in the dehydrogenation of alkanes at temperatures as high as 250 °C, without decomposition. However, even at this reaction temperature the system did not match the performance of complexes **25** and **28**, nor in yield of the terminal olefin or in the TONs.

13.4.2
Ir-POCOP Pincer Compounds

In spite of the success that Ir—PCP pincer compounds have enjoyed in the development of more efficient and promising alkane dehydrogenation systems, one caveat of using these complexes has been the sometimes difficult or tedious synthesis of the pincer ligands and complexes. An answer to this problem was

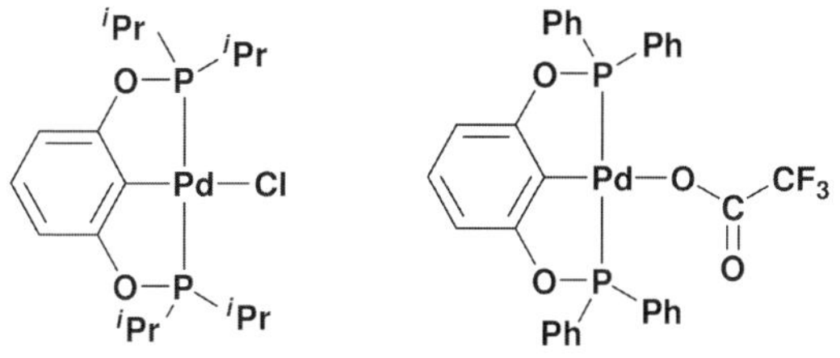

13.4 Alkane Dehydrogenation Catalyzed by Ir Pincer Complexes

Scheme 13.18

Morales-Morales & Jensen — [phosphinite PCP Pd–Cl complex with ^{i}Pr groups]

Bedford — [phosphinite PCP Pd complex with Ph groups and trifluoroacetate]

provided by Morales-Morales and coworkers [38] and by Bedford and colleagues [39], almost simultaneously, in the year 2000 when they independently reported the synthesis of the first PCP phosphinite ligands and their palladium derivatives (Scheme 13.18).

Since then, these ligands and their complexes have become increasingly important due to the fact that they exhibit the same characteristics of robustness and thermal stability and, in most of cases, an enhanced reactivity compared to their phosphine counterparts. Thus, phosphinite PCP pincer complexes have been used intensely during the most recent era of catalytic processes and reactions involving the activation of aliphatic C—H bonds [40].

The first reports regarding transfer dehydrogenation of alkanes using iridium–phosphinite PCP pincer complexes were made by Morales-Morales and colleagues [41] and by Brookhart and coworkers [42] in 2004. Morales-Morales, Jensen and colleagues reported the use of an iridium–phosphinite PCP pincer complex (**30**) (Scheme 13.19) for both the transfer dehydrogenation of COA in the presence of *tert*-butylethylene (tbe) and the acceptorless dehydrogenation of *n*-undecane. Here, the reactivity of these complexes was found to be slightly superior to that of the analogous phosphine derivative [IrH$_2${C$_6$H$_3$-2,6-(CH$_2$P^{t}Bu$_2$)$_2$}$_2$] (**25**) with similar yields and TONs, whilst exhibiting the same problems of product inhibition and isomerization (Scheme 13.20).

In 2004, Brookhart and coworkers [42] synthesized a series of new phosphinite *p*-XPCPIrHCl pincer complexes (Scheme 13.21). In comparison with their phosphine counterparts, where the air- and moisture-sensitive hydridochloride compound must be converted into the even more sensitive dihydride species by treatment with LiBHEt$_3$. In this case, generation of the catalytically active species is made *in situ* by reacting complexes (**31** to **36**) with NaO^{t}Bu in the reaction mixture (Scheme 13.22). Under these reaction conditions, the transfer dehydrogenation of COA with tbe to form cyclooctene (COE) and *tert*-butylethane (TBA) was easily accomplished. Under comparable conditions, these new catalysts (**31–36**) are almost one order of magnitude more active in terms of turnover frequency (TOF), TONs and substrate conversion than the benchmark catalyst [IrH$_2${C$_6$H$_3$-2,6-(CH$_2$P^{t}Bu$_2$)$_2$}$_2$] (**25**). A further dehydrogenation of COE to form 1,3-COD was accomplished by transfer dehydrogenation in the presence of tbe at a high COE concentration. Moreover,

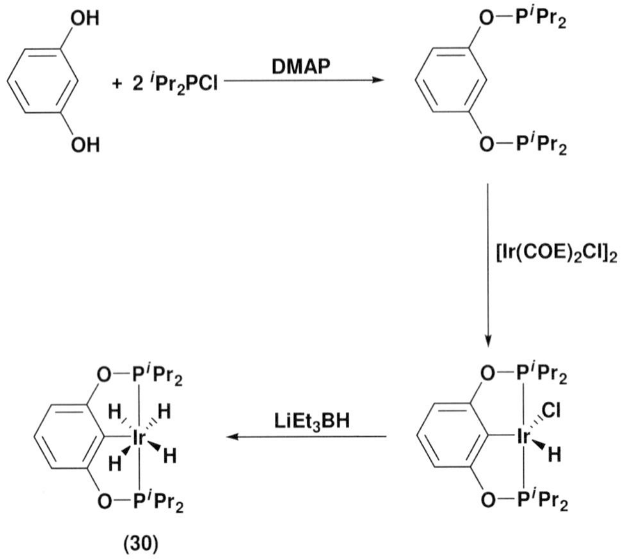

Scheme 13.19

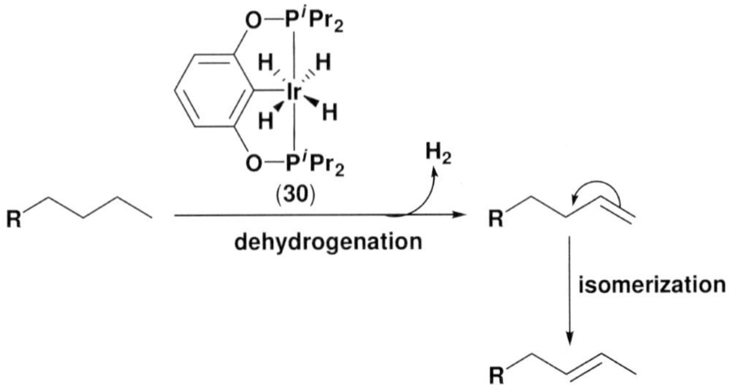

Scheme 13.20

in the absence of another hydrogen acceptor, COE itself serves as the hydrogen acceptor, giving rise to a disproportionation of COE into COA and 1,3-COD, which is further transformed into o-xylene and ethylbenzene at temperatures as low as 200 °C. However, the disproportionation of COE into 1,3-COD and COA at 200 °C is only operative at relatively low COE:catalyst ratios (ca. 450:1).

A major problem associated with the above-described systems is that of product inhibition of the catalyst and isomerization processes, when a certain yield of α-olefin has been attained. It would be desirable, therefore, to have a catalyst (or a combination of catalysts) capable of using and/or draining the newly formed double bonds in such a way that the olefin would cause neither of these problems.

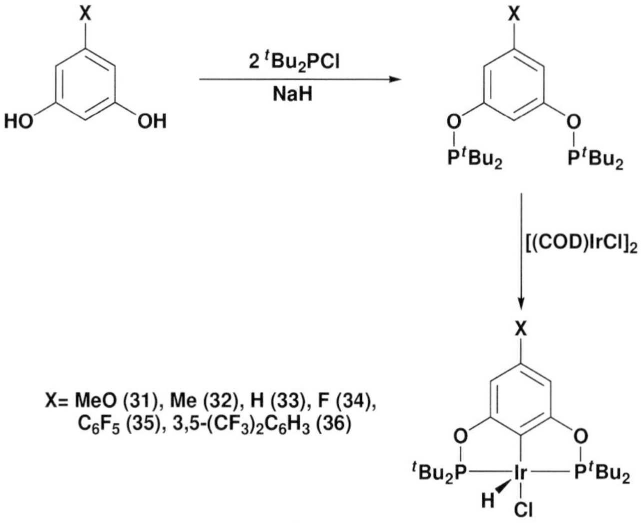

13.4 Alkane Dehydrogenation Catalyzed by Ir Pincer Complexes

Scheme 13.21

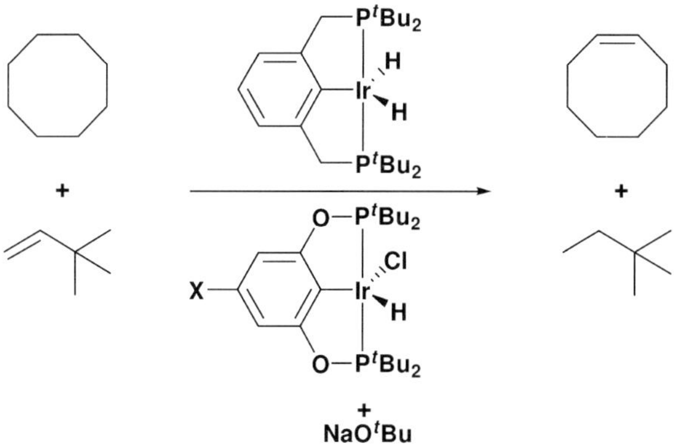

X = MeO (31), Me (32), H (33), F (34), C$_6$F$_5$ (35), 3,5-(CF$_3$)$_2$C$_6$H$_3$ (36)

Scheme 13.22

The situation would be even more interesting if a tandem process that incorporated the dehydrogenation of linear alkanes to form α-olefins and the further formation of other high-value products, was possible. An answer to this problem was provided recently by the joint efforts of Goldman and Brookhart, who described the combined application of an iridium–phosphinite PCP pincer complex with a Schrock-type metathesis catalyst [43]. The concept was simple: an acceptorless dehydrogenative catalyst for linear alkanes was used to produce the corresponding α-olefin, followed by an olefin-metathesis catalyzed by the Schrock-type catalyst. In this way, consumption of the recently formed α-olefin would allow the dehydrogenation catalyst to maintain its function, while the metathesis catalyst would consume all of the α-olefin produced; the result would be a synergistic system.

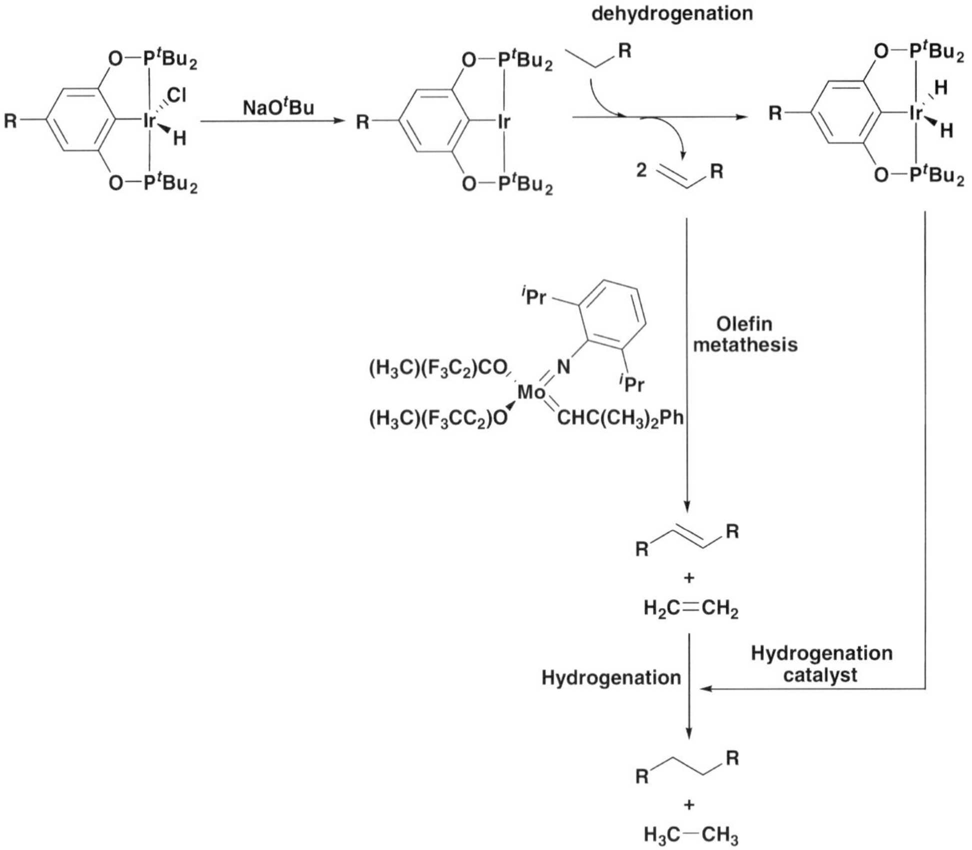

Scheme 13.23

Unfortunately, the phosphine analogue was found to be more effective that the phosphinite derivative, due to a more rapid isomerization to the internal olefin when using the latter catalyst. However, the system would be completely selective for the linear product (n-alkane) (Scheme 13.23).

Although this process has shown much promise, decomposition of the olefin metathesis catalyst appears to limit the conversion; nonetheless, it is expected that a more robust and compatible olefin metathesis catalyst will yield higher TONs.

More recently, the same principle was applied by the same authors to cyclic alkanes for catalytic ring expansion, contraction and metathesis–polymerization (Scheme 13.24) [44]. By using the tandem dehydrogenation–olefin metathesis system shown in Scheme 13.23, it was possible to achieve a metathesis–cyclo-oligomerization of COA and cyclodecane (CDA). This afforded cycloalkanes with different carbon numbers, predominantly multiples of the substrate carbon number; the major products were dimers, with successively smaller proportions of higher cyclo-oligomers and polymers.

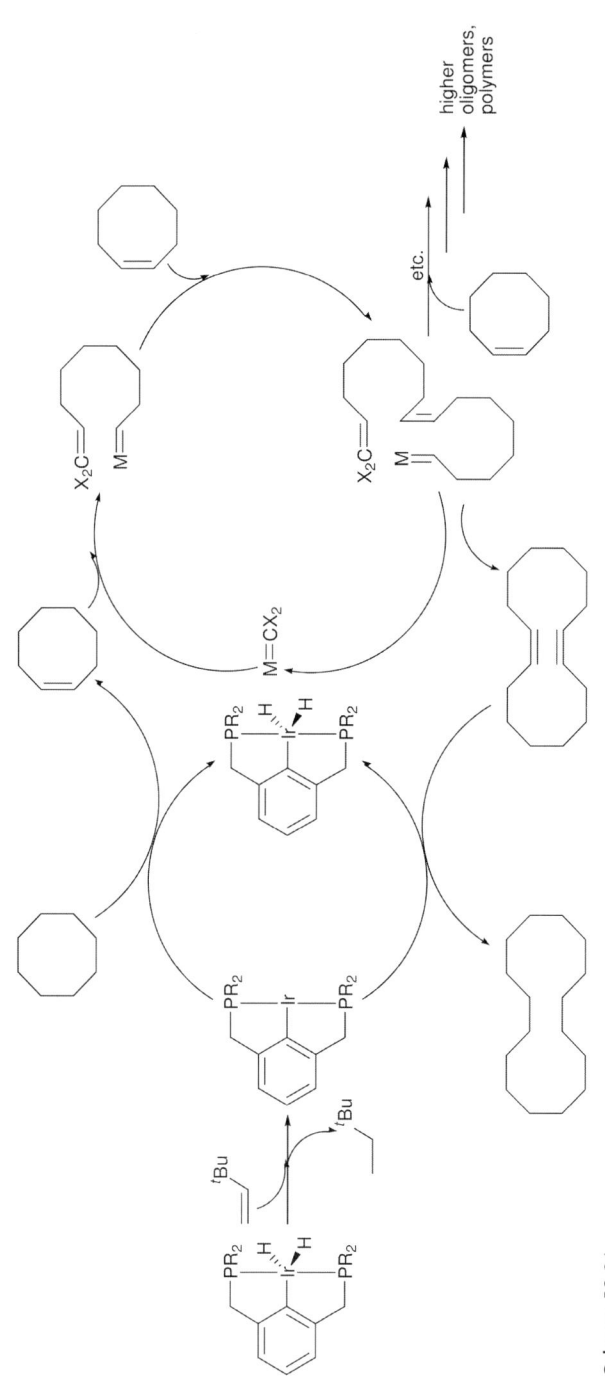

Scheme 13.24

13.5
Final Remarks

The details of iridium chemistry presented in this chapter represent a clear reflection of the major importance of this metal in our understanding of C—H activation, and in the further development of efficient catalytic systems for alkane dehydrogenation. It would appear today that the potential of the iridium species used in these processes might have reached maturity, with multiple applications such as tandem catalytic processes having emerged as potential methods in the production of high-value materials. In this respect, iridium PCP pincer compounds may prove to be industrially feasible for alkane dehydrogenation, not only to provide an extremely versatile backbone but also to allow fine-tuning of the species to create the best possible catalysts. Whilst the remarkable thermal stability and selectivity of these species invites further exploitation with other substrates, their compatibility with other catalytic systems may lead to the discovery of other tandem, catalytic processes. As the diminishing levels of the world's oil reserves continue to cause concern for future energy sources, the efficient use of both current and as-yet unexploited reserves becomes fundamental. Indeed, it is in this area where the use of homogeneous catalysis in environmentally friendly yet atom-economic processes – as exemplified by the high reactivity and selectivity of iridium complexes – will undoubtedly emerge in the development of new technologies. Undoubtedly, while the chemistry of alkane dehydrogenation continues to grow, the iridium species will serve as regular companions in these developments.

Acknowledgments

The support and enthusiasm of the former and current group members and colleagues at the authors' institution are gratefully acknowledged. The research studies described in this chapter were supported by CONACYT (F58692) and DGAPA-UNAM (IN227008).

References

1 Kleinman, J.P. and Dubeck, M. (1963) *Journal of the American Chemical Society*, **85**, 1544.
2 Chatt, J. and Davidson, J.M. (1965) *Journal of the Chemical Society*, 843.
3 See for instance: (a) Kakiuchi, F. and Chatani, N. (2003) *Advanced Synthesis Catalysis*, **345**, 1077.
(b) Goj, L.A. and Gunnoe, T.B. (2005) *Current Organic Chemistry*, **9**, 671.
4 Crabtree, R.H. (1995) *Chemical Reviews*, **95**, 987.
5 (a) Arndtsen, B.A., Bergman, R.G., Mobley, T.A. and Peterson, T.H. (1995) *Accounts of Chemical Research*, **28**, 154;
(b) Kakiuchi, F. and Murai, S. (1999) *Activation of Unreactive Bonds and Organic Synthesis* (ed. S. Murai), Springer-Verlag, Berlin, Chapter 3, p. 56;
(c) Jones, D.W. (2000) *Science*, **287**, 1942.
6 See for instance: (a) Crabtree, R.H. (1985) *Chemical Reviews*, **85**, 245;
(b) Crabtree, R.H. (2004) *Journal of Organometallic Chemistry*, **689**, 4083;

(c) Bergman, R.G. (2007) *Nature*, **446**, 391;
(d) Labinger, J.A. and Bercaw, J.E. (2002) *Nature*, **417**, 507;
(e) Crabtree, R.H. (2001) *Journal of the Chemical Society - Dalton Transactions*, 2437.

7 Goldberg, K.I. and Goldman, A.S. (eds) (2004) *Activation and Functionalization of C—H Bonds*, ACS Symposium Series 885, American Chemical Society, Washington, DC.

8 (a) Bennett, M.A. and Milner, D.L. (1967) *Journal of the Chemical Society D - Chemical Communications*, 581;
(b) Vaska, L. (1968) *Accounts of Chemical Research*, **1**, 335.

9 Cotton, F.A., Frenz, B.A. and Hunter, D.L. (1974) *Journal of the Chemical Society D - Chemical Communications*, 755.

10 (a) Janowicz, A.H. and Bergman, R.G. (1982) *Journal of the American Chemical Society*, **104**, 352;
(b) Janowicz, A.H. and Bergman, R.G. (1983) *Journal of the American Chemical Society*, **105**, 3929;
(c) Joyano, J.K., McMaster, A.D. and Graham, W.A.G. (1983) *Journal of the American Chemical Society*, **105**, 7190;
(d) Jones, W.D. and Feher, F.J. (1984) *Journal of the American Chemical Society*, **106**, 1650;
(e) Periana, R.A. and Bergman, R.G. (1984) *Organometallics*, **3**, 508.

11 (a) Sundarum, K.M., Shreehan, M.M. and Olszewski, E.F. (1991) *Kirk-Othmer Encyclopedia of Chemical Technology*, Volume **9**, 4th edn (eds J.I. Kroschwitz and M. Howe-Grant), Wiley-Interscience, New York, p. 877;
(b) Tullo, A.H. (2001) *Chemical and Engineering News*, **79**, 18.

12 Crabtree, R.H., Mihelcic, J.M. and Quirk, J.M. (1979) *Journal of the American Chemical Society*, **101**, 7738.

13 (a) Crabtree, R.H., Felkin, H., Khan, T. and Morris, G.E. (1977) *Journal of Organometallic Chemistry*, **141**, 205, and references therein; (b) Crabtree, R.H. (1979) *Accounts of Chemical Research*, **12**, 331.

14 Crabtree, R.H., Mellea, M.F., Mihelcic, J.M. and Quirk, J.M. (1982) *Journal of American Chemical Society*, **104**, 107.

15 (a) Baudry, D., Ephritikhine, M., Felkin, H. and Holmes-Smith, R. (1983) *Journal of the Chemical Society D - Chemical Communications*, 788;
(b) Baudry, D., Ephritikhine, M., Felkin, H. and Zakrzewski, J. (1984) *Tetrahedron Letters*, **25**, 1283.

16 (a) Burk, M.J., Crabtree, R.H., Parnell, C.P. and Uriarte, R.J. (1984) *Organometallics*, **3**, 816; (b) Crabtree, R.H., Parnell, C.P. and Uriarte, R.J. (1987) *Organometallics*, **6**, 696.

17 (a) Burk, M.J., Crabtree, R.H. and McGrath, D.V. (1985) *Journal of the Chemical Society D - Chemical Communications*, 1829;
(b) Burk, M.J. and Crabtree, R.H. (1987) *Journal of the American Chemical Society*, **109**, 8025.

18 (a) Felkin, H., Fillebeen-Khan, T., Gault, Y., Holmes-Smith, R. and Zakrzewski, J. (1984) *Tetrahedron Letters*, **25**, 1279;
(b) Felkin, H., Fillebeen-Khan, T., Holmes-Smith, R. and Yingrui, L. (1985) *Tetrahedron Letters*, **26**, 1999.

19 (a) Nomura, K. and Saito, Y. (1988) *Journal of the Chemical Society D - Chemical Communications*, 161;
(b) Nomura, K. and Saito, Y. (1989) *Journal of Molecular Catalysis A - Chemical*, **54**, 57.

20 (a) Sakakura, T., Sodeyama, T. and Tanaka, M. (1989) *New Journal of Chemistry*, **13**, 737;
(b) Sakakura, T., Abe, F. and Tanaka M. (1991) *Chemistry Letters*, 359.

21 (a) Maguire, J.A., Boese, W.T. and Goldman, A.S. (1989) *Journal of the American Chemical Society*, **11**, 7088;
(b) Maguire, J.A. and Goldman, A.S. (1991) *Journal of the American Chemical Society*, **113**, 6706;
(c) Maguire, J.A., Petrillo, A. and Goldman, A.S. (1992) *Journal of the American Chemical Society*, **114**, 9492.

22 (a) Fujii, T. and Saito, Y. (1990) *Journal of the Chemical Society D - Chemistry Communications*, 757;
(b) Fujii, T., Higashino, Y. and Saito, Y. (1993) *Journal of the Chemical Society - Dalton Transactions*, 517.

23 Aoki, T. and Crabtree, R.H. (1993) *Organometallics*, **12**, 294.

24 Anton, D.R. and Crabtree, R.H. (1983) *Organometallics*, **2**, 855.

25 Ortiz, J.V., Havels, Z. and Hoffmann, R. (1984) *Helvetica Chimica Acta*, **67**, 1.
26 Belli, J. and Jensen, C.M. (1996) *Organometallics*, **15**, 1532.
27 Moulton, C.J. and Shaw, B.L. (1976) *Journal of the Chemical Society - Dalton Transactions*, 1020.
28 Gupta, M., Hagen, C., Flesher, R.J., Kaska, W.C. and Jensen, C.M. (1996) *Chemical Communications*, 2083.
29 Wang, K., Goldman, M.E., Emge, T.J. and Goldman, A.S. (1996) *Journal of Organometallic Chemistry*, **518**, 55.
30 Leitner, W. and Six, C. (1997) *Chemische Berichte-Recueil*, **130**, 555.
31 (a) Gupta, M., Hagen, C., Kaska, W.C., Cramer, R.E. and Jensen, C.M. (1997) *Journal of the American Chemical Society*, **119**, 840;
(b) Gupta, M., Kaska, W.C. and Jensen, C.M. (1997) *Chemical Communications*, 461;
(c) Lee, D.W., Kaska, W.C. and Jensen, C.M. (1998) *Organometallics*, **17**, 1;
(d) Gómez-Benítez, V., Redón, R. and Morales-Morales, D. (2003) *Revista de la Sociedad Química de México*, **47**, 124.
32 (a) Jensen, C.M. (1999) *Chemical Communications*, 2443;
(b) Thomson, D.T. (1998) *Platinum Metals Reviews*, **42**, 71;
(c) Ray, A., Zhu, K., Kissin, Y.V., Cherian, A.E., Coates, G.W. and Goldman, A.S. (2005) *Chemical Communications*, 3388;
(d) Ray, A., Kissin, Y.V., Zhu, K., Goldman, A.S., Cherian, A.E. and Coates, G.W. (2006) *Journal of Molecular Catalysis A - Chemical*, **256**, 200.
33 Liu, F., Pak, E.B., Singh, B., Jensen, C.M. and Goldman, A.S. (1999) *Journal of the American Chemical Society*, **121**, 4086.
34 Liu, F. and Goldman, A.S. (1999) *Chemical Communications*, 655.
35 Xu, W., Rosini, G.P., Gupta, M., Jensen, C.M., Kaska, W.C., Krogh-Jespersen, K. and Goldman, A.S. (1997) *Chemical Communications*, 2273.
36 (a) Li, S. and Hall, M.B. (1999) *Organometallics*, **18**, 5682;
(b) Niu, S. and Hall, M.B. (1999) *Journal of the American Chemical Society*, **121**, 3992;
(c) Krogh-Jespersen, K., Czerw, M., Kanzelberger, M. and Goldman, A.S. (2001) *Journal of Chemical Information and Computer Sciences*, **41**, 56;
(d) Krogh-Jespersen, K., Czerw, M., Summa, N., Renkema, K.B., Achord, P.A. and Goldman, A.S. (2002) *Journal of the American Chemical Society*, **124**, 11404;
(e) Krogh-Jespersen, K., Czerw, M. and Goldman, A.S. (2002) *Journal of Molecular Catalysis A - Chemical*, **189**, 95.
37 Haenel, M.W., Oevers, S., Angermund, K., Kaska, W.C., Fan, H-J. and Hall, M.B. (2001) *Angewandte Chemie - International Edition*, **40**, 3596.
38 Morales-Morales, D., Grause, C., Kasaoka, K., Redón, R., Cramer, R.E. and Jensen, C.M. (2000) *Inorganica Chimica Acta*, **300-302**, 958.
39 Bedford, R.B., Draper, S.M., Scully, P.N. and Welch, S.L. (2000) *New Journal of Chemistry*, **24**, 745.
40 Morales-Morales, D. and Jensen, C.M. (eds) (2007) *The Chemistry of Pincer Compounds*, Vol. **151**, Elsevier, Amsterdam, The Netherlands, Chapter 9.
41 Morales-Morales, D., Redón, R., Yung, C. and Jensen, C.M. (2004) *Inorganica Chimica Acta*, **357**, 2953.
42 Göttker-Schnetmann, I., White, P. and Brookhart, M. (2004) *Journal of the American Chemical Society*, **126**, 1804.
43 Goldman, A.S., Roy, A.H., Huang, Z., Ahuja, R., Schinski, W. and Brookhart, M. (2006) *Science*, **312**, 257.
44 Ahuja, R., Kundu, S., Goldman, A.S., Brookhart, M., Vicente, B.C. and Scott, S.L. (2008) *Chemical Communications*, 253.

14
Transformations of (Organo)silicon Compounds Catalyzed by Iridium Complexes

Bogdan Marciniec and Ireneusz Kownacki

14.1
Introduction

Numerous reactions of organic compounds catalyzed by transition-metal complexes have been explored during the past 50 years, including hydrogenation, hydroformylation, the Wacker process, olefin metathesis and the Ziegler–Natta polymerization of olefins. nevertheless, among the catalytic transformations of (organo)silicon compounds only hydrosilylation reactions are well known as processes of industrial importance [1–5]. During the past two decades, some other reactions of silicon compounds catalyzed by transition-metal complexes have been revealed and have undergone spectacular development. Among these are the dehydrogenative silylation and bis-silylation of alkenes and alkynes by hydrosilanes and disilanes, respectively, the silylative coupling of alkenes with vinylsilanes, metathesis of silicon-containing alkenes and coupling of the C—H of alkenes and arenes with organosilicon compounds, as well as the dehydrocoupling of hydrosilanes, silylformylation and silylcarbonylation of a variety of organic compounds (for reviews, see Refs [5–12]).

Most of the above reactions occur via a mechanism involving intermediates with a metal–silicon bond (i.e. silicometallics) and a metal–hydrogen bond, accompanied (or sided) only occasionally by compounds containing metal–carbon bonds (i.e. organometallics) that are characteristic of the key intermediates of transition-metal-catalyzed transformations of organic compounds (for recent reviews, see Refs [11, 13]).

Platinum complexes have been mainly used in the hydrosilylation of carbon–carbon bonds, and ruthenium complexes in the metathesis and silylative coupling of olefins with vinylsilanes. Most of these processes (except for olefin metathesis) may also proceed efficiently in the presence of rhodium and iridium complexes.

This chapter describes the application of iridium-catalyzed reactions in the synthesis of molecular and macromolecular organosilicon compounds and related silicon derivatives. Some mechanistic implications are introduced which illustrate the specific catalytic activation of organic and silicon compounds by iridium

Iridium Complexes in Organic Synthesis. Edited by Luis A. Oro and Carmen Claver
Copyright © 2009 WILEY-VCH Verlag GmbH & Co. KGaA, Weinheim
ISBN: 978-3-527-31996-1

complexes in comparative analysis, particularly with the respective rhodium complexes.

14.2
Hydrosilylation and Dehydrogenative Silylation of Carbon–Carbon Multiple Bonds

Most research and industrial syntheses based on hydrosilylation are carried out in the presence of platinum complexes with chloroplatinic acid and other d^8-Pt(II) and d^{10}-Pt(0) complexes as precursors. Platinum catalysts, supported on either carbon or silica, have also appeared to be effective, especially in some industrial processes [1–4]. The high catalytic activity of platinum and other transition-metal complexes comprising an unsaturated compound (e.g. olefin) and silicon hydride, gives rise to various side reactions involving processes of olefins (isomerization, hydrogenation, polymerization) and/or reactions of silicon hydrides (redistribution and dehydrocoupling), as well as reactions involving both substrates (such as the dehydrogenative silylation of olefins and acetylenes). The latter system may be useful as a synthetic method to produce finally unsaturated compounds.

14.2.1
Hydrosilylation and Dehydrogenative Silylation of Alkenes

Unlike Pt catalysts, the triad complexes of iron and cobalt catalyze competitively both dehydrogenative silylation and hydrosilylation [11]. The reaction can proceed via a complex containing the σ-alkyl and σ-silylalkyl ligands (Scheme 14.1).

The concurrent β-H transfer from the two ligands to the metal is a key step for two alternative reactions, namely hydrosilylation and/or dehydrogenative silylation [6, 11].

In contrast to the Pt(0) and Pt(II) complexes and the corresponding Rh(I) and Rh(III) complexes, the iridium complexes have rarely been employed as hydrosilylation catalysts [1–4]. Iridium–phosphine complexes with d^8 metal configuration – for example, $[Ir(CO)Cl(PPh_3)_2]$ (Vaska's complex) and $[Ir(CO)H(PPh_3)_3]$ – were first tested some 40 years ago in the hydrosilylation of olefins. Although they underwent oxidative addition with hydrosilanes (simultaneously to Rh(I) com-

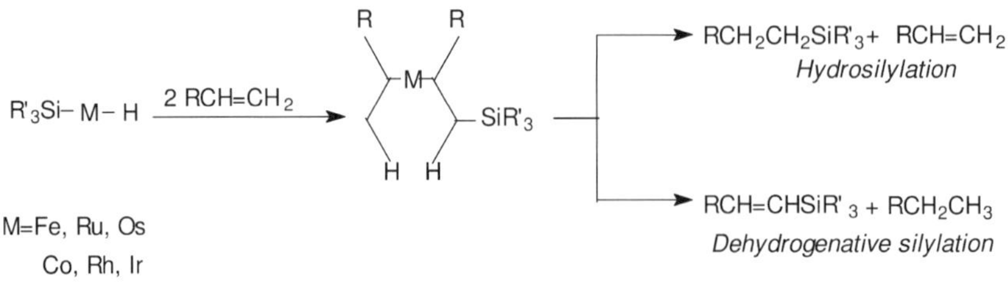

Scheme 14.1 Organometallic intermediate competitive hydrosilylation and dehydrogenative silylation of alkenes.

plexes) to yield respective Ir(III) adducts, they appeared to be inactive towards hydrosilylation [14]. Instead, such adducts readily dissociate, eliminating the chlorosilane molecule, according to Equation 14.1:

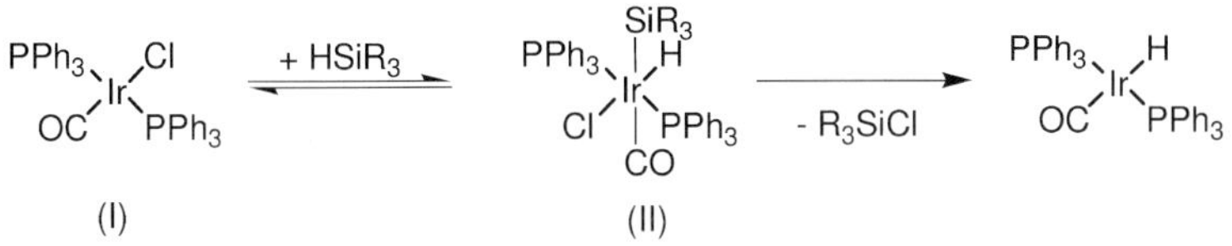

$$\underset{(I)}{\text{PPh}_3\text{-Ir(Cl)(CO)(PPh}_3)} + \text{HSiR}_3 \rightleftharpoons \underset{(II)}{\text{PPh}_3\text{-Ir(SiR}_3)(H)(Cl)(PPh}_3)(CO)} \xrightarrow{-\text{R}_3\text{SiCl}} \text{PPh}_3\text{-Ir(H)(CO)(PPh}_3)$$

(14.1)

For this reason, phosphine-free complexes (e.g. [{Ir(μ-Cl)(coe)$_2$}$_2$] and [{Ir(μ-Cl)(cod)$_2$}$_2$]), as well as chloroligand-free complexes such as [{Ir(μ-OMe)(cod)$_2$}$_2$] and [Ir{μ-(OSiMe$_3$)(cod)}$_2$], were found to be basic precursors of all iridium-catalyzed hydrosilylation and dehydrogenative silylations. [IrH$_2$(SiEt)$_3$(cod)(L)] (L = PPh$_3$ or AsPh$_3$), synthesized from [{Ir(μ-OMe)(cod)$_2$}$_2$], appeared to be active catalysts for the dehydrogenative silylation of ethylene and 1-hexene (Scheme 14.2) [15].

Ir(I), Ir(II) and Ir(V) complexes stabilized by an O-donor ligand (e.g. [Ir(coe)(triso)] and [Ir(C$_2$H$_4$)$_2$(triso)] (triso = tridentate tris(diphenyloxosphoranyl)methanides) are effective catalysts for the dehydrogenative silylation and hydrosilylation of ethylene [16–18].

A family of cationic catalysts such as [Ir(cod)(PCy$_3$)Py]$^+$[PF$_6$]$^-$ [19] and zwitterionic [20] Ir(I) complexes have been tested in the hydrosilylation of styrene, and represent an effective class of the hydrosilylation catalysts to yield predominantly β-adduct accompanied by α-adduct and traces of unsaturated products (see Table 14.1).

The hydrosilylation of ethylene by the early-late transition-metal heterodinuclear complexes [CpTa(μ-CH$_2$)$_2$Ir(CO)$_2$] has been studied mainly in a bid to recognize the mechanism of reaction, which occurs via a predominant alkene/Ir—H insertion pathway over a minor insertion of ethylene into the Ir—Si bonds [21].

The complexes of the composition [{Ir(μ-X)(diene)}$_2$], where X = halogen, OH, OMe (e.g. [IrCl(CO)(cod)]) appeared to be very effective catalysts for the hydrosilylation of allyl chloride by trialkoxy- and alkylalkoxy-silanes [22]. Other iridium complexes have been subsequently reported as catalysts for the synthesis of silane

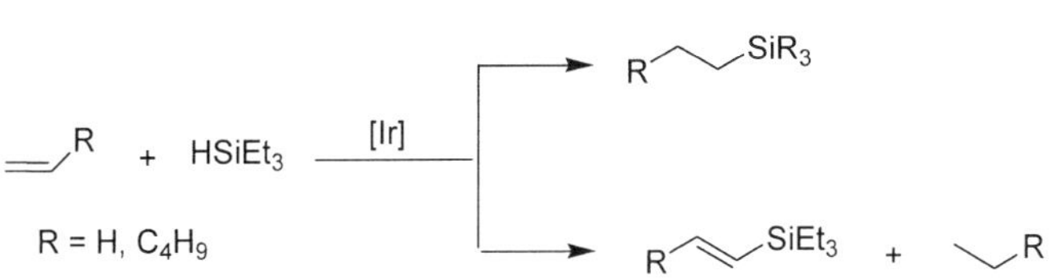

Scheme 14.2 Competitive hydrosilylation and dehydrogenative silylation of ethylene and 1-hexene.

Table 14.1 Addition of triethylsilane to styrene.

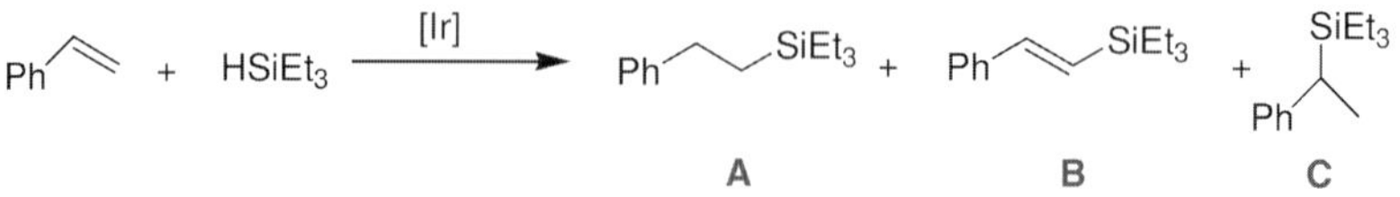

Ph−CH=CH$_2$ + HSiEt$_3$ →[Ir] Ph−CH$_2$−CH$_2$−SiEt$_3$ (A) + Ph−CH=CH−SiEt$_3$ (B) + Ph−CH(SiEt$_3$)−CH$_3$ type product with Ph−C(SiEt$_3$)=... (C)

Catalyst	Solvent	Temperature (°C)	Yield (A/B/C) (%)
[Ir(cod)(PCy$_3$)(Py)]$^+$[PF$_6$]$^{-a}$	DCE	60	>99 (51/40/7)
[Ir(cod)(κ^2-P,N-3-P^{i}Pr$_2$-2-NMe$_2$-indene]$^+$[PF$_6$]$^{-b}$	DCE	60	66 (66/<1/<1)
	THF	60	>99 (91/1/8)
	THF	24	35 (35/<1/<1)
[Ir(cod)(κ^2-P,N-3-P^{i}Pr$_2$-2-NMe$_2$-indenide]^b	DCE	60	>99 (99/<1/<1/<1)
	DCE	24	86 (86/<1/<1/<1)
	THF	60	98 (94/2/<1)
	THF	24	88 (88/<1/<1)
	Toluene	60	64 (55/2/<1)

Reaction conditions: ratio: styrene : Et$_3$SiH = 5 : 1.
a Ref. [19].
b Ref. [20].

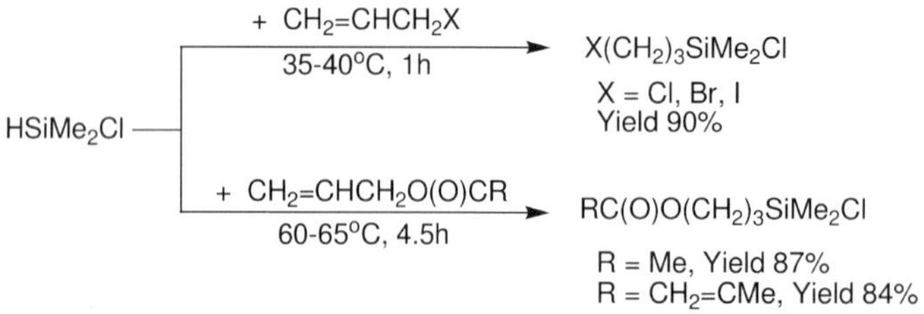

HSiMe$_2$Cl
+ CH$_2$=CHCH$_2$X, 35–40 °C, 1 h → X(CH$_2$)$_3$SiMe$_2$Cl
X = Cl, Br, I
Yield 90%

+ CH$_2$=CHCH$_2$O(O)CR, 60–65 °C, 4.5 h → RC(O)O(CH$_2$)$_3$SiMe$_2$Cl
R = Me, Yield 87%
R = CH$_2$=CMe, Yield 84%

Scheme 14.3 Hydrosilylation of allyl derivatives with dimethylchlorosilane.

coupling agents via hydrosilylation of olefins with alkoxysubstituted and other hydrosilanes [23].

[{Ir(μ-X)(cod)}$_2$], complexes (where X = Cl, Br, I) appeared to be effective catalysts for the synthesis of 3-halopropyldimethylchlorosilane and 3-dimethylchlorosilylpropyl esters via hydrosilylation according to the equation shown in Scheme 14.3 [24–26].

14.2.2
Application of Hydrosilylation in Polymer Chemistry

Iridium–siloxide complexes [Ir(cod)(PCy$_3$)(OSiMe$_3$)] and [Ir(CO)(PCy$_3$)$_2$(OSiMe$_3$)] were examined in the crosslinking of silicones via hydrosilylation to be effective catalysts for the model homogeneous hydrosilylation of vinyltris(trimethylsiloxy)-silanes with heptamethyltrisiloxane, as well as for crosslinking of the commercial polysiloxane system [27]. The curing process catalyzed by iridium complexes pro-

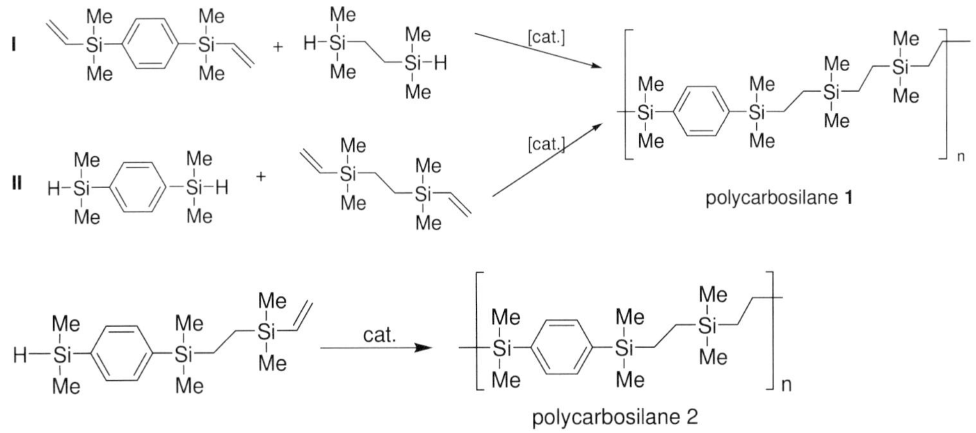

Scheme 14.4 Synthesis of poly(phenylene-silylene-ethylene)s via hydrosilylation polymerization process.

ceeds at a higher temperature (ca. 200 °C) than the process catalyzed by the Karstedt-catalyst–DAM system, but does not require an inhibitor (diallylmaleate; DAM) to maintain the low viscosity of the reaction mixture at room temperature for several days. Silicone-based polymers and polymeric compositions with special applications (e.g. LED devices) can be also synthesized by hydrosilylation. It was reported recently that, low or moderately branched vinyl-siloxane oligomers and polymers, as well as polymer films, were prepared via a hydrosilylation reaction catalyzed by transition-metal complexes including also [IrCl(PPh$_3$)$_3$], [{Ir(μ-Cl)(cod)}$_2$] and iridium 2,4-pentadionate [28].

Linear polycarbosilanes and polycarbosiloxanes – especially those containing arylene units in the chain – have specific physico-chemical properties which can be applicable in heat-resistant materials [29–31]. Phenylene–silylene–ethylene-polymers, which may serve as potential substrates for applications as membrane materials are usually obtained in the presence of platinum catalysts [32], although other transition-metal complexes have also been tested in this process.

Rhodium and iridium siloxide complexes of the general formula [{M(μ-OSiMe$_3$)(cod)}$_2$] and [M(cod)(PCy$_3$)(OSiMe$_3$)] (M = Rh and Ir) have been used successfully as catalysts of hydrosilylation polymerization occurring according to Scheme 14.4. The reactions of both types lead to linear poly(phenylene–silylene–ethylene)s, and although the best results have been achieved using a rhodium catalyst, iridium siloxides are also effective catalysts for these processes in exemplary reactions (see Table 14.2).

14.2.3
Hydrosilylation and Dehydrogenative Silylation of Alkynes

The transition-metal-catalyzed hydrosilylation of alkynes remains one of the most common routes for the synthesis of vinyl-substituted silicon compounds to yield three possible products, β-E, β-Z and α (Equation 14.2):

Table 14.2 Molecular weights and polydispersity indexes of polycarbosilanes 1 obtained via reactions I and II.

Catalyst	M_w	M_w/M_n	Yield (%)
Via reaction I			
[Pt$_2${(CH$_2$=CHSiMe$_2$)$_2$O}$_3$]	14 100	1.88	88
[{Rh(μ-OSiMe$_3$)(cod)}$_2$]	20 200	2.04	95
[Rh(cod)(OSiMe$_3$)(PCy$_3$)]	16 400	2.26	82
[{Ir(cod)(μ-OSiMe$_3$)}$_2$]	10 500	1.93	90
[Ir(cod)(OSiMe$_3$)(PCy$_3$)]	14 000	2.10	74
Via reaction II			
[Pt$_2${(CH$_2$=CHSiMe$_2$)$_2$O}$_3$]	16 800	1.63	66
[{Rh(μ-OSiMe$_3$)(cod)}$_2$]	6200	2.32	78
[Rh(cod)(PCy$_3$)(OSiMe$_3$)]	8300	2.87	77
[{Ir(μ-OSiMe$_3$)(cod)}$_2$]	4100	1.96	83

Reaction conditions: toluene, 110 °C, 24 h.
[HSi≡]:[CH$_2$=CHSi≡]:[cat.] = 1:1:10^{-5} (for [Pt] catalyst) or 1:1:10^{-3} for [Rh] and [Ir] catalysts.

$$R_3SiH + R\equiv\!\!=\!\!-H \xrightarrow{[M]} R\!\!-\!\!\text{=}\!\!-SiR_3 + R\!\!-\!\!\text{=}\!\!-SiR_3 + \text{=}\!\!<\!\!{SiEt_3 \atop R}$$
$$\beta\text{–E} \qquad \beta\text{–Z} \qquad \alpha$$

(14.2)

The highly stereoselective hydrosilylation of terminal alkynes to obtain almost exclusively the β-Z-adduct, is observed when the O-donor ligands such as tridentate tris(diphenyloxosphoranyl)methanides (triso) and (acac) ligands are used in Ir(I), Ir(III) and Ir(V) complexes – that is, [Ir(coe)$_2$(triso)], [Ir(C$_2$H$_4$)$_2$(triso)], [Ir(H)$_2$(SiPh$_2$Me)$_2$(triso)], [Ir(C$_2$H$_4$)(H)(SiPh$_3$)(triso)] [4, 16–18, 33], and [Ir(acac)(H)(SiEt$_3$)(PCy$_3$)] [34].

Computational and catalytic studies of the hydrosilylation of terminal alkynes have been very recently reported, with the use of [{Ir(μ-Cl)(Cl)(Cp*)}$_2$] catalyst to afford highly stereoselectively β-Z-vinylsilanes with high yields (>90%) [35]. E-isomers can be also found among the products, due to subsequent Z → E isomerization under the conditions employed. The catalytic cycle is based on an Ir(III)–Ir(V) oxidative addition and direct reductive elimination of the β-Z-vinylsilane. Other iridium complexes have been found to be active in the hydrosilylation of phenylacetylene and 1-alkynes; for example, when phenylacetylene is used as a substrate, dehydrogenative silylation products are also formed (see Scheme 14.5 and Table 14.3).

$$Ph\!\!-\!\!\equiv + HSiEt_3 \xrightarrow{[Ir]} \begin{array}{l} Ph\!-\!\!\equiv\!\!-SiEt_3 + Ph\!\!-\!\!\text{=} \\ Ph\!-\!\!\equiv\!\!-SiEt_3 + H_2 \end{array}$$

Scheme 14.5 Dehydrogenative silylation of phenylacetylene.

14.2 Hydrosilylation and Dehydrogenative Silylation of Carbon–Carbon Multiple Bonds

Table 14.3 Hydrosilylation of alkynes.

Catalysts	HSiR$_3$	R'C≡CH (R')	Yield (%)	Ratio E/Z/α/DH	Conditions	Reference
[Ir(CO)(η^2-C$_8$H$_{14}$)(η^5-C$_9$H$_7$)]	HSi^{i}Pr$_3$	PhC≡CH	–	1/4/–/–	60°C, 120h, benzene	[36]
[Ir(CO)(η^2-C$_8$H$_{14}$)(η^5-C$_5$H$_5$)]	HSi^{i}Pr$_3$	PhC≡CH	–	1/4/–/–	60°C, 96h, benzene	
[Ir(L)$_2$(triso), L–coe, C$_2$H$_4$]	HSiR$_3$	PhC≡CH C$_5$H$_{11}$C≡CH	80 39	1/128/–/– 1/19/–/– E/Z/α/DH (%)	25°C, 20h, CD$_2$Cl$_2$	[18]
[IrBr(CO)(dppe)]	HSiEt$_3$	PhC≡CCH$_3$	54	17/37/–/–	65°C, 54h, CH$_2$Cl$_2$	[37]
[Ir$_2$Cl(cod)$_2$(TIME^{Bu})]Cl	HSiMe$_2$Ph	^{n}BuC≡CH	44	29/51/20/–	60°C, 24h, [cat.] = 0.1%, CDCl$_3$	[38]
			46	24/57/19/–	60°C, 72h, [cat.] = 0.1%, CDCl$_3$	
[Ir(cod) (η^5-Ind)]	PhMe$_2$SiH	n-C$_6$H$_{13}$C≡CH	96	94/4/2/–	80°C, 24h, toluene	[39]
[Ir(cod)(pzpy)]	HSiEt$_3$	PhC≡CH	–	42/19/7/19	60°C, 24h, CH$_2$Cl$_2$	[40]
[Ir(NCCH$_3$) (cod) (PMe$_3$)]BF$_4$	HSiEt$_3$	PhC≡CH	100	9/74/2/8	50°C, 1h, CH$_2$Cl$_2$	[41]
			90	21/50/3/10	50°C, 2h, CH$_2$Cl$_2$	
			47	27/35/6/8		
[Ir(NCCH$_3$) (TFB) (P^{i}Pr$_3$)]BF$_4$	HSiEt$_3$	PhC≡CH	100	24/47/5/12	50°C, 1h, CH$_2$Cl$_2$	
			74	23/45/6/13	50°C, 3h, CH$_2$Cl$_2$	
[IrH$_2$(NCCH$_3$)$_3$(P^{i}Pr$_3$)]BF$_4$	HSiEt$_3$	PhC≡CH	100	26/35/9/15	50°C, 1.3h, CH$_2$Cl$_2$	
			61	25/37/7/13	50°C, 0.2h, CH$_2$Cl$_2$	
[Ir(CO)$_2$ (C$_2$Ph) (PCy$_3$)] [Ir(C$_2$Ph)(TFB)(PCy$_3$)]	HSiEt$_3$ HSiEt$_3$	PhC≡CH PhC≡CH	88 90	26/38/7/11 14/66/1/6	60°C, CH$_2$Cl$_2$, argon	[42]

Data compiled with Table 14.3 show that neutral iridium complexes catalyze mainly the addition of an Si—H bond into a C—C triple bond in organic reactants with good yields. For example, iridium–indenyl catalyzed reactions give efficiently Z-2-silylstyrene as well as for complex [Ir(L)$_2$(triso)] (L = coe, C$_2$H$_4$) [18]. The same product was formed exclusively, although in the case of [Ir(cod)(η^5-Ind)] [39] reversed selectivity was observed, E-isomer was formed with selectivity of 94%. Cationic iridium complexes also catalyze the conversion of terminal alkynes with substituted silanes to silylated derivatives, there being more active than neutral complexes. In the presence of a cationic iridium species the formation of silylated alkyne derivatives was also observed [41, 43]. Exceptionally, [Ir(cod)(pzpy)] and iridium alkynyl complexes [Ir(CO)$_2$(C$_2$Ph)(PCy$_3$)], [Ir(C$_2$Ph)(TFB)(PCy$_3$)] were found to catalyze the formation of silylacetylene derivatives. The participation of each compound in a reaction mixture is not solely dependent on the catalyst used; rather, an excess of acetylene derivatives favors dehydrogenative silylation to yield the olefin and corresponding alkynylsilane [44].

The dehydrogenative silylation of alkynes may dominate under optimum conditions to afford the corresponding silylalkynes in good yields and with high selectivity when the [Ir$_4$(CO)$_{12}$] + PPh$_3$ system is used [45]. The use of 2 equiv. of phenylacetylene in the presence of [Ir$_4$(CO)$_{12}$] (1 mol%) + PPh$_3$ (12 mol%) gave the corresponding silylphenylacetylene in 96% yield and 97% selectivity. This method is applicable for a variety of terminal acetylenes RC≡CH (where R = Ph, C$_6$H$_{13}$, cyclohexyl and ^{t}Bu). The dehydrogenative silylation of phenylacetylene with triethylsilane is accompanied by the generation of molecular hydrogen, and the formation of styrene is also observed.

In order to prevent phenylacetylene hydrogenation, and to eliminate the evolution of hydrogen, other unsaturated compounds such as diethylfumarate, diethylmaleate and diphenylacetylene were added as a hydrogen scavengers [45].

14.3
Asymmetric Hydrosilylation of Ketones and Imines

Iridium complexes are known to be generally less active in hydrosilylation reactions when compared to rhodium derivatives, although iridium-based catalysts with bonded chiral carbene ligands have been used successfully in the synthesis of chiral alcohols and amines via hydrosilylation/protodesilylation of ketones [46–52] and imines [53–55]. The iridium-catalyzed reaction of acetophenone derivatives with organosubstituted silanes often gives two products (Equation 14.3):

$$R-C(=O)-CH_3 + SiH_nR'_{4-n} \xrightarrow{[Ir]} R-CH(OSiH_{n-1}R'_{4-n})-CH_3 \;\; (I) \;\; + \;\; R-C(OSiH_{n-1}R'_{4-n})=CH_2 \;\; (II) \quad (14.3)$$

R = Aryl, R' = Alkyl, Aryl

Neutral iridium(I) complexes [Ir(Cl)(cod)(L)] [46] consisting of the chiral carbene ligands L1 and L2 have been shown as active catalysts for the asymmetric hydrosilylation of acetophenone with Ph$_2$SiH$_2$, giving >99% yield of **I** at −20 °C in 16 h at one catalyst loading. However, only low enantiomeric excess (ee)-values of *sec*-phenethyl alcohol were obtained for the iridium catalytic systems used – that is, 15% ee for L1 and 2% ee for L2.

L1   L2   L3

R = ^{i}Pr, Bn

Iridium(I) precursors [Ir(cod)(L)] with bidentate N-heterocyclic carbene ligands L3 appeared slightly less active in the hydrosilylation of acetophenone with diphenylsilane than did the similar rhodium complexes, giving respectively yields of 85% of **I** and 15% of **II** for the ^{i}Pr substituent, and 83% of **I** and 17% of **II** for the benzyl moiety, after 2 h reaction at room temperature [47]. However, when carbene ligands of type L3 were used a significant increase in the ee-value of the *sec*-phenethyl alcohol *R* isomer of up to 60% was observed.

Other carbene iridium complexes (C1, C2) were also applied as catalysts of this reaction, but their catalytic activity was very low (only 10% yields of **I** for C1 and 58% for C2 were obtained) [48].

C1   C2

Catalytic systems based on the commonly used iridium precursor [{Ir(μ-Cl)(cod)}$_2$] and diferrocenyl dihalcogenides of L4 and L5 type were also studied in the asymmetric hydrosilylation of acetophenone, giving a relatively high yield of *sec*-phenetyl alcohol silyl ether (**I**) and a moderate ee of one stereoisomer [49].

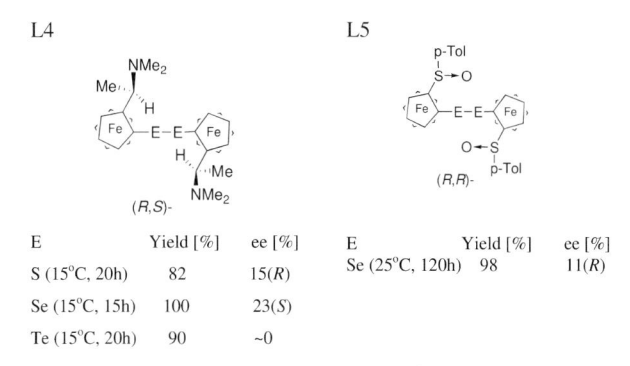

L4 (R,S)-

E	Yield [%]	ee [%]
S (15°C, 20h)	82	15(R)
Se (15°C, 15h)	100	23(S)
Te (15°C, 20h)	90	~0

L5 (R,R)-

E	Yield [%]	ee [%]
Se (25°C, 120h)	98	11(R)

The cationic complex [Ir(CO)(κ³-N,N,N-(S,S)-^{i}Pr-pybox)][PF$_6$] [50] was also found to be catalytically active in the addition of Ph$_2$SiH$_2$ to acetophenone, with complete conversion of the ketone into the corresponding silyl ether (**I**) at room temperature after 72 h of reaction. However, desilylation of the product (**I**) led to racemic 1-phenylethanol, which means that the reduction took place without asymmetric induction.

The enantiomerically pure dithiourea derivative L6 has been successfully applied to the iridium-catalyzed [{Ir(μ-Cl)(cod)}$_2$] asymmetric hydrosilylation of acetophenone, as well as of various alkylaryl ketones [51]. In the case of acetophenone, when the iridium catalytic system ([Ir]/L6 was used with a 10-fold excess of L6 (R = Ph), an encouraging enantioselectivity was achieved (74% ee), together with a yield of 30% at 50 °C after 24 h.

L6

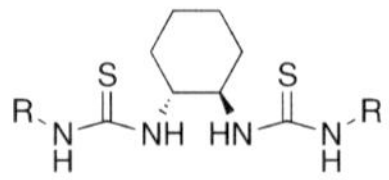

R = Ph

When a threefold excess of L6 versus iridium was used, the yield of hydrosilylation/protodesilylation products was up to 53%, with 55% ee (Scheme 14.6).

The cationic iridium–hydrido-silylene complex [IrCp*(H)(=SiPh$_2$)(PMe$_3$)][B(C$_6$F$_5$)$_4$] has also been recently reported to effectively catalyze the reaction of substituted silanes with ketones (Equation 14.4) [52].

R = Me; R' = H   yield 44%
R = Ph, R' = H   yield 54%

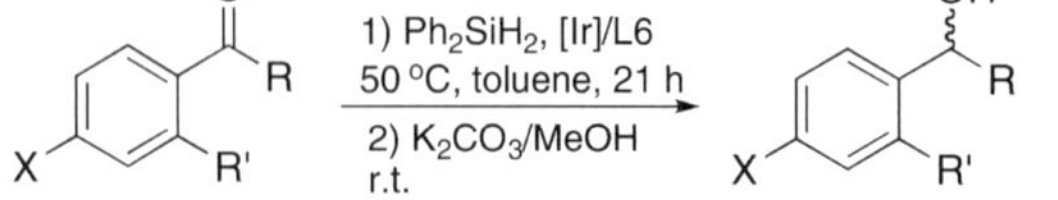

(14.4)

1) Ph$_2$SiH$_2$, [Ir]/L6
50 °C, toluene, 21 h
2) K$_2$CO$_3$/MeOH
r.t.

R	R'	X	Yield (%)	e.e. (R) (%)
CH$_3$	H	H	50	55
CH$_3$	H	CF$_3$	24	13
CO$_2$CH$_3$	H	H	53	0
(CH$_2$)$_2$CH$_3$	H	H	23	9
CH(CH$_3$)$_2$	H	H	6	12
C(CH$_3$)$_3$	H	H	2	0
-(CH$_2$)$_3$-		H	34	28

Scheme 14.6

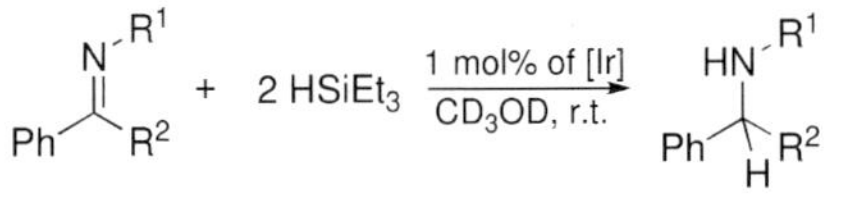

14.3 Asymmetric Hydrosilylation of Ketones and Imines

$$\underset{Ph}{\overset{N^{R^1}}{\|}}_{R^2} + 2\,HSiEt_3 \xrightarrow[CD_3OD,\,r.t.]{1\,mol\%\,of\,[Ir]} \underset{Ph}{\overset{HN^{R^1}}{\underset{H}{|}}}_{R^2}$$

	Yield (%) (Time)
$R^1 = Ph, R^2 = H$	100 (~10 min)
$R^1 = Ph, R^2 = Me$	100 (~10 min)
$R^1 = {}^nBu, R^2 = H$	66 (~27 min)

Scheme 14.7

As mentioned above, iridium complexes are also active in the formation of amines via the hydrosilylation/protodesilylation of imines. In the presence of 2 equiv. of $HSiEt_3$, the cationic complex $[Ir\{bis(pyrazol-1-yl)methane\}(CO)_2][BPh_4]$ (C4) catalyzes the reduction of various imines, including N-alkyl and N-aryl imines and both aldimines and ketimines. Excellent conversions directly to the amine products were achieved rapidly at room temperature in a methanol solution (Scheme 14.7) [53].

Cyclic imines can be also transformed to the corresponding amines.

The above-mentioned iridium complex, C4, as well as $[\{Ir[bis(1\text{-methylimidazol-}2\text{-yl})methane]\,(CO)_2\}][BPh_4]$ (C5), also appeared very efficient in a one-pot tandem hydroamination/hydrosilylation reaction of 4-pentyn-1-amine with $HSiEt_3$ to form 2-methylpyrroline, and then subsequently 1-(triethylsilyl)-2-methylpyrrolidine with an essentially quantitative yield (Equation 14.5) [54].

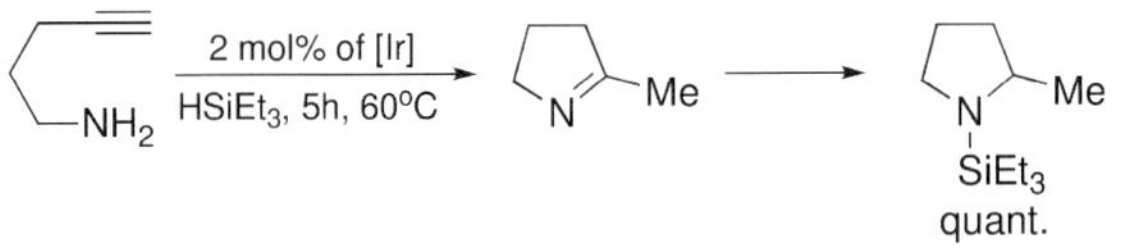

$$\underset{NH_2}{\overset{\equiv}{\diagdown}} \xrightarrow[HSiEt_3,\,5h,\,60°C]{2\,mol\%\,of\,[Ir]} \overset{}{\underset{N}{\diagdown}}\!\!-Me \longrightarrow \underset{\underset{SiEt_3}{N}}{\diagdown}\!\!-Me \quad\text{quant.} \tag{14.5}$$

Chiral oxazolinylphosphines were used as effective ligands for the iridium-catalyzed asymmetric hydrosilylation of imines to afford the corresponding sec-amines with high enantioselectivities (up to 89% ee) after hydrolysis in almost quantitative yields (Equation 14.6) [55]. The following derivatives as efficient ligands were used (see also Table 14.4):

L7: ferrocenyl oxazoline with PPh$_2$, N–S–Ph

L8: phenyl oxazoline with PPh$_2$, N–S–Ph

$$\underset{R^3}{\overset{N^{R^1}}{\|}}_{R^2} + 2\,HSiEt_3 \xrightarrow{[Ir]} \xrightarrow{H^+} \underset{R^3}{\overset{HN^{R^1}}{\underset{H}{|}}}_{R^2} \tag{14.6}$$

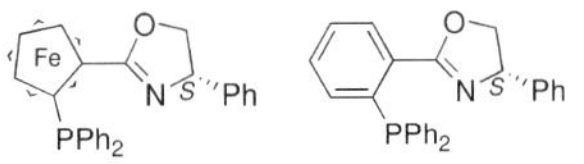

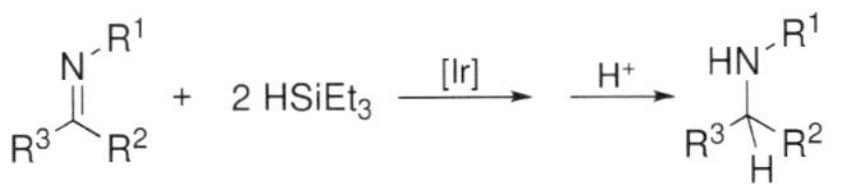

Table 14.4 Formation of amines via iridium-catalyzed hydrosilylation and subsequent protodesilylation.

Metal/Ligand	Imine	Reaction time (h)	Amine	Yield (%)	e.e. (%) (config.)
Ir/L7	2-Ph-pyrroline	20	2-Ph-pyrrolidine	>95	85 (S)
Ir/L7	2-Ph-tetrahydropyridine	100	2-Ph-piperidine	18	7 (S)
Ir/L7	Ph(Me)C=N-Me	60	Ph-CH(Me)-NHMe	56	89
Ir/L8		48		24	16
Ir/L7	Ph(Me)C=N-Ph	60	Ph-CH(Me)-NHPh	25	23

Ir = [{Ir(μ-Cl)(cod)}$_2$]; Ph$_2$SiH$_2$ = 2 mmol; imine = 1 mmol; Ir = 0.005 mmol; L = 0.01 mmol; Et$_2$O, 0 °C.

14.4
Transformation of Organosilicon Compounds in the Presence of Carbon Monoxide

14.4.1
Hydroformylation of Vinylsilanes

Iridium complexes were also found to be effective catalysts for the hydroformylation of vinylsilanes. The study by Crudden and Alper [56] on the catalytic activity of IrCl$_3$ and iridium complexes proved its efficiency in hydroformylation reaction on the example of triethylvinylsilane. Hydrated iridium trichloride, with the addition of AgPF$_6$ or AgBF$_4$, under optimum conditions (100 °C, 700 psi CO, 100 psi H$_2$) and after preactivation, catalyzes the conversion of Et$_3$SiCH=CH$_2$ to an appropriate aldehyde. The complexes tested, namely [Ir(cod)(BPh$_4$)], [Ir(cod)$_2$][BF$_4$], [{Ir(CO)$_3$Cl}$_n$] and [{Ir(μ-Cl)(cod)}$_2$], with the exception of Vaska's complex, were effective catalysts for the hydroformylation of triethylvinylsilane, with or without preactivation. The iridium complexes showed a completely opposite selectivity to that of the rhodium complexes, giving the linear aldehyde with a yield of 90–100% of the total aldehyde product (Scheme 14.8). This approach represents a convenient alternative to the existing methodology, which requires the addition of a vast excess of tertiary phosphines.

Iridium(I) siloxide complexes – that is [{Ir(μ-OSiMe$_3$)(cod)}$_2$] and [Ir(cod)(PCy$_3$)(OSiMe$_3$)] – have also been studied in the transformation of various vinylsilanes under pressure of syn-gas, giving mixtures of silylaldehydes accompanied by the hydrogenation product (Scheme 14.9) [57].

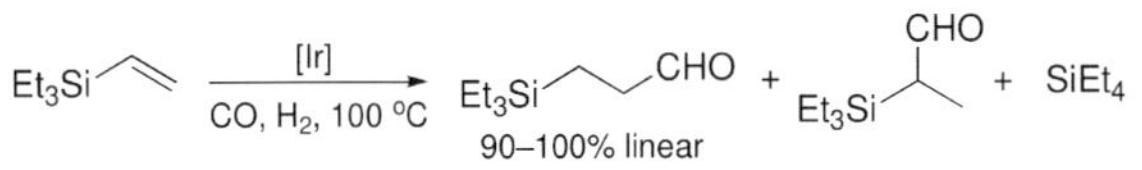

$$Et_3Si\diagup\!\!\!\diagdown \xrightarrow[CO,\ H_2,\ 100\ ^\circ C]{[Ir]} Et_3Si\diagup\!\!\diagdown\!\!\diagup CHO + Et_3Si\diagdown\!\!\diagup^{CHO} + SiEt_4$$
90–100% linear

Catalyst	Conversion [%]	Yield (Selectivity) of linear product [%]	Yield of Hydrogenation [%]
IrCl$_3$/AgBF$_4$[a]	92	59(98)	16
[Ir(cod)$_2$][BF$_4$]	83	80(97)	15
[Ir(cod)(BPh$_4$)][b]	100	73(94)	0

Reaction conditions: 3.2 mmol substrate, 1.8% catalyst (0.058 mmol), 1 M in dry distilled solvent, 100°C, 700 psi of CO, 100 psi of H$_2$, 3h; [a] - 1.5-2 equiv of AgBF$_4$, pre-activation 160°C, 1h C$_6$H$_6$/CHCl$_3$; [b] – isolated yield, CHCl$_3$

Scheme 14.8

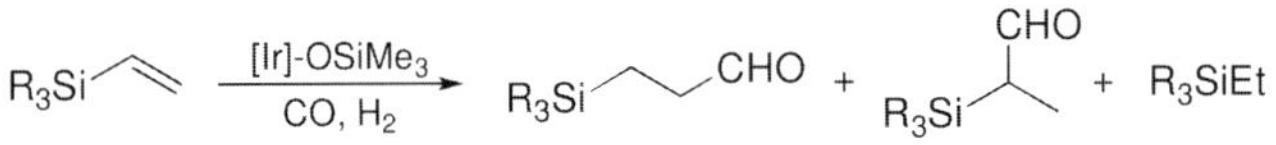

$$R_3Si\diagup\!\!\!\diagdown \xrightarrow[CO,\ H_2]{[Ir]-OSiMe_3} R_3Si\diagup\!\!\diagdown\!\!\diagup CHO + R_3Si\diagdown\!\!\diagup^{CHO} + R_3SiEt$$

Complex	R$_3$	Yield [%]		
		R$_3$SiCH$_2$CH$_2$CHO	R$_3$SiCH(CHO)CH$_3$	R$_3$SiEt
[{Ir(μ-OSiMe$_3$)(cod)}$_2$][a]	Me$_3$	14	11	75
	Me$_2$Ph	18	10	72
[Ir(cod)(PCy$_3$)(OSiMe$_3$)][b]	Me$_3$	38	12	44
	Me$_2$Ph	52	9	34

Reaction conditions: 80 °C, 10 atm H$_2$/CO (1:1), [Ir] = 6.7x10^{-3} M, [a] – 1h, [b] – 3h

Scheme 14.9

14.4.2
Silylcarbonylation of Alkenes and Alkynes

When substituted silanes are used instead of hydrogen, the process is referred to as silylformylation or silylcarbonylation. Only rhodium complexes catalyze the transformation of unsaturated compounds to silylaldehydes via the silylformylation reaction. Iridium complexes also are able to catalyze the simultaneous incorporation of substituted silanes and CO into unsaturated compounds, although during the reaction other types of product are formed. In the presence of [{IrCl(CO$_3$)}$_n$] and [Ir$_4$(CO)$_{12}$]) the alkenes react with trisubstituted silanes and CO to give enol silyl ethers of acyl silanes [58] according to Scheme 14.10.

Iridium siloxide complexes show a similar activity. Catalytic tests performed in the presence of [{Ir(μ-OSiMe$_3$)(cod)}$_2$], with the use of trimethylvinylsilane and dimethylphenylsilane as reactants [59], gave the same type of silicon derivatives as those obtained by Murai and coworkers [58], but the siloxide iridium precursor used appeared to be more efficient under milder conditions. When the [Ir(cod)(PCy$_3$)(OSiMe$_3$)] was used rather than the binuclear iridium siloxide complex, Z-Me$_3$SiCH$_2$CH=CHOSiMe$_2$Ph was obtained exclusively [59].

The iridium cluster [Ir$_4$(CO)$_{12}$], which was used successfully as an olefin silylcarbonylation catalyst, also appeared effective in the formation of nitrogen hetero-

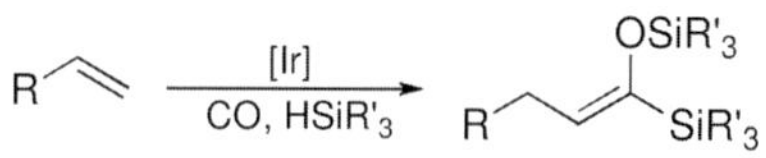

Scheme 14.10

$$R\diagup\!\!\!\!\diagdown \xrightarrow[\text{CO, HSiR'}_3]{[Ir]} R\diagup\!\!\diagdown\!\!\overset{\text{OSiR'}_3}{\underset{\text{SiR'}_3}{\diagup\!\!\diagdown}}$$

R	Yield (%)(E/Z ratio)
H	67(66/34)
Ph	50(72/28)
Me$_3$Si	73(73/27)
BuOCH$_2$	75(68/32)
Me$_3$SiCH$_2$	53(79/21)
Me$_3$SiOCH$_2$	67(67/33)
(EtO)$_2$CH	58(73/27)
NCCH$_2$	45(65/35)
(furyl)-CH$_2$CH$_2$	56(65/35)

Reaction conditions: [alkene] : [HSiR'$_3$] : [Ir] = 10 : 1 : 5x10^{-2}, 50 bars of CO, C$_6$H$_6$, 140°C, 48h.

Scheme 14.11

Silylcarbonylation with [Ir$_4$(CO)$_{12}$], CO, MeCN using HSiMe$_2^t$Bu:

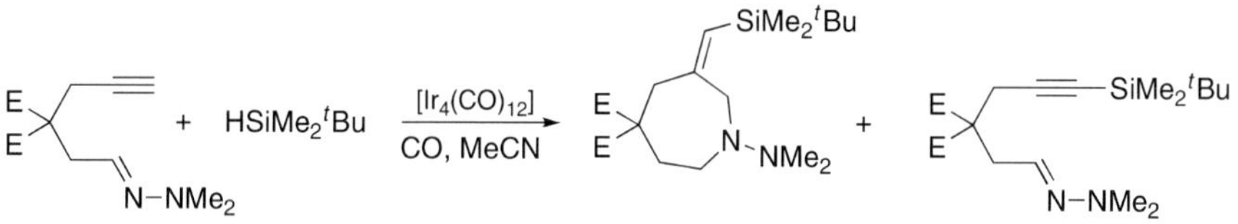

E	Yield of cyclic product (%)
H	16
COOEt	53

cyclic derivatives via the silylcarbonylation of acetylene hydrazones (Scheme 14.11) [60].

The binuclear precursor [{Ir(μ-Cl)(cod)}$_2$] has been also studied in the carbonylative silylcarbocyclization of enynes occurring according to Equation 14.7:

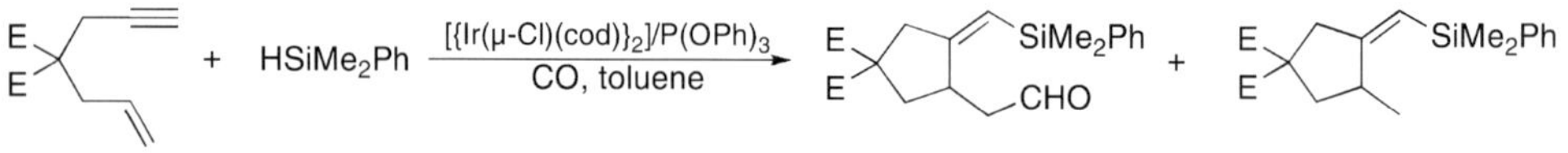

Enyne + HSiMe$_2$Ph $\xrightarrow[\text{CO, toluene}]{[\{Ir(\mu\text{-Cl})(cod)\}_2]/P(OPh)_3}$ cyclopentane products with SiMe$_2$Ph and CHO groups

(14.7)

The tests performed under various conditions showed only a moderate selectivity of the iridium complex in this process. Under optimized conditions (20 bar CO, 80 °C, 18 h) only 75% of the desired product (aldehyde) was obtained [61].

Iridium–phosphine complexes were found to be efficient carbonylative alkyne–alkene coupling catalysts [62]. Although frequently applied in other transformations, the dimeric complex [{Ir(μ-Cl)(cod)}$_2$] appeared to be a very active catalyst in the coupling of silylated diynes with CO [63], giving bicyclic products with a carbonyl moiety (Scheme 14.12).

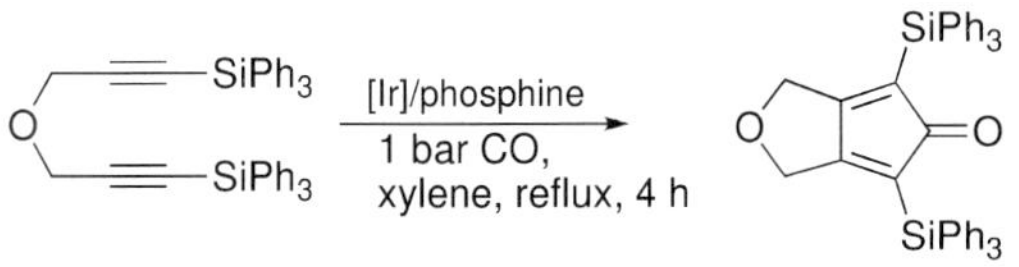

[Scheme 14.12 reaction: alkyne-diyne with two SiPh3 groups, [Ir]/phosphine, 1 bar CO, xylene, reflux, 4 h → bicyclic cyclopentenone product with two SiPh3 groups]

Ligand	Yield (%)	Ligand	Yield (%)
none	18	dppe	37
4PPh$_3$	49	dppp	52
4P(4-F-C$_6$H$_4$)$_3$	42	dppb	0
4P(2-furyl)$_3$	10	dppf	14

Scheme 14.12

14.5
Silylation of Aromatic Carbon–Hydrogen Bonds

The direct silylation of arenes through C—H bond activation provides an attractive route for the synthesis of useful aromatic compounds [64]. Vaska's complex was the first of the iridium catalysts to be reported for activation of the C—H bond in benzene by Si—H of pentamethyldisiloxane to yield phenylsubstituted siloxane [65]. However, a very attractive method for the aromatic C—H silylation with disilanes has been recently reported by the groups of Ishiyama and Miyaura [66–68].

The reaction of tetrafluorodi-sec-butyldisilane of many arenes catalyzed by Ir(I) complexes generated from [{Ir(μ-OMe)(cod)}$_2$] and 2,9-diisopropyl-1,10-phenantroline (dipphen) results in the formation of the relevant corresponding arylfluorosilanes in high yields and with excellent regioselectivities (Equation 14.8).

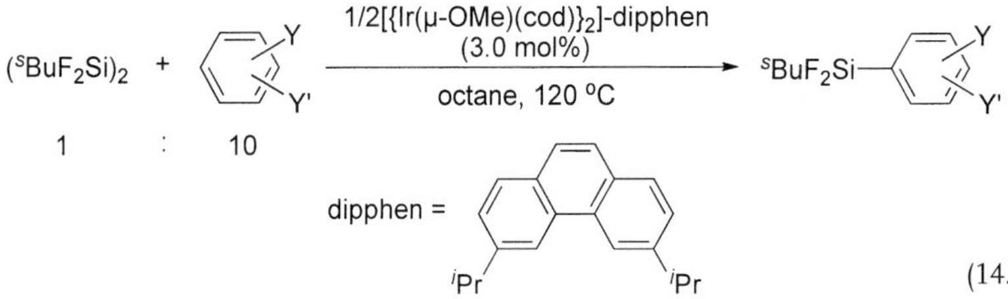

$$(^sBuF_2Si)_2 + \text{ArH} \xrightarrow[\text{octane, 120 °C}]{1/2[\{Ir(\mu\text{-OMe})(cod)\}_2]\text{-dipphen (3.0 mol\%)}} {}^sBuF_2Si\text{-Ar}$$

1 : 10

dipphen = [2,9-diisopropyl-1,10-phenanthroline structure with ^{i}Pr groups] (14.8)

The aromatic silylation of five-membered heteroarenes under the same conditions (catalyst, temperature, solvent) also proceeded in regioselective fashion. Both, thiophene and furane derivatives are exclusively silylated at the α-position, but 1-triisopropylsilyl-pyrrole and -indole each produce selectively β-silyl products (Equations 14.9 and 14.10).

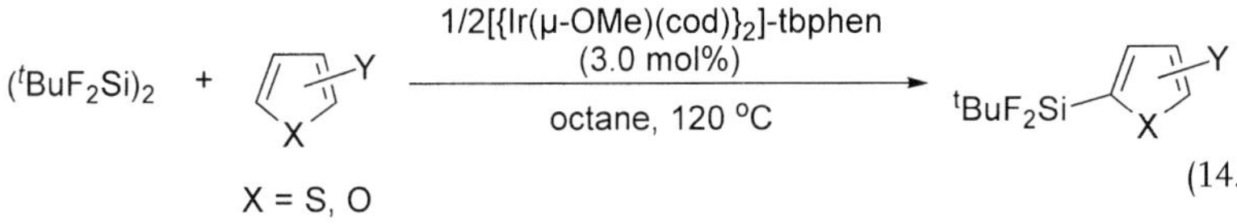

$$(^tBuF_2Si)_2 + \text{heteroarene} \xrightarrow[\text{octane, 120 °C}]{1/2[\{Ir(\mu\text{-OMe})(cod)\}_2]\text{-tbphen (3.0 mol\%)}} {}^tBuF_2Si\text{-heteroarene}$$

X = S, O (14.9)

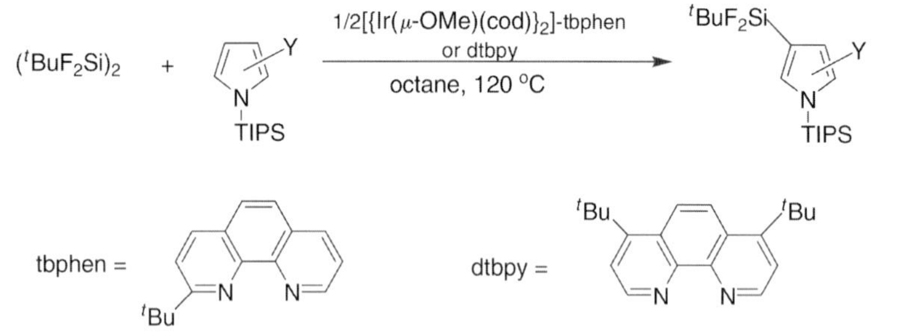

(14.10)

Fluorosilylsubstituted aryl derivatives were found to be useful reagents for carbon–carbon bond formation via palladium-catalyzed cross-coupling with aryl halides in the presence of fluoride anions as Si—C bond activator in dimethylformamide (DMF), as well as rhodium-catalyzed 1,4-addition to α,β-unsaturated ketones in the presence of a fluoride anion source (Equation 14.11) [66, 69, 70].

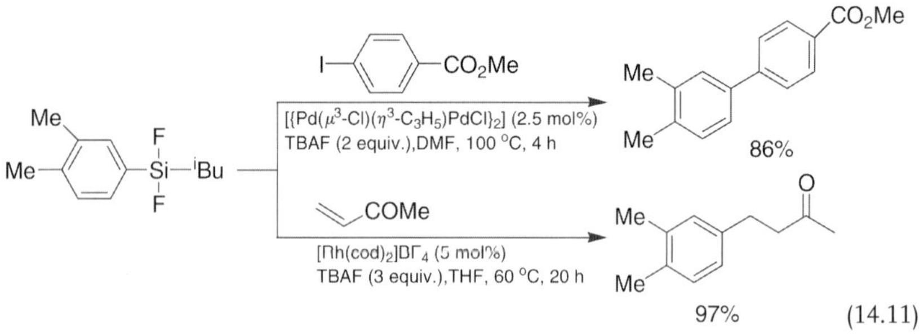

(14.11)

14.6
Silylation of Alkenes with Vinylsilanes

During the past two decades, within the series of our studies, we have developed a silylative coupling reaction of olefins with vinylsubstituted silicon compounds which takes place in the presence of transition-metal complexes (e.g. ruthenium and rhodium) that initially contain or generate M—H and M—Si bonds (for reviews, see Refs [5] and [6]). The reaction involves activation of the =C—H bond of olefins and cleavage of the =C—Si bond of vinylsilane. The reaction, which is catalyzed by complexes of the type [{M(μ-OSiMe$_3$)(cod)}$_2$] (where M = Rh, Ir) occurs according to Equation 14.12 [71, 72].

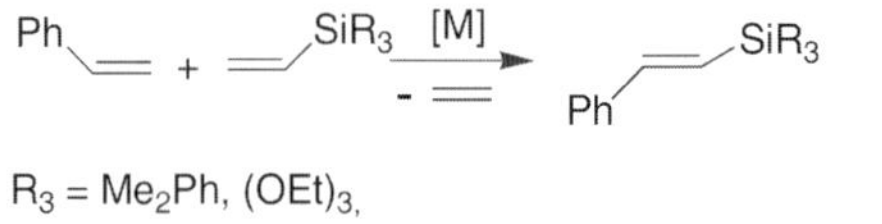

(14.12)

R_3 = Me$_2$Ph, (OEt)$_3$.

However, in the presence of iridium siloxide and bulky substituents at the silicon in vinylsilanes instead of silylative coupling, the codimerization (hydrovinylation) of styrene and vinylsilanes occurs according to Equation 14.13:

$$\text{Ph-CH=CH}_2 + \text{CH}_2\text{=CH-SiR}_3 \xrightarrow{[\{Ir(\mu\text{-OSiMe}_3)(cod)\}_2]} \text{Ph-CH=CH-CH}_2\text{-SiR}_3 \quad (14.13)$$

$R_3 = (O^tBu)_3, (OSiMe_3)_3$

14.7
Alcoholysis and Oxygenation of Hydrosilanes

The alcoholysis of hydrosilanes (or the O-silylation of alcohols) represents a crucial way of synthesizing the silyl ethers that are among the most widely used protective groups of the hydroxyl functionality in organic synthesis (Equation 14.14) [73].

$$R'_3SiH + ROH \longrightarrow R'_3SiOR + H_2 \quad (14.14)$$

Many transition-metal complexes have been reported as catalysts of this reaction, including [Ir(μ-Cl)(coe)$_2$]$_2$ [74] and [IrH$_2$(solv.)(PPh$_3$)][SbF$_6$] [75]. The latter catalyst appeared to be a very active and highly selective. The hydroxyl group can be selectively silylated, even in the presence of other potentially reactive C=C and C=O groups. The order of relative reactivities of alcohol isomers is: secondary alcohol > primary alcohol > tertiary alcohol.

Iridium cationic complexes have recently been used as highly effective catalysts for the regioselective di- and tri-silylation of simple glycopyranosides with *tert*-butyldimethylsilane [76].

Other cationic Ir(I) complexes containing the bidentate ligands, such as bis(1-pyrazolyl)methane (BPM) and bis(3,5-dimethyl-1-pyrazolyl)methane (dmBPM), that is [Ir(BPM)(CO)$_2$]$^+$[BPh$_4$]$^-$ and [Ir(dmBPM)(CO)$_2$]$^+$BPh$_4$]$^-$, were also reported to be effective catalysts for the alcoholysis of binary and tertiary silanes [77].

From among the many transition-metal complexes which have been extensively studied as catalysts of oxygenation of the Si—H bond, a commercially available, air-stable [{Ir(μ-Cl)(coe)}$_2$] complex was shown to be a highly efficient catalyst of the oxygenation of organosilanes to silanols, performed under essentially neutral and mild conditions (at room temperature) and even in the presence of water [78].

14.8
Isomerization of Silyl Olefins

As mentioned above, isomerization of the carbon–carbon (—C=C—) bond is a concurrent and/or consecutive reaction of many transformations of silicon

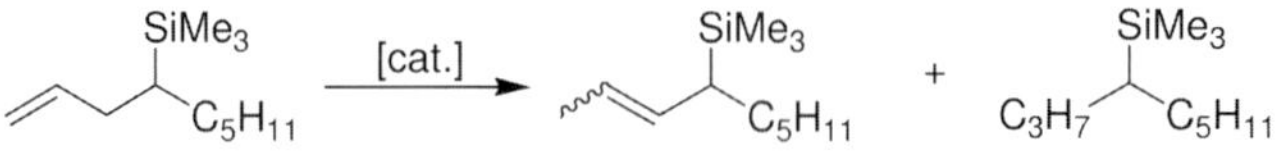

Scheme 14.13

$SiMe_3$-CH₂-CH=CH₂ chain with C_5H_{11} → [cat.] → product **1** ($SiMe_3$, C_5H_{11}) + product **2** ($SiMe_3$, C_3H_7, C_5H_{11})

Catalyst	Conversion (%)	Ratio of products (E)1/(Z)1/2
[Ir(cod)(PPh₂Me)₂][PF₆][a]	100	86/9/6
[Ir(cod)(PPh₂Me)₂][PF₆][b]	100	87/6/8
[Ir(cod)(Py)(PCy₃)][PF₆][b]	76	74/10/16
[Ir(cod)(Py)(PCy₃)][PF₆][c]	83	76/15/9
[Ir(cod)(DPPB)][PF₆][b]	13	77/0/24
[Ir(cod)(PPh₃)₂][PF₆][b]	100	97/0/3

[a] temp. 20 °C; 2 h [b] temp. -20 °C, 2 h; [c] temp. -20 °C, 23 h

compounds with olefins, catalyzed by transition-metal complexes. The cationic iridium complex prepared via the hydrogenation of $[Ir(cod)_2][PF_6]/2PR_3$ was found to serve as an excellent catalyst of the stereoselective isomerization (often exceeding 99%) of primary allyl silyl ethers to (E)-enol ethers and secondary allyl ether to (Z)-enol ethers (Equation 14.15) [79, 80].

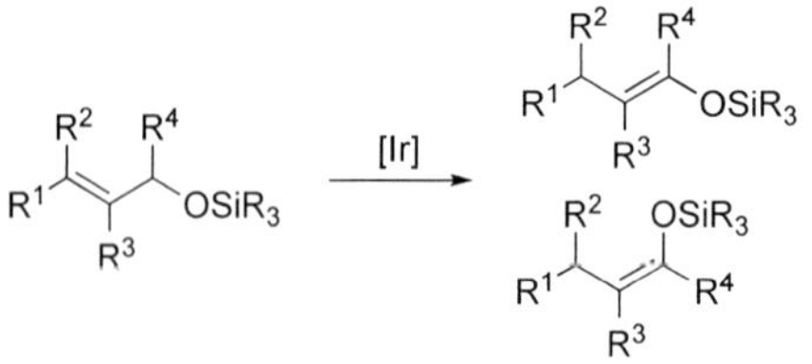

(14.15)

[Ir] = [Ir(cod)(PMePh₂)₂][PF₆]/(H₂ or catecholborane)
[Ir(cod)₂][PF₆]-2R₃P/H₂
$R^1 = C_3H_7$, R^2, R^3 = H, SiR_3 = Si^tBuMe_2, R^4 = Ph

A similar cationic complex, for example $[Ir(cod)(PPh_2Me)_2]^+[PF_6]^-$ was reported to be an excellent catalyst for the isomerization of various 3-silyloxy-1-propenylboronates under mild conditions (room temperature) [81].

On the other hand, the cationic complex $[Ir(cod)(PR_3)_2]^+[PF_6]^-$ appeared a very good catalyst for migration of the C=C bond to be applied for the regiocontrolled synthesis of allylsilane from olefin silylation at a remote sp³ carbon (Scheme 14.13) [82].

14.9
Addition of Silylacetylenes ≡C—H Bond into Imines

Iridium-catalyzed reactions of organosilicon derivatives with various organic compounds involving C—C bond formation have been found suitable for the synthesis

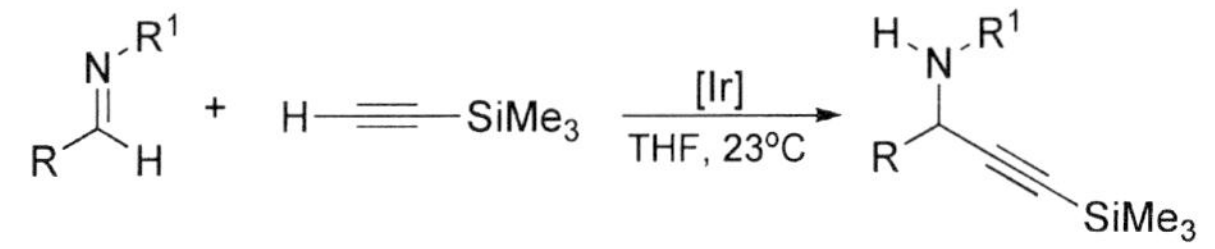

Scheme 14.14

R/R^1	Ph/Bn	^{t}Bu/Bn	Pr/Bn	^{i}Pr/Bn	4-Br-Ph/Bn	Ph/4-MeO-Ph
Yield (%)	76	84	69	65^a	54	85

of silyl-functionalized organic compounds. The iridium binuclear complexes [{Ir(μ-Cl)(cod)}$_2$] effectively catalyzes the addition of trimethylsilylacetylene to aldimines via alkynyl C—H bond activation, yielding secondary amines (Scheme 14.14) [83].

Products of a similar type have been obtained in good yields in a tandem reaction occurring via the condensation of aldehydes with primary or secondary amines and an iridium-mediated double addition of ethynyltrimethyl silane to aldamine formed in the condensation step [84].

Binuclear iridium(I) complexes – that is [{Ir(μ-Cl)(cod)}$_2$], [{Ir(μ-OMe)(cod)}$_2$], [{Ir(μ-Ph)(cod)}$_2$] – were found to be efficient catalysts of ethynyltrimethylsilane addition into quinoline and isoquinoline derivatives (see Equations 14.16 and 14.17) [85].

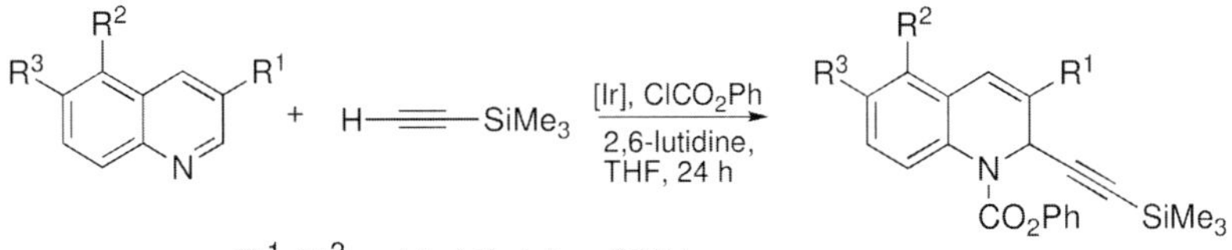

R^1-R^3 = H, Yield = 85%
R^1 = Me; R^2, R^3 = H, Yield = 79% (14.16)

The presence of other groups for example Br, CN, NO$_2$, OMe, OH gives the yields of 13 to 80% (Equation 14.16).

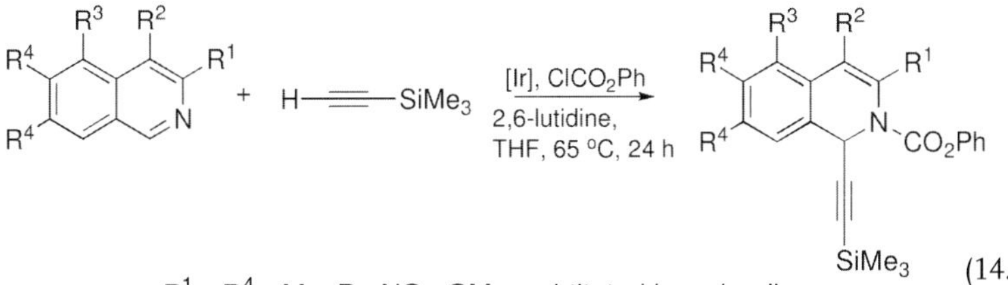

R^1 – R^4 = Me, Br, NO$_2$, OMe, subtituted isoquinodines, yield = 61–82% (Equation 14.16) (14.17)

14.10
Conclusions

Although, among the catalytic conversions of (organo)silicon compounds, hydrosilylation remains the most commercially used process, a variety of other reactions catalyzed by transition-metal complexes have been identified and developed during the past two decades.

Whereas, platinum complexes are used predominantly as efficient catalysts in the hydrosilylation of carbon–carbon multiple bonds, cobalt and iron triad complexes play a crucial role in the catalysis of other processes, such as the hydrosilylation of C=O and C=N, dehydrogenative silylation, silylcarbonylation, and silylation with vinylsilanes and disilanes.

Among the latter group, iridium complexes (though less common than rhodium) and perhaps also ruthenium play crucial roles in many of the above-mentioned transformations of silicon compounds, leading to the creation of silicon–carbon bonds. Examples include the hydrosilylation or dehydrogenative silylation of alkenes and alkynes, the hydroformylation of vinylsilanes, and the silylformylation of alkynes as well as activation of the sp^2C—H of arenes (by disilanes) and alkenes (by vinylsilanes).

Other silicon derivatives containing Si—X—C bonds (where X is O and/or N) can be successfully prepared by using iridium-catalyzed reactions such as the asymmetric hydrosilylation of ketones and amines, the silylcarbonylation of alkenes, and the alcoholysis of Si—H bonds. Indeed, oxygenation of the latter bond to silanol also proceeds smoothly in the presence of iridium compounds.

By contrast, the isomerization of silyl olefins and addition of silylacetylenes ≡C—H bond into imines catalyzed by iridium complexes appears to serve as a suitable route for the synthesis of silylfunctionalized organic compounds. Hence, the acquisition of experimental data on catalysis by iridium complexes in silicon chemistry may be regarded as an initial stage in the quest for catalytic processes leading to the synthesis of other p-block (e.g. B, Ge, Sn, P)–carbon bond-containing compounds.

Abbreviations

coe	cyclooctene
cod	1,5-cyclo-octadiene
triso	tris(diphenyloxosphoranyl)methanides
DAM	diallylmaleate
LED	light-emitting diode
TFB	tetrafluorobenzobarrelene
^{i}Pr-pybox	2,6-bis[4'S-4'-isopropyloxazolin-2'-yl]pyridine
dppe	1,2-bis(diphenylphosphino)ethane
dppp	1,3-bis(diphenylphosphino)-propane
dppb	1,4-bis(diphenylphosphino)butane

dppf	1,10-bis-(diphenylphosphino)ferrocene
dipphen	2,9-diisopropyl-1,10-phenanthroline
tbphen	2-*tert*-butyl-1,10-phenanthroline
dtbpy	4,7-di-*tert*-butyl-1,10-phenanthroline
BPM	bis(1-pyrazolyl)methane
dmBPM	bis(3,5-dimethyl-1-pyrazolyl)methane
pzpy	3-(2-pyridyl)-pyrazol-1-yl
TBAF	tetrabutylamonium fluoride
η^5-Ind	indenyl

References

1 Ojima, I., Li, Z. and Zhu, J. (1998) *The Chemistry of Organic Silicon Compounds* (eds Z. Rappoport and Y. Apeloig), John Wiley & Sons, Ltd, Chichester, Chapter 29.

2 Marciniec, B. (1992) *Comprehensive Handbook on Hydrosilylation*, Pergamon Press, Oxford.

3 Marciniec, B. (2002) *Applied Homogeneous Catalysis with Organometallic Compounds* (eds B. Cornils and W.A. Herrmann), Wiley-VCH Verlag GmbH, Weinheim, Chapter 2.6.

4 Marciniec, B. (2002) *Silicon Chemistry*, **1**, 155–75.

5 Brook, M.A. (2000) *Silicon in Organic Organometallic and Polymer Chemistry*, John Wiley & Sons, Ltd, New York, Chapter 9.

6 Reichl, J.A. and Berry, D.H. (1999) *Advances in Organometallic Chemistry*, **43**, 197–265.

7 Marciniec, B. (2000) *Applied Organometallic Chemistry*, **14**, 527–38.

8 Marciniec, B. and Pietraszuk, C. (2003) *Handbook of Metathesis* (ed. R. Grubbs), Wiley-VCH Verlag GmbH, Weinheim, Chapter 2.13.

9 Marciniec, B. and Pietraszuk, C. (2003) *Current Organic Chemistry*, **7**, 691–735.

10 Marciniec, B. and Pietraszuk, C. (2004) Synthesis of silicon derivatives with ruthenium catalysts, in *Topics in Organometallic Chemistry*, Vol. **11** (ed. P. Dixneuf), Springer-Verlag, Berlin, pp. 197–248.

11 Marciniec, B. (2005) *Coordination Chemistry Reviews*, **249**, 2374–90.

12 Marciniec, B., Pietraszuk, C., Kownacki, I. and Zaidlewicz, M. (2005) Vinyl- and arylsilicon, germanium, and boron compounds, in *Comprehensive Organic Functional Group Transformations II*, Vol. **1** (eds A.R. Katrytzky and R.J.K. Taylor), Elsevier, Amsterdam, pp. 941–1024.

13 Marciniec, B., Pawluc, P. and Pietraszuk, C. (2007) Inorganometallic Chemistry, in *Inorganic and Bio-Inorganic Chemistry, Encyclopedia of Life Support Systems* (ed. I. Bertini), Developed under the Auspices of UNESCO, EOLSS Publishers Co. Ltd, Oxford, UK www.eols.net.

14 Cundy, C.S. Kingston, B.M. and Lappert, M.F. (1973) *Advances in Organometallic Chemistry*, **11**, 253–311.

15 Fernandez, M.J., Esterueles, M.A., Jimenez, M.S. and Oro, L.A. (1986) *Organometallics*, **5**, 1519–20.

16 Tanke, R.S. and Crabtree, R.H. (1991) *Organometallics*, **10**, 415–18.

17 Fernandez, M.J. and Oro, L.A. (1988) *Journal of Molecular Catalysis*, **45**, 7–15.

18 Tanke, R.S. and Crabtree, R.H. (1990) *Journal of the American Chemical Society*, **112**, 7984–9.

19 Cipot, J., Ferguson, M.J. and Stradito, M. (2006) *Inorganica Chimica Acta*, **359**, 2780–5.

20 Cipot, J., McDonald, R., Ferguson, M.J., Schatte, G. and Stradito, M. (2007) *Organometallics*, **26**, 594–608.

21 Hostetler, M.J., Butts, M.D. and Bergman, R.G. (1993) *Organometallics*, **12**, 65–75.

22 Quirk, J.M. and Kanner, B. (1987) US Patent 4658050.
23 Nashiwaki, A. and Kiyomori, A. (2003) Japanese Patent 2003 096086.
24 Kropfgans, F. and Frings, A. (1996) European Patent 0709392 A1.
25 Kubota, T. and Yamamoto, A. (1995) Japanese Patent 1993 270278.
26 Baumann, F. and Hoffmann, M. (2006) Patent DE 102004052424.
27 Kownacki, I., Marciniec, B., Macina, A., Rubinsztajn, S. and Lamb, D. (2007) *Applied Catalysis A. General*, **317**, 53–7.
28 Khanarian, G. and Pedicini, A. (2007) European Patent 1801163 A1.
29 Kotani, J., Tsumura, M., Iwahara, T. and Hirose, T. (1995) European Patent 0661331 A2.
30 Jones, R.G., Ando, W. and Chojnowski, J. (eds) (2000) *Silicon-Containing Polymers*, Kluwer Academic, Dordrecht.
31 Rickle, G.K. (1994) *Journal of Applied Polymer Science Part A – Polymer Chemistry*, **51**, 605–15.
32 Pawluc, P., Marciniec, B., Kownacki, I. and Maciejewski, H. (2005) *Applied Organometallic Chemistry*, **19**, 49–54.
33 Tanke, R.S. and Crabtree, R.H. (1990) *Journal of the Chemical Society D – Chemical Communications*, 1056–7.
34 Esteruelas, M.A., Lahoz, F.J., Oñate, E., Oro, L.A. and Rodríguez, L. (1996) *Organometallics*, **15**, 823–34.
35 Sridevi, V. S., Fan, W.Y. and Leong, W.K. (2007) *Organometallics*, **26**, 1157–60.
36 Szajek, L.P. and Shapley, J.R. (1994) *Organometallics*, **13**, 1395–403.
37 Dield, L.D. and Ward, A.J. (2003) *Journal of Organometallic Chemistry*, **681**, 91–7.
38 Mas-Marza, E., Poyatos, M., Sanau, M. and Peris, E. (2004) *Inorganic Chemistry*, **43**, 2213–19.
39 Miyake, Y. Isomura, E. and Iyoda, M. (2006) *Chemistry Letters*, **35**, 836–7.
40 Martínez, A.P., Fabra, M.J., García, M.P., Lahoz, F.J., Oro, L.A. and Teat, S.J. (2005) *Inorganica Chimica Acta*, **358**, 1635–44.
41 Martin, M., Sola, E., Torres, O., Plou, P. and Oro, L.A. (2003) *Organometallics*, **22**, 5406–17.
42 Esteruelas, M.A., Olivan, M. and Oro, L.A. (1996) *Organometallics*, **15**, 814–22.
43 Jun, C.H. and Crabtree, R.H. (1993) *Journal of Organometallic Chemistry*, **447**, 177–87.
44 Vicent, C., Viciano, M., Mas-Marzá, E., Sanaú, M. and Peris, E. (2006) *Organometallics*, **25**, 3713–20.
45 Shimizu, R. and Fuchikami, T. (2000) *Tetrahedron Letters*, **41**, 907–10.
46 Herrmann, W.A., Baskakov, D., Herdtweck, E., Hoffman, S.D., Bunlaksananusorn, T., Rampf, F. and Rodefeld, L. (2006) *Organometallics*, **25**, 2449–56.
47 Chianese, A.R. and Crabtree, R.H. (2005) *Organometallics*, **24**, 4432–6.
48 Chen, T., Liu, X.-G. and Shi, M. (2007) *Tetrahedron*, **63**, 4874–80.
49 Nishibayashi, Y., Segawa, K., Singh, J.D., Fukuzawa, S., Ohe, K. and Uemura, S. (1996) *Organometallics*, **15**, 370–9.
50 Cuervo, D., Diez, J., Gamasa, M.P., Gimeno, J. and Paredes, P. (2006) *European Journal of Inorganic Chemistry*, 599–608.
51 Karame, I., Tommasino, M.L. and Lemaire, M. (2003) *Journal of Molecular Catalysis A – Chemical*, **196**, 137–43.
52 Klei, S.R., Don Tilley, T. and Bergman, R. (2002) *Organometallics*, **21**, 4648–61.
53 Field, L.D., Messerle, B.A. and Rumble, S.L. (2005) *European Journal of Organic Chemistry*, 2881–3.
54 Field, L.D., Messerle, B.A. and Wren, S.L. (2003) *Organometallics*, **22**, 4393–5.
55 Takei, I., Nishibayshi, Y., Arikawa, Y., Uemura, S. and Hidai, M. (1999) *Organometallics*, **18**, 2271–4.
56 Crudden, C.M. and Alper, H. (1994) *Journal of Organic Chemistry*, **59**, 3091–7.
57 Mieczynska, E., Trzeciak, A.M., Ziólkowski, J.J., Kownacki, I. and Marciniec, B. (2005) *Journal of Molecular Catalysis A – Chemical*, **237**, 246–53.
58 Chatani, N., Ikeda, S., Ohe, K. and Murai, S. (1992) *Journal of the American Chemical Society*, **114**, 9710–11.
59 Kownacki, I., Marciniec, B., Szubert, K. and Kubicki, M. (2005) *Organometallics*, **24**, 6179–83.
60 Chatani, N., Yamaguchi, S., Fukumoto, Y. and Murai, S. (1995) *Organometallics*, **14**, 4418–20.

61 Maerten, E., Delerue, H., Queste, M., Nowicki, A., Suisse, I. and Agbossou-Niedercorn, F. (2004) *Tetrahedron: Asymmetry*, **15**, 3019–22.
62 Shibata, T. and Takagi, K. (2000) *Journal of the American Chemical Society*, **122**, 9852–3.
63 Miyake, Y., Isomura, E. and Iyoda, M. (2006) *Chemistry Letters*, **35**, 836–7.
64 Kakiuchi, F. and Chatani, N. (2003) *Advanced Synthesis Catalysis*, **345**, 1077–101.
65 Gustavson, W.A., Epstein, P.S. and Curtis, M.D. (1982) *Organometallics*, **1**, 884–5.
66 Ishiyama, T., Sato, K., Nishio, Y. and Miyaura, N. (2003) *Angewandte Chemie – International Edition*, **42**, 5346–8.
67 Saiki, T., Nishio, Y., Ishiyama, T. and Miyaura, N. (2006) *Organometallics*, **25**, 6068–73.
68 Ishiyama, T., Sato, K., Nishio, Y., Saiki, T. and Miyaura, N. (2005) *Chemical Communications*, 5065–7.
69 Sakakura, T., Tokunaga, Y., Sodeyama, T. and Tanaka, M. (1987) *Chemistry Letters*, 2375–7.
70 Ishikawa, M., Naka, A. and Ohshita, J. (1993) *Organometallics*, **12**, 4987–92.
71 Marciniec, B., Kownacki, I. and Kubicki, M. (2002) *Organometallics*, **21**, 3263–70.
72 Marciniec, B., Kownacki, I., Kubicki, M., Krzyzanowski, P., Walczuk, E. and Blazejewska-Chadyniak, P. (2003) *Perspectives in Organometallic Chemistry* (eds C.G. Screttas and B.R. Steele), Royal Society of Chemistry, Cambridge, pp. 253–64.
73 Lukevics, E. and Dzintara, M. (1985) *Journal of Organometallic Chemistry*, **295**, 265–315.
74 (a) Blacburn, S.N., Haszeldine, S.N., Parish, R.V. and Setchfi, J.H. (1980) *Journal of Organometallic Chemistry*, **192**, 329–38;
(b) Dwyer, J., Hilal, H.S. and Parish, R.V. (1982) *Journal of Organometallic Chemistry*, **228**, 191–201.
75 Luo, X.L. and Crabtree, R.H. (1989) *Journal of the American Chemical Society*, **111**, 2527–35.
76 Chung, M.K. and Schlaf, M. (2005) *Journal of the American Chemical Society*, **127**, 18085–92.
77 Field, L.D., Messerle, B.A., Rehr, M., Soler, L.P. and Hambley, T.W. (2003) *Organometallics*, **22**, 2387–95.
78 Lee, Y., Seomoon, D., Kim, S., Han, H., Chang, S. and Lee, P.H. (2004) *Journal of Organic Chemistry*, **69**, 1741–3.
79 Ohmura, T., Shirai, Y., Yamamoto, Y. and Miyaura, N. (1998) *Chemical Communications*, 1337–8.
80 Ohmura, T., Yamamoto, Y. and Miyaura, N. (1999) *Organometallics*, **18**, 413–16.
81 Yamamoto, Y., Miyairi, T., Ohmura, T. and Miyaura, N. (1999) *Journal of Organic Chemistry*, **64**, 296–8.
82 Matsuda, I., Kato, T., Sato, S. and Izumi, Y. (1986) *Tetrahedron Letters*, **27**, 5747–50.
83 Fischer, C. and Carreira, E.M. (2001) *Organic Letters*, **3**, 4319–21.
84 Sakaguchi, S., Mizuta, T., Furuwan, M., Kubo, T. and Ishii, Y. (2004) *Chemical Communications*, 1638–9.
85 Yamazaki, Y., Fujita, K. and Yamaguchi, R. (2004) *Chemistry Letters*, **33**, 1316–17.

15
Catalytic Properties of Soluble Iridium Nanoparticles
Jackson D. Scholten and Jaïrton Dupont

15.1
Introduction

It is well known that transition-metal nanoparticles are only kinetically stable. Hence, when they are freely dissolved in solution they must be stabilized in order to prevent their agglomeration, whereby they diffuse together and coalesce, eventually to form the bulk metal – a situation which is thermodynamically favored [1, 2]. The stabilization of soluble metal nanoparticles can be achieved by using stabilizing agents that provide electrostatic and/or steric protection; examples include water-soluble polymers, quaternary ammonium salts, surfactants and/or polyoxoanions [3–8]. The stabilizing agents also play an important role in controlling the diameter, shape and size distribution of the metal nanoparticles, as well as their surface properties – and hence their catalytic properties. The ideal nanoparticle/stabilizer combination should exhibit a synergistic effect that enhances both the activity and durability of the catalyst (by providing a stable metal nanoparticle environment), but without causing any loss of the catalytic properties. Usually, soluble and stable Ir(0) nanoparticles are easily synthesized by the reduction of Ir(I) or Ir(III) compounds in the presence of classical surfactants, typically polyoxoanions associated with tetrabutylammonium salt or ionic liquids that provide steric/electrostatic stabilization modes.

15.2
Synthesis of Soluble Iridium Nanoparticles

15.2.1
Polyoxoanions

Soluble and stable iridium nanoparticles (3.0 ± 0.4 nm diameter) have been prepared by reduction of the polyoxoanion-supported Ir(I) complex $(n\text{-Bu}_4\text{N})_5\text{Na}_3$ [(COD)Ir($P_2W_{15}Nb_3O_{62}$)] (COD = 1,5-cyclo-octadiene) with molecular hydrogen in

Iridium Complexes in Organic Synthesis. Edited by Luis A. Oro and Carmen Claver
Copyright © 2009 WILEY-VCH Verlag GmbH & Co. KGaA, Weinheim
ISBN: 978-3-527-31996-1

acetone solution and in the presence of Bu_4N^+ salt [9, 10]. These nanoparticles have been characterized by several methods, including transmission electron microscopy (TEM), elemental analysis, fast-atom bombardment mass spectroscopy, electron diffraction, IR and UV-visible spectroscopy, electrophoresis and ultracentrifugation solution molecular-weight measurements. The average composition of the synthesized nanoparticles is $[Ir(0)_{\sim 900}(P_4W_{30}Nb_6O_{123}^{-16})_{\sim 60}](n\text{-}Bu_4N)_{\sim 660}Na_{\sim 300}$, and it was noted that the polyoxoanion is in its oxidized state and Nb—O—Nb bridged is in its aggregate form $(P_4W_{30}Nb_6O_{123}^{-16})$. Nanoparticle stabilization is provided by the polyoxoanion $(P_4W_{30}Nb_6O_{123}^{-16})$ and its associated cationic Bu_4N^+ species, which form a protective layer surrounding the nanoparticles, thus preventing agglomeration. Starting from the same precursor $(n\text{-}Bu_4N)_5Na_3[(COD)Ir(P_2W_{15}Nb_3O_{62})]$, iridium nanoparticles of 2.0 ± 0.3 nm mean diameter have also been prepared in the presence of cyclohexene. The elemental analysis of the nanoclusters indicates that their chemical composition is $[Ir(0)_{\sim 300}(P_4W_{30}Nb_6O_{123}^{-16})_{\sim 33}](n\text{-}Bu_4N)_{\sim 300}Na_{\sim 233}$.

15.2.2
Surfactants

A simple and general method for the preparation of surfactant-free, thiol-functionalized iridium nanoparticles was reported by Ulman and coworkers in 1999 [11]. The synthesis consisted of a reduction of the dihydrogen hexachloroiridate (IV) $H_2IrCl_6 \cdot H_2O$ precursor by lithium triethylborohydride ('super-hydride') in the presence of octadecanethiol ($C_{18}H_{37}SH$) in tetrahydrofuran (THF) (Scheme 15.1). The obtained iridium nanoparticles were crystalline with fcc (face-centered cubic) packing, and showed a wider size distribution with diameters ranging from 2.25 to 4.25 nm.

In 2005, Stowell and Korgel [12] developed the synthesis of iridium nanoparticles by reduction of the precursor [(methylcyclopentadienyl)(COD)Ir] with hexadecanediol in the presence of different capping ligands such as oleic acid/oleylamine, tetraoctylammonium bromide (TOAB), tetraoctylphosphonium bromide (TOPB) and trioctylphosphine (TOP). Iridium nanoparticles with diameters varying between 10 and 100 nm and synthesized with a TOP-capping ligand were considered to be the poorest quality, with a wide polydispersion and irregular shapes. Reduction of the organometallic precursor in the presence of TOAB-capping ligand resulted in crystalline and size-monodispersed iridium nanoparticles with diameters in the range of 1.5–3.0 nm, whereas when using the TOPB-capping ligand the nanoparticles have diameters of 2.0–5.0 nm. The latter nanoparticles

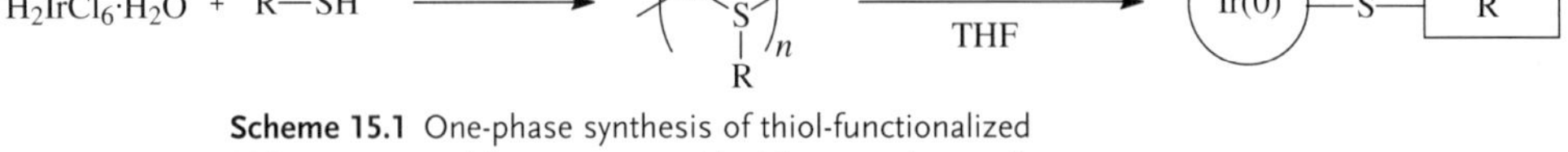

Scheme 15.1 One-phase synthesis of thiol-functionalized iridium nanoparticles as proposed by Ulman and coworkers.

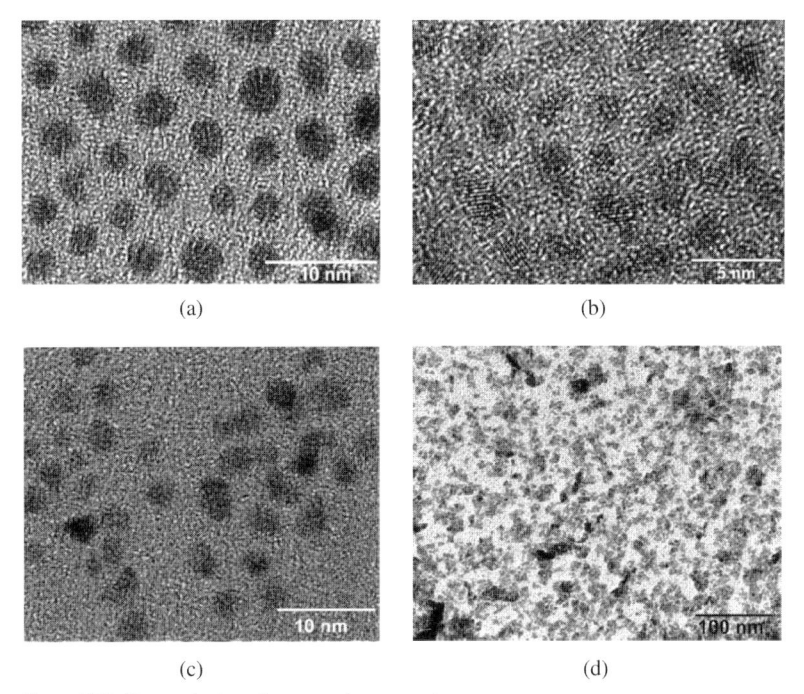

Figure 15.1 Transmission electron microscopy images showing iridium nanoparticles prepared in four different capping ligands. (a) Oleic acid/oleylamine; (b) TOAB; (c) TOPB; (d) TOP. (Reproduced with permission from Ref. [12]; © 2005 American Chemical Society).

were crystalline, but more polydisperse, when compared to those produced with the TOAB ligand. However, the best results were obtained for iridium nanoparticles (diameter ~4.0 nm) prepared in the oleic acid/oleylamine capping ligand, on the basis of their crystallinity, narrow size and shape-distribution, as observed by TEM (Figure 15.1).

An interesting method to produce water-soluble iridium nanoparticles was proposed by Chaudret and coworkers [13]. Here, aqueous soluble iridium nanoparticles were synthesized by the chemical reduction of iridium trichloride with sodium borohydride in an aqueous solution of the surfactant N,N-dimethyl-N-cetyl-N-(2-hydroxyethyl)ammonium chloride (Scheme 15.2). The precursor reduction was assisted by sonication, while the gradual conversion of Ir(III) ions to Ir(0) nanoparticles was followed using UV spectroscopy. The use of a molar surfactant:Ir ratio of 10 proved sufficient to obtain stable aqueous soluble iridium nanoparticles; however, if the molar surfactant:Ir ratio used was <10 then agglomeration was observed in solution after several days. TEM analysis of the iridium nanoparticles revealed a monodispersed size distribution and a mean diameter of 1.9 ± 0.7 nm (Figure 15.2).

$$\text{IrCl}_3 \xrightarrow[\text{H}_2\text{O, room temperature}]{\text{NaBH}_4, \text{ surfactant, ultrasound irradiation}} [\text{Ir}(0)]_n$$

Surfactant: HO–CH$_2$CH$_2$–N$^+$(CH$_3$)$_2$–(CH$_2$)$_n$–CH$_3$ Cl$^-$

Scheme 15.2 System employed by Chaudret and coworkers to prepare soluble iridium nanoparticles in an aqueous medium.

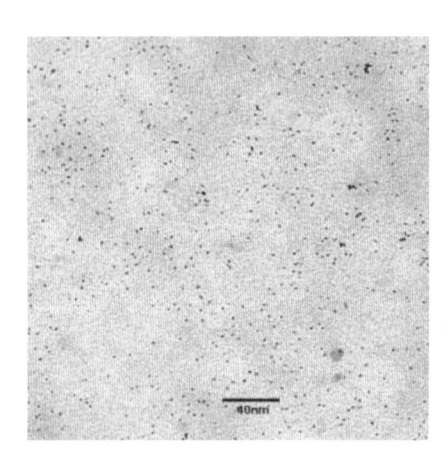

Figure 15.2 Transmission electron microscopy image of iridium nanoparticles of 1.9 ± 0.7 nm in diameter (400 particles counted) prepared in the presence of the surfactant N,N-dimethyl-N-cetyl-N-(2-hydroxyethyl)ammonium chloride. (Reproduced with permission from Ref. [13]; © 2004 Wiley-VCH).

15.2.3
Imidazolium Ionic Liquids

Imidazolium-based ionic liquids (ILs) have been used extensively as media for the formation and stabilization of transition-metal nanoparticles [14–17]. These 1,3-dialkylimidazolium salts (Figure 15.3) possess very interesting properties: they have a very low vapor pressure, they are nonflammable, have high thermal and electrochemical stabilities, and display different solubilities in organic solvents [18–20].

R_1; R_2 = Me, Et, n-Bu
X = PF$_6$, BF$_4$, CF$_3$SO$_3$ (OTf), (CF$_3$SO$_2$)$_2$N (NTf$_2$), EtSO$_4$

Figure 15.3 The chemical structures of the general imidazolium-based ionic liquids.

In particular, the ILs differ from classical ammonium salts in one very important aspect, namely that they possess preorganized structures, mainly through hydrogen bonds [21–23]. Such bonds induce structural directionality, in contrast to the classical salts where the aggregates display charge-ordering structures. The ILs have been used with great success in the synthesis of iridium nanoparticles, as

$$[\text{Ir(COD)Cl}]_2 \xrightarrow[\text{IL, 75 °C}]{\text{H}_2} [\text{Ir}(0)]_n$$

Scheme 15.3 Preparation of soluble iridium nanoparticles from *in situ* reduction of the organometallic precursor [Ir(COD)Cl]$_2$ in imidazolium ionic liquids.

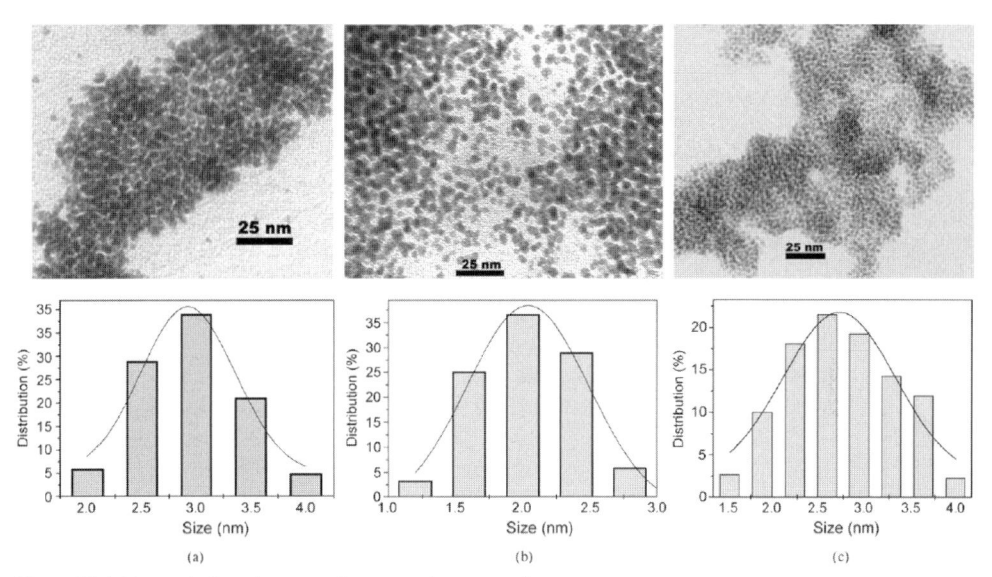

Figure 15.4 Transmission electron microscopy images and size-distribution histograms (300 particles counted) for iridium nanoparticles prepared in: (a) BMI·BF$_4$; (b) BMI·PF$_6$; and (c) BMI·CF$_3$SO$_3$. (Reproduced with permission from Ref. [25]; © 2006 Elsevier).

they may provide electrostatic stabilization. In this context, reduction of the organometallic precursor [Ir(COD)Cl]$_2$ by molecular hydrogen in 1-*n*-butyl-3-methylimidazolium hexafluorophosphate (BMI·PF$_6$) afforded iridium nanoparticles (Scheme 15.3) with irregular shapes and a monomodal distribution and a mean diameter of 2.0 ± 0.4 nm [15, 24].

The typical *in situ* reduction of the precursor [Ir(COD)Cl]$_2$ by molecular hydrogen under the same reaction conditions have been also performed in 1-*n*-butyl-3-methylimidazolium trifluoromethanesulfonate (BMI·CF$_3$SO$_3$) and 1-*n*-butyl-3-methylimidazolium tetrafluoroborate (BMI·BF$_4$) [25]. The iridium nanoparticles prepared in BMI·CF$_3$SO$_3$ and BMI·BF$_4$ ILs, as previously observed with BMI·PF$_6$, display irregular shapes with a monomodal size distribution (Figure 15.4). Mean diameters in the range of 2–3 nm were estimated with *in situ* TEM and small-angle X-ray scattering (SAXS) analyses of the Ir(0) nanoparticles soluble in the ionic liquids, and by X-ray diffraction (XRD) of the isolated material. The mean diameters of iridium nanoparticles synthesized in the three ILs, as estimated by TEM, SAXS and XRD, are summarized in Table 15.1.

Table 15.1 Comparison of the mean diameters of iridium nanoparticles as determined by TEM, SAXS and XRD techniques [25].

Sample	TEM (nm)	SAXS (nm)	XRD (nm)
[Ir(0)] BMI·PF$_6$	2.0 ± 0.4	2.8 ± 0.4	2.1 ± 0.5
[Ir(0)] BMI·CF$_3$SO$_3$	2.6 ± 0.6	2.4 ± 0.3	2.5 ± 0.3
[Ir(0)] BMI·BF$_4$	2.9 ± 0.4	3.0 ± 0.3	2.6 ± 0.4

Notably, a good agreement was found between the TEM, SAXS and XRD methods for determining the mean relative diameters of the nanoparticles (assuming a spherical shape) in the ILs (see Table 15.1). Moreover, studies of the isolated iridium nanoparticles with X-ray photoelectron spectroscopy (XPS) showed the presence of Ir—F bonds for those samples prepared in BMI·BF$_4$ and BMI·PF$_6$. In addition, in the case of the sample obtained in BMI·PF$_6$ there was also a very small contribution of phosphorus. Of note, no chloride was detected in these samples. In the case of Ir(0) prepared in BMI·CF$_3$SO$_3$, only the contributions of Ir—Ir and Ir—O bonds were observed in the Ir 4f spectra. Clearly, the F and P (BMI·BF$_4$ and BMI·PF$_6$) signals indicate that the isolated nanoparticles contain residues of the ILs, although no other impurities were detected within the sensitivity range of the technique used. The XPS signals of the Ir 4f region for the three samples are shown in Figure 15.5.

In addition, the sample prepared with BMI·PF$_6$ was submitted to Ar$^+$ sputtering followed by further XPS analysis. After sputtering, the F 1s signal was eliminated as well as the Ir—F component in the Ir 4f region (which displays mainly the Ir—Ir component), showing that only the external surface iridium atoms were bounded to F (Figure 15.6). These results indicated strongly that, besides the presence of an Ir—O layer, it was the effective interaction of the IL with the metal surface that may have been responsible for stabilization of the nanoparticles.

XPS measurements showed clearly the interactions of the IL with the metal surface, that occurs through F (for BF$_4^-$ and PF$_6^-$) or O (for CF$_3$SO$_3^-$) of the anions, demonstrating the formation of an IL protective layer surrounding the iridium nanoparticles. Additional extended X-ray absorption fine structure (EXAFS) analyses also provided evidence for interaction of the IL liquid with the metal surface.

SAXS analysis indicated the formation of an IL layer surrounding the metal particles, with an extended molecular length of approximately 2.8–4.0 nm that was dependent on the type of anion. This result suggests the presence of a semi-organized anionic species composed of supramolecular aggregates of the type $[(BMI)_{x-n}(X)_x]^{n-}$, as is usually observed in solid, liquid, gas-phase and solution structural organization of imidazolium ILs. This multilayer is most likely composed of anions located immediately adjacent to the nanoparticle surface, providing Coulombic repulsion, and countercations that provide the charge balance, that is, quite close to DLVO (Derjaugin–Landau–Verwey–Overbeek) -type stabilization.

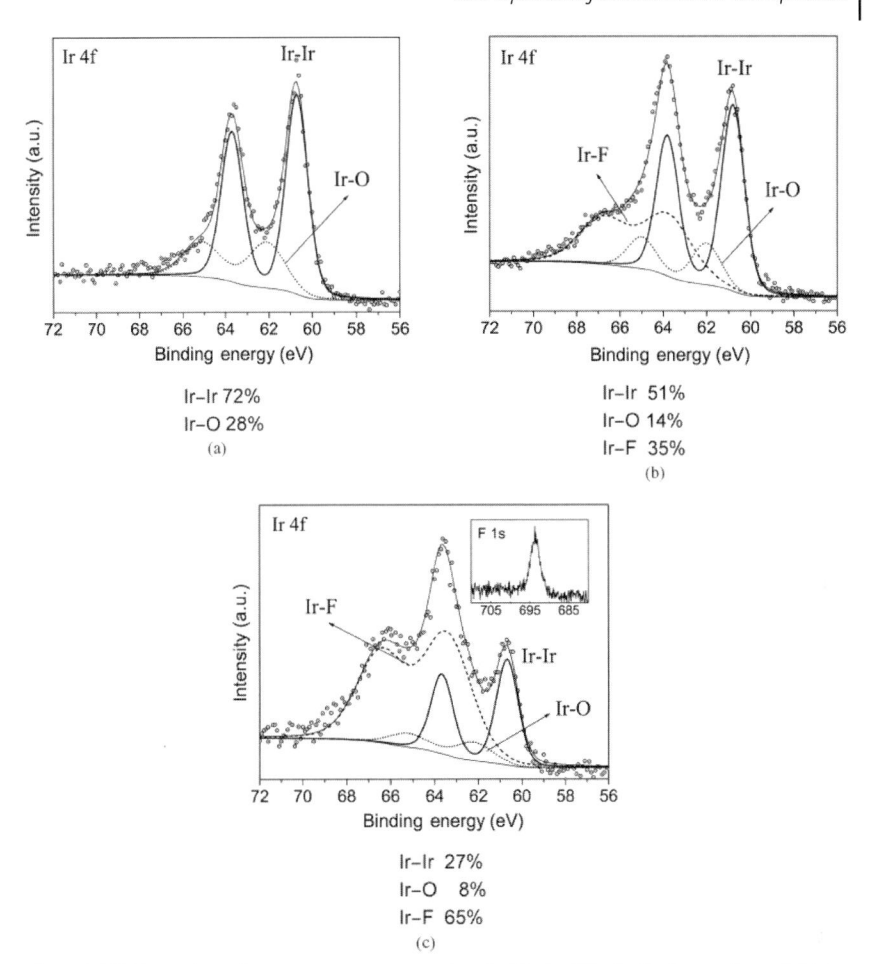

Figure 15.5 X-ray photoelectron spectroscopy of the Ir(0) nanoparticles prepared with: (a) BMI·CF$_3$SO$_3$; (b) BMI·BF$_4$; and (c) BMI·PF$_6$, showing the Ir 4f region with the fitting results. The Ir 4f doublet presents up to three components corresponding to Ir—Ir (solid line), Ir—O (dotted line) and Ir—F bonds (dashed line). The inset of panel (c) shows the F 1s signal observed in that case. The relative contributions for each component are also listed for all samples. (Reproduced with permission from Ref. [25]; © 2006 Elsevier).

However, the DLVO model cannot completely explain the stabilization properties of imidazolium ILs towards the Ir(0) nanoparticles as it treats counterions as mono-ionic point charges and was not designed to account for sterically stabilized systems. Together with the electrostatic stabilization provided by the intrinsic high charge of the IL, a steric type of stabilization can also be envisaged. This is due to the presence of anionic and cationic supramolecular aggregates of the type $[(BMI)_x(X)_{x-n}]^{n+}[(BMI)_{x-n}(X)_x]^{n-}$, where BMI is the 1-$n$-butyl-3-methylimidazolium cation and X is the anion.

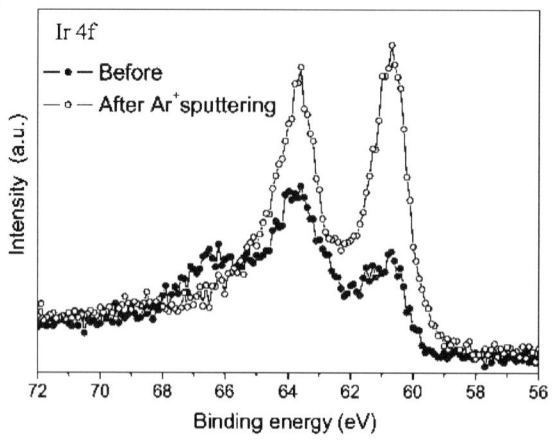

Figure 15.6 X-ray photoelectron spectroscopy of the Ir 4f region for Ir(0) nanoparticles prepared in BMI·PF$_6$ before and after Ar$^+$ sputtering. Ar$^+$ sputtering eliminates the outermost layers of the particles, which results in the Ir 4f region with mainly Ir—Ir bond components. (Reproduced with permission from Ref. [25]; © 2006 Elsevier).

Scheme 15.4 Deuterium/hydrogen (D/H) exchange reaction in ionic liquids promoted by Ir(0) nanoparticles.

However, hydrogen/deuterium (H/D) labeling and ^{2}H NMR studies of nanoparticles formed by the reaction of [Ir(COD)(MeCN)$_2$]BF$_4$ dissolved in BMI·NTf$_2$/acetone in the presence of 1,8-dimethylaminonaphthalene with D$_2$ at 22 °C showed that ILs with hydrogens mainly in the C$_2$ position of the imidazolium ring formed N-heterocyclic carbenes (NHCs) with Ir(0) nanoparticles [26]. These NHC species, when attached to a metal surface, may provide an enhanced stabilization for transition-metal nanoparticles in ILS. It was also observed independently, by using ^{1}H NMR and ^{2}H NMR experiments, that the D/H exchange only occurred after the *in situ* formation of Ir(0) nanoparticles during the hydrogenation of cyclohexene using [Ir(COD)Cl]$_2$ as catalyst precursor and molecular hydrogen in [BMI]-d$_3$·NTf$_2$ (Scheme 15.4). Of note, the D/H exchange reaction was seen to occur preferentially at the less-acidic C$_4$ and C$_5$ imidazolium positions [27].

The D/H exchange occurred mainly after complete consumption of the alkene, and no D-incorporated alkane was detected, which indicated that the coordinated NHC was easily displaced by the alkene and that these carbenes were less strongly bounded to the metal surface than was seen with mononuclear metal compounds [28]. These results strongly suggested that the imidazolium cations reacted with the nanoparticle surface preferentially as aggregates of the type $\{[(DAI)_x(X)_{x-n}]^{n+}[(DAI)_{x-n}(X)_x]^{n-}\}_n$ (where DAI is the 1,3-dialkylimidazolium cation and X the anion), rather than as isolated imidazolium cations.

15.3
Kinetic Studies of Iridium Nanoparticle Formation: The Autocatalytic Mechanism

Using the well-defined system of polyoxoanion/Bu_4N^+-stabilized iridium nanoparticles [9, 29] as a model for the studies, Finke and coworkers [30] proposed a method that attempted to explain the formation and growth of transition-metal nanoparticles. This indirect method is based on an autocatalytic mechanism that considers a *nucleation* step in which a precursor A is converted to a zero-valent nuclei B with a rate constant k_1, and a second step that considers the *autocatalytic surface growth* of the metal nanoparticles where species B catalyzes its own formation with a rate constant k_2 (Scheme 15.5).

(a) $A \xrightarrow{k_1} B$

(b) $A + B \xrightarrow{k_2} 2B$

(c) $\alpha [B + \text{alkene} + H_2] \xrightarrow{\text{fast}} B + \text{alkane}]$

(d) $[A + \alpha \text{ alkene} + \alpha H_2] \xrightarrow{k_{obs}} B + \alpha \text{ alkane}]$

Scheme 15.5 The pseudo-elementary step concept proposed by Finke and coworkers to monitor transition-metal nanoparticle formation.

In this autocatalytic mechanism the olefin hydrogenation step is considered faster than the nucleation and growth steps. When the olefin hydrogenation is a rapid process, the equations can be deduced in terms of the two constants, k_1 and k_2, the values of which can be obtained from kinetic Equation 15.1:

$$[cyclohexene]_t = \frac{\frac{k_1}{k_2} + [cyclohexene]_0}{1 + \frac{k_1}{k_2 [cyclohexene]_0} e^{(k_1 + k_2 [cyclohexene]_0)t}} \quad (15.1)$$

A more detailed description of the use of a pseudo-elementary step for the treatment of hydrogenation kinetic data and derivatization of the kinetic equations can be found elsewhere [30–32].

This kinetic equation is applied to the observed kinetic curves obtained in cyclohexene hydrogenation (model reaction) following the molecular hydrogen consumption. Of note, the present kinetic equation provides the value of k_{2obs} and not k_2. However, the real value of the rate constant k_2 can be obtained easily using the relationship: $k_2 = k_{2obs} \times S/C$, where S/C is the substrate/catalyst molar ratio (the catalyst is given as the number of metallic moles employed).

An interesting additional experiment to follow the iridium nanoparticles formation was demonstrated by Watzky and Finke [30], who used a direct method of monitoring by gas–liquid chromatography (GLC) the evolution of cyclo-octane from the conversion of the precursor $(Bu_4N)_5Na_3[(COD)Ir(P_2W_{15}Nb_3O_{62})]$ into metallic iridium nanoparticles under molecular hydrogen. The curve derived from the loss of the precursor presents a sigmoid shape, and fits well using the same kinetic equation (Equation 15.1). This results in similar rate constants k_1 and k_2 to those obtained from the hydrogenation curve in the same experiment (within experimental error and after introduction of required mathematical correction factors). The rate constants k_1 and k_2 were estimated as: k_1 hydrogenation = 1.8 (±0.2) × $10^{-3} h^{-1}$, k_2 hydrogenation(corrected) = 2.5 (±0.3) × $10^3 M^{-1} h^{-1}$; k_1 GLC = 2.8 (±1.8) × $10^{-3} h^{-1}$, k_2 GLC(corrected) = 2.3 (±0.2) × $10^3 M^{-1} h^{-1}$. The large error observed for rate constant k_1 when obtained by GLC was attributed to the lower precision of this technique when compared to hydrogenation measurements.

Moreover, the effects of other added substances such as water, acetic acid, olefin, polyoxoanion stabilizer $[Bu_4N]_9P_2W_{15}Nb_3O_{62}$, temperature and hydrogen pressure were also investigated for the formation of iridium nanoclusters.

Hence, two methods are available that can be applied to follow nanoparticles formation and growth: (i) an *indirect method* that utilizes the consumption of molecular hydrogen pressure versus time; and (ii) a *direct method* that follows the loss of precursor by the 1:1 conversion of its cyclo-octadiene ligand to cyclo-octane by GLC measurements. The mechanism developed by Watzky and Finke suggests that the nanoparticles act as 'living-metal polymers'–a concept that could be used to obtain particles with defined sizes simply by adding the appropriate amounts of catalyst precursors [32].

Recently, Hornstein and Finke [33] have included a third step in the autocatalytic mechanism, known as *bimolecular aggregation* (B + B → C; rate constant k_3), where B is the active catalyst and C is the deactivated catalyst. However, these authors discovered a more general four-step mechanism by adding a fourth step into the autocatalytic mechanism [34]; this was called the *autocatalytic agglomeration step (bulk metallic formation)*, and involves a double autocatalysis which considers that small particles B agglomerate with the larger-bulk particles C (B + C → 1.5C; rate constant k_4). The accepted modern autocatalytic mechanism to investigate the formation of general transition-metal nanoparticles is constituted by four steps, as indicated in Scheme 15.6.

$$A \xrightarrow{k_1} B \quad \text{Nucleation}$$

$$A + B \xrightarrow{k_2} 2B \quad \text{Surface growth}$$

$$B + B \xrightarrow{k_3} C \quad \text{Bimolecular aggregation}$$

$$B + C \xrightarrow{k_4} 1.5C \quad \text{Bulk metallic formation}$$

Scheme 15.6 The accepted autocatalytic mechanism for monitoring the formation of transition-metal nanoparticles, as proposed by Finke and coworkers.

15.3 Kinetic Studies of Iridium Nanoparticle Formation: The Autocatalytic Mechanism

In the same context, the autocatalytic mechanism was successfully applied to the formation of iridium nanoparticles dispersed in imidazolium-based ILs [25, 35]. The iridium nanoparticles formation was followed in a typical hydrogenation of 1-decene with *in situ* reduction of the organometallic precursor [Ir(COD)Cl]$_2$ dissolved in ILs and molecular hydrogen. Typically, the kinetics curves obtained from the hydrogen consumption were treated using the pseudo-elementary step and fitted by the integrated rate equation (Equation 15.1), for nucleation (A → B, k_1) and autocatalytic surface growth (A + B → 2B, k_2). The kinetics curves were sigmoid, and well fitted by the cited rate equation, thus giving a good indication for the formation of metallic iridium nanoparticles. Attempts to fit the kinetic curves by including the third step (bimolecular aggregation; B + B → C, rate constant k_3) and the recently discovered double autocatalytic mechanism that consider a fourth step (bulk metallic formation; B + C → 1.5C, rate constant k_4) failed, which suggested that agglomeration is not important in these cases. The experimental data and calculated curves using the autocatalytic mechanism for the hydrogenation of 1-decene in different ionic liquids are shown in Figure 15.7.

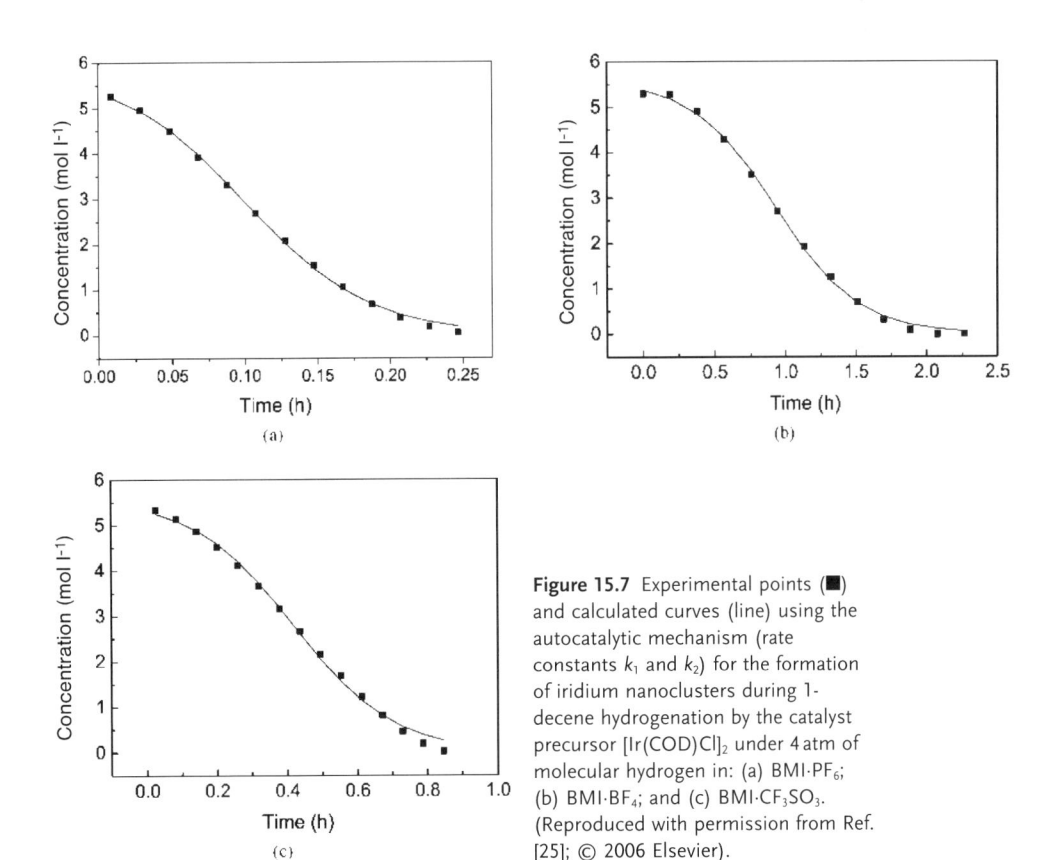

Figure 15.7 Experimental points (■) and calculated curves (line) using the autocatalytic mechanism (rate constants k_1 and k_2) for the formation of iridium nanoclusters during 1-decene hydrogenation by the catalyst precursor [Ir(COD)Cl]$_2$ under 4 atm of molecular hydrogen in: (a) BMI·PF$_6$; (b) BMI·BF$_4$; and (c) BMI·CF$_3$SO$_3$. (Reproduced with permission from Ref. [25]; © 2006 Elsevier).

15.4
Catalytic Applications of Soluble Iridium Nanoparticles

During recent years there has been a considerable increase in the number of reports relating to catalytic applications of transition-metal nanoparticles [2, 36, 37]. The major use of these nanoparticles in catalysis is based on their intrinsic electronic properties and the subsequent physical-chemical properties that lie between those of the smallest element from which the nanoparticles can be created, and those of the bulk material [38, 39].

Indeed, in many cases the creation of soluble nanoparticles has provided singular catalytic activities/selectivities that differ from those expected for both molecular (single-site) and heterogeneous (multi-site) catalysts [40, 41]. As a result, iridium nanoparticles have attracted much interest in terms of their catalytic performance in the hydrogenation of olefins, ketones and aromatic compounds.

The so-called 'soluble heterogeneous catalysts' $Ir(0)_{\sim 300}$ nanoparticles (synthesized by the reduction of polyoxoanion-supported Ir(I) precursors in acetone solution under molecular hydrogen) have shown full conversion in cyclohexene hydrogenation at 22 °C; this compares well with the corresponding heterogeneous catalyst $Ir(0)/Al_2O_3$, with a turnover frequency (TOF) [mol product/(mol exposed Ir(0) atoms × h)] of $3200 \pm 1000 \, h^{-1}$. By comparison, a commercial 7.9% dispersed $Ir(0)/\gamma-Al_2O_3$ had a verified TOF of $3950 \pm 1000 \, h^{-1}$, while a value of $1740 \pm 250 \, h^{-1}$ was observed with Exxon's catalyst 80% dispersed $Ir(0)/\eta-Al_2O_3$ [29, 42]. Moreover, the iridium nanoclusters have been shown to retain their catalytic activities (up to a total of 18 000 turnovers) over 10 days in solution before deactivation (this was superior to the TOF of 10 000 reported by Moiseev and coworkers [43] for soluble palladium clusters). Such catalytic activities are comparable to those of the classical heterogeneous catalysts when operating under the same reaction conditions.

The acid-assisted hydrogenation of neat acetone by Ir(0) nanoparticles was also studied by Ozkar and Finke [44]. This system consisted of an *in situ* reduction of $[Ir(COD)Cl]_2$ in neat acetone under 40 psi of H_2 at 22 °C. The acid production derives from reduction of the precursor to form 1 equiv. of HCl per Ir(I) ion reduced. Total acetone hydrogenation was observed in 9 h, presenting selectivities of 0.95 equiv. in 2-propanol, 0.025 equiv. of diisopropyl ether (this product arising from the H^+-catalyzed condensation of 2-propanol) and 0.025 equiv. of water, with an initial TOF of $1.9 \, s^{-1}$. When a 'catalyst lifetime' test was performed over 32 h, a total of 16 400 catalytic turnovers was achieved, although some catalyst deactivation due to aggregation and formation of the bulk metal was observed. The catalyst was also found to be more selective in the presence of molecular sieves (selectivity of 100% in 2-propanol and 188 000 total turnovers over 110 h prior to catalyst deactivation due to aggregation), most likely due to the nanoparticles being supported on the sieves and thus avoiding aggregation. Control experiments also showed that assistance from the acid was essential in order for the acetone hydrogenation to take place.

Using BMI·PF$_6$ IL as the reaction medium for acetone hydrogenation, Finke's group showed that, under mild reactions conditions, ILs may act as nanoparticles catalyst poisons. The addition of only 0.1 equiv. of the IL was shown to poison the formation of nanoclusters from the precursor [Ir(COD)Cl]$_2$ at 22 °C and 2.7 atm (2.7 × 10^3) of H$_2$ [45].

Chaudret and coworkers [13] described the application of their previously prepared aqueous soluble iridium nanoparticles in biphasic arene hydrogenation under mild conditions. This catalytic system was considered to be very useful for the hydrogenation of several benzene derivatives, with rapid conversions but no hydrogenolysis products. When the recycling and catalyst lifetimes were investigated for the substrates anisole, toluene and p-cresol, the TOFs remained virtually unchanged after three runs for each substrate, indicating high stability of the soluble aqueous iridium nanoclusters. An analysis of the stereochemistry of hydrogenated monocyclic compounds also showed the major product in all cases to be the *cis* diastereomer. Some results for the hydrogenation of aromatic compounds in biphasic media, using aqueous soluble iridium nanoparticles as catalytic phase under mild conditions, are summarized in Table 15.2.

Iridium nanoparticles prepared in the presence of different capping ligands [12] were used in the catalytic hydrogenation of 1-decene (substrate : Ir mass ratio = 1000) under 3 psi of molecular hydrogen at 75 °C. Whereas, iridium nanoparticles prepared in the presence of oleic acid/oleylamine and trioctylphosphine (TOP) did not exhibit any catalytic activity towards 1-decene hydrogenation, those nanoparticles synthesized in the presence of TOAB and TOPB did effect such catalysis. Moreover, the TOPB-coated nanoparticles exhibited the highest TOFs. In order to calculate the TOFs it was assumed that all surface atoms were active, even though it was expected that some sites would remain unavailable for catalysis due to the bonded capping ligand. The TOFs obtained for 1-decene hydrogenation using iridium nanoparticles prepared in TOAB and TOPB capping ligands were 4 s^{-1} and 270 s^{-1}, respectively.

This difference was attributed to the ligand binding strength, as the Ir—N bond has a shorter equilibrium bond length when compared to Ir—P. This suggests that the Ir—N bond presents a larger binding energy and, consequently, a stronger attachment to the nanoparticle surface. As discussed by the present authors, this proposal is in agreement with the iridium nanoparticle diameters observed by TEM (~1.5 nm for TOAB-capped, ~5 nm for TOPB-capped nanoparticles, respectively). It is considered, therefore, that a large proportion of the iridium atoms are exposed to the TOPB-capping ligand in order to provide catalytic hydrogenation.

These results show clearly the effect on the catalytic activity of 1-decene hydrogenation when the capping ligand is attached to the metal surface of the iridium nanoparticles. In this situation, the agent used must be strong enough to bind to the metal surface and provide a good stabilization, but weak enough so as to permit access of the substrates to the metal catalytic site.

Interestingly, the authors noted an increasing catalytic activity during a recycling reaction conducted with TOAB-coated iridium nanoparticles, over four cycles, with

Table 15.2 Biphasic hydrogenation of arenes using soluble aqueous iridium nanoparticles as catalyst [13].[a]

Entry	Substrate	Product (yield %)	Time (h)[b]	TOF (h^{-1})[c]
1	Anisole	Methoxycyclohexane (100)	0.75	400
2	Phenol	Cyclohexanol (100)	1.0	300
3	Benzene	Cyclohexane (100)	0.8	375
4	Toluene	Methylcyclohexane (100)	0.8	375
5	Ethyl benzoate	Ethyl cyclohexanoate (100)	1.9	157
6	Styrene	Ethylcyclohexane (100)	1.16	344
7	o-Xylene	1,2-Dimethylcyclohexane cis (95), trans (5)	1.65	181
8	m-Xylene	1,3-Dimethylcyclohexane cis (85), trans (15)	1.4	214
9	p-Xylene	1,4-Dimethylcyclohexane cis (80), trans (20)	1.2	250
10	o-Cresol	2-Methylcyclohexanol cis (90), trans (10)	1.5	200
11	m-Cresol	3-Methylcyclohexanol cis (80), trans (20)	1.2	250
12	p-Cresol	4-Methylcyclohexanol cis (78), trans (22)	1.0	300

a Experimental conditions: substrate:Ir molar ratio = 100, catalyst (1.85 × 10^{-5} mol), N,N-dimethyl-N-cetyl-N-(2-hydroxyethyl)ammonium chloride (1.85 × 10^{-4} mol), water (20 ml), 40 bar hydrogen pressure, 20 °C, stirring rate 1000 min^{-1}.
b Determined by GC measurement.
c TOFs considered as mol H$_2$ mol Ir^{-1} h^{-1}.

successive TOF-values of 4, 13, 50 and 124 s^{-1}. This increase in catalytic activity is due to ligand desorption occurring at the metal surface during the recycle reactions. However, in a fifth reaction cycle the catalytic activity decreased considerably (TOF 38 s^{-1}). This reduction can be explained due to aggregation of the iridium nanoparticles after catalysis, as observed by TEM (with desorption of the stabilizer capping ligand having led to slight particle aggregation during the previous four cycles). The same behavior was observed in recycling reactions of nanoparticles

Table 15.3 Biphasic hydrogenation reactions by Ir(0) nanoparticles in BMI·PF$_6$ under 4 atm of H$_2$ (constant pressure) at 75 °C and olefin/Ir = 1200 molar ratio [24].

Entry	Substrate	Time (h)	Conversion (%)[b]
1	1-Decene	0.5	100
2[a]	1-Decene	0.5	56
3	Styrene	1.0	63
4	Cyclohexene	3.2	100
5	Methyl methacrylate	17	100
6	4-Vinylcyclohexene	1.0	91[c]
7	4-Vinylcyclohexene	4.0	100[d]

a Reaction performed in CH$_2$Cl$_2$ (in absence of ionic liquid).
b Conversion determined using GC.
c Conversion and selectivity in 4-ethylcyclohexene.
d Conversion in ethylcyclohexane.

prepared with TOPB-coated ligand, with catalytic activity increasing over five cycles but decreasing in the sixth.

The *in situ* reduction of the precursor [Ir(COD)Cl]$_2$ dispersed in BMI·PF$_6$ at 75 °C and under 4 atm of H$_2$ provided a suitable medium for the synthesis of Ir(0) nanoparticles, and represents an ideal system for the biphasic hydrogenation reactions of several olefins (Table 15.3) [24]. Of note, the TOF observed for this system (6000 h^{-1} at 1200 rpm and 75 °C) was considerably higher than those obtained under biphasic conditions by classical transition-metal catalyst precursors in ILs under similar reaction conditions [46–48].

The black solution containing IL and Ir(0) nanoparticles (~2.0 nm in diameter, as determined by TEM) can also be recycled in catalytic hydrogenation reactions. The catalytic activity of these soluble iridium nanoparticles was also compared to that of the Crabtree catalyst ([Ir(COD)(PCy$_3$)py]PF$_6$) in BMI·PF$_6$. These nanoparticles maintain an efficient activity for at least seven recycles, whereas the Crabtree catalyst suffers a significant reduction in activity during recycling reactions.

The iridium nanoclusters prepared in ILs can also catalyze the hydrogenation of arenes [15]. Such reactions can be performed by isolated Ir(0) nanoparticles (solventless system), redispersed in either an imidazolium IL (biphasic system) or in acetone (homogeneous system). As expected, those reactions performed in solventless and homogeneous conditions required a shorter reaction time for complete substrate conversion compared to reactions in BMI·PF$_6$, a fact which can be explained by considering the nature of the biphasic reactions that are controlled by mass-transfer process.

Moreover, Ir(0) nanoparticles recovered after benzene hydrogenation using *solventless* conditions can be reused for at least seven runs without any significant loss in catalytic activity. The results of arene hydrogenations catalyzed by iridium nanoparticles in different media are listed in Tables 15.4 and 15.5.

The competitive hydrogenation of alkyl-substituted arenes was also performed with Ir(0) nanoparticles [49]. Using toluene as a standard substrate, several toluene/benzene and toluene/monoalkylbenzene hydrogenation experiments were conducted in order to determine the selectivity constants of the transition-metal nanoparticles. These selectivity constants can be used to predict the relative reactivity of any other couple of monoalkylbenzenes. A series of initial reaction

Table 15.4 Benzene hydrogenation by Ir(0) nanoparticles in different systems under 4 atm of molecular hydrogen (constant pressure) at 75 °C [15].

Entry	System	Benzene:Ir[a]	Time (h)[b]	TOF (h^{-1})[c]
1	BMI·PF$_6$	250	5	50 (88)
2	BMI·PF$_6$	500	10	50 (88)
3	BMI·PF$_6$	1200	14	85 (150)
4	Solventless	250	2	125 (219)
5	Solventless	500	7	71 (124)
6	Acetone	500	2.5[d]	200 (351)

a Benzene:Ir molar ratio.
b Time required for 100% conversion, determined by GC.
c TOF calculated considering total metal (mol cyclohexane mol metal^{-1} h^{-1}); TOF values in parentheses were corrected for exposed metal.
d Total benzene conversion and 82% of acetone reduction to isopropanol.

Table 15.5 Hydrogenation of arenes by Ir(0) nanoparticles in different systems under 4 atm of molecular hydrogen (constant pressure) at 75 °C [15].

Entry	System	Arene	Arene:Ir[a]	Time (h)	Conversion (%)[b]	TOF (h^{-1})[c]
1	BMI·PF$_6$	Toluene	1200	25	93	44 (78)
2	BMI·PF$_6$	p-Xylene	500	18	86[d]	24 (42)
3	Solventless	p-Xylene	250	12	100[e]	21 (36)
4	Solventless	Methylbenzoate	250	18	92	13 (22)
5	Solventless	Acetophenone	250	16	100[f]	15 (27)
6	Solventless	Anisole	250	18	74[g]	10 (18)

a Arene:Ir molar ratio.
b Conversion determined by GC.
c TOF calculated considering total metal (mol cyclohexane mol metal^{-1} h^{-1}); TOF values in parentheses were corrected for exposed metal.
d Cis:trans 5:1.
e Cis:trans 3:1.
f 42% ethylcyclohexane and 58% 1-cyclohexyl-1-ethanol.
g 84% methoxycyclohexane and 16% cyclohexane.

rate constants obtained in competitive hydrogenation reactions catalyzed by Ir(0) nanoparticles can be correlated with the Taft equation, indicating that the relative reaction constants for the alkyl substituents can be expressed only by steric factors and are thus independent of any other factors. This competitive method was shown to be suitable for estimating the selectivity constants for couples of alkylbenzenes, and can also be used as a chemical probe for quantitative comparison of the selectivity and steric influence imposed by different types of metal catalysts.

Iridium nanoparticles prepared in imidazolium-based ILs have been also used in the catalytic hydrogenation of ketones under mild conditions [50]. Firstly, cyclohexanone was chosen as the model substrate to optimize the reaction conditions (temperature, hydrogen pressure, catalyst concentration). Initially, isolated Ir(0) nanoparticles were tested in a solventless system for the hydrogenation of cyclohexanone; the preliminarily results are listed in Table 15.6.

As expected, cyclohexanone hydrogenation performed in an IL has a longer reaction time than in solventless conditions. Where using iridium nanoparticles dispersed in an IL, the biphasic hydrogenation of cyclohexanone could be performed at least 15 times, without any considerable loss in catalytic activity; this contrasted with the use of nanoparticles in solventless conditions, when the catalytic activity begins to decline after the third cycle. The standard experimental conditions established for the hydrogenation of other carbonyl compounds were 75 °C, 4 atm of H_2 and a molar substrate:Ir ratio of 250.

The iridium nanoparticles showed good activity for the hydrogenation of both cyclic and acyclic saturated ketones, including an additional experiment for benzaldehyde (Table 15.7, entry 7). The main results relating to hydrogenation reactions performed under solventless conditions are listed in Table 15.7.

Table 15.6 Cyclohexanone hydrogenation by iridium nanoparticles in solventless conditions [50].

Entry	Ketone:Ir[a]	Temperature (°C)	Pressure (atm)[b]	Time (h)[c]	TOF (h^{-1})[d]
1	250	30	4	4.5	55
2	250	75	4	2.0	125
3	250	100	4	2.0	125
4	250	75	2	2.0	125
5	250	75	6	1.2	208
6	500	75	4	4.0	125
7	750	75	4	8.0	94
8[e]	250	75	4	17.5	14

a   Ketone:Ir molar ratio.
b   Constant pressure of H_2.
c   Time for total conversion determined by GC.
d   Mol cyclohexanol mol Ir^{-1} h^{-1}.
e   Ir(0) nanoparticles redispersed in BMI·PF_6.

Table 15.7 Hydrogenation of ketones by Ir(0) nanoparticles (substrate : ratio = 250) in solventless condition under 4 atm of H$_2$ (constant pressure) at 75 °C and the additional experiment for benzaldehyde [50].

Entry	Substrate	Product	Time (h)	Conversion (%)[a]	TOF (h^{-1})[b]
1	cyclopentanone	cyclopentanol	4.0	100[c]	62.5
2	2-pentanone	2-pentanol	2.5	96	96
3	methyl isopropyl ketone	methyl isopropyl carbinol	2.5	96	96
4	methyl tert-butyl ketone	methyl tert-butyl carbinol	3.7	100	68
5	ethyl 2-oxopropanoate	ethyl 2-hydroxypropanoate	2.5	98	98
6	acetone	isopropanol	2.0	95	119
7	benzaldehyde	benzyl alcohol	15.0	100	17

a Conversions determined by GC analyses.
b Mol product mol Ir^{-1} h^{-1}.
c 88% cyclopentanol and 12% bicyclopentyl ether.

When using the same reaction conditions, a high selectivity for C=C double bond reduction was observed in the hydrogenation of α,β-unsaturated ketones such as cyclohexenone and isophorone. For cyclohexenone hydrogenation, the selectivities measured at 90% cyclohexenone conversion corresponded to 83% of cyclohexanone, 16% of cyclohexanol and 1% of cyclohexane (hydrogenolysis product). In the case of isophorone, 91% selectivity in the saturated ketone was achieved at 75% isophorone conversion. Interestingly, there was no evidence for the formation of unsaturated alcohols in either reaction, even at low substrate conversions, which suggests that the reduction of C=C double bond is much faster than of the C=O bond. It also strongly suggests that any alcohols present in the medium are derived from the hydrogenation of cyclic saturated ketone formed during the reaction.

The hydrogenation of acetophenone by Ir(0) nanoparticles was poorly selective in terms of reduction of the aromatic ring over the carbonyl group. For example, only 35% selectivity in cyclohexylmethylketone was observed at 50% acetophenone conversion (ethylcyclohexane derived from the hydrogenolysis of the alcohol was also observed). Thus, there was no preferred coordination of the aromatic ring or carbonyl group on the metal surface. These results were in opposition to those obtained for benzaldehyde (additional experiment, Table 15.7, entry 7), where the hydrogenation was highly selective to the carbonyl group.

Iridium nanoparticles also catalyze the hydrogenation of benzylmethylketone, with high selectivity in reduction of the aromatic ring (92% selectivity in saturated ketone, 8% in saturated alcohol at 97% benzylmethylketone conversion). This preferential coordination of the aromatic ring can be attributed to steric effects that make carbonyl coordination difficult. Therefore, metallic iridium nanoparticles prepared in ILs may serve as active catalysts for the hydrogenation of carbonyl compounds in both solventless and biphasic conditions.

The kinetics of the 1-decene hydrogenation reaction catalyzed by Ir(0) nanoparticles in $BMI \cdot PF_6$ was investigated in detail [35]. 1-Decene hydrogenation by the Ir(0) nanoparticles follows the classical monomolecular surface reaction mechanism, $v = k_c K[S]/1+K[S]$, where k_c is the kinetic constant for the formation of the products and K corresponds to the adsorption constant. The adsorption constant calculated for 1-decene adsorption in the surface of iridium nanoparticles in the ionic liquid ($K = 1.32 \pm 0.20\,M^{-1}$) is independent of the hydrogen concentration, even at low pressure. The catalytic constant (k_c) is almost identical ($0.45 \pm 0.06\,min^{-1}$) under hydrogen pressures $\geq 4\,atm$, but at low hydrogen pressures (2 atm) the hydrogenation rate underwent a significant decrease. This indicated that, at pressures <4 atm the hydrogen solubility in ionic liquid follows a mass transfer-controlled process and, under the kinetic conditions, the surface of the catalytic nanoparticles is not fully saturated. At higher hydrogen pressures the hydrogen concentration in the ionic liquid has no influence on the reaction rate due to the saturation on the nanoparticles' surface.

It follows therefore that, under hydrogen pressures $\geq 4\,atm$, k_c and K are independent of the hydrogen concentration; this indicates a zero-order dependence on hydrogen pressure, such that the reaction depends only on the concentration of 1-decene in the ionic liquid.

15.5
Conclusions

Soluble and stable Ir(0) nanoparticles of small size and narrow range of mean diameter may be easily prepared using a simple reduction of Ir(I) or Ir(III) compounds by molecular hydrogen or hydrides, and in the presence of stabilizing agents such as surfactants, polyoxoanions or imidazolium salts. The surface properties of these Ir(0) nanoparticles depend heavily on the nature of the stabilizing agent and the solvent (water, organic or ionic liquid). Indeed, good capping ligands – which stabilize robust nanocrystals with very narrow size distributions – are almost inactive catalysts for the hydrogenation reactions. In contrast, metal nanoparticle-containing stabilizers that bind less strongly to the metal surface than do other anions generate a higher catalytic activity. The hydrogenation of simple alkenes by Ir(0) nanoparticles dispersed in ILs appears to depend on steric hindrance at the C=C double bond, and follows the same trend observed with classical iridium complexes in homogeneous conditions (terminal > disubstituted > trisubstituted > tetrasubstituted). The process of alkene hydrogenation by Ir(0) nanoparticles in $BMI \cdot PF_6$ follows the classical monomolecular surface reaction mechanism, $v = k_c K[S]/1+K[S]$. Finally, these soluble iridium nanoparticles serve as effective catalysts for the single-phase or multi-phase hydrogenation of alkenes, aromatic compounds and ketones.

References

1 Gelesky, M.A., Umpierre, A.P., Machado, G., Correia, R.R.B., Magno, W.C., Morais, J., Ebeling, G. and Dupont, J. (2005) *Journal of the American Chemical Society*, **127**, 4588–9.
2 Migowski, P. and Dupont, J. (2007) *Chemistry – A European Journal*, **13**, 32–9.
3 Aiken, J.D. and Finke, R.G. (1998) *Journal of the American Chemical Society*, **120**, 9545–54.
4 Aiken, J.D. and Finke, R.G. (1999) *Journal of the American Chemical Society*, **121**, 8803–10.
5 Pan, C., Pelzer, K., Philippot, K., Chaudret, B., Dassenoy, F., Lecante, P. and Casanove, M.J. (2001) *Journal of the American Chemical Society*, **123**, 7584–93.
6 Pellegatta, J.L., Blandy, C., Colliere, V., Choukroun, R., Chaudret, B., Cheng, P. and Philippot, K. (2002) *Journal of Molecular Catalysis A – Chemical*, **178**, 55–61.
7 Schulz, J., Roucoux, A. and Patin, H. (1999) *Chemical Communications*, 535–6.
8 Reetz, M.T. and Lohmer, G. (1996) *Chemical Communications*, 1921–2.
9 Lin, Y. and Finke, R.G. (1994) *Journal of the American Chemical Society*, **116**, 8335–53.
10 Aiken, J.D., Lin, Y. and Finke, R.G. (1996) *Journal of Molecular Catalysis A – Chemical*, **114**, 29–51.
11 Yee, C.K., Jordan, R., Ulman, A., White, H., King, A., Rafailovich, M. and Sokolov, J. (1999) *Langmuir*, **15**, 3486–91.
12 Stowell, C.A. and Korgel, B.A. (2005) *Nano Letters*, **5**, 1203–7.
13 Mevellec, V., Roucoux, A., Ramirez, E., Philippot, K. and Chaudret, B. (2004) *Advanced Synthesis Catalysis*, **346**, 72–6.
14 Scheeren, C.W., Machado, G., Teixeira, S.R., Morais, J., Domingos, J.B. and Dupont, J. (2006) *The Journal of Physical Chemistry B*, **110**, 13011–20.
15 Fonseca, G.S., Umpierre, A.P., Fichtner, P.F.P., Teixeira, S.R. and Dupont, J. (2003) *Chemistry – A European Journal*, **9**, 3263–9.
16 Silveira, E.T., Umpierre, A.P., Rossi, L.M., Machado, G., Morais, J., Soares, G.V., Baumvol, I.L.R., Teixeira, S.R., Fichtner, P.F.P. and Dupont, J. (2004) *Chemistry – A European Journal*, **10**, 3734–40.
17 Machado, G., Scholten, J.D., de Vargas, T., Teixeira, S.R., Ronchi, L.H. and Dupont, J. (2007) *International Journal of Nanotechnology*, **4**, 541–63.
18 Dupont, J., de Souza, R.F. and Suarez, P.A.Z. (2002) *Chemical Reviews*, **102**, 3667–91.
19 Fei, Z.F., Geldbach, T.J., Zhao, D.B. and Dyson, P.J. (2006) *Chemistry – A European Journal*, **12**, 2123–30.
20 Wasserscheid, P. and Welton, T. (eds) (2003) *Ionic Liquids in Synthesis*, Wiley-VCH Verlag GmbH, Weinheim.
21 Tsuzuki, S., Tokuda, H., Hayamizu, K. and Watanabe, M. (2005) *The Journal of*

Physical Chemistry B, **109**, 16474–81.
22 Hardacre, C., Holbrey, J.D., McMath, S.E.J., Bowron, D.T. and Soper, A.K. (2003) *Journal of Chemical Physics*, **118**, 273–8.
23 Dupont, J. and Suarez, P.A.Z. (2006) *Physical Chemistry Chemical Physics*, **8**, 2441–52.
24 Dupont, J., Fonseca, G.S., Umpierre, A.P., Fichtner, P.F.P. and Teixeira, S.R. (2002) *Journal of the American Chemical Society*, **124**, 4228–9.
25 Fonseca, G.S., Machado, G., Teixeira, S.R., Fecher, G.H., Morais, J., Alves, M.C.M. and Dupont, J. (2006) *Journal of Colloid and Interface Science*, **301**, 193–204.
26 Ott, L.S., Cline, M.L., Deetlefs, M., Seddon, K.R. and Finke, R.G. (2005) *Journal of the American Chemical Society*, **127**, 5758–9.
27 Scholten, J.D., Ebeling, G. and Dupont, J. (2007) *Dalton Transactions*, 5554–60.
28 Bacciu, D., Cavell, K.J., Fallis, I.A. and Ooi, L.L. (2005) *Angewandte Chemie – International Edition*, **44**, 5282–4.
29 Lin, Y. and Finke, R.G. (1994) *Inorganic Chemistry*, **33**, 4891–910.
30 Watzky, M.A. and Finke, R.G. (1997) *Journal of the American Chemical Society*, **119**, 10382–400.
31 Widegren, J.A., Aiken, J.D., Ozkar, S. and Finke, R.G. (2001) *Chemistry of Materials*, **13**, 312–24.
32 Watzky, M.A. and Finke, R.G. (1997) *Chemistry of Materials*, **9**, 3083–95.
33 Hornstein, B.J. and Finke, R.G. (2004) *Chemistry of Materials*, **16**, 139–50.
34 Besson, C., Finney, E.E. and Finke, R.G. (2005) *Chemistry of Materials*, **17**, 4925–38.
35 Fonseca, G.S., Domingos, J.B., Nome, F. and Dupont, J. (2006) *Journal of Molecular Catalysis A – Chemical*, **248**, 10–16.
36 Roucoux, A., Schulz, J. and Patin, H. (2002) *Chemical Reviews*, **102**, 3757–78.
37 Lu, A.H., Salabas, E.L. and Schuth, F. (2007) *Angewandte Chemie – International Edition*, **46**, 1222–44.
38 Bell, A.T. (2003) *Science*, **299**, 1688–91.
39 Rolison, D.R. (2003) *Science*, **299**, 1698–701.
40 Jansat, S., Gomez, M., Philippot, K., Muller, G., Guiu, E., Claver, C., Castillon, S. and Chaudret, B. (2004) *Journal of the American Chemical Society*, **126**, 1592–3.
41 Valden, M., Lai, X. and Goodman, D.W. (1998) *Science*, **281**, 1647–50.
42 Aiken, J.D. and Finke, R.G. (1999) *Journal of Molecular Catalysis A – Chemical*, **145**, 1–44.
43 Vargaftik, M.N., Zagorodnikov, V.P., Stolarov, I.P., Moiseev, I.I., Kochubey, D.I., Likholobov, V.A., Chuvilin, A.L. and Zamaraev, K.I. (1989) *Journal of Molecular Catalysis A – Chemical*, **53**, 315–48.
44 Ozkar, S. and Finke, R.G. (2005) *Journal of the American Chemical Society*, **127**, 4800–8.
45 Ott, L.S., Campbell, S., Seddon, K.R. and Finke, R.G. (2007) *Inorganic Chemistry*, **46**, 10335–44.
46 Suarez, P.A.Z., Dullius, J.E.L., Einloft, S., DeSouza, R.F. and Dupont, J. (1996) *Polyhedron*, **15**, 1217–19.
47 Chauvin, Y., Mussmann, L. and Olivier, H. (1996) *Angewandte Chemie – International Edition*, **34**, 2698–700.
48 Dullius, J.E.L., Suarez, P.A.Z., Einloft, S., de Souza, R.F., Dupont, J., Fischer, J. and De Cian, A. (1998) *Organometallics*, **17**, 815–19.
49 Fonseca, G.S., Silveira, E.T., Gelesky, M.A. and Dupont, J. (2005) *Advanced Synthesis Catalysis*, **347**, 847–53.
50 Fonseca, G.S., Scholten, J.D. and Dupont, J. (2004) *Synlett*, 1525–8.

Index

a

acetonitrile 21 ff, 77
acetophenone 80 f, 91 f, 115
acrylate dimerization 252 f
AHA (alkyne hydroamination) 150 ff
– intermolecular 152 ff
– intramolecular 152 ff
– proposed mechanism 156
alcohols 108 ff, 123 ff, 126 ff, 138 ff
– addition to terminal alkynes 262 f
alkylation of ketones 258 ff
aldehydes 137 f
– in three-component coupling 249 ff
alkene metathesis 305
alkylation 258 ff
N-alkylation of amines 260 ff
– of amines with alcohols 123 ff
– of ketones 258 ff
– of methylene compounds 131 f
– of secondary alcohols 130 f
alkyne dimerization 247 ff
– proposed mechanism 248 f
alkyne hydroamination (AHA) 150 ff
– intermolecular 152 ff
– intramolecular 152 ff
– proposed mechanism 156
alkyne trimerization 248 f
alkynes 345
– addition of alcohols 262 f
– addition of water 262 f
– aromatic homologation 266 f
– dimerization, see also alkyne dimerisation 247 ff
– in three-component coupling 249 ff
– trimerization, see also alkyne trimerisation 247 ff
allylic alkylation 208
– allylation of enamines 223
– allylic etherification 239
– allylic substitution 256 f

amidation 128ff, 232 f
– amination of allylic alcohols 228 f
– asymmetric 207 ff
– catalyst activation 225 ff
– conditions 213 ff
– decarboxylative 223 f, 232 f
– DFT calculations 211
– intramolecular amination 227
– proposed mechanism 213 ff
– reactivity 208 ff
– regioselectivity 208 ff, 257
– selectivity 227, 236
– stereoselectivity 209, 257 f
– steric course 210 f
– synthesis of biologically active compounds 240 ff
– using a glycine equivalent 221 f
– using aliphatic nitro compounds 220 f
– using aryl zinc compounds 224
– using enolates 217 ff, 222 f
– using N-nucleophiles 225 ff, 234 f
– using O-nucleophiles 235 ff
– using pronucleophiles 229 ff
– using silanolates 238 ff
amines 123 ff
– addition to terminal alkynes 262 f
– aliphatic 225 ff
– in three-component coupling 249 ff
aniline 147 f, 163 f
arsines 63 ff
arylation 263 f
asymmetric allylic substitution, see also allylic substitution 207 ff
asymmetric transfer hydrogenation (ATH) 70

b

benzaldehyde 62 ff, 80 f, 101
benzylidene acetone 59

N-benzylideneaniline   34 ff
bipy   71, 102, 121
bis(diphenylphosphino)ethane (ddpe)   177, 274 f
bis(diphenylphosphino)methane (ddpm)   177
1,1'-bis(diphenylphosphino)ferrocene (dppf)   274 f
bisimidazoles   75 ff
(bis)oxazolines   88 ff
bond activation
– B—H bonds   169 ff
– C—C bonds   314 ff
– C—E bonds   313 ff
– C—H bonds   251, 263, 296 ff, 307 ff, 321 ff, 330
– C—halide bonds   313 f
– C—O bonds   314
– C—P bonds   328 f
– Ir—H bonds   311
– N—H bonds   159 ff
boron-addition   169 ff
borylation   181 ff, 186
– catalyst efficiency   185
– of aromatic C-H bonds   264 f
– proposed mechanism   184
– regioselectivity   184
butatriene   247 f
tert-butyl-acetylene   27 ff

c

caproaldehyde   62 ff
carbonylation   191 ff
– Cativa process   200 ff
– Celanese process   195
– DFT studies   198 ff
– iridium based   196 ff
– iridium-platinum based   202 ff
– migratory CO insertion   198 ff, 202 ff
– Monsanto process   192 ff
– proposed mechanism   196 f, 201
– rhodium based   192 ff
– role of the cocatalyst   203
– water gas shift (WGS) reaction   197 ff
catalysis
– application in fine chemicals industry   1 ff
– biocatalytic   99, 102 f
– catalysts cost   3
– catalytic performance, see also TOF, TON   3 f
– heterogeneous   99 ff
– homogeneous   55 ff
– hydrogenation, see hydrogenation   4 ff
– of Ir-NHC complexes   49 ff
– supported   99 ff
catalyst activity   57 ff, 303 f, 325 ff
– selectivity of nanoparticles   375 ff
catalyst precursor   16 ff, 183, 365 ff
catalytic performance   3
– of nanoparticles   375 ff
catechol   170 ff
Cativa process   200 ff
Celanese process   195
C—H activation   19 ff, 32ff, 40 ff, 49, 181 ff
– intramolecular   42
– selectivity   182
C—H borylation   181 ff
– selectivity   182
chalcone   59
chelate effect   45
chelating ligands
– aminophosphine   72 ff
– bipy   71
– bipyrimidine   111
– (bis)oxazolines   88 ff
– diamines   67 ff, 71 ff, 89 f, 111
– hybrid HHC-oxazoline ligands   51 f
– NC-chelating ligands   77 ff
– N-heterocyclic carbenes (NHCs)   39 ff, 50 ff, 73 ff
– NN-chelating ligands   83 ff, 89 f
– NNN-chelating ligands   86 ff
– NS-chelating ligands   93 ff
– phen   71
– phosphinooxazolines   257
– Pincer ligands   77 ff
– PN ligands   9, 23, 59
– PS ligands   68 ff
– Schiff-bases   83 ff
– TIME (1,1,1-[tris(3-alkylimidazole-2-ylidene)methyl]ethane)   45
– TIMEN (1,1,1-[tris(3-alkylimidazole-2-ylidene)methyl]amine)   45
CHIRAPHOS   81 ff, 280
C—N bond formation   123 ff
coupling reactions
– allylic substitution   256 f
– C—C   12, 130 ff, 305 ff, 341 ff, 358 ff
– C—Si coupling   341 ff
– iridium catalyzed   249 ff
– Suzuki-Myaura   183
– three-component coupling   249 ff
– three-component coupling, proposed mechanism   250 f

– Tishchenko reaction   137 f
– yielding vinyl ethers   253 ff
Crabtree's catalyst   15 ff
crotonaldehyde   100
cyclization
– of amino alcohols   126 ff
– of diols   135 ff
– of primary amines   127 f
cycloaddition   273 ff
– [2+2+1]   279 ff
– [2+2+2]   274 ff
– [4+2]   284
– [5+1]   284
– chemoselectivity   276
– enantioselective   277 ff
– Pauson-Khand reaction   279 ff
– regioselectivity   283
– solid-phase   276
cycloisomerization   285 ff
– proposed mechanism   286
cyclooctadiene   16 ff, 70 ff, 151 ff, 172, 304 f, 343 f, 366
cyclysation, *see also* cycloaddition, cycloisomerization   287, 288 ff
cymene   90

d
ddpm (bis(diphenylphosphino)methane)   177
dehalogenation   313 f
dehydrogenation   298, 305 ff, 321 ff, 329 ff
– dehydrogenative silylation   342 ff
– hydrogen acceptor   327 ff, 331 ff
– isomerization   331, 334
– of alkanes   298 f, 321 ff, 335 f
– of ammonia borane   315 f
– of ketones   299
– photochemical   325 ff
– proposed mechanism   299 ff, 328 f
DIAPHOX   217, 219
diazabutadienes   73 f
diboration of unsaturated C-C bonds   180 f
dimerization of primary alcohols   262
DKR (dynamic kinetic resolution)   118 ff
dppe (bis(diphenylphosphino)ethane)   177, 274 f
dppf (1,1'-bis(diphenylphosphino)ferrocene)   274 f
dynamic kinetic resolution   118 ff

e
β-elimination   22
esters   137 f

f
farnesol   11
fine chemicals   1 ff, 296
fluoride   150

h
N-heterocyclic carbene (NHC) ligands   39 ff, 180 f
– abnormal NHCs (aNHCs)   46 ff
– chelating bis-NHCs   43 ff
hydroamination
– alkyne hydroamination (AHA)   150 ff
– Ir complexes   156 ff
– of alkynes   145, 150 ff
– of allenes   145 ff
– olefin hydroamination (OHA)   146 ff
– reactions of Ir complexes   156 ff
hydroboration
– *anti*-Markovnikov-addition product   177
– catalyzed by lanthanum complexes   178
– diastereoselectivity   173 ff
– diboration   180 ff
– of alkenes   172 ff, 176
– of internal alkenes   177 ff
– proposed mechanism   175 ff
– regioselectivity   174 ff
– rhodium catalyzed   175 ff
– *trans*   179 f
hydroformylation of organosilicon compounds   352 ff
hydrogen acceptor   298 ff, 327 ff
hydrogen storage   315 f
hydrogen transfer reactions   298 ff
– alkylation of amines   123 ff
– C—C bond formation   130 ff
– C—O bond formation   135 ff
– cyclization of amino alcohols   126 ff
– oxidation of alcohols   108 ff, 138 ff
– proposed mechanism   126, 133, 139
hydrogenation   298 ff, 324 f, 377 ff
– *N*-alkylation of amines   123 ff
– α-keto-acids   122 f
– asymmetric   65 ff, 112 ff
– asymmetric transfer hydrogenation (ATH)   70, 81 ff
– biphasic   378 ff
– chemoselective   56 ff, 66 ff
– C—N bond formation   123 ff
– enantioselective   4 ff, 57, 64 ff, 70 ff, 81 ff, 91 ff, 113 ff, 117,
– in aqueous media   66, 80 f, 116 f, 119 ff
– of 1-alkynes   22, 26 ff
– of alcohols   108 ff

- of alkaloids   92 ff
- of N-benzylideneaniline   34 ff
- of tert-butyl-acetylene   27 ff
- of C=C bonds   8 ff, 21ff, 50 ff, 57, 92,
- of C=N bonds   4 ff, 34 ff, 50 ff, 85 f, 113, 123 ff
- of C=O bonds   50 ff, 55 ff, 65 ff, 83 ff, 99 ff, 113 ff, 381 ff
- of nitroarenes   12 f
- of phenylacetylene   25 ff
- of propene   24 ff
- of quinolines   112 ff
- of styrene   30 ff
- pH-dependant   120 ff
- transfer hydrogenation (TH)   69 ff, 107 ff, 113 ff, 299, 305 ff, 333 f
- using nanoparticles   377 ff
hydrosilylation   17 ff, 341 ff
- asymmetric   52
- in polymer chemistry   344 f
- isomerization   357 f
- of alkynes   345 f
- of imines   348 ff
- of ketones   52, 348 ff
- of styrene   343 f
- stereoselectivity   346 ff
hydroxylamine derivatives   234 ff

i

imidazolium salts   43 ff, 98, 368
- bisimidazolium salts   48 ff
- in the synthesis of nanoparticles   368
- stabilisation of nanoparticles   368
imines   112
indoles   152 ff
industrial applications   6 f, 9ff
- carbonylation   191 ff
- Cativa process   200 ff
- Celanese process   195
- methanol carbonylation   191 ff
- Monsanto process   192 ff
innocent ligand   16 ff
insertion   22
iridium complexes
- 14-electron species   329
- alkinyl   311
- alkyl   160 ff
- allyl   25
- amino   156 ff
- arene   31 ff
- aryl   29 ff, 107 ff
- as precatalysts   183
- as precursors for nanoparticles   365 ff
- asymmetric catalysts   51 f, 85 ff, 113 f

- bipyrimidine   111
- boryl   169ff
- carbonyl   45 ff, 62, 77, 95 f, 151, 198 ff
- compared to La complexes   178
- compared to Pd complexes   207 f, 210, 226
- compared to Pt complexes   342
- compared to Re complexes   327 ff
- compared to Rh complexes   115 f, 172 f, 178, 182 ff, 198 ff, 202, 210, 283, 342
- Cp*   107 ff
- dihydrido   15 ff, 259, 302, 325 f
- dinuclear   45 f, 148, 326 f, 343, 354, 359
- diphosphine   4 ff, 148 f
- ethylene   146 ff
- fluoride   165 f
- for C—H bond activation   322 ff
- for coupling reactions, *see also* coupling reactions   247 ff
- for the synthesis of biologically active compounds   242
- hydrido   159 ff, 251, 301, 310
- hydroamination   156 ff
- hydroxo   120ff
- Ir(I) complexes   36, 59 ff, 66 ff, 73, 146 ff, 185, 196 ff, 251, 274 ff, 302 f, 310 f, 343 f
- Ir(III) complexes   29 ff, 36 f, 67 ff, 74ff, 146 ff, 159 ff, 181 ff, 196 ff, 211, 252, 287 f, 302 f, 309 ff, 343 f
- Ir(V) complexes   329, 343 f
- Ir=Ir double bond   78 f
- isomerization   20 ff, 28
- NHC   43 ff, 110, 151, 349 ff
- octahedral   308
- phosphino   159 f, 274 f
- Pincer-type   295 ff, 329 ff
- trihydrido   78
- trinuclear   45 f
- triphosphine   15 ff
- water-soluble   66, 80 f, 116 f
- zwitterionic   178

j

Josiphos   7

k

α-keto-acids   122
ketones   108 ff, 113 ff, 258 ff

l

γ-lactams   233 f
ligands
- acetonitrile   21 ff, 33 ff, 110

- allyl   25
- amines   156 ff
- aminophosphine   72 ff
- ammonia   159
- arsines   63
- aryl   29 ff, 107 ff
- asymmetric diferrocenyl ligands   98 f
- bidentate   357
- bipy   71, 102, 121
- bipyrimidine   111
- (bis)oxazolines   88 ff
- boryl   172 ff
- capping ligands   366 ff
- carbonyl   171 ff, 202 ff
- CHIRAPHOS   81 ff, 280
- Cp* (pentamethylcyclopentadienyl)   107 ff
- cymene   90
- diamines   67 ff, 71 ff, 89 f, 115
- DIAPHOX   217, 219
- diphosphine   164, 249
- dppe   177
- dppm   177
- N-heterocyclic carbenes (NHCs)   39 ff, 50 ff, 73 ff, 98 f, 110, 151, 180 f, 349 ff
- hybrid HHC-oxazoline ligands   51 f
- innocent ligand   16 ff
- Josiphos   7
- NC-chelating ligands   77 ff
- NN-chelating ligands   83 ff, 89 f
- NNN-chelating ligands   86 ff
- NS-chelating ligands   93 ff
- oxazolinylphosphines   351
- pentamethylcyclopentadienyl (Cp*)   107 ff
- phen   71
- phosphines   56 ff, 62ff, 70, 81 ff, 148 f, 176 ff, 183, 257
- phosphinooxazolines   211 f, 257
- phosphonic acid   66
- phosphoramidites   12, 212 ff, 257
- PHOX   16
- Pincer ligands   72 ff, 77 ff158 ff, 164 ff, 295 ff, 329 ff
- PN ligands   9, 23, 59, 151
- polydentate ligands   95 f, 148 f
- PROPHOS   81 ff
- PS ligands   68 ff
- pybox   12, 350
- pyridine   74 ff
- QUINAP   176
- S,S-DIOP   59
- Schiff-bases   83 ff
- terdentate ligands see also pincer ligands   295 ff

- TIME (1,1,1-[tris(3-alkylimidazole-2-ylidene)methyl]ethane)   45
- TIMEN (1,1,1-[tris(3-alkylimidazole-2-ylidene)methyl]amine)   45
- TIPPS   70 ff
- tridentate ligands, see also pincer ligands   346 ff
- TsCYDN   115 f

m

methanol carbonylation   191 ff
Monsanto process   192 ff

n

nanoparticles
- catalytic application   75ff, 375 ff
- formation kinetics   373 ff
- monodisperse   366
- polydisperse   367
- solulable iridium nanoparticles   365 ff
- stabilisation   365 ff, 370 ff
- synthesis   365 ff
nickel complexes   322
O-nucleophiles
- alkoxides   237
- hydroxylamine derivatives   237 f
- phenolates   235 ff

o

OHA (olefin hydroamination)   146 ff
- proposed mechanism   155 f
Oppenauer-type oxidation   108 ff

p

Pauson-Khand reaction   279 ff
pentamethylcyclopentadienyl (Cp*)   107 ff
Pfalz's catalyst   16
phen   71
phenylacetylene   25 ff, 157 f, 311 f
phosphines   177, 179, 257, 355
phosphinooxazolines (PHOX)   211f, 257
phosphoramidites   212 ff, 257
Pincer ligands   72 ff, 295 ff, 302 ff
- ECE   295
- NCN   295
- PCP   295 ff, 302 ff, 314 f
- PCP   329 ff, 333 ff
- PNP   308 ff
- POCOP   332 ff
- SCS   295
piperidine   150
PN ligands   9
polycarbosilanes   344 f

polydentate ligands
– aminophosphine   72 ff
– bipy   71
– bipyrimidine   111
– (bis)oxazolines   88 ff
– diamines   67 ff, 71 ff, 89 f, 111
– N-heterocyclic carbenes (NHCs)   39 ff, 50 ff, 73 ff
– hybrid HHC-oxazoline ligands   51 f
– NC-chelating ligands   77 ff
– NN-chelating ligands   83 ff, 89 f
– NNN-chelating ligands   86 ff
– NS-chelating ligands   93 ff
– phen   71
– phosphinooxazolines   257
– Pincer ligands   77 ff
– PN ligands   9, 23, 59
– PS ligands   68 ff
– Schiff-bases   83 ff
– TIME (1,1,1-[tris(3-alkylimidazole-2-ylidene)methyl]ethane)   45
– TIMEN (1,1,1-[tris(3-alkylimidazole-2-ylidene)methyl]amine)   45
polyoxoanions   365 ff
precatalysts   183
pronucleophiles   229
PROPHOS   81 ff
pybox   12, 217, 219, 350

q

QUINAP (1-(2-diphenylphosphino-1-naphtyl)   176
quinolines   112

r

reductive amination   7
reductive elimination   22, 44
rhenium complexes   322, 325 f
rhodium complexes   178

ring closing metathesis (RCM)   233 f, 239
ring rearrangement methathesis (RRM)   242
ruthenium complexes   322

s

silanolates   238 ff
silylation   17 ff, 356 ff
– of aromatic C-H bonds   264 f
– of C-H bonds   355 ff
silylcarbonylation   353 ff
surfactants   366 ff

t

TEM (transmission electron microscopy)   367 ff
TIME (1,1,1-[tris(3-alkylimidazole-2-ylidene)methyl]ethane)   45
TIMEN (1,1,1-[tris(3-alkylimidazole-2-ylidene)methyl]amine)   45
TIPPS   70 ff
Tishchenko reaction   137 f
γ-tocotrienyl acetate   11
TOF (turnover frequency)   297, 376 ff
Tolman's electronic parameter   40
TON (turnover number)   303 ff, 330, 336
transfer hydrogenation (TH)   69 ff
– in water   80 f
triazole-based NHCs   50 ff
TsCYDN   115 f

v

Vaska's complex   283 f, 342 f

w

water gas shift (WGS) reaction   197 ff
Wilkinson's Catalyst   15

z

zinc compounds   224